FUNDAMENTALS OF SOIL ECOLOGY

FUNDAMENTALS OF SOIL ECOLOGY

THIRD EDITION

David C. Coleman

Mac A. Callaham, Jr.

D. A. Crossley, Jr.

Academic Press is an imprint of Elsevier
125 London Wall, London EC2Y 5AS, United Kingdom
525 B Street, Suite 1800, San Diego, CA 92101-4495, United States
50 Hampshire Street, 5th Floor, Cambridge, MA 02139, United States
The Boulevard, Langford Lane, Kidlington, Oxford OX5 1GB, United Kingdom

British Library Cataloguing-in-Publication Data
A catalogue record for this book is available from the British Library

Library of Congress Cataloging-in-Publication Data
A catalog record for this book is available from the Library of Congress

ISBN: 978-0-12-805251-8

For Information on all Academic Press publications
visit our website at https://www.elsevier.com/books-and-journals

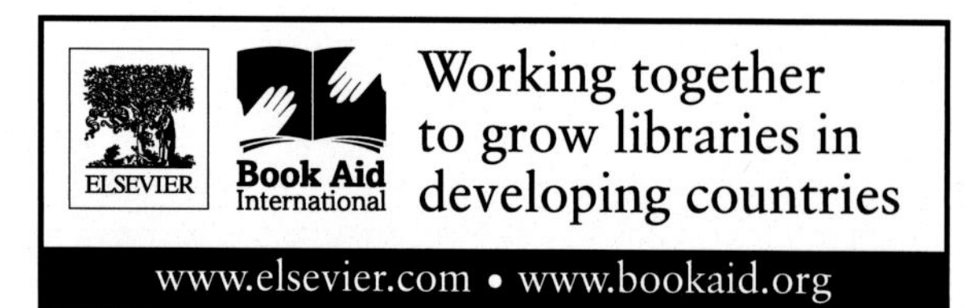

Publisher: Candice Janco
Acquisition Editor: Anneka Hess
Editorial Project Manager: Emily Thomson
Production Project Manager: Vijayaraj Purushothaman
Cover Designer: Christian Bilbow
Nematode photo on cover: Mitchell Pavao-Zuckerman

Typeset by MPS Limited, Chennai, India

Contents

Preface

This book, in its two previous editions, has spanned more than a generation: 21 years and 13 years, respectively. The field of Soil Ecology has moved ahead at an exponentially increasing rate.

In this Third Edition, we have noted the rapid expansion of taxonomic knowledge of prokaryotes, both Archaea and Bacteria. Perhaps even more progress has been made in the understanding of mycorrhizal interactions with their plant hosts and decomposing substrates as well (Chapters 2 and 3: Primary Production Processes in Soils: Roots and Rhizosphere Associates, and Secondary Production: Activities of Heterotrophic Organisms—Microbes). However, we feel that a strong emphasis in this book on the holistic nature of root-soil-microbial–faunal interactions will encourage the soil ecologist to continue to include the faunal side of the equation even more so than one generation ago. Thus the faunal chapter (Chapter 4: Secondary Production: Activities of Heterotrophic Organisms—The Soil Fauna) is by far the most extensive of any of our chapters. The role of soil biodiversity globally is of increasing interest; we cover that and responses of soils to global climate change in Chapters 7 and 8, Soil Biodiversity and Linkages to Soil Processes, and Future Developments in Soil Ecology.

With the continuing interest shown by our readers in the final chapter (Chapter 9: Laboratory and Field Exercises in Soil Ecology) on laboratory and field studies in our field, we have updated and expanded its coverage significantly. We hope this book, with its conjoint consideration of biology, chemistry, and physics of soil systems, will continue to be useful in courses in ecology, forestry, agronomy, and soil science in general.

We extend our appreciation to the long list of graduate and postdoctoral students listed in the Preface to the Second Edition, and also include recent students and colleagues involved with the Soil Biology and Ecology course offered at the University of Georgia, United States. David Coyle and Uma Nagendra deserve special thanks for assisting with instruction of the course in 2013 and 2015, respectively.

One major difference that readers will note in this third edition is the addition of many new color photographs throughout the text. These photos came from many sources, but four individuals were particularly generous with their images. Foremost among these is Andy Murray whose in situ photos of collembolans and other mesofauna are unparalleled in quality in our opinion. David Walter provided spectacular images of mites, and Bruce Snyder and Roberto Carrera-Martínez provided images of various macrofauna and other items of interest throughout.

We owe a special debt of gratitude to Melanie Taylor who provided logistical and intellectual support throughout the revision process, and who provided some text in Chapter 9, Laboratory and Field Exercises in Soil Ecology. We owe further gratitude to

Shela Mou whose dogged attention to detail has produced the most accurate literature cited list yet presented in the three editions.

As ever, we extend our deepest gratitude to our families for their continued patience, forbearance, and support, as we worked on the text.

Last, but not least, we express the considerable thanks owed to our recently deceased colleagues, Amyan Macfadyen, Dennis Parkinson, John Waid, and Gregor Yeates, who were pioneers in integrative soil ecology, and whose contributions to the discipline may readily be seen throughout this volume.

David C. Coleman, Mac A. Callaham, Jr. and D.A. Crossley, Jr.
Athens, GA, United States, 24 May 2017

CHAPTER

1

Introduction to Soil: Historical Overview, Soil Science Basics, and the Fitness of the Soil Environment

1.1 WHY SOIL ECOLOGY?

A new student of soil ecology might reasonably ask the question that leads this chapter! We felt that for the third edition of Fundamentals of Soil Ecology, we should present our argument for a continued global effort toward the careful study of soil and its ecological properties. We suggest that there are at least four compelling, and somewhat interconnected, reasons for the study of soil ecology, but there are likely many more.

1. Perhaps the most compelling reason to study soil ecology in the 21st century is humanity's dependence upon the products of soil. Current projections of human population on Earth suggest that there will be between 9.6 and 12.3 billion people on earth by the year 2100, and that this growth will continue well into the 22nd century (Gerland et al., 2014). Clearly, all of these billions will rely on food produced on agricultural soils, but the ability of the finite resource of Earth's soils to sustainably provide the service of food production is not completely clear. Understanding the consequences of agricultural expansion or intensification will require a deep understanding of soil and its inhabitants.
2. Soil has been, and remains one of the final frontiers in biodiversity research. As a reservoir of biodiversity, soil offers an amazing opportunity to study the phenomenon of biological diversity. Because soil is by its nature immobile, it must evolve, in place, a biological community that is theoretically capable of dealing with the full set of past climatic/disturbance conditions that may have ever been experienced in a particular location. From cold and dry Antarctic soils to warm and wet tropical soils, soil everywhere harbors bacteria, fungi, and animals, which act and interact in ways that allow ecosystems to function. However, given the dramatic rate of change in climatic conditions currently underway on earth, it is unknown whether the soil biotic communities of the world's ecosystems will have the capacity to adapt and/or evolve to novel combinations of climate, vegetation, and soil. How will soil communities

DOI: http://dx.doi.org/10.1016/B978-0-12-805251-8.00001-6

respond to changes associated with human forcing of atmospheric processes? The answer will stand as a test of the importance of biodiversity to ecosystem function and the capacity of such diversity to buffer ecosystems against global change, and will be critical with regard to a continuous, sustainable supply of food and fiber for human consumption.

3. Soil is an excellent place to study ecology. In spite of the inconvenient fact that soil is a stubbornly opaque medium, it still provides opportunities to pursue nearly all the subdisciplines of terrestrial ecology: Population, Community, Ecosystem, Stoichiometry, Trophic, Agro-, Disease, Microbial, Restoration, etc. All these avenues of ecological inquiry find a home in soils, and each will be discussed throughout the text. We hope the reader will see that soil–plant–animal ecosystems provide fertile intellectual ground to plow for ecologists of all stripes (Fig. 1.1).

4. Soil ecology is fun! Even at very early ages, children become fascinated with playing in the dirt. Watch any group of children in the outdoors for long enough, and the play will take a turn toward the soil. A hole will be dug, a root excavated, a worm or beetle-grub discovered. Such discoveries elicit strong emotions whether wonder, excitement, or disgust. Children study soil in a semiscientific way, seeking for the perfect clay texture for mud-pies, or for aggregate structure that yields clods suitable for throwing at the neighbor or sibling. All this is correctly called "play," but if the adult scientist can maintain a modicum of childlike inquisitiveness about soil and its many inhabitants, the work of soil ecology can sometimes feel like play, and is undeniably fun!

1.2 THE HISTORICAL BACKGROUND OF SOIL ECOLOGY

The "roots" of human understanding of soil biology and ecology can be traced into antiquity and probably even beyond the written word. We can only imagine hunter–gatherer societies attuned to life cycles of plant roots, fungi, and soil animals important to their diets, their welfare or their cultures, and particularly to environmental conditions favorable to such organisms. Indeed, early agriculture must certainly have

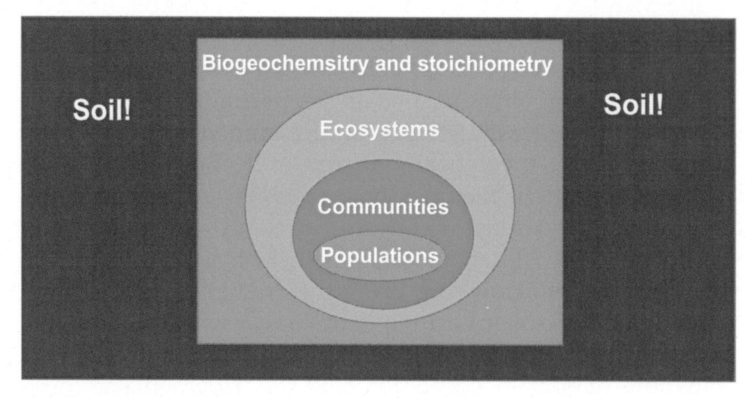

FIGURE 1.1 The full range of opportunities for ecological study within soil.

developed, at least in part, from a practical knowledge of soils and their physical and biological characteristics.

Soil is so fundamental to human life that it has been reflected for millennia in our languages. The Hebrew word for soil is *adama*, from which comes the name Adam—the first man of the Semitic religions who, in these traditions, was formed from clay (Hillel, 1991). Early civilizations had obvious relationships with soils. The Mesopotamian region encompasses present-day Iraq and Kuwait, occupying the valley of the Euphrates and Tigris rivers from their origin as they come out from the high tablelands and mountains of present-day Armenia to their mouth at the Persian Gulf. It had one of the earliest recorded civilizations, the Sumerian, dating from about 3300 years BCE (Hillel, 1991). An inventory taken in the time of the early Caliphates showed 12,500,000 acres (nearly 5,100,000 hectares) under cultivation in the southern half of Mesopotamia (Whitney, 1925). With many centuries of irrigation, this so-called hydraulic civilization was plagued with problems of siltation and salinization, which was written about at the time of King Hammurabi (1760 BCE) (Hillel, 1991). An impressive sequence of civilizations waxed and waned over the millennia: Sumerian, Akkadian, Babylonian, and Assyrian, as cultivation shifted from the lower to central and upper regions of Mesopotamia. Siltation and salinization continue to beset present-day civilizations that practice extensive irrigation-based agriculture with examples worldwide (Rengasamy, 2006).

To the east of Mesopotamia, past the deserts of southern Iran and of Baluchistan, lies the Indus River Valley. Another irrigation-based civilization developed here, probably under the influence of the Mesopotamian civilization. The Indus River civilization probably encompassed a total land area far exceeding that of either Sumeria or Egypt; little is known about it. No written records have been discovered, but its fate, like that of the Sumerian, succumbed to environmental degradation, exacerbated by the extensive deforestation which occurred to provide fuel to bake the bricks used in construction (Hillel, 1991). The bricks in Mesopotamian cities were sun-baked, similar to the adobe style of construction used in the deserts of the southwestern United States.

In contrast, the Egyptian civilization persisted more or less in place, as a result of the annual floods of the Nile River, which renewed soil fertility in vast areas along the river's length as it flowed northward. Over the millennia, from 1 to 3 million people lived along the Nile, and produced enough grain to export wheat and barley to many countries around the Mediterranean rim. Now that the population is some 30 times greater, it must import some foodstuffs and is economically in questionable condition, in spite of the vast areas being irrigated with water from the Aswan high dam.

The ancient Chinese concept of fundamental elements included earth, air, fire, water, and moon. In the Yao dynasty from 2357 to 2261 BCE, the first attempt was made at soil classification surveying. The Emperor established nine classes of soils in as many provinces of China, with a taxation system based upon this system. These classes included the yellow and mellow soils of Young Chow (Shensi and Kansu); the red, clayey, and rich soils of Su Chow (Shantung, Kiangsu, and Anhwei); the whitish, rich salty soils of Tsing Chow (Shantung); the mellow, rich, dark, and thin soils of Yu Chow (Honan); the whitish and mellow soils of Ki Chow (Chili and Shansi); the black and rich soils of Yen Chow (Chili and Shantung); the greenish and light soils of Liang Chow (Szechuan and Shensi); and the miry soils of King Chow (Hunan and Hupeh) and Yang Chow (Kiangsu)

(Whitney, 1925). This system reflects a sophisticated knowledge within early Chinese civilization of soils and their relationship with plant growth. Interestingly, in recognition of the importance of biological activity in soils, the ancient Chinese termed earthworms as "angels of the soil" (Blakemore, 2002).

The Greeks believed there were four basic elements: earth, air, fire, and water; and Aristotle, understanding the role of earthworms in organic matter decomposition, considered earthworms to be the "intestines of the earth" (Edwards and Lofty, 1977). In Greek mythology, the champion of mortal men, Heracles, defeated Antaeus, perhaps symbolically ushering in the age of man's dominion over the soil. A wild and dangerous giant, Antaeus, who was the son of Gaia (goddess of the earth), was invincible so long as he maintained physical contact with the soil, but Heracles recognized this and lifted the giant into the air and squeezed him to death in a bear hug. The Greeks and Romans also had a clear differentiation of the productive capacities of different types of soils. They referred to good soils as "fat," and soils of lower quality as "lean" (Whitney, 1925). For the Roman writers, "humus" referred to soil or earth. Virgil (79–19 BC), in his *Georgics*, named the loamy soil *pinguis humus* and used the words *humus*, *solum*, or *terra* more or less interchangeably for the notions of soil and earth. Columella in the 1st century AD noted, "wheat needs two feet of good *humus*" (Feller, 1997; italics added).

The word *humus* seems to have entered the European scientific vocabulary in the 18th century. Thus in Diderot and d'Alembert's Encyclopaedia (vol. 8) in 1765: "Humus, natural history, this Latin word is often borrowed by naturalists (even into French) and denotes the mold, the earth of the garden, the earth formed by plant decomposition. It refers to the brown or darkish earth on the surface of the ground. Refer to the mold or vegetable mold" (translation in Feller, 1997).

By the beginning of the 19th century, the leading authorities with a biological view of soils were Leeuwenhoek, Linnaeus, and other pre-Darwinians, and then Darwin himself (1837, 1881), who "fathered" the modern era. Müller (1879, 1887), cited in Feller (1997), laid the groundwork for the present-day scientific bases of the different forms of humus, and even included a general survey of soil genetic processes in cold and temperate climates. Müller developed terms for the three humus types—Mull, Mor, and MullartigerTorf—the latter equivalent to Moder. *Mull* is mold and *Torf* is peat in Danish. Thus MullartigerTorf is *mold peat* in Danish, and it is viewed as an intermediate form between the two extremes (see Feller, 1997 for more details on the history of these fascinating substances).

The first scientific view of soils as natural bodies that develop under the influence of climate and biological activity acting on geological substrates arose in Russia with the work of Dokuchaev and his followers (Feller, 1997; Zonn and Eroshkina, 1996), in Europe with Müller's (1887) descriptions of soil horizon development (Tandarich et al., 2002), and in England with Darwin's observations on textural sorting through animal activities that resulted in soil horizonation (Johnson and Schaetzl, 2015). In any case, the ecological basis of the Russian tradition is clear in the words of Glinka (1927; cited in Jenny, 1941), a disciple of Dokuchaev, whose view of soil included "... not only a natural body with definite properties, but also its geographical position and surroundings, i.e., climate, vegetation, and animal life." This Russian perspective predates the formal statement of the ecosystem concept by several decades (Tansley, 1935).

During this early period of theoretical development across the Atlantic, soil science in the United States was more concerned with practical matters of agriculture, such as soil productivity and crop growth (Tandarich et al., 2002) and, later, on restoration of soils badly degraded from poor management (e.g., the "dust bowl" in the Great Plains and the severely eroded croplands of the southeastern United States) (Sutter, 2015). It was not until the 1920s that ideas of pedogenesis gained wide recognition in the United States. Within the next decade, Jenny (1941) published his classic work on soil formation, drawing heavily from Dokuchaev's ideas to synthesize pedological and ecological perspectives into the concept of a "... soil system [that] is only a part of a much larger system ... composed of the upper part of the lithosphere, the lower part of the atmosphere, and a considerable part of the biosphere." He formulated this concept into the now famous "fundamental equation of soil-forming factors":

$$s = f(cl, o, r, p, t, \ldots)$$

where s refers to the state of a body of soil at a point in time; f refers to function; cl to climate; o to organisms; r to relief or topography; p to parent material; and t to time. Jenny, probably more than any North American soil scientist of his era, emphasized the importance of the biota in and upon soils. His last major work, *The Soil Resource* (1980), is now a classic in the literature on ecosystem ecology.

Since Jenny's work, research in soil ecology has experienced a "renaissance" as the significance of biological activity in soil formation, organic matter dynamics, and nutrient cycling have become widely recognized. The post-World War II scientific boom was an important impetus for science generally, including soil science. In the United States, the Atomic Energy Commission (later the Department of Energy), through the national laboratories, funded soil biology in relation to nutrient and radioisotope recycling in soil systems (Auerbach, 1958); more recently, the National Science Foundation's Division of Environmental Biology and the United States Department of Agriculture (USDA) National Research Initiative in Soils and Soil Biology have supported a wide array of research in soil ecology. The International Biological Program (IBP) on the international scene greatly expanded methodologies in soil ecology and increased our knowledge of ecological energetics and soil biological processes (Coleman, 2010; Golley, 1993).

In a concise review of almost a century's work by pedologists and soil scientists, Johnson and Johnson (2010) offered the following passage with regard to a simple, overarching definition of soil in the broadest possible sense, which we endorse here as being perhaps the most succinct expression of these ideas to date:

> So, how should we define soil insofar as it technically is the 'skin' of landforms, an integument of planets, and in the case of Earth primarily biophysically-biochemically produced with an epidermal biomantle? The best scientific definition is one that is most useful and comprehensible to a spectrum of scientists and the lay public. It should be simply expressed, scientifically sound, easily explainable, and cover all cases (Johnson et al. 1997), and ideally have universality. Being applicable to soil on Earth and on all other generally lithic-composed planets and planetoids, a definition that fits these criteria is: ***Soil is substrate at or near the surface of Earth and similar bodies altered by biological, chemical, and/or physical agents and processes*** (emphasis added).

In summary, all of these developments and advances in knowledge, from the ancient to the modern, have led to a vast literature upon which is based our current understanding of the soil beneath our feet and the vital role that this living milieu plays in sustaining life on a thin, dynamic, fragile planetary crust.

1.3 WATER AS A CONSTITUENT OF SOIL

> The occurrence of water is, moreover, not less important and hardly less general upon the land. In addition to lakes and streams, water is almost everywhere present in large quantities in the soil, retained there mainly by capillary action, and often at greater depths. ***(Henderson, 1913).***

Lawrence J. Henderson, a noted physical chemist and physiologist, published a book (*The Fitness of the Environment*, 1913), which was a landmark among books on biological topics. Henderson's thesis is that one substance, water, is responsible for the characteristics of life, and the biosphere as we know it. The highly bipolar nature of water, with its twin hydrogen bonds, leads to a number of intriguing characteristics (e.g., high specific heat), which have enabled life in the thin diaphanous veil of the biosphere (Lovelock, 1979, 1988) to extend and proliferate almost endlessly through the air, water, soil, and several kilometers into the earth's mantle (Whitman et al., 1998).

A central fact of soil science is that certain physicochemical relationships of matter in all areas of the biosphere are mediated by water. Thus soil, which we normally think of as opaque and solid, from the wettest organic muck soil to the parched environs of the Atacama, Kalahari, Gobi, or Mojave deserts, is dominated by the amount and availability of water.

Consider water in each of its phases—solid, liquid, and gaseous:

1. *Solid:* In aquatic ecosystems, water freezes from the top down, because it has its greatest density at 4°C. This allows for organismal activity to continue at lower depths and in sediments as well. In soil, the well-insulated nature of the soil materials and water with its high specific heat means that there is less likelihood of rapid freezing. Water expands when it freezes. In more polar climates (and in some temperate ones), soil can be subjected to "frost heaving," which can be quite disruptive, depending on the nature of the subsurface materials.
2. *Liquid:* Water's high specific heat of 1 calorie per gram per degree Celsius increase in temperature has a significant stabilizing influence in bodies of water and soil (Hadas, 1979). The effect of the high specific heat is to reduce fluctuations in temperature. The location of the liquid, in various films, or in empty spaces, has a marked influence on the soil biota.
3. *Vapor:* It is somewhat counterintuitive but true that the atmosphere within air-dry soil (gravimetric water content of 2% by weight) has a relative humidity of 98%. The consequences of this humidity for life in the soil are profound. Most soil organisms spend their lives in an atmosphere saturated with water. Many soil animals absorb and lose water through their integuments, and are entirely dependent upon saturated atmospheres for their existence.

From the pragmatic viewpoint of the soil physicist, we can consider aqueous and vapor phases of water conjointly. Following a moisture release curve, one can trace the pattern of water, in volume and location in the soil pore spaces, in the following manner (Vannier, 1987). Starting with freestanding, or gravitational, water at saturation, the system is essentially subaquatic (Fig. 1.2). With subsequent evaporation and plant transpirational water losses from the soil, the freestanding water disappears, leaving some capillary-bound water (Fig. 1.3), which has been termed the edaphic system. Further evaporation then occurs, resulting in the virtual absence of any capillary water, leaving only the adsorbed water at a very high negative water tension (Fig. 1.4).

The implications of this complex three-dimensional milieu are of fundamental importance for a very diverse biota. Vannier (1973) proposed the term "porosphere" for this intricate arrangement of sand, silt, clay, and organic matter. Primitive invertebrates first successfully undertook the exploitation of aerial conditions at the beginning of the Paleozoic era (Vannier, 1987). This transition probably took place via the soil medium, which provided the necessary gradient between the fully aquatic and aerial milieus. This water-saturated environment, so necessary for such primitive, wingless (Apterygote) forms as the Collembola, or springtails (Fig. 1.2), is equally important for the transient life-forms such as the larval forms of many flying insects, including Diptera and Coleoptera. In addition, many of the micro- and mesofauna (described in Chapter 4, Secondary Production: Activities and Functions of Heterotrophic Organisms—The Soil Fauna) could be considered part of the "terrestrial nannoplankton" (Stout, 1963). Stout included all of the water-film inhabitants, namely: bacteria and yeasts, protozoa, rotifers, nematodes,

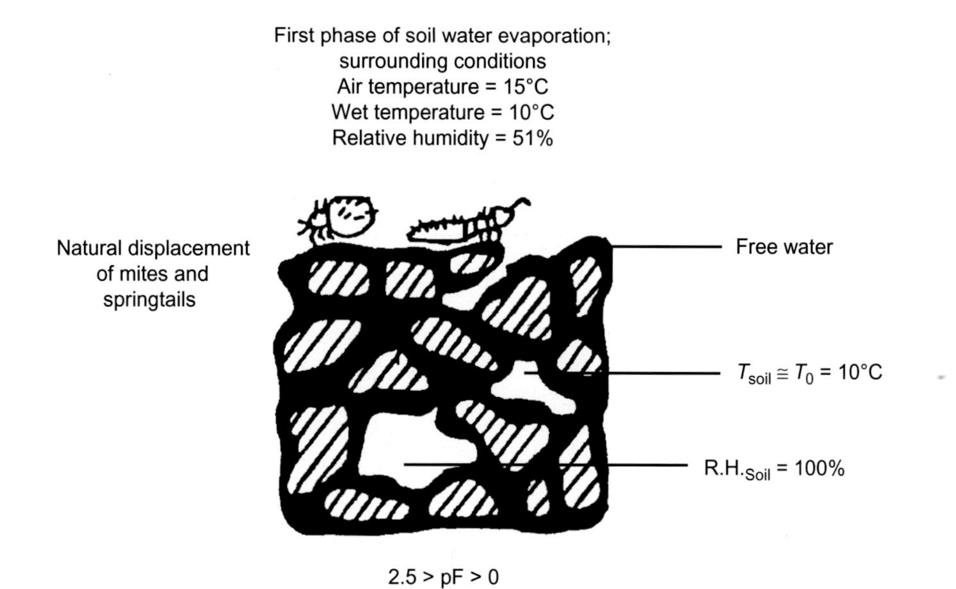

FIGURE 1.2 Gravitational moisture (the subaquatic system) in the soil framework. pF = −log cm H_2O suction; R.H. = relative humidity; 2.5pF = field capacity. *Source: From Vannier, G., 1987. The porosphere as an ecological medium emphasized in Professor Ghilarov's work on soil animal adaptations. Biol. Fert. Soil. 3, 39 – 44.*

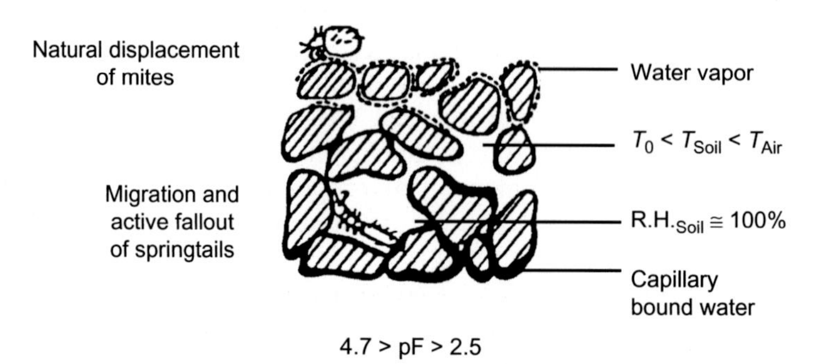

FIGURE 1.3 Capillary moisture (the edaphic system) in the soil framework. 4.7pF = −5 mPa; 2.5pF = −0.03 mPa = field capacity. *Source: From Vannier, G., 1987. The porosphere as an ecological medium emphasized in Professor Ghilarov's work on soil animal adaptations. Biol. Fert. Soil. 3, 39 − 44.*

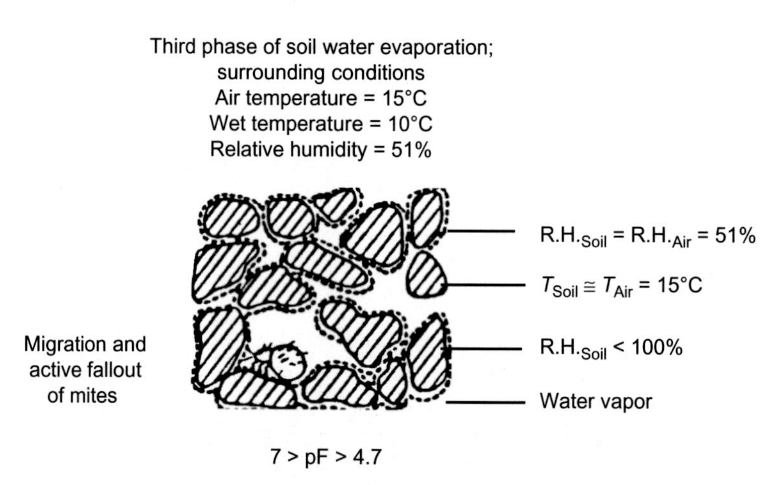

FIGURE 1.4 Adsorptional moisture (the aerial system) in the soil framework. 7pF = −1000 mPa; 4.7pF = −5 mPa; permanent wilting point = −1.5 mPa = 4.18pF. *Source: From Vannier, G., 1987. The porosphere as an ecological medium emphasized in Professor Ghilarov's work on soil animal adaptations. Biol. Fert. Soil. 3, 39 − 44.*

copepods, and enchytraeids (the small oligochaetes also called potworms). Raoul Francé, a German sociologist, made analogies between aquatic plankton and the small- and medium-sized organisms that inhabit the water films and water-filled pores in soils, terming them: "Das Edaphon" (Francé, 1921).

As noted in Figs. 1.2–1.4, there is a marked difference in moisture requirements of some of the soil microarthropods. Thus another major group, the Acari, or mites, are often able to tolerate considerably more desiccation than the more sensitive Collembola. In both cases, the microarthropods make a gradual exit from the soil matrix as the desiccation sequence described earlier continues.

Other organisms, more dependent on the existence of free water or water films, include the protozoa and nematoda, the life histories and feeding characteristics of which are covered in Chapter 4, Secondary Production: Activities and Functions of Heterotrophic Organisms—The Soil Fauna. In a sense, the very small fauna, and the bacteria they feed upon, exist in a qualitatively different world from the other fauna, or from fungi. Both larger fauna and fungi move in and out of various water films through various pores, which are less than 100% saturated with water vapor, with comparative ease (Hattori, 1994).

In conclusion, this overview of soil physical characteristics and their biological consequences notes the following: "For a physicist, porous bodies are solids with an internal surface that endows them with a remarkable set of hygroscopic properties. For example, a clay such as bentonite has an internal surface in excess of 800 $m^2 g^{-1}$, and a clay soil containing 72 percent montmorillonite possesses an internal surface equal to 579 $m^2 g^{-1}$. The capacity to condense gases on free walls of capillary spaces (the phenomenon of adsorption) permits porous bodies to reconstitute water reserves from atmospheric water vapor" (Vannier, 1987). Later, we will address the phenomenon of adsorption in other contexts, ones that are equally important for soil function as we know it.

1.4 ELEMENTAL CONSTITUTION OF SOIL

Many elements are found within the earth's crust, and most of them are in soil as well. However, a few elements predominate. These are hydrogen, carbon, oxygen, nitrogen, phosphorus, sulfur, aluminum, silicon, and alkali and alkaline earth metals. Various trace elements or micronutrients are also biologically important as enzyme cofactors, and include iron, cobalt, nickel, copper, magnesium, manganese, molybdenum, and zinc.

A more functional and esthetically pleasing approach is to define soil as predominantly a sand–silt–clay matrix, containing living (biomass) and dead (necromass) organic matter, with varying amounts of gases and liquids within the matrix. In fact, the interactions of geological, hydrological, and atmospheric (Fig. 1.5) facets overlap with those of the biosphere, leading to the union of all, overlapping in part in the pedosphere. Soils, in addition to the three geometric dimensions, are also greatly influenced by the fourth dimension of time, over which the physicochemical and biological processes occur.

1.5 HOW SOILS ARE FORMED

Soils are the resultant of the interactions of several factors—climate, organisms, parent material, and topography (relief)—all acting through time (Jenny, 1941, 1980) (Fig. 1.6). These factors affect major ecosystem processes (e.g., primary production, decomposition, and nutrient cycling), which lead to the development of ecosystem properties unique to that soil type, given its previous history. Thus characteristics such as cation-exchange capacity, texture, structure, and organic matter status are the outcomes of the aforementioned processes operating as constrained by the controlling factors. Different arrays of processes may predominate in various ecosystems (see Fig. 1.6).

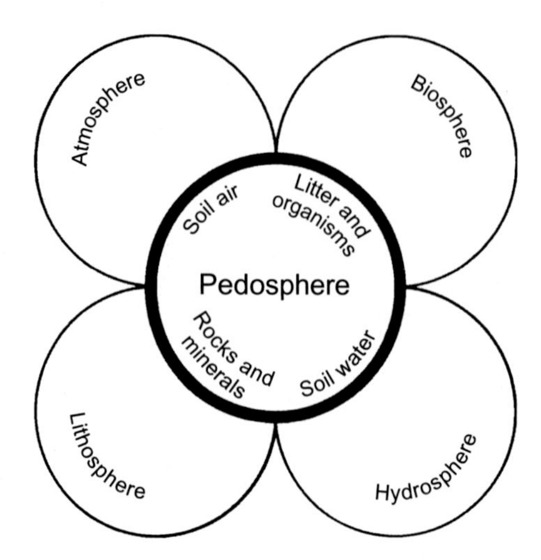

FIGURE 1.5 The pedosphere, showing interactions of abiotic and biotic entities in the soil matrix. *Source: From FitzPatrick, E.A., 1984. Micromorphology of Soils. Chapman and Hall, London (FitzPatrick, 1984).*

1.6 PROFILE DEVELOPMENT

The abiotic and biotic factors noted earlier lead to certain chemical changes down through the top few decimeters of soil (Fig. 1.7A, B). In many soils, particularly in more mesic or moist regions of the world, there is leaching and redeposition of minerals and nutrients, often accompanied by a distinct color change (profile development). Thus as one descends through the profile from the air-litter surface, one passes through the litter (L), fermentation (F), and humification (H) zones (O_i, O_e, and O_a, respectively), then reaching the mineral soil surface, which contains the preponderant amount of organic matter (A horizon). The upper portion of the A horizon is termed the topsoil, and under conditions of cultivation, the upper 12–25 cm is called the plow layer or furrow slice. This is followed by the horizon of maximum leaching, or eluviation, of silicate clays, Fe, and Al oxides, known as the E horizon. The B horizon is next, with deeper dwelling organisms and somewhat weathered material. This is followed by the C horizon, the unconsolidated mineral material above bedrock. The solum includes the A, E, and B horizons plus some of the cemented layers of the C horizon. All these horizons are part of the regolith, the material that overlies bedrock. More details on soil classification and profile formation are given in soil textbooks, such as Russell (1973) and Brady and Weil (2000).

The work of the soil ecologist is made somewhat easier by the fact that the top 10–15 cm of the A horizon, and the L, F, and H horizons (O_i, O_e, and O_a) of forested soils contain the majority of plant roots, microbes, and fauna (Coleman et al., 1983; Paul and Clark, 1996). Hence a majority of the biological and chemical activities occur in this layer. Indeed, a majority of microbial and algal-feeding fauna, such as protozoa (Elliott and Coleman, 1977; Kuikman et al., 1990) and rotifers and tardigrades (Leetham et al., 1982), are within 1 or 2 cm of the surface. Microarthropods are most abundant usually in the top 5 cm of forest soils (Schenker, 1984) or grassland soils (Seastedt, 1984a), but are occasionally more abundant at 20–25 cm and even 40–45 cm at certain times of the year in

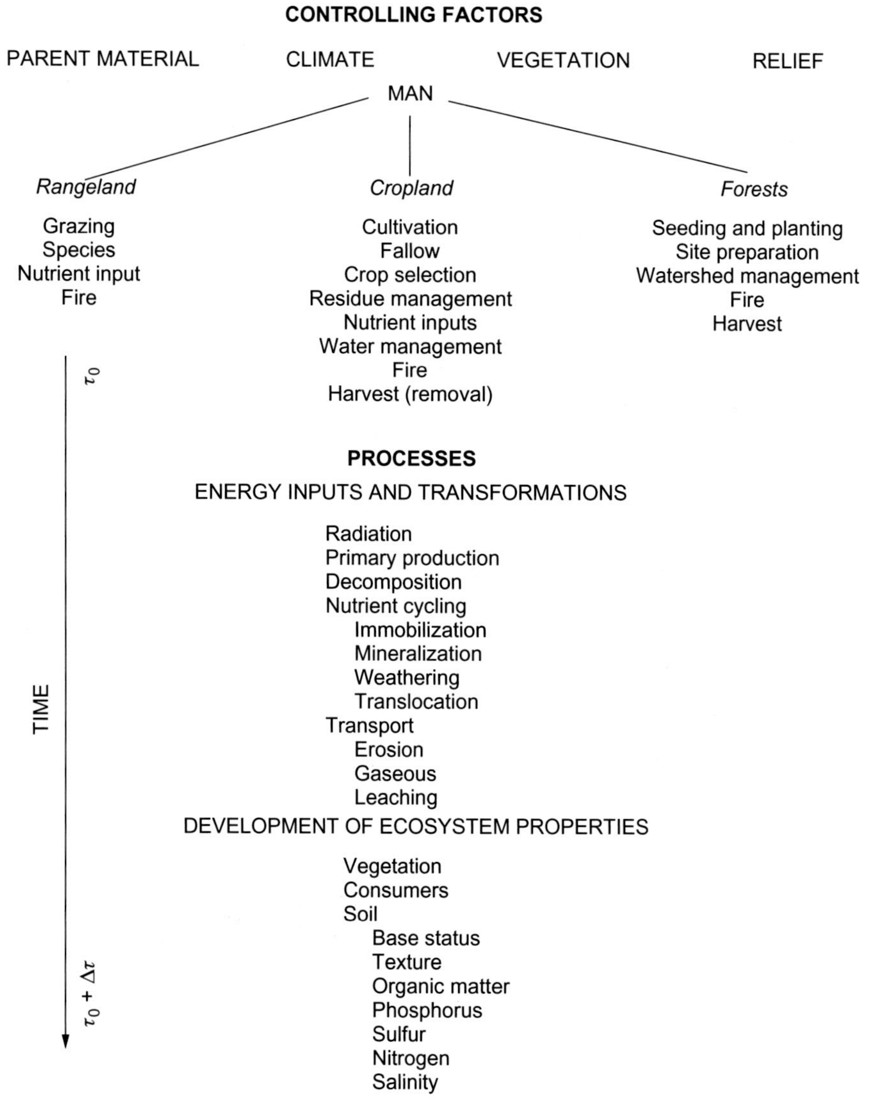

FIGURE 1.6 Soil-forming factors and processes, interaction over time. *Source: From Coleman, D.C., Reid, C.P.P., Cole, C.V., 1983. Biological strategies of nutrient cycling in soil systems. Adv. Ecol. Res. 13, 1 – 55, modified from Jenny, H., 1980. The Soil Resource: Origin and Behavior. Ecological Studies 37, Springer-Verlag, New York.*

tallgrass prairie (O'Lear and Blair, 1999). This region may be "primed," in a sense, by the continual input of leaf, twig, and root materials, as well as algal and cyanobacterial production and turnover in some ecosystems, while soil mesofauna such as nematodes and microarthropods may be concentrated in the top 5 cm. Significant numbers of nematodes may be found at several meters' depth in xeric sites such as deserts in the American Southwest (Freckman and Virginia, 1989).

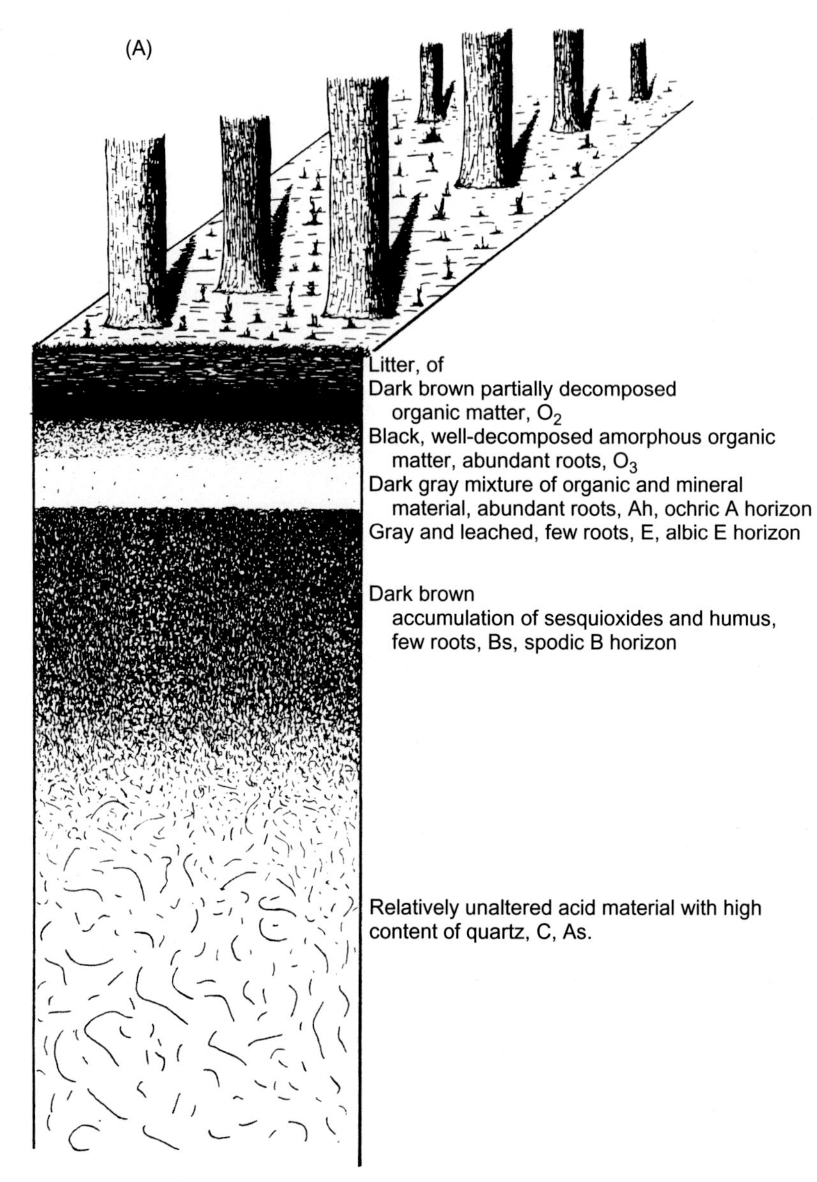

FIGURE 1.7 (A) Diagram of a Podzol (spodosol in North American soil taxonomy) profile with minerals accumulating in subsurface horizons. This is the characteristic soil of coniferous forests. (B) Diagram of a Cambisol profile, with the organic matter well mixed in the A horizon; due to faunal mixing there is no mineral accumulation in subsurface horizons. This is the characteristic soil of the temperate deciduous forests. *Source: From FitzPatrick, E.A., 1984. Micromorphology of Soils. Chapman and Hall, London.*

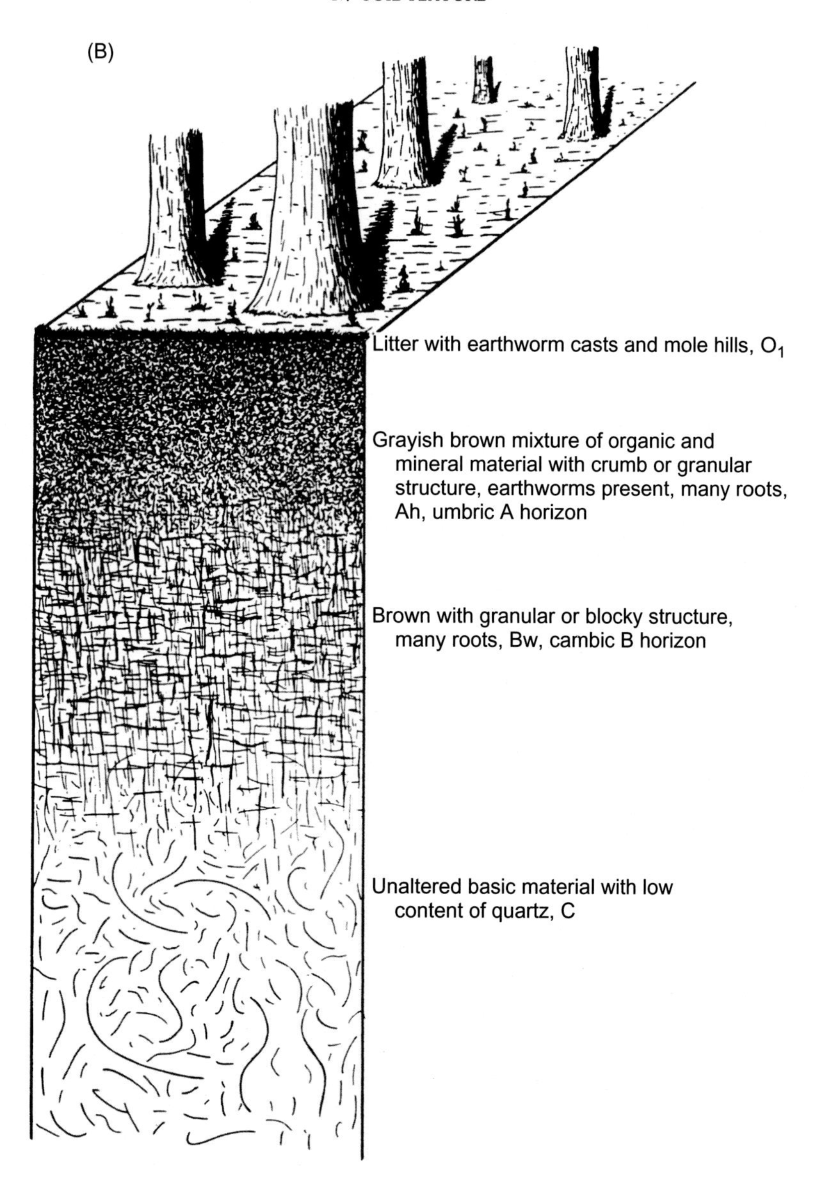

FIGURE 1.7 (Continued).

1.7 SOIL TEXTURE

Historically, texture was a term used to describe the workability of an agricultural soil. A heavy, clay soil required more effort (horsepower) to till than a lighter, sandy loam (Russell, 1973). A more quantifiable approach is to characterize soils in terms of the sand, silt, and clay present, which are ranged on a spectrum of light–intermediate–heavy or sandy–silt–clay. The array of textural classes (Fig. 1.8) shows percentages of sand, silt, and clay, and the resulting soil types such as sandy, loamy, or clayey soils.

The origin and mineralogical composition of mineral particles in soil is a most interesting and complex one. The particles are in two major categories: (1) crystalline minerals derived from primary rock, and (2) those derived from weathering animal and plant residues. The microcrystalline forms comprised calcium carbonate, iron, or aluminum oxides, or silica.

The clay fraction, so important in imparting specific physical properties to soils, to microbial life, and to plant activity via nutrient availability, comprised particles less than 2 μm in diameter. Unlike the sand–silt minerals, clays are weathered forms of primary minerals, and hence they are referred to as secondary minerals. Coarse clay particles (0.5 μm) often are derived from quartz and mica; finer clays (0.1 μm) are clay minerals or weathered products of these (such as hydrated ferric, aluminum, titanium, and manganese oxides).

No matter what size the particle is, microorganisms in unsaturated soil exist in a world dominated by the presence of extensive surfaces. There seems to be a general advantage to microbes living at these interfaces in terms of enhanced nutrient concentrations and the potential to use many of the physical substrates themselves as energy or nutrient sources. The thickness of water films in unsaturated conditions allows the microbes little option except to adhere to the surfaces (Mills, 2003). We discuss some of the microbial dynamics and interactions with soil organic matter in Chapter 3, Secondary Production: Activities and Functions of Heterotrophic Organisms—Microbes.

The roles of coarse and fine clays in organic matter dynamics are under intensive scrutiny in several laboratories around the world (Oades and Waters, 1991; Six et al., 1999). It is possible that labile (i.e., easily metabolized) constituents of organic matter are preferentially adsorbed onto fine clay particles and may be a significant source of energy for the soil microbes (Anderson and Coleman, 1985). For more information on the environmental attributes of clays, see Hillel (1998).

1.8 CLAY MINERAL STRUCTURE

The clay minerals in soil are in the form of layer-lattice minerals, and are made up of sheets of hydroxyl ions or oxygen. The clay minerals fall into two groups: (1) those with three groups of ions lying in a plane (the 1:1 group of minerals), and (2) those with four groups of ions lying in a plane (the 2:1 group of minerals). The type mineral of the 1:1 group is kaolinite, which typically has a very low charge on it. In contrast, the 2:1 type mineral, for example illite, carries an appreciably higher negative charge per unit weight than the kaolin group. More detailed information on the clay particles, their composition, and charges upon them is given in Theng (1979) and Oades et al. (1989).

A key concern to the soil ecologist is the extremely high surface area found per gram of clay mineral. Surface areas can range from 50 to 100 $m^2\ g^{-1}$ for kaolinitic clays, from 300 to 500 $m^2\ g^{-1}$ for vermiculites, and from 700 to 800 $m^2\ g^{-1}$ for well-dispersed smectites (Russell, 1973). These impressively large surface areas can play a pivotal role in adsorbing

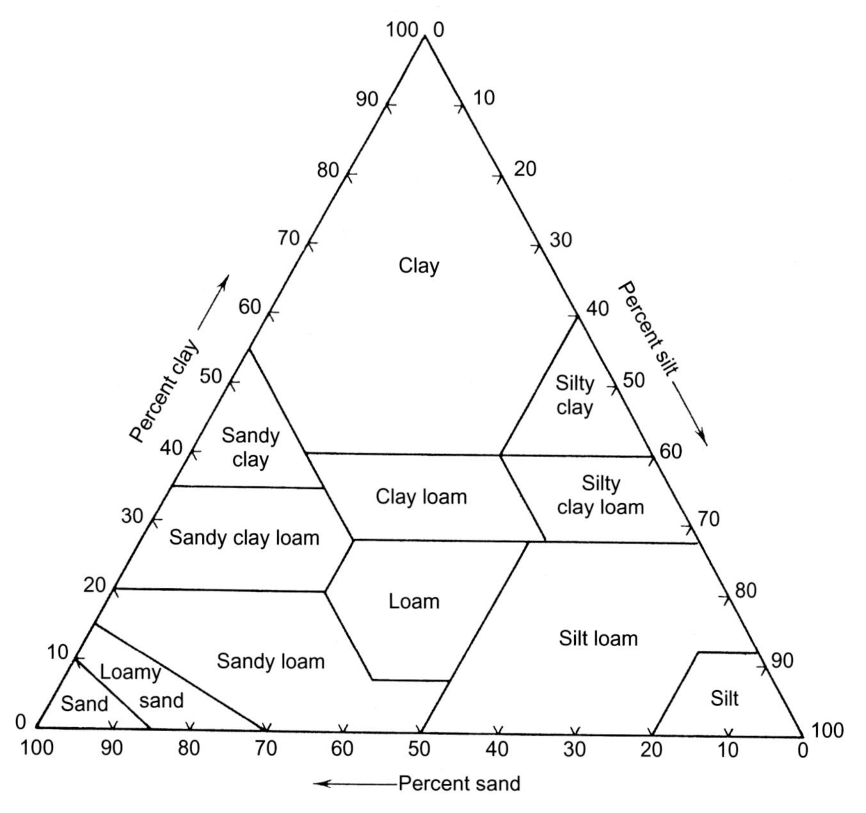

FIGURE 1.8 The soil texture triangle. Using this diagram, the textural name of a soil may be determined from a mechanical analysis. The points corresponding to the percentages of sand, silt, and/or clay present in a given soil sample are located on their respective axes. Lines are then projected inward, based on the relative abundance of these particles (only two of the three must be known). The name of the compartment in which the two lines intersect is the class name of the soil in question. *Source: Modified from Buckman and Brady (1970).*

and desorbing inorganic and organic constituents in soils, and have only recently been treated in an appropriately analytical fashion as an integral part of the soil nutrient system (Oades et al., 1989; Tisdall and Oades, 1982).

1.9 SOIL STRUCTURE

Structure refers to the ways in which soil particles are arranged or grouped spatially. The groupings may occur at any size level on a continuum from either extreme of what are nonstructural states: single grained (such as loose sand grains) or massive aggregates of aggregates (large, irregular solid).

An additional aspect of aggregates, their stabilization once they are formed, is significant for soil ecology. Stabilization is the result of various binding agents. Plant and microbial polysaccharides and gums serve as binding agents (Cheshire, 1979; Cheshire et al., 1984; Harris et al., 1964). A variety of other organic compounds act as binding agents (Cheshire, 1979), and some biological agents such as roots and fungal hyphae (Tisdall, 1991; Tisdall and Oades, 1979, 1982) play a similar role.

The implications of soil structure refer not only to the particles but also extend to the pore spaces within the structure, as noted earlier. Indeed, it is the nature of the porosity that exists in a well-structured soil that leads to the most viable communities within it. This in turn has strong implications for ecosystem management, particularly for agroecosystems (Elliott and Coleman, 1988). There is a very active area of research in soil ecology related to dynamics of micro- and macroaggregates, in relationship to drying–wetting cycles and tillage management. Denef et al. (2001) measured marked differences in aggregate formation and breakdown as a function of amount of bacterial and fungal activity in soils with ^{13}C-labeled crop residues. They traced differences in fine intraaggregate Particulate Organic Matter (POM) to variations in wetting and drying regimes versus those soils not experiencing such environmental fluctuations. We discuss the aggregate formation process further in Chapter 3, Secondary Production: Activities and Functions of Heterotrophic Organisms—Microbes on microbes and their effects on ecosystems.

Several types of structural forms are found in soils. The four major types are platelike, prismlike, blocklike, and spheroidal (Fig. 1.9). All of these are "variations on a theme," as it were, of a fundamental unit of soil aggregation: the ped. A ped is a unit of soil structure, such as an aggregate, crumb, prism, block, or granule, formed by natural processes. This is distinguished from a clod, which is artificial or man-made (Brady and Weil, 2000). Soils may have peds of differing shapes, in surface and subsurface horizons. These are the result of differing temperature, moisture, and chemical and biological conditions at various levels in the soil profile.

Another concept is helpful in soil structure: the pedon. This is an area, from 1 to 10 m^2, under which a soil may be fully characterized. Later in the book, we will consider the arrangement of soil units in a landscape, and in an entire region. Next, we will examine some of the causes for the formation, or genesis, of soil structure.

Input of organic matter to soil is one of the major agents of soil structure. The organic matter comes from both living and dead sources (roots, leaves, microbes, and fauna). Various physical processes, such as deformation and compression by roots and soil fauna, and freezing–thawing or wetting–drying, also have significant influences on soil structure. It is generally recognized that plant roots and humus (resistant organic breakdown products) play a major role in the formation of aggregates (Elliott and Coleman, 1988; Paul, 2015). However, bacteria and fungi and their metabolic products play an equally prominent role in promoting granulation (Cheshire, 1979; Foster, 1985; Griffiths, 1965). We will explore organic matter dynamics in the sections on soil biology.

The interaction of organic matter and mineral components of soils has a profound influence on cation-adsorption capabilities. The interchange of cations in solution with cations on these surface-active materials is an important phenomenon for soil fertility. The capacity of soils to adsorb ions (the cation-exchange capacity) is due to the sum of exchange sites on both organic matter and minerals. However, in most soils, organic matter has the higher exchange capacity (number of exchange sites). For a more extensive account, see Paul (2015).

There is a hierarchical nature to the ways in which soil structure is achieved, and it reflects the biological interactions within the soil matrix (Elliott and Coleman, 1988; Six

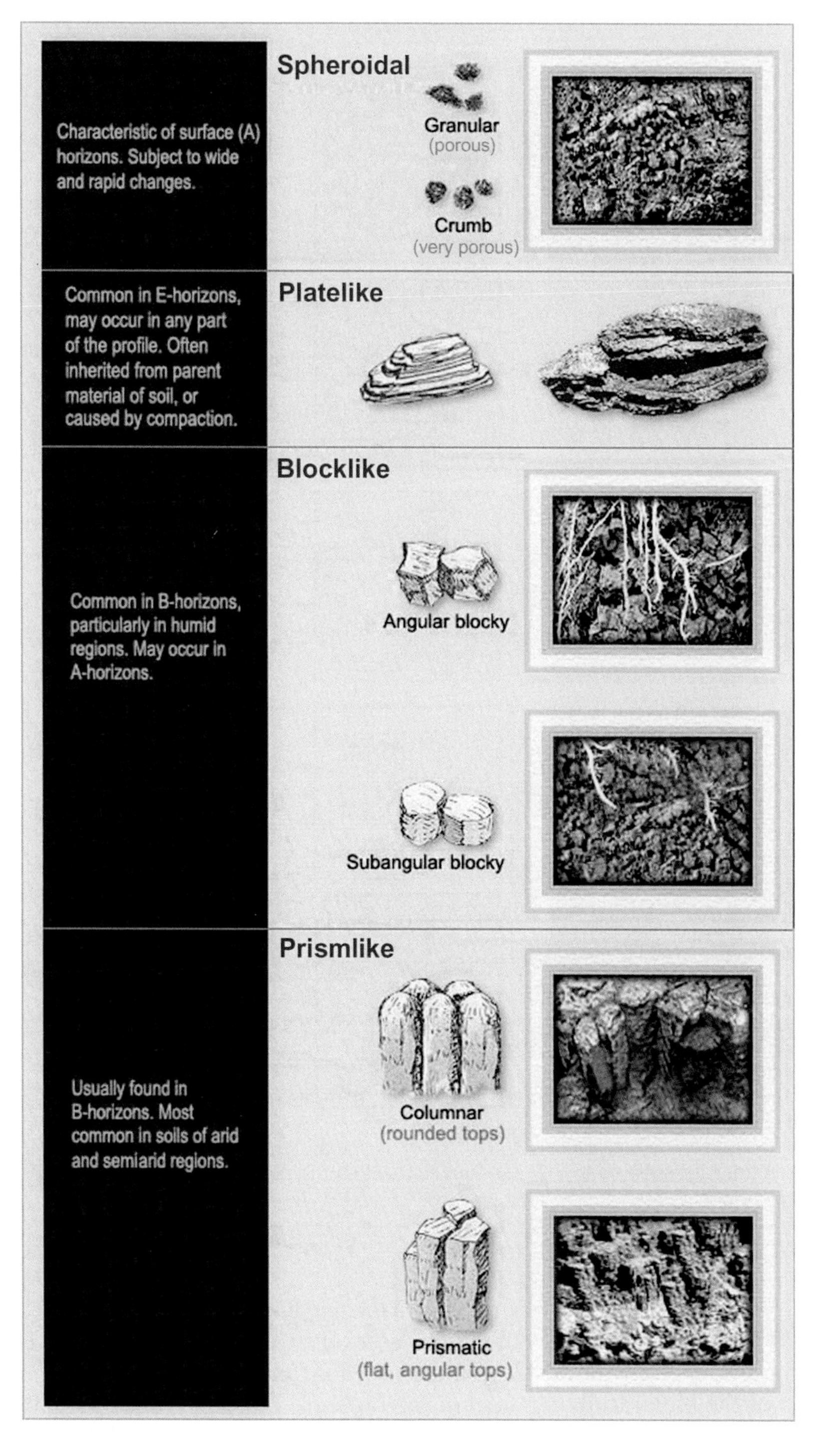

FIGURE 1.9 Various structural types found in mineral soils. Their location in the profile is sugested. In arable topsoils, a stable granular structure is prized. *Source: From Brady, N.C., 1974. The Nature and Properties of Soils, eighth ed. MacMillan, New York (Brady, 1974).*

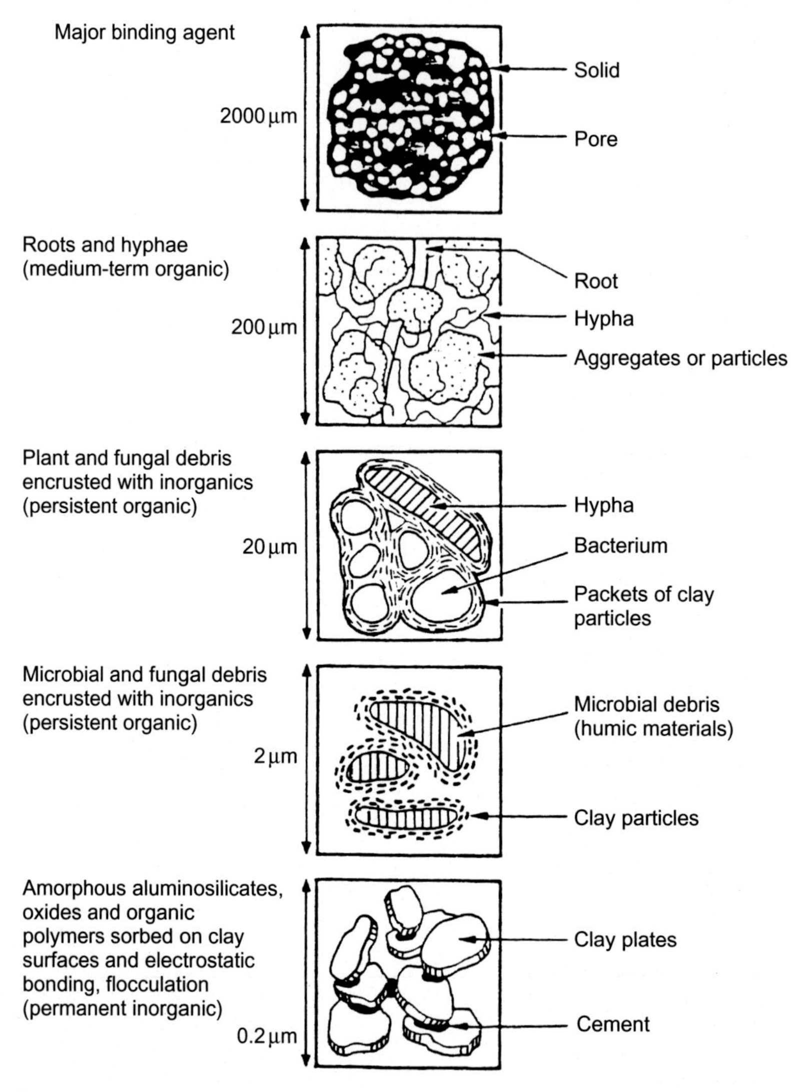

FIGURE 1.10 Soil microaggregates, across five orders of magnitude, beginning at the level of clay particles, through plant and fungal debris, up to a 2-mm diameter soil crumb. *Source: From Tisdall, J.M., Oades, J.M., 1982. Organic matter and waterstable aggregates in soils. Eur. J. Soil Sci. 33, 141 – 163.*

et al., 2002a). Several Australian researchers (Oades, 1984; Tisdall and Oades, 1982; Waters and Oades, 1991) have noted how the processes of structuring soils extend over many orders of magnitude, from the level of the individual clay platelet to the ped in a given soil. For most of the biologically significant interactions, one can consider changes across a range of at least six orders of magnitude from $<0.01\ \mu m$ to <1 cm (Tisdall and Oades, 1982) (Fig. 1.10). Not all soils are aggregated by biological agents; for heavily weathered

Oxisols with kaolinite-oxide clays, there seems to be no hierarchy of organization below 20 μm, because only physicochemical forces predominate there (Oades and Waters, 1991). Studies in our Horseshoe Bend agroecosystem project at the University of Georgia have uncovered significant differences between tillage regimes (conventional, moldboard plowing vs no-tillage, direct drilling of the seeds into the soil). The aggregates in the 53–106 and 106–250 μm categories are most affected by fungal growth and proliferation, reflecting physical binding and the increased amounts of acid-hydrolysable carbohydrates, which are more prevalent in the no-tillage treatments as compared with the bacteria-dominated conventional tillage systems (Beare et al., 1994a, 1994b, 1997).

It is the interactions between physical, chemical, and biological agents in soils that are so fascinating, complex, and important to consider as we increase the intensity of management of terrestrial ecosystems, or alter their usage in response to increased human concerns about their use, and also strive for effective sustainability of them worldwide (Coleman et al., 1992, 1998). Indeed, Lavelle (2000) observed that soil ecology can be considered to have arisen from the convergence of three major approaches: (1) the development of enormous databases on communities of microorganisms and invertebrates and their energy budgets via the International Biological Program (e.g., Petersen and Luxton, 1982); (2) the placement of decomposition processes on center stage, bridging soil chemistry with soil biology (Swift et al., 1979); and (3) an appreciation of the effects of soil organisms on soil structure, including the influence of macrofauna as ecosystem engineers (Bal, 1982, Jones et al., 1994). We would add a fourth dimension, that of soils and sediments as repositories or libraries of DNA. In Siberia, several permafrost cores dating from 10,000 to 400,000 years old have yielded at least 19 different plant taxa, as well as megafauna sequences of mammoth, bison, and horse (Willerslev et al., 2003). Temperate cave sediments from New Zealand yielded 29 taxa characteristic of the prehuman environment, including two species of ratite moas. These genetic records of paleoenvironments will add to our understanding of past ecosystem structure and, possibly, function.

1.10 SOILS AS SUPPLIERS OF ECOSYSTEM SERVICES

Soils are large repositories of mineral and organic wealth, available for both the use and misuse by civilizations on this planet (Hillel, 1991). Levels of soil carbon have dropped by as much as 50% after 50–100 years of intensive farming in the North American Great Plains (Haas et al., 1957). Similar concerns were expressed about loss of organic matter and erosion of soils in the Mediterranean region at the time of Plato in the third century BCE, as noted earlier (Whitney, 1925).

An example of the monetary value of what soils provide is given by the costs of raising crops in intense nonsoil conditions using hydroponic culture. Construction of a modern hydroponics system in the United States, including pumps and sophisticated computer control systems, costs upward of $850,000 per hectare (FAO 1990, cited by Daily et al., 1997). Soils also play significant roles in the regulation of global greenhouse gases such as carbon dioxide, methane, and nitrous oxides (Schimel and Gulledge, 1998). As we present in detail in later chapters, the cleansing and recycling role that soils play in processing organic wastes and recycling nutrients constitutes one of the major benefits provided "free" to humanity and all the biota (outside the market economy) but worth literally trillions of dollars per year as one of the major ecosystem services (Costanza et al., 1997) on Earth.

1.11 SUMMARY

The physical properties of the soil are the production of continued interactions between soil biota and their abiotic milieu. Water, the "universal solvent," exerts a strong influence on the biota because many of the biota are adapted to life in a saturated atmosphere. The interplay between liquid and gaseous phases of water, in turn, is largely determined by pore size. The arrangement of particles in soils (the porosphere) is an important determinant for the ecology of the soil microbes (Archaea, bacteria, fungi) and fauna.

Soil formation—the product of climate, organisms, parent material, and topography, over time—leads to various soil types. Profile development and soil texture are the product of interactions of these factors. The capabilities for nutrient retention, important for primary producers in all soils, are affected by both mineral content and soil organic matter, with organic matter usually having the higher number of exchange sites. The aggregate structure of soils is biologically mediated in many soil types. Soils play major roles in both recycling matter and nutrients, as well as being important sources and sinks of global greenhouse gases. It is apparent that soil ecology is being considered much more centrally in ecological studies and in ecosystem management as well.

CHAPTER

2

Primary Production Processes in Soils: Roots and Rhizosphere Associates

2.1 INTRODUCTION

Lotka (1925), in his classic overview of ecological function, considered the system-level features of carbon gain or anabolism, and system-level losses of reduced carbon or catabolism. This chapter is concerned with the primary sources of organic carbon inputs to soils, or system anabolism. These inputs have a major impact on nutrient (N, P, and S) dynamics and soil food web function, as will be shown in Chapter 5, Decomposition and Nutrient Cycling and Chapter 6, Soil Foodwebs: Detritivory and Microbivory in Soils.

How can we best address the problems of measurement of primary production? Some ecological studies have declared that accurate measurements of belowground inputs to ecosystems are virtually insurmountable, and assumed that belowground production equals that of production aboveground for total net primary production (NPP) (Fogel, 1985). This rule-of-thumb is clearly inadequate, and often very wrong (Vogt et al., 1986). One of our objectives in this chapter is to address the processes and principles underlying primary production, and to indicate where the "state of the art" is now, and is likely to be, over the next several years. A wide range of new techniques is now available. We anticipate that information on and our understanding of belowground NPP will continue to increase.

2.2 THE PRIMARY PRODUCTION PROCESS

In the process of carbon reduction, there is a net accumulation of sugars, or their equivalents, in the organism's tissues. The costs of photosynthesis are extensively treated by plant physiologists, and are out of the purview of this book. Other costs, related to

Fundamentals of Soil Ecology
DOI: http://dx.doi.org/10.1016/B978-0-12-805251-8.00002-8

movement of the photosynthates within the plant, and allocation to symbiotic associates, are significant to the plant and to the ecosystem, and will be considered further on.

Gross primary production minus plant respiration yields NPP. NPP is the resultant of two principal processes: (1) increases in biomass, and (2) losses due to organic detritus production, which follows from or is dependent on the biomass production (Fogel, 1985). The detritus production includes leaves, branches, bark, inflorescences, seeds, and roots. Additional losses are traceable to exudation, volatilization, leaching, and herbivory (Cheng, 1996; Cheng et al., 1993; Jones et al., 2009).

Measurement of aboveground components is at times tedious, but fairly complete in many studies (see reviews by Persson, 1980; and Swank and Crossley, 1988). In contrast, measurement of belowground production processes has been fraught with errors and many difficulties. However, the total allocation of NPP belowground is often 50% or greater (Coleman, 1976; Fogel, 1985, 1991; Harris et al., 1977; Kuzyakov and Domanski, 2000) (Table 2.1). A sizable portion of the total production is contributed by fine roots, which often have a high turnover rate, of weeks to months (Table 2.2), which may be closely linked to nitrogen availability on a seasonal basis (McCormack et al., 2015; Nadelhoffer et al., 1985, 1992; Publicover and Vogt, 1993). In addition to production of fibrous root tissues, there are accompanying inputs of soluble compounds, namely organic acids, sugars, and other compounds. All of these have a considerable impact on rhizosphere, the zone of soil immediately surrounding the root, comprised of root secretions, exfoliations, and the microbial communities contained therein (Curl and Truelove, 1986; Hiltner, 1904) processes. We cover these later in the chapter.

TABLE 2.1 Annual Production ($Mg \cdot ha^{-1}$) of Fine Roots (<2 mm) and Root Production as Percentage of Total NPP in Different Ecosystems

Ecosystem	Age (Years)	% Contribution	Production
CONIFEROUS FOREST			
Douglas fir	55	73	4.1–11.0
Loblolly pine	?	?	8.6
Scots pine	14	60	3.5
DECIDUOUS FOREST			
Liriodendron	80?	40	9.0
Oak–maple	80	?	5.4
HERBACEOUS			
Corn	<1	25	1.2–4.2
Soybean	<1	25	0.6
Tallgrass prairie	?	50	5.1

From Fogel, R., 1985. Roots as primary producers in below-ground ecosystems. In: Fitter, A.H., Atkinson, D., Read, D. J., Usher, M.B. (Eds.), Ecological Interactions in Soil: Plants, Microbes and Animals. Blackwell, Oxford, pp. 23–36 (Fogel, 1985) and Coleman, D.C., Andrews, R., Ellis, J.E., Singh, J.S., 1976. Energy flow and partitioning in selected man-managed and natural ecosystems. Agro-Ecosystems, 3, 45–54

TABLE 2.2 Annual Losses (Due to Consumption and Decomposition) of Fine Root Biomass in Different Forests

Ecosystem	Loss (% total)
DECIDUOUS FOREST	
European beech	80–92
Oak	52
Liriodendron	42
Walnut	90
CONIFEROUS FOREST	
Douglas fir	40–47
Scots pine	66

From Fogel, R., 1985. Roots as primary producers in below-ground ecosystems. In: Fitter, A.H., Atkinson, D., Read, D.J., Usher, M.B. (Eds.), Ecological Interactions in Soil: Plants, Microbes and Animals. Blackwell, Oxford, pp. 23–36.

2.2.1 Zones of Root Growth and Activity

Roots are spatially diverse and temporally variable in their function. Root exudation may be considered to be in several zones, going from the tip inward: in its broadest sense, this release of organic C is often termed rhizodeposition (Jones et al., 2004). The term rhizodeposition includes a wide range of processes by which C enters the soil including: (1) root cap and border cell loss, (2) death and lysis of root cells (cortex, root hairs, etc.), (3) flow of C to root-associated symbionts living in the soil (e.g., mycorrhizas), (4) gaseous losses, (5) leakage of solutes from living cells (root exudates), and (6) insoluble polymer secretion from living cells (mucilage; Fig. 2.1) (Jones et al., 2009).

When comparing across ecosystems, one needs to be aware of marked differences in root morphology and distribution, that is, root architecture (Fitter, 1985, 1991). Thus, wheat roots in a Kansas field are not markedly different in size, with primary and secondary laterals arising from root initials. In contrast, coniferous tree roots are often comprised of long, supporting lateral roots, and short roots, which do the primary job of water and nutrient absorption. Ecologists often use a rather simple, pragmatic classification approach: (1) fine roots, <2 mm dia., and (2) structural roots, >2 mm dia. (Fogel, 1991).

2.3 METHODS OF SAMPLING

There are several methods for sampling roots, many of which have been reviewed by Böhm (1979). They may be generally classified into two principal approaches (Upchurch and Taylor, 1990): (1) destructive (sampling soil cores or monoliths), and (2) nondestructive, or observational, using rhizotrons or borescopes, termed minirhizotrons (Cheng et al., 1990; Pregitzer et al., 2002; Upchurch and Taylor, 1990).

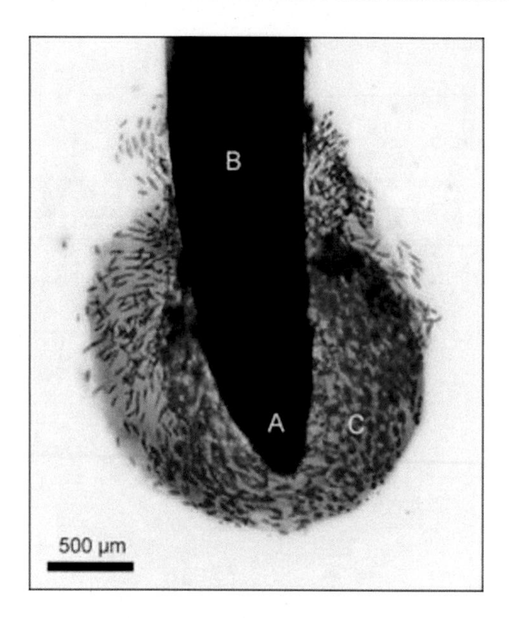

FIGURE 2.1 Light microscope image showing the large amount of mucilage (*blue* halo surrounding the root) and border cells production in a *Zea mays* L. root tip. Labels indicate the root quiescent center (A), the main root elongation zone (B), and the mucilage halo in which the border cell are embedded (C). The mucilage is stained with *aniline blue. Source: From Jones, D.A., Nguyen, C., Finlay, R.D., 2009. Carbon flow in the rhizosphere: carbon trading at the soil-root interface. Plant Soil 321, 5–33.*

2.3.1 Destructive Techniques

2.3.1.1 The Harvest Method

This method involves taking samples, usually as soil cores, dry-sorting the organic material, or rinsing it free by use of water or other flotation media, then sieving, sorting, and obtaining dry mass values. For sorting and categorizing roots, three factors need to be considered: root diameter, spatial distribution, and also temporal distribution (Fogel, 1985). Much of the existing data have been derived from thousands of cores, washed, sorted, and analyzed by legions of weary researchers. Some of these data have been truly informative and worth the effort. Other efforts, perhaps a majority of the published papers, have limited value. In the course of measuring root production by the harvest method, scientists often use what is known as the "peak-trough" calculation, in which the peaks and valleys of root standing crops through the course of a growing season as represented on a graph are successively added or subtracted about some general mean level. Unfortunately, there can be a fairly frequent occurrence of no net changes in root biomass, perhaps as often as 30% of the time in grasslands studies (Singh et al., 1984), known as zero-sum years, which have no net production, as the increases in production are canceled out by those periods which show decreases. These problems were reviewed by Singh et al. (1984) (Fig. 2.2). They extensively analyzed a grassland root production data set, looking for effects of sample (replicate number) size, and sampling frequency, coming to the conclusion that fairly frequently, perhaps in 3 years out of 10, one could expect to measure no significant increments to growth, using the harvest method. In addition, they compared the amount of NPP that one would expect from the peak-trough harvest method, and multiple-year-based computer simulation model of root production and turnover. They found that the peak-trough method at times overestimated the "true" or simulated root

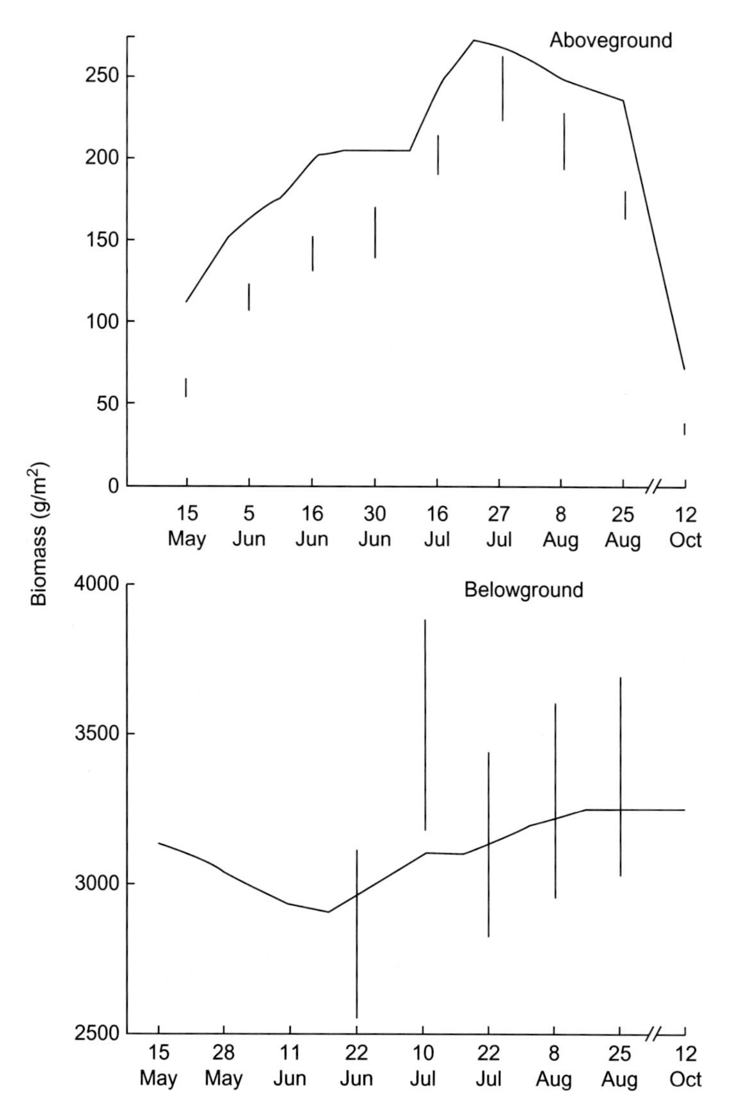

FIGURE 2.2 Comparison of values for aboveground and belowground biomass (g/m^2) predicted by a simulation model with data collected in the field. The curves represent output of the model; vertical bars are means of field data plus and minus one standard error. *Source: From Singh, J.S., Lauenroth, W.K., Hunt, H.W., Swift, D.M., 1984. Bias and random errors in estimators of net root production: a simulation approach. Ecology 65, 1760–1764.*

production by as much as 150%, due to widely varying means, leading to spuriously high "production" values. The simulated production is not more "real" than the data, of course, but they raised the question that perhaps the peak-trough method, as applied usually, may lead often to some significant overestimates of root production rates.

Considerable information is available on fine root production (FRP) in forested ecosystems. Nadelhoffer and Raich (1992) compiled 59 published estimates of annual net FRP from 43 forest sites worldwide. They compared four techniques used by investigators: (1) sequential core method (calculated as differences in means of fine root biomass between sampling periods summed across growing seasons); (2) maximum–minimum method (simpler than method 1), in that it uses only the difference between annual minimum and maximum fine root biomass to estimate FRP; (3) ingrowth core method, similar to the method of Steen (1984, 1991; Brassard et al., 2011) cited later; and (4) the N budget method (based on annual measures of net N mineralization in soil and net N flux into aboveground tissues). Annual N allocation to fine roots is calculated from the difference between net N mineralization and net N fluxes into aboveground tissues. FRP is then calculated as the product of annual N allocation to fine roots and the C:N ratio in fine roots (Nadelhoffer et al., 1985). Interestingly, Total Root Allocation (TRA) = root respiration + root production, was predicted from aboveground litterfall carbon with an $r^2 = 0.36$, $P < .02$. While this method leaves almost 2/3 of the variability in root production unaccounted for, the other methods are even less reliable in predicting FRP.

2.3.1.2 *Isotope-Dilution Method*

Another approach, using the ^{14}C-dilution technique, has been used to determine belowground biomass turnover in grasslands. Milchunas et al. (1985) performed a pulse labeling of plants for a few hours, then followed the time course of new ^{12}C label incorporated into both soluble and structural tissues of the root systems a few weeks to months later. They then calculated the subsequent production, on the assumption that any tissues lost would have a constant ratio of $^{14}C/^{12}C$ in the structural tissues. An additional step was to include ^{14}C incorporated into plant cell walls between sampling times 1 and 2. This greatly reduced the errors of the estimates (Table 2.3). Milchunas et al. (1985) found that the grass roots were continuing to mobilize additional amounts of Carbon-14 from storage tissues in the grasses, but then made further measurements of the labeled plants to adequately account for the translocated Carbon-14. Other researchers, notably Caldwell and Camp (1974), have used the isotope-dilution technique with considerable success.

2.3.1.3 *Root-Ingrowth Technique*

The root-ingrowth technique (Brasssard et al., 2011; Steen, 1984, 1991) involves the removal of long cores of soil, sieving the soil free of roots, and then replacing the root-free soil into nylon tubular mesh bags, with a mesh size of 5–7 mm. The mesh bags are inserted by drawing them over a plastic tube, and the pipe plus mesh bag on the outside is inserted into the hole in the field soil. The soil is tamped down in 5 cm increments, as the pipe is gradually withdrawn, leaving the mesh bag in position in the hole. Care must be taken to have a bulk density similar to that in the surrounding matrix. After the soil mesh bags are placed in their respective holes, roots are allowed to regrow into the bags. The bags are then recovered at various intervals, and the living and recent dead root biomass measured (Hansson et al., 1991; Steen, 1991). The principal assumptions are that growth into the root-free soil is the same as the root production would have been in the normal, undisturbed soil. Some concerns one might have about this technique are: was the bulk density of the soil in the mesh tubes identical to that in the surrounding soil? Also, were any significant soil aggregates broken in the soil sieving process, which might alter rates of root growth in the bags? Larger soil aggregates might be left intact, if they do not contain any roots (Steen, 1991). What effect is caused by higher water contents in soil volumes without living roots? Advantages of the technique include being able to get a clear, more accurate measure of production of roots over discrete time intervals. Also, it is possible to obtain information on decomposition of dead roots by placing fine-mesh (<0.1 mm) cloth bags into the soil cylinders, and determine the loss rates of dead roots simultaneously with

TABLE 2.3 Comparison of Increments in the Belowground Biomass of Blue Grama and Wheat as Determined by Complete Harvest and $^{14}C/^{12}C$ Dilution Techniques[a]

Cell-Wall Carbon					Belowground Production[b]				
					Uncorrected			Corrected for ^{14}C Incorporation	
Time (days)	^{12}C(g)	$^{14}C(g \times 10^{-4})$[c]	$R(\times 10^{-4})$[d]	$R_c(\times 10^{-4})$[e]	Harvest	$(R_1C_1/R_2C_2-1)B_1$[f]	Error (%)[g]	$(R_1C_1/R_2C_2-1)B_1$[f]	Error (%)[g]
BLUE GRAMA									
0.15	0.70	4.37	6.23						
5	0.90	22.56	25.07	10.58	3.52	−0.65	−118	3.48	−1
8	0.95	33.48	35.26	15.70	3.64	1.22	−66	3.56	−2
11	1.19	34.57	28.99	16.21	2.70	0.53	−80	2.73	1
25	1.40	39.67	28.25	18.60	2.15	0.78	−64	2.14	0
40	2.13	48.04	22.52						
WHEAT									
0.15	0.53	1.82	3.46						
5	0.52	5.46	10.46	2.52	3.45	1.02	−70	3.53	2
8	0.72	6.91	9.60	3.19	2.96	1.25	−58	2.97	0
11	0.77	7.64	9.99	3.52	2.73	1.35	−50	2.68	−2
25	1.26	8.73	6.95	4.02	1.31	0.51	−56	1.33	2
40	2.02	9.83	4.87	4.53	−0.20	−0.35	75	−0.15	−25
62	2.17	10.19	4.70						

[a]*Values are expressed on an ash-free, dry-weight basis.*
[b]*Increment is grams of total root (structural plus labile) between that sampling time and the last sampling time.*
[c]*^{14}C = (DPM sample) (60s/min) [Ci(3.7 × 10^{10}DPS)] (0.22442g/Ci).*
[d]*$R = {}^{14}C/^{12}C$; $R_1 = {}^{14}C/^{12}C$ for that sampling time; and $R_2 = {}^{14}C/^{12}C$ for the last sampling time.*
[e]*$R_C = R_2$ corrected for the amount of ^{14}C incorporated into structural material between that sampling time and the last sampling time.*
[f]*C = % cell wall at time 1 or 2; B_1 = biomass (structural plus labile) at time 1.*
[g]*(Dilution−harvest)/harvest × 100.*
Modified from Milchunas, D.G., Lauenroth, W.K., Singh, J.S., Cole, C.V., Hunt, H.W., 1985. Root turnover and production by ^{14}C dilution: implications of carbon partitioning in plants. Plant Soil 88, 353–365.

measurement of new root ingrowth over time (Titlyanova, 1987). However, see papers by Reid and Goss (1982), Cheng and Coleman (1990), and Cheng et al. (2014) for comments about live root and organic matter interactions. We will address some of these concerns further, in the course of discussing decomposition processes in soils.

2.3.1.4 *Nondestructive Techniques*

Rhizotrons and minirhizotrons: over the last few years, there has been a great resurgence of interest in observational, nondestructive techniques for studying root-related processes. Several review volumes present a detailed discussion of minirhizotron and rhizotron usage: Box and Hammond (1990), Fahey et al. (1999), and Taylor (1987). In essence, the rhizotron approach involves installing a large glass plate in an observation gallery, and then measuring the growth of roots against the glass, over time (Davis et al., 2004; Fogel and Lussenhop, 1991) (Fig. 2.3). Using this technique, one can follow a large part of a given root population visible through the glass over various time periods. The disadvantages are that the soil profile must be recreated, and retamped to an equivalent bulk density, or mass per unit volume of soil, which closely approximates that of the surrounding soil. It is also only a small fraction of an entire field or forest.

Minirhizotrons, on the other hand, have a smaller amount of surface area in one place, being tubular (5–7 cm.dia.), and are placed, as are the rhizotrons, at a 20–25 degree angle from the vertical (Fig. 2.4). However, being light and readily handled, they enable extensive replication in any given plot, experimental treatment or entire field site tubes may be of either glass or a durable plastic, such as polycarbonate. For example, Cheng et al. (1990) used 12 minirhizotron tubes in each replicate, and 2 replicates per treatment (conventional tillage and no-tillage) in a study of sorghum root growth and turnover in a southeastern agroecosystem. Other studies have followed the dynamics of soil mesofauna, namely collembola, in fields under various crops in Michigan agroecosystems (Snider et al., 1990). A number of precautions should be employed in the usage of minirhizotrons, so as to avoid artifacts of placement. For example, total root biomasses can be underestimated in the top 7–10 cm if inadequate care is taken to shield the top of the minirhizotron tubes from transmitted light. Also, adequate soil/tube contact needs to be ensured by careful drilling and smoothing of the bored hole (preferably using a hydraulic coring apparatus), as noted by Box and Johnson (1987). If there is some open space between the outer tube surface and the soil, roots may respond as if this is a major soil crack, and preferentially grow along it (van Noordwijk et al., 1993). To handle the large amounts of data and images obtained using minirhizotrons, it is necessary to use image analysis programs, such as those described by Hendrick and Pregitzer (1992), Pregitzer et al. (2002), and Smucker et al. (1987). With the advent of digital analysis techniques and image storage on CD-ROM, the literally millions of bits of information per soil/root image can be manipulated and analyzed reasonably promptly and efficiently. Caution must be taken, however, to ensure that the material used for the tube has a minimal effect on the roots being observed. Withington et al. (2003) compared minirhizotron data for glass, acrylic, and butyrate tubes in an apple orchard, and acrylic and butyrate tubes in a study with six forest tree species. Root phenology and morphology were generally similar among tubes. Root survivorship varied markedly between hardwood and conifer species, however, probably due to hydrolysis by fungi interacting with the plastic tubes. Comparison of root

FIGURE 2.3 University of Michigan Soil Biotron. (A) View of tunnel and aboveground laboratory from the south. (B) View from the west. Note *white* pine stump left after logging and burning in about 1917. (C) Interior of tunnel showing window bays covered with insulated shutters. (D) Close-up of glass, wire-reinforced, 6-mm-thick windowpane. Note fungal rhizomorphs. The wire grid is about 2 cm by 2 cm. *Source: From Fogel, R. Lussenhop, J., 1991. The University of Michigan Soil Biotron: a platform for soil biology research in a natural forest. In: Atkinson, D. (Ed.), Plant Root Growth an Ecological Perspective. Blackwell, Oxford, pp. 61–73.*

standing crop data from cores with minirhizotron standing crop showed a closer match with the acrylic than the butyrate data. Glass was considered to be the most inert, but one-third of the glass tubes were lost due to breakage over the winter in the Pennsylvania site.

Frank et al. (2002), using minirhizotrons, measured significant increases in root growth in nine higher elevation (1635–2370 m.) mixed-grass grazing-lands in Yellowstone

FIGURE 2.4 An auger jig system used to install angled minirhizotron tubes. *Source: From Mackie-Dawson and Atkinson, 1991.*

National Park. They found that large migratory herds of elk, bison, and pronghorn, by their grazing, stimulated aboveground, belowground, and whole-grassland productivity by 21%, 35%, and 32%, respectively. This feedback effect, which was demonstrated earlier by Dyer and Bokhari (1976) and McNaughton (1976), will be addressed further in system-level effects considered in Chapter 5, Decomposition and Nutrient Cycling.

In a study of seven minirhizotron data sets, Crocker et al. (2003) substituted root numbers for root lengths using a regression technique. Linear regression models fitted between root length and root number for production and mortality of a wide range of tree species from subtropical to boreal conditions. Treatments yielded r^2 values ranging from 0.79 to 0.99, indicating that changes in root numbers can be used to predict root length dynamics reliably. Slope values for mean root segment length ranged from 2.34 to 8.38 mm per root segment for both production and mortality. Crocker et al. (2003) caution that the quantitative relationship between root lengths and numbers must be established for a particular species-treatment combination, but it will save much time required to quantify root dynamics.

Additional approaches to calculation of FRP and turnover continue to appear in the scientific literature. Gaudinski et al. (2001) used one-time measurements of radiocarbon (^{14}C) in fine roots (<2 mm dia.) from three temperate forests in the eastern United States: a coniferous forest in Maine, a mixed hardwood forest in Massachusetts and a loblolly pine plantation in South Carolina. Roots were sampled as either mixed live and dead, and others as live roots. Using accelerator mass spectrometry (AMS) to analyze very small samples, Gaudinski et al. (2001) found that root tissues are derived from recently fixed carbon, and the storage time prior to allocation to root growth is <2 years and more likely < 1 years. Live roots in the organic horizons are made of C fixed 3–8 years ago, versus roots in mineral B horizons, with 11–18 years mean age. This spatial component to root age has not been measured before, and has important implications in calculating more realistic carbon budgets for terrestrial ecosystems. This assessment of mean root ages in forest tree roots is in marked contrast to the more rapid turnover times as noted earlier in the studies using minirhizotrons and other direct means of observation. It does not negate the findings of the observational studies, but emphasizes the need to be aware of the wide range of age of fine roots in the entire soil profile.

The field of root production and turnover studies is increasing in quantity and quality as well. Thus redefining what constitute fine roots improves understanding of belowground contributions to terrestrial biosphere processes. McCormack et al. (2015) developed a framework of shorter lived absorptive roots and a longer lived transport fine root pool. Their new calculations showed a decrease of FRP and turnover, using their redefined categories, which represents 22% of terrestrial NPP. This is a significant drop by 30% from older, larger estimates. Thus there is a real need to take a closer look at fine roots as newly defined, and mycorrhizal extramatrical mycelial contributions to total C inputs in soils. One must also keep in mind the caveats as presented by Jones et al. (2009), namely that much of the carbon inputs into soil derive from root exudates and exfoliates.

In a detailed process study using seedlings of wheat (*Triticum aestivum*), Kaiser et al. (2015) examined the fates of pulses of ^{13}C-labeled phospho- and neutral lipid fatty acids, tracing in situ flows of recently assimilated C of $^{13}CO_2$ moved through arbuscular mycorrhiza (AM) into root and hyphal associated microbial communities. Using nanoscale secondary ion mass spectrometry (nano-SIMS), they found intraradical hyphae were significantly enriched compared to other root cortex areas within 8 hours after labeling. A significant and fresh proportion of ^{13}C photosynthates was delivered via the AM pathway and was utilized by different microbial groups, compared to C directly released by the roots. This study by Kaiser et al. (2015) serves as an object lesson in following the timing and pattern of release of photosynthates in the rhizosphere.

2.4 ADDITIONAL SOURCES OF PRIMARY PRODUCTION

An additional contribution to NPP comes from algal populations in the surface few millimeters, or on the soil surface, itself. By measuring CO_2 fixation by cyanobacteria and algae on the surface of intact cores taken from an agroecosystem, Shimmel and Darley (1985) calculated that approximately 39 g carbon was fixed per m^2 per year. This is a small proportion (5%) of total NPP for the study site, the conventional-tillage agricultural system in Georgia (noted earlier), which averaged 800 g m^{-2} y^{-1} aboveground NPP. The type of organic matter, and the amount that may feed directly into detritivorous fauna could be of importance, beyond the total production figures on an annual basis. For example, cryptogamic crusts can be significant agents of nitrogen fixation, providing inputs of nitrogen and carbon in nutrient-poor aridlands ecosystems (Belnap, 2002; Coe et al., 2012; Evans and Belnap, 1999).

2.5 SYMBIOTIC ASSOCIATES OF ROOTS

From the earliest origins of a land flora, over 400 million years ago, there has been a structural/functional interaction, between plant roots and AM (Fig. 2.5A) as shown in the fossil record (Malloch et al., 1980; Pirozynski and Malloch, 1975). Probably the earliest land plants arose in the Ordovician and were similar to current-day hornworts and

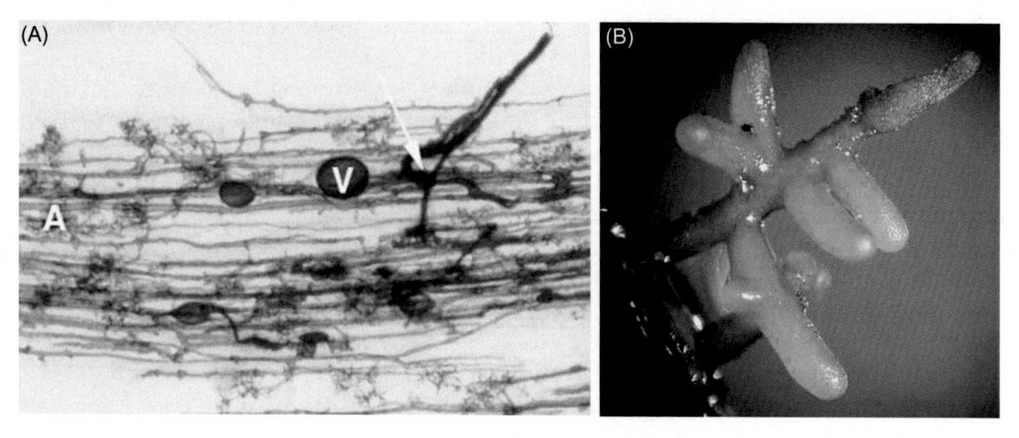

FIGURE 2.5 (A) A microscopic view of a root colonized by arbuscular mycorrhizal fungi (AMF). V indicates vesicles, A indicates arubuscules, and the arrow shows the fungal entry point. (B) A root tip colonized by ectomycorrhizal fungi. *Source: Photos courtesy of mycorrhizas.info and Dr. J. Walker.*

liverworts (Redeker, 2002). Today's hornworts and liverworts do form associations with AM fungi, but because they do not have true roots, they do not meet all the criteria of AM (Read et al., 2000). Most families of terrestrial plants have mycorrhizal symbionts, with two families, the Cruciferae and Chenopodiaceae being conspicuous exceptions (Allen, 1991). Indeed, recent analyses have shown that zygomycetous fungi colonize a wide range of lower land plants (hornworts, many hepatics, lycopods, Ophioglossales, Psilotales (horsetails), and Gleicheniaceae) (Read et al., 2000). These associations are structurally analogous to mycorrhizas, but their functions remain to be determined (Read et al., 2000).

The ectomycorrhizae, or ectomycorrhizal (ECM) (Fig. 2.5B), are prevalent in several tree families, such as the Fagaceae (including the beeches and oaks), and also the Pinaceae within the Conifers. ECM arose relatively recently, only 160 million years ago, in the Cretaceous (St. John and Coleman, 1983).

After examining the structures of the principal types of mycorrhizae, we will consider information on carbon costs to the plant, as well.

2.6 MYCORRHIZAL STRUCTURE AND FUNCTION

Seven different categories of mycorrhizal symbiosis have been distinguished on the basis of their morphological characteristics and the fungal and plant species involved (Finlay, 2008). AM is the most ancient and widespread form. AM, the so-called endomycorrhiza, are characterized by structures within root cells, called arbuscules, because they grow and ramify, tree-like, within the cell (Fig. 2.5A). They are members of the Phycomycete fungi. Most, but not all AM (two exceptions are known in the family Endogonaceae) also have storage structures known as vesicles, which store oil-rich products. AM send out hyphae for several centimeters (a maximum of 6–10) into the surrounding soil, and are instrumental in facilitating nutrient uptake, particularly phosphate ions (Allen, 1991). AM are known only as obligate mutualists (i.e., the root provides carbon, and the mycorrhiza tap an enhanced pool of mineral nutrients), and have not been cultured yet apart from their host roots. The AM fungi are included within the Order Glomeromycota, with over 282 SSU rRNA gene virtual taxa described with all Glomeromycota sequences (Oepik et al., 2010).

Because AM hyphae will grow out from the germinating chlamydospore toward root surfaces, responding to soluble compounds, principally strigolactones, they are considered to have slight saprophytic competence (Azcón-Aguilar and Barea, 1995). Interestingly, a similar sort of signaling system is involved with roots and rhizobia, with flavonoids playing a primary role. Thus flavonoids, released by the plant roots, signal to rhizobia in the rhizosphere which then produce nodulating factors (Nod factors) that are recognized by the plant (Oldroyd, 2013). Some rhizosphere microorganisms seem to stimulate AM germination and mycelial growth, functioning either by detoxifying or removing inhibitors from the growth medium, or utilizing self-inhibiting compounds from the AM fungus, enabling more growth than would be possible under axenic conditions (Azcón-Aguilar and Barea, 1995). For AMF, perception of strigolactone exudation into the rhizosphere promotes spore germination and hyphal branching. AMF produce Myc factors, including lipochitooligosaccharides (LCOs), signals that activate the symbiosis signaling pathway in the root. AMF invasion involves an infection peg from the hyphopodium that allows fungal hyphal growth into the root epidermal cell (Fig. 2.6) (Oldroyd, 2013).

Paleobotanical and molecular sequence data suggest that the first land plants formed associations with Glomalean fungi from the Glomeromycota about 460 million years ago (Redecker et al., 2000). Thus, these symbioses occurred some 300–400 million years before the appearance of root nodule symbioses with nitrogen-fixing bacteria. AM symbioses can be formed with a very wide range of plant species, as many as 250,000. Only 150–200 species of AM fungi have so far been distinguished on the basis of morphology, but DNA-based studies suggest the true diversity of these symbionts may be very much higher (Fitter, 2005; Santos-González et al., 2007). The symbiosis is characterized by highly branched fungal structures, arbuscules, which grow intracellularly without penetrating the host plasmalemma.

Ericoid mycorrhiza are formed in three plant families, the Ericaceae, Empetraceae, and Epacridaceae, all belonging to the order Ericales. These plants grow principally as dwarf shrubs in upland and lowland heaths and other nutrient-impoverished areas, such as the understory vegetation of boreal forests. Ericoid mycorrhizas also occur in warm Mediterranean climate zones in chaparral vegetation systems throughout the world, suggesting that nutritional, rather than climatic, factors determine their distribution. Around 3400 plant species form this type of mycorrhizal association with various fungi from the Ascomycota. Ericaceous mycorrhiza are noted for the ability to facilitate direct uptake of organic N in low pH environments, and some of them produce proteases to enhance the N uptake without going through any external mineralization in the soil solution (Bending and Read, 1996). As may often be the case with residues derived from ericaceous sclerophylls, the essential elements (N and P) will be masked by skeletal materials, namely lignin or its breakdown products. In these circumstances, the ability of the mycorrhizal fungus to produce lignase or phenol oxidase activities and thus expose the nutrient-

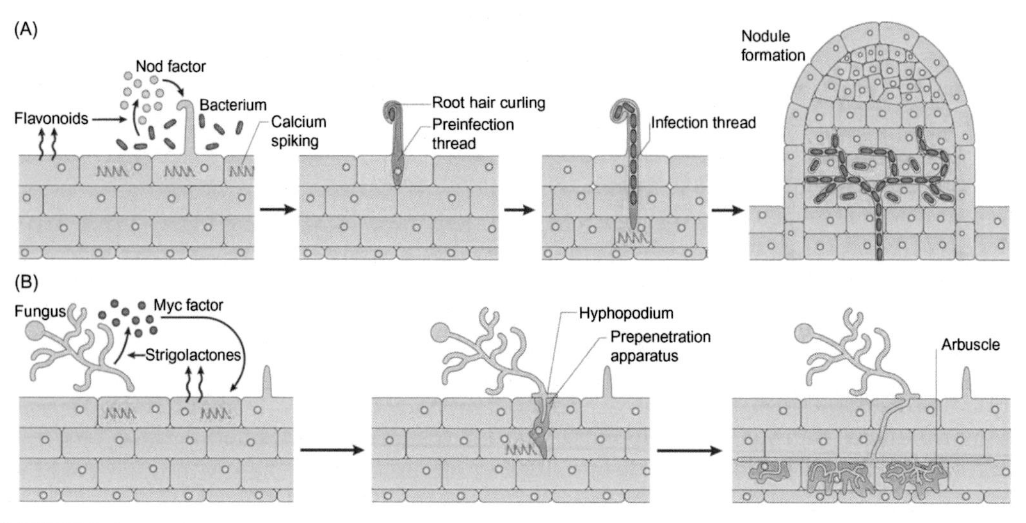

FIGURE 2.6 (A) Flavonoids released by the plant signal to rhizobia in the rhizosphere, which in turn produce nodulation factors (Nod factors) recognized by the plant. This activates the symbiosis signaling pathway, leading to calcium oscillations in epidermal and cortical cells preceding their colonization. Rhizobia enter the plant root via root hair cells that grow around and trap the bacteria inside a root hair curl. Infection threads (invasive invaginations of the plant cell) allow invasion of the rhizobia into the root tissue. Nodule meristem forms in the root cortex below the site of bacterial infection. Infection threads grow toward the emergent nodules and ramify within the nodule tissue. Rhizobia may remain inside the infection threads, but more often, the bacteria are released into membrane-bound compartments inside the cells of the nodule where N-fixation occurs. (B). Strigolactone release by the plant root signals to AMF in the rhizosphere. This promotes spore germination and hyphal branching. AMF produce mycorrhizal factors (Myc factors), including LCOs and, possibly, signals that activate the symbiosis signaling pathway in the root, leading to calcium oscillations. AMF invasion involves an infection peg from the hyphopodium that allows fungal hyphal growth into the root epidermal cell. The fungus colonizes the plant root cortex through intercellular hyphal growth. Arbuscules are formed in inner root cortical cells from the intercellular hyphae. *Source: Modified from Oldroyd, G.E.D., 2013. Speak, friend, and enter: signalling systems that promote beneficial symbiotic associations in plants. Nat. Rev./Microbiol. 11, 252–263.*

containing substrates would be just as important as production of the enzymes, for example, phosphatases and proteases that are directly involved in nutrient release. Also see comments on Future Directions in Mycorrhizal Research, at the end of this chapter. For information on these, and other less-common mycorrhiza, refer to Allen (1991, 1992), Finlay (2008), Read (1991), and Smith and Read (1997).

Many long-lived perennial plants and trees form a third type of mycorrhiza, ectomycorrhiza. ECM, ectomycorrhiza, are significantly different in physiology and ecology. These are principally Basidiomycetes, and proliferate between cells, not inside them, as was the case for AM. An obvious morphological alteration occurs, with formation of the mantle and Hartig net (a combination of epidermal cells and ECM fungal tissues) on the exterior of the root (Fig. 2.5B). ECM send hyphae out literally several meters into the surrounding soil. The hyphae aid in nutrient uptake, including inorganic and some organic N + P compounds (Read, 1991). The hyphae constitute a significant proportion of carbon allocated to belowground NPP in coniferous forests (Vogt et al., 1982). The reproductive

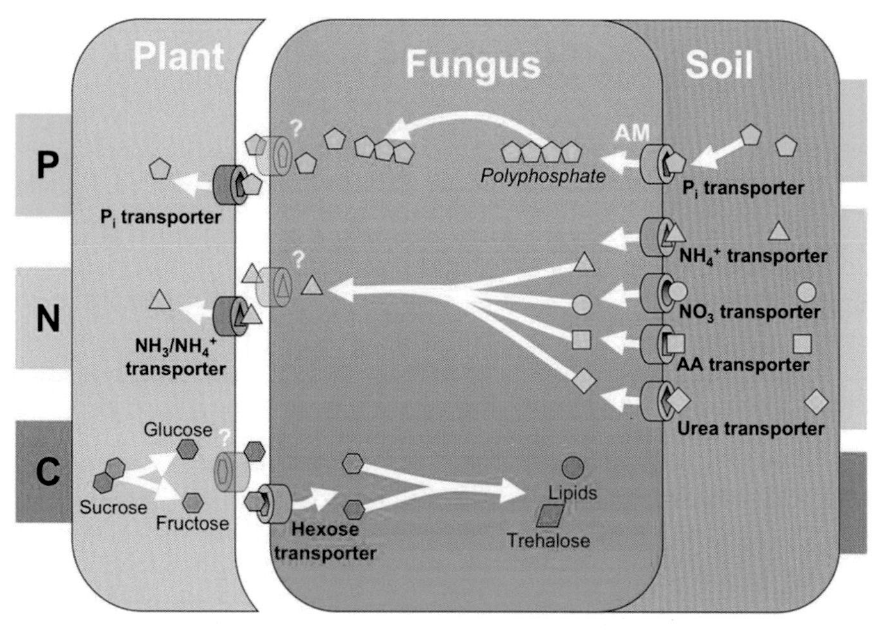

FIGURE 2.7 Scheme summarizing the main nutrient exchange processes in EM and AM symbiosis. Emphasis is placed on the translocation of phosphorus (P), nitrogen (N), and carbon (C) at the soil–fungus and fungus–plant interface. Inorganic P and mineral or organic forms of N, such as NH_4^+, NO_3^-, and AA, are taken up by specialized transporters located on the fungal membrane in the extraradical mycelium. NH_3/NH_4^+ and P_i (the latter originated in AM fungi from the hydrolysis of polyphosphate) are imported from the symbiotic interface to the plant cells through selective transporters. Hexose transporters import plant-derived carbon to the fungus, whereas transporter proteins involved in the export of nutrients from either the plant or fungus have not been identified yet. This questions whether such processes indeed result from active, protein-mediated transport or involve passive export mechanisms. *AA*, amino acids. *Source: From Bonfante, P., Genre, A., 2010. Mechanisms underlying beneficial plant fungus interactions in mycorrhizal symbiosis. Nat. Commun. 1, 48.*

structures of ECM are the often-observed mushrooms in oak or pine forests. ECM will form resting stages, or sclerotia, which are cordlike bundles of hyphae, which can persist for years. ECM, unlike AM, often can be cultured apart from their host plants. Some ECM have considerable decomposing capabilities, and can obtain a portion of their reduced carbon from decomposing substrates, that is, leaf litter. The main nutrient exchange processes in AM and ECM symbioses have been summarized by Bonfante and Genre (2010) (Fig. 2.7).

The fungi are predominantly from the Basidiomycota and Ascomycota and as many as 10,000 fungal species and 8000 plant species may be involved, globally (Taylor and Alexander, 2005). Although this represents only a small fraction of the total number of terrestrial plants, these species often form the dominant components of forest ecosystems occupying a disproportionately large area. The plant species involved are usually trees or shrubs from cool, temperate boreal, or montane forests, but also include arctic-alpine dwarf shrub communities, Mediterranean/chaparral vegetation, and many species in the Dipterocarpaceae and leguminous Caesalpinoideae in tropical forests. The fungi do not penetrate the host cells, and the symbiosis is characterized by the presence of a fungal mantle or sheath around each of the short roots, as well as a network of intercellular hyphae penetrating between the epidermal and cortical cells, the so-called Hartig net. Like the arbuscules in AM, this interface is an effective way of increasing the surface area of contact between the fungus and its plant host. The mantle is usually connected to a more or less well-developed extraradical mycelium (ERM) which may extend for many centimeters from the root into the soil. This ERM may form a significant fraction of the total microbial biomass in forest soils and estimates of 20–200 kg ha^{-1} y^{-1} ECM production have been made (Ekblad et al., 2013). Carbon inputs to the soil would be half of these dry mass values.

In most types of mycorrhizal symbiosis the fungal symbionts depend upon their autotrophic plant hosts to supply carbon, however in the orchid mycorrhiza formed by orchids this dependency may be reversed. The family Orchidaceae is the largest in the plant kingdom and estimated to contain 30,000 species. Although most orchids have green leaves and are autotrophic when fully established, about 100 species are achlorophyllous as adults and all pass through a germination and early development phase when they are dependent on an external supply of nutrients and organic carbon. They have minute dust-like seeds with no reserves, and are initially entirely dependent upon the supply of carbon and nitrogen from fungi. In total, more than 400 plant species are achlorophyllous and described as "mycoheterotrophic," obtaining their carbon from fungi. The fungi colonizing these plants were originally thought to be effective saprotrophs or parasites, but DNA-based studies of the fungi have now shown that most of them are mycorrhizal fungi simultaneously colonizing other autotrophic plants (Leake, 2004).

The Monotropaceae are mycoheterotrophic plants that form a fifth type of mycorrhizal association, monotropoid mycorrhiza. Structurally, these are like ectomycorrhizas, often with a well-developed mantle but with a more superficial Hartig net, with single hyphae growing into the epidermal cells, forming peg-like structures. The achlorophyllous monotropoid plants are wholly dependent on the fungi for reduced carbon and soil nutrients, and it appears that the carbon comes indirectly from autotrophic host plants that are also attached to the same ECM mycelium (Finlay, 2008). Two other types of mycorrhizal association have been distinguished, the arbutoid mycorrhiza and ectendomycorrhiza. The former associations are formed between fungi that are normally ECM and plants in the genera Arbutus, Arctostaphylos, and the family Pyrolaceae, where intracellular fungal penetration occurs. Ectendomycorrhizas have features of both ectomycorrhizas and endomycorrhizas, a sheath that may be reduced or absent, a Hartig net that is usually well developed, but also intracellular penetration. Some of the fungi that form ectendomycorrhizas with Pinus and Larix also form ectomycorrhizas with other tree species. Ectendo-, arbutoid, and monotropoid mycorrhizas can, to some extent, be seen as more specialized cases of the general mycorrhizal habit (Finlay, 2008).

2.6.1 Commonalities between Rhizobia and Mycorrhiza for Colonizing Roots

The signaling pathways between plants and fungi and colonizing bacterial symbionts have been identified in detail. Thus root leakage of strigolactones into soil and their rapid hydrolysis is instrumental in signaling root proximity in the rhizosphere. At least seven genes (SYM genes) are known to be required for both AM and rhizobium-legume symbioses to occur (Bonfante and Genre, 2010). This is a very complex sequence, with calcium channels and transporters involved.

It is important to note that mycorrhizal fungi and nitrogen-fixing bacteria are responsible for from 5%–20% (in grassland and savanna) to 80% (in temperate and boreal forests) of all nitrogen and up to 75% of phosphorus that is acquired by plants annually (Van der Heijden et al., 2008). An innovative study by Hobbie et al. (2002) examined the flows of ^{13}C and ^{14}C to basidiomycete sporocarps, needles, and litter in a Western Oregon forest soil, using AMS of 1–2 mg samples of soils and tissues. Mycorrhizal associations were indicated by very young (0–2 years) age of ^{14}C, whereas the saprotrophic genera averaged 10 years in radiocarbon age (Hobbie et al., 2002). With analytical tools now at hand, Hobbie et al. (2002) suggest that needle and fungal carbohydrates should be analyzed for ^{14}C content separately from needle and fungal protein. They predict that protein will be significantly older than carbohydrates when amino acids are taken up directly from the soil, versus the faster flow of C from photosynthates to the current crop of needles in litter.

2.7 ECOSYSTEM-LEVEL CONSEQUENCES OF ECTOMYCORRHIZAL FUNCTION

ECM form fungal mats in the surface soils of Douglas Fir (*Pseudotsuga menziesii*) and probably other coniferous forests. The mats contain higher microbial biomass, with two to three times more soil mesofauna (collembola, mites, and nematodes) feeding on them (Cromack et al., 1988). The fungal mats play an important role in buffering against Fe and Al activity in these acidic soils, on the west slope of the Cascade Mountains in the Pacific Northwest of the United States (Entry et al., 1992). As Fe and Al activity in soil increases, calcium oxalate crystals formed by the fungal mats dissociate, releasing calcium, and chelating the iron and aluminum. Entry et al. (1991) also measured greater litter decomposition rates and greater mineralization of N and P in ECM mat soils.

Alternative methods for determining EM fungal community makeup have been developed using molecular techniques. Landeweert et al. (2003) used a basidiomycete-specific primer pair (ITS1F-ITS4B) to amplify fungal internal transcribed spacer from total DNA extracts of the soil horizons, followed by an amplified basidiomycete DNA cloning and sequencing procedure, to identify the EM fungi present. The soil samples were from four distinct horizons of a spodosol profile, under coniferous (Norway spruce (*Picea abies*) and Scots pine (*Pinus sylvestris*)) vegetation. By identifying basidiomycete mycelium in soil, the EM fungal community was analyzed in a novel fashion, excluding EM root tips from the analysis. Landeweert et al. (2003) sampled from the O layer (0–2 cm thick), and E horizon (from 3 to 18 cm), an enriched eluvial or B horizon (18–35 cm) and parent material, C (>40 cm). They found 16 of the 25 total operational taxonomic units exclusively in the deeper mineral soil, or B horizon. The authors suggested that these distributions might be the result of somewhat higher amounts of C and higher pH existing deeper in the profile. This analysis demonstrates the need to consider the full suite of EM fungi present, and is a cautionary note that the entire profile should be considered when determining species richness in a given site.

Interplays between ECM and AM fungi under field conditions are of great interest to soil ecologists. Brzostek et al. (2015) experimentally reduced belowground C supply to soils via tree girdling, and contrasted responses in control and girdled plots for three consecutive growing seasons. They hypothesized that decreases in belowground C supply would have stronger effects in plots dominated by ECM trees rather than AM trees. In ECM-dominated plots, girdling decreased the activity of enzymes that break down soil organic matter (SOM) by c. 40%, indicating that, in control plots, C supply from ECM roots primes microbial decomposition. In AM-dominated plots, girdling had little effect on SOM-degrading enzymes, but increased the decomposition of AM leaf litter by c. 43%, suggesting that, in control plots, AM roots may intensify microbial competition for nutrients.

Over long-term, successional time spans, carbon sequestration is related to mycorrhizal fungal community shifts during long-term succession in boreal forests. Cord-forming ECM and ericaceous MF play opposing roles in belowground C storage. Thus, there is high production and turnover of ECM early on, then the cord-formers decline, and ericoid MF dominate, facilitating production of humic material via formation of melanized hyphae, thus resisting decomposition (Clemmensen et al., 2015).

Plant nutrient-acquisition strategies change with soil age. Cluster-bearing (root) species tend to dominate on severely P-impoverished, ancient soils, where P sensitivity is relatively common (Lambers et al., 2008). Cluster roots release carboxylates into the rhizosphere, thus solubilizing poorly available nutrients (such as P) within the soil. In a follow-up study, Lambers et al. (2010) studied plant mineral nutrition in ancient landscapes in Western Australia. They found that high plant species diversity on old, climatically buffered, infertile soils is linked to functional diversity for nutritional strategies. This contrasts with young, frequently disturbed, fertile landscapes, which have a somewhat lower plant species diversity.

The effects of season on belowground transfers to ECM fungi are very pronounced in a number of forested systems in boreal regions. Högberg et al. (2010) noted a >500% increase in belowground allocation of plant C in late season (August) versus a much lower allocation earlier in the summer (June). Labeled C was found in fungal fatty acid biomarkers and rarely in bacteria. There were significant transfers of the biomarkers to Collembola, but not to either Acari or Enchytraeids in the site.

2.8 ACTINORHIZA

Another symbiotic associate with roots plays an important role in many forested ecosystems worldwide. It is the actinorhiza, an actinomycete (filamentous, branching, gram-positive bacteria) that forms nodules and fixes dinitrogen in a fashion analogous to that used by rhizobia. A majority of actinorhizal species are pioneers on nitrogen-poor, open sites (Baker and Schwintzer, 1990). The dominant actinorhizal genus is *Frankia*, occurring on roots of 8 plant families, encompassing 24 genera and some 230 species of dicotyledons. Prominent actinorhizal plant families and genera are: Betulaceae, on 47 *Alnus* species; Casuarinaceae, on 16 spp. of *Casuarina*, and 54 spp. of *Allocasuarina*; Myricaceae, 28 spp. of *Myrica*, and Rhamnaceae, 31 spp. of *Ceanothus* (Russo, 2005). These genera are widespread in ecosystems on all continents except Australia, and in Oceania (e.g., Australia and New Zealand) (Baker and Schwintzer, 1990).

2.9 CARBON ALLOCATION IN THE ROOT/RHIZOSPHERE

Looking at the root/soil system as a whole, what is the totality of the resources involved, and how are these resources allocated, under various conditions of stress and soil type?

Several reviewers (Badri and Vivanco, 2009; Cheng, 1996; Cheng et al., 1993; Coleman, 1976; Coleman et al., 1983; Fogel, 1985, 1991; Kuzyakov, 2002; Martin and Kemp, 1986) have noted that from 20% to 50% more carbon enters the rhizosphere from root exudates and exfoliates (sloughed cells and root hairs) than actually is present as fibrous roots, at the end of a growing season. This was determined in a series of experiments, using Carbon-14 as a radiotracer of the particulate and soluble carbon (Barber and Martin, 1976; Shamoot et al., 1968). In fact, the mere change from a hydroponic medium to a sand medium was enough to double the amount of labile carbon as an input to the medium. This difference was attributed to the abrasion of roots against sand particles. In addition the root/rhizosphere microflora has the potential to act as a sizable carbon sink (Helal and Sauerbeck, 1991; Wang et al., 1989), which can double the losses to soil, as well. This is convincing proof that the combined belowground system: roots, microbes, soil, and fauna is governed by source–sink relationships, just as are intact plants (i.e., roots and shoots).

Extensive amounts and complexities of carbon compounds are elaborated in the rhizosphere (Cheng et al., 1993; Foster, 1988; Foster et al., 1983; Kilbertus, 1980; Lee and Foster, 1991; Rovira et al., 1979). The extent to which this exuded carbon is integral to root and rhizosphere function is of great interest to ecologists. Nitrogen-fixing bacteria residing in the rhizosphere and the release of their N to the plant can be stimulated by root exudates (Rao et al., 1998, cited in Jones et al., 2003). There are numerous direct and indirect positive and negative effects of carbon flows in the rhizosphere, that encompass a wide array of symbiotic associations and trophic and biochemical interactions (Jones et al., 2003, Figs. 2.8 and 2.9). Although the potential for rhizodeposition-driven N_2 fixation in the soil is small in comparison to inorganic and symbiotic fixation inputs, it may be of importance in N-limited ecosystems (Jones et al., 2003). The boundary layer between roots and soil, the so-called mucigel (Jenny and Grossenbacher, 1963), is jointly contributed by microbes and root surfaces. Studies of the root tip and capsule components have been most informative about the roles of signal molecules that are exuded at subnutritional rates in soil. One of the key components involved is the border cells. These cells are lost from the root cap at a rate regulated by the root and secrete compounds that alter the environment of and gene expression in soil microorganisms and fauna (Farrar et al., 2003). These root tip capsule components include high MW mucilage secreted by the root cap, as well as cell-wall breakdown products resulting from separation of literally thousands of border cells from each other and the root cap.

Much research has been expended on methods to separate total CO_2 efflux into that from microbial respiration from SOM and that from roots and rhizosphere microbial populations. Methods employed have been summarized into three broad categories

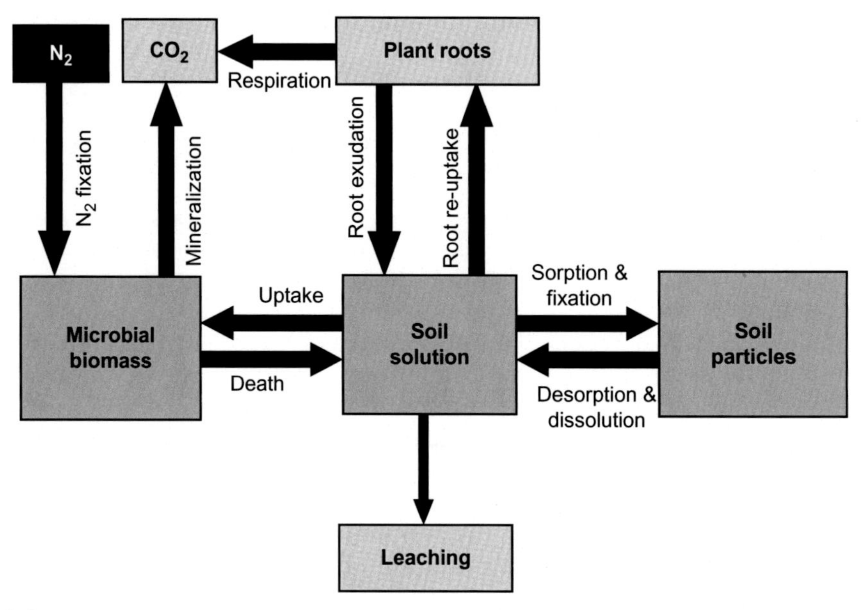

FIGURE 2.8 Schematic representation of the major carbon fluxes and pools in the rhizosphere. *Source: From Jones, D.L., Farrar, J., Giller, K.E., 2003. Associative nitrogen fixation and root exudation—what is theoretically possible in the rhizosphere? Symbiosis 35, 19–38.*

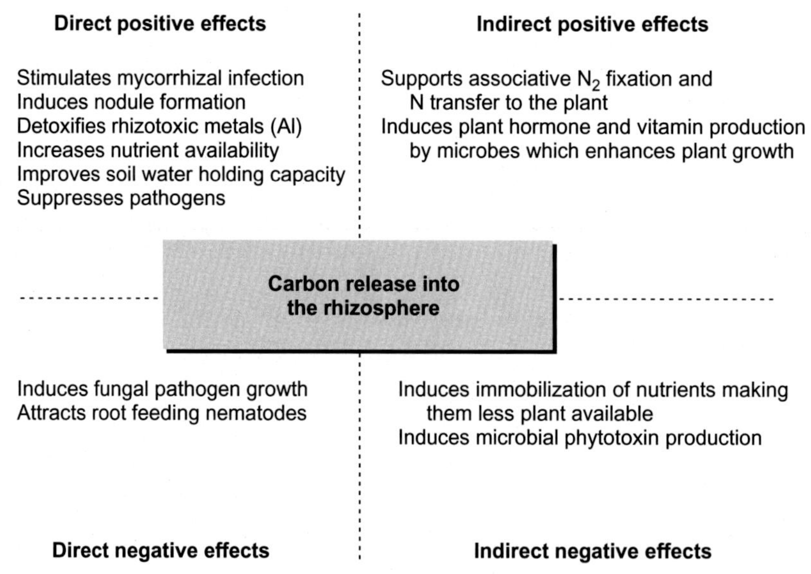

FIGURE 2.9 Schematic representation of the positive and negative direct and indirect effects of root exudates on plant growth. *Source: From Jones, D.L., Farrar, J., Giller, K.E., 2003. Associative nitrogen fixation and root exudation—what is theoretically possible in the rhizosphere? Symbiosis 35, 19–38.*

(Hanson et al., 2000): component integration, root exclusion, and isotopic approaches. Component integration entails separation of the constituent soil components involved in respiring CO_2 (i.e., roots, sieved soil, and litter) followed by measurement of the specific rates of CO_2 outputs from each component part (Coleman, 1973; Trumbore, 1995). The root exclusion method estimates root respiration indirectly by subtracting the soil respiration without roots from the soil respiration with roots (Anderson, 1973; Edwards, 1991). Isotope methods refer to the use of either radioisotopic^{14}C (Cheng et al., 1993; Horwath et al., 1994), or stable ^{13}C isotopes to trace the origin of soil respiration (Andrew et al., 1999; Robinson and Scrimgeour, 1995; Trumbore, 2006). Isotope methods have a significant advantage over the component integration and root exclusion methods because they permit researchers to partition C between rhizosphere respiration and SOM decomposition in situ, and avoiding the effects of soil disturbance.

Other studies, using real-time monitoring of Carbon-11 under laboratory conditions (Wang et al., 1989), found that mycorrhizal infection nearly doubled the "sink strength" of the roots, hence there was greater carbon flow of translocated photosynthate in studies of African Panicum grasses (the same species studied by McNaughton, 1976 and McNaughton et al., 1998), when compared with noninoculated control plants. Some elegant field studies in Scottish grassland soils near Edinburgh have demonstrated significant flows of carbon from roots to mycorrhiza. Within 21 hours of pulse labeling a grassland sward in the field with $^{13}CO_2$, between 3.9% and 6.2% passed through the external mycelium of the AM fungal symbionts to the atmosphere (Johnson et al., 2002). This is the first in the field verification of similar results measured using pot experiments. Additional recent pot experiments have exposed mycorrhizal plants to fossil (C-14 depleted) carbon dioxide and collected samples of ERM hyphae over the following 29 days. Analyses of their carbon-14 content by AMS revealed that most ERM hyphae of AM fungi live, on average, 5 to 6 days (Staddon et al., 2003). This high turnover rate indicates the existence of a large and rapid mycorrhizal pathway of carbon in the soil carbon cycle.

There is an equally important flow of nitrogen to and from AM fungi and organic sources in soils. Using two species of *Glomus* species, Leigh et al. (2009) found up to 33% of nitrogen in organic patch material was transferred to the plant by *Glomus intraradices*. Note this is a gross, not net, transfer. The N cycling and AMF phenomenon has been reviewed more generally by Veresoglou et al. (2012), who noted that experimental studies are still in their nascent phases, and encouraged more investigators to pursue studies in this important area.

2.10 FUTURE DIRECTIONS FOR RESEARCH ON ROOTS AND MYCORRHIZAL FUNCTION AND BIODIVERSITY

As researchers and government agencies become ever more interested in and concerned about "sustainability" and long-term management of ecosystems, they will require much more information on system-level carbon allocation and energetics of these ecosystems. A recent example is related to concerns about carbon sequestration and ecosystem carbon cycling. Root/mycorrhizal interactions have been found to be diagnostic of significant differences in potential seedling relative growth rate (RGR) (Cornelissen et al., 2001). Plant species with ericoid mycorrhiza showed consistently low RGR, low foliar N and P concentrations, and poor litter decomposability. Species with ectomycorrhiza had an intermediate RGR, higher foliar N and P, and intermediate to poor litter decomposability; plant species with AM showed comparatively high RGR, high foliar N and P, and fast litter decomposition. The incorporation of mycorrhizal associations into functional-type classifications should prove useful in assessing plant-mediated controls on carbon and nutrient cycling.

Several studies of mycorrhizal function have noted the ways in which mycorrhiza facilitate nutrient uptake from a wide range of sources. For example, Perez-Moreno and Read (2001) measured an enhanced recycling of N and P nutrients from the necromass of nematodes in cultures with *Betula pendula*. In nonmycorrhizal treatments, the uptake of N and P was slightly greater, but with mycorrhiza of *Paxillus involutus* (an ECM) present, the nutrient uptake was significantly higher still. ECM plants (e.g., Lodgepole pine = *Pinus contorta*) and mycorrhizal fungi such as *Amanita muscaria* have been demonstrated to utilize organic forms of nitrogen directly (Abuzinadah and Read, 1989; Finlay et al., 1992). In addition, ericoid mycorrhizae can produce extracellular enzymes that mineralize nitrogen from protein–tannin complexes (Bending and Read, 1996; Leake and Read, 1989). In a number of ecosystems worldwide that have ericaceous or heath vegetation, this direct pathway, "short circuiting" the microbial mineralization pathway would enable the ericaceous plants to survive, indeed thrive, in the absence of adequate N from usual mineralization rates (Northup et al., 1995). Note that newly available analytical techniques such as AMS (see Staddon et al., 2003, earlier) have allowed detection of small quantities of C other than from atmospheric sources, such as the direct uptake of amino acids, as noted earlier, and also from anaplerotic (dark fixation) pathways, the latter providing as much as 3% of total C in ECM or saprophytic tissues (Hobbie et al., 2002).

As noted earlier, the vast majority of plants have mycorrhizal associates. Little is known yet of the species richness of mycorrhiza, particularly of the arbuscular, or AM type. Until recently, AM mycorrhiza were assumed to have little host-specificity, and to generally colonize a wide range of possible host species. In an intensive study of the plant–AM fungal interactions within a 1-ha old field in North Carolina, Bever et al. (2001) found that, rather than the initial estimate (in 1992) of 11 AM species in this field, they now have isolated at least 37 species, with one-third of them previously unrecorded. The ecological preference ranges of each species are quite different, reflecting significantly different optima for temperature, moisture content, host, and phenological phases of the plants in the field. The implications of these varied interactions for plant diversity are very large, and the subject of increasing interest among plant community ecologists, as the extent of mycorrhizal growth and uptake of labile carbon may affect the overall plant community makeup in interesting, hitherto unthought-of fashions (Bever, 1999).

The molecular identification of AM has increased greatly in the last decade. By taking samples from within growing roots, amplifying, and producing 18S rDNA sequences, Redeker et al. (2000) and Redeker (2002) found a phylogenetically deep divergence of lineages within the Glomales, one of the principal groups within the AM fungi. In addition, two or more species were found to be cooccurring within the same root, indicating the probable existence of complex interactions of the fungi involved. The possibility for various species of AM fungi to become active at various times of the year is an intriguing one for future research on root-mycorrhizal fungal ecology (Redeker, 2002).

Until recently, there have been few estimates of the amount of carbon inputs into soil from mycorrhizal fungi. Ekblad et al. (2013) estimated a dry matter deposition range of 20–200 kg ha^{-1} y^{-1} for a wide array of ecosystems dominated by ECM fungi. This would be approximately 10–100 kg ha^{-1} $y^{-1}$of carbon, a significant proportion of the total allocated belowground in many forested ecosystems.

Seasonal effects of root growth and turnover inevitably lead to differences in rate of production of mycorrhizal fungi, one would think. Studying seasonal changes in root growth in a boreal pine forest soil, Högberg et al. (2010) measured a five-fold increase of belowground allocation of plant C in late season (August) versus early season (June). Labeled C was traced into fungal fatty acid biomarkers (rarely in bacteria) and in Collembola, not in Acari or Enchytraeids (Högberg et al., 2010).

Over longer time spans, mycorrhizal fungal networks provide a facilitating mechanism for growth of plant seedlings in forested ecosystems, and possibly ones in grasslands as well. In a study of 21 seedling species in the field, Van der Heijden and Horton (2009) measured enhanced seedling growth in 48% of the cases, negative effects in 25%, and no effects for 27%.

Carbon sequestration patterns relative to the mycorrhizal community shift during long-term successional processes in boreal forests. Cord-forming ECM and ericoid MF play opposite roles in belowground carbon storage. There is high production of the former early on, then cord-formers decrease, and ericaceous MF dominate, facilitating production of humic material via formation of melanized hyphae, thus resisting decomposition (Clemmensen et al., 2015).

Studies of the several-year-long effects of elevated carbon dioxide effects on soils subsystems have been studied by several research groups, using such tools as the Free Air Carbon Enrichment (FACE) experiments (Drigo et al., 2010). They developed a conceptual model (Fig. 2.10) of flows in a mycorrhiza–plant–soil system, with flows into protists and nematodes viewed as indirect effects due to grazing and herbivory. Subsequently, Cheng et al. (2012) measured increases in organic C decomposition due to AM activity under elevated levels of carbon dioxide.

In a later study, Cheng et al. (2014) studied the rhizosphere priming effect (RPE) which is often very large. Thus, RPE can lead to from 50% reduction to a 380% increase on SOM decomposition rates under field conditions. Using a case study of the Duke FACE experiment, they found that the performance of the PhotoCent (Cheng et al., 2014) model was significantly improved, with an RPE induced 40% increase in SOM decomposition rate for the elevated CO_2 treatment. This demonstrates the value of incorporating RPE into future ecosystem models.

In addition to plant–fungal interactions, there are additional interactions with underground fauna to be considered. Klironomos and Kendrick (1995) found that the hyphae of AM are generally less palatable than the hyphae of soil-borne conidial fungi. Such species-specific differences in palatability have been observed in ECM fungi as well (Schultz, 1991). We consider feeding behaviors in Chapter 4, Secondary Production: Activities and

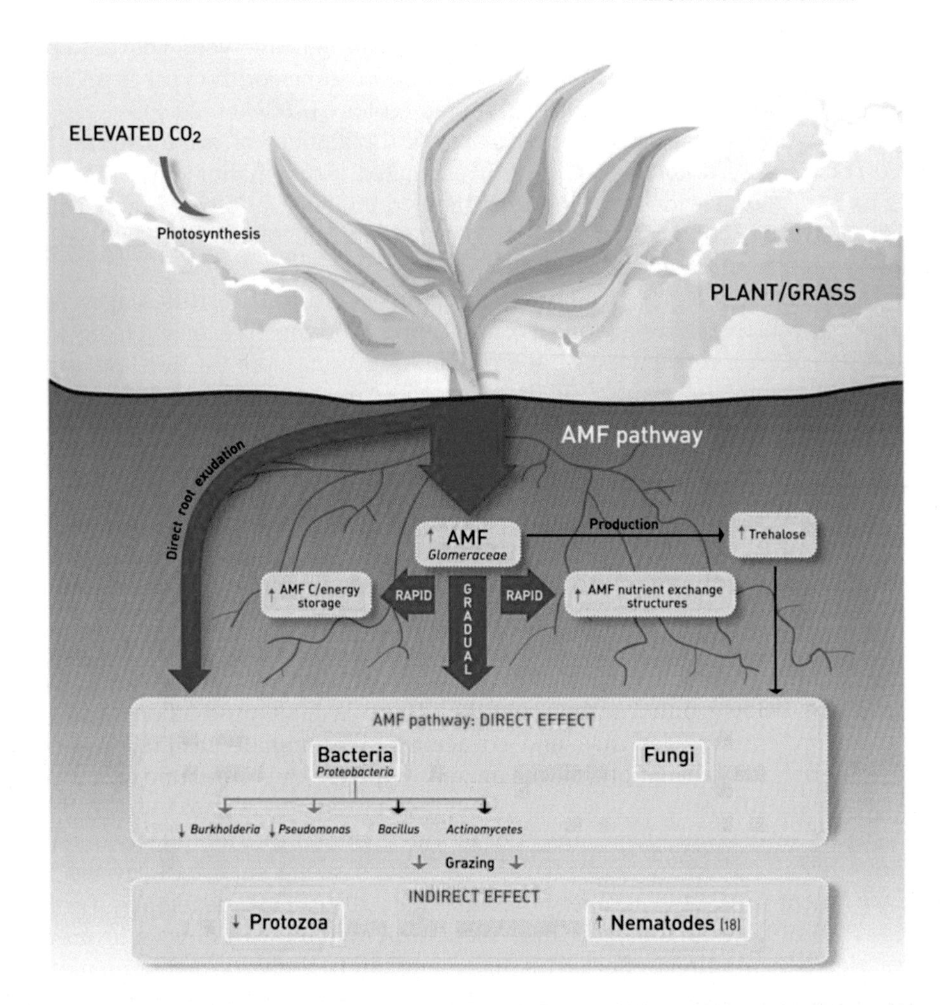

FIGURE 2.10 Conceptual model of C flow in mycorrhizal plant–soil systems summarizing the observed effects of elevated CO_2 atmospheric concentrations on soil communities. *Brown* arrows indicate increases and decreases in the respective community sizes, as determined by real-time PCR and lipid analyses, as well as changes in community structure and carbon flow. Absence of an arrow indicates no significant change in the community size or structure. *Red* arrows indicate no effect of increased C availability because of elevated CO_2 on the Actinomycetes spp. and Bacillus spp. communities. The mechanism and magnitude of the C flow along the soil–food web are indicated by the *green* arrows. *Source: Modified from Drigo et al. (2010).*

Functions of Heterotrophic Organisms—The Soil Fauna and system-level impacts in Chapters 5, Decomposition and Nutrient Cycling and 6, Soil Foodwebs: Detritivory and Microbivory in Soils.

The tools and analytical skills are at hand; it is now necessary to proceed with as much care in the assessment and measurement of belowground processes as has heretofore been given to aboveground processes.

2.11 SUMMARY

Primary production processes constitute the principal biochemical motive force for all subsequent activities of heterotrophs in soils. The inputs come in two directions: from aboveground onto the soil surface as litter, and belowground, as roots, which contribute exudates and exfoliated cells while the roots are alive, and then as root litter, once the roots die.

A wide range of direct measurements of root production and turnover are now in use. These include various nondestructive techniques, including rhizotrons and minirhizotrons, and destructive techniques, including soil coring, and isotopic-labeling of roots, followed by destructive sampling at specified time intervals to determine dynamics, for example, over an entire growing season.

Of equal importance to roots themselves are their generally more efficient physiological extensions, the root–fungus mutualistic association, mycorrhiza. At a cost of 5%–30% of the total photosynthate translocated belowground, mycorrhiza assist in obtaining inorganic nutrients, water, and in some cases, organic nutrients over a much wider range of the soil volume than roots alone. This symbiotic association has a significant effect on other biota, namely microbes and fauna, which inhabit all soil systems.

CHAPTER

3

Secondary Production: Activities of Heterotrophic Organisms—Microbes

3.1 INTRODUCTION

We will now consider system-level catabolism (cata-bolos = breaking-down activity), or dissipation and transformation of energy. The transfer of energy from the primary producers into organisms farther along the food chain supports a wide range of heterotrophs. The production of new body tissues from primary production is called secondary production. If the plant food sources are living, the linkages are called a grazing food chain. Conversely, if the contributions from net primary production are dead, the sequence is termed a detrital food chain. These different food chains (or food webs) are often termed "green food webs" for grazing on live plant matter, and "brown food webs" for detrital systems (e.g., Zou et al., 2016).

This difference in green and brown food chains has some impact on system function, in that the feedback effects on the living tissues are direct, rather than indirect, as in the detrital pathway. Soil food chains and webs are discussed further in Chapter 6, Soil Foodwebs: Detritivory and Microbivory in Soils.

The array of energy dissipators, or heterotrophs, in soil is incredibly diverse. The size range goes from <1 μm in length (bacteria) to the largest fossorial mammals, such as aardvarks or badgers, and giant earthworms, which reach 2 m in length (Lee, 1985). Larger entities include ant and termite colonies, considered by some as "superorganisms" (Emerson, 1956). Larger yet are supercolonies of one organism of uniform genetic material, such as the extended mycelium of an *Armillaria* fungus in Oregon, which extended over more than 9.6 km^2 (Ferguson et al., 2003).

All heterotrophs, of whatever size or volume, are involved in ingesting organic carbon and associated nutrients and assimilating them into carbohydrates, lipids, and proteins. Using a portion for production of new body tissue, an extensive amount (40% or more) of the chemical bond energy is lost as metabolic heat and evolved CO_2. The initially synthesized plant carbohydrate (or its equivalent) is catabolized according to the general formula: $C_6\ H_{12}\ O_6 + 6O_2 \rightarrow 6CO_2 + 6H_2O$; the more general formulation thus becomes: $C_n\ H_{2n}\ O_n + nO_2 \rightarrow nCO_2 + nH_2O$. Stoichiometrically, this is the reverse of photosynthesis,

Fundamentals of Soil Ecology
DOI: http://dx.doi.org/10.1016/B978-0-12-805251-8.00003-X

which was discussed in Chapter 2, Primary Production Processes in Soils: Roots and Rhizosphere Associates.

3.2 COMPOUNDS BEING DECOMPOSED

There are literally thousands of chemical and biochemical compounds involved in catabolism. Viewed in an ecological context, however, they can be classified into two functional categories: (1) primary compounds, those which are directly derived from plant, microbial, or animal tissues, and (2) secondary compounds, those which are produced as a result of organic matter/mineral interactions, usually resulting in small or large chemical changes in chemical bonds or degree of aromaticity.

Both categories comprise a few major types (or groups) of compounds: soluble or labile, versus relatively insoluble (in water) nonlabile or resistant compounds. Compounds in the former category include organic acids, amino acids, and simple sugars. Compounds in the latter category include lignin, cellulose, cutins, and waxes. One should also consider biochemical versus biological bond types, as defined by McGill and Cole (1981) and expanded upon in Zechmeister-Boltenstern et al. (2015). These reflect the differences between ester linkages, designated R–C–O–O–R, which yield energy when broken, and the carbonyl C–N, C–P, or C–S bonds, which require energy to be cleaved, yielding nutrients to the microbes (Newman and Tate, 1980).

3.3 MICROBIAL ACTIVITIES IN RELATION TO CATABOLISM IN SOIL SYSTEMS

The principal "players" in the decomposition process are the microbial populations, i.e., the bacteria, Archaea, fungi, and viruses. The bacteria and fungi are as biochemically diverse as they are diverse in phylogeny. Bacteria, currently considered to encompass from 60 to 80 phyla and a number that will undoubtedly grow as new habitats are sampled and organisms sequenced, are probably the most speciose array of organisms on earth (Youssef et al., 2015; Tiedje et al., 2001; Torsvik and Øvreås, 2002). In addition, they are undoubtedly the most numerous organisms, and have been estimated to number from 4 to 6×10^{30} cells, with a sizable proportion, >70%, being in the subsurface, which includes the earth's mantle to 4 km in depth (Whitman et al., 1998). Numbers of bacteria in soils of all biomes were estimated to be 2.5×10^{29} cells, with some of the larger quantities in desert scrub and savanna lands. The foregoing counts translate into 2×10^{9} cells g^{-1} in the top meter, and 1×10^{8} cells g^{-1} in the 1–8 m soil depth, with numbers in forest soils being markedly lower (Whitman et al., 1998).

In soils away from the rhizosphere, the environment for bacteria is usually stressful. A majority of bacteria exist in this low nutrient condition and may be starving (Morita, 1997). For now, one should note that, although some bacteria can double every 20 minutes or less in growth media in the laboratory, they may undergo only 2–3 divisions per year, on average, in soil under field conditions, due to the extreme limitations of available reduced carbon substrates. This energetic limitation is considered in detail in Chapter 6,

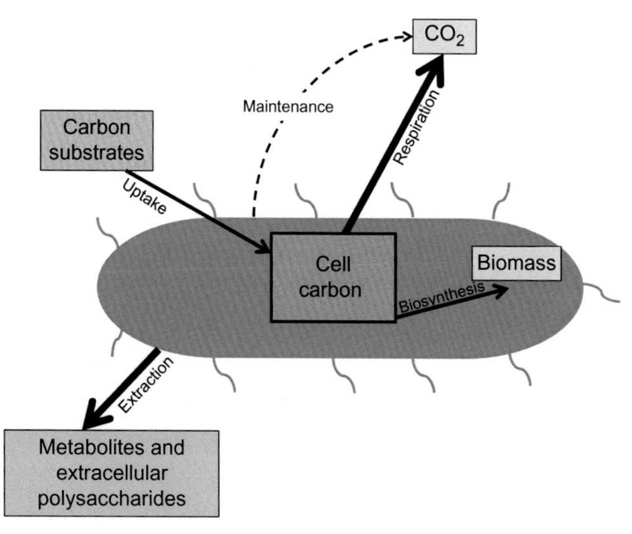

FIGURE 3.1 Flow of carbon in a bacterial cell. *Source: Redrawn from Scow, K.M., 1997. Soil microbial communities and carbon flow in agroecosystems. In: Jackson, L.E. (Ed.), Ecology in Agriculture. Academic Press, San Diego, California, pp. 367–413.*

Soil Foodwebs: Detritivory and Microbivory in Soils. A pictorial representation of bacterial carbon flow is given in Fig. 3.1 (Scow, 1997). Note the aforementioned flows to both growth and maintenance respiration. The latter requirement becomes limiting to many bacteria under conditions of nutrient limitation. In anoxic or low-redox microsites within soil aggregates and faunal guts, decomposition via microbial fermentation or anaerobic respiration with nitrate or other electron acceptors can occur. Decomposition linked to aerobic respiration would occur in regions with higher levels of oxygen, such as on the exteriors of aggregates.

Many genera of prokaryotes, including both bacteria and archaea, have evolved the highly important biochemical trait of "fixing" (rupturing the triple covalent bonds of) dinitrogen, making it available as ammonium for plant or microbial uptake (Postgate, 1987). This has important ramifications for nitrogen and phosphorus cycling and interactions with soil organic matter (SOM, Falkowski et al., 2008; Giller, 2001; Stewart et al., 1990), although the estimated total amount of nitrogen entering terrestrial ecosystems via this pathway remains uncertain (Stocker et al., 2016; Vitousek et al., 2013).

An additional factor to consider early on in our examination of soil nutrient cycling processes is the strong competition that exists between roots and microorganisms for nitrogen. There is a very strong temporal aspect to this competition. Analyzing 424 data pairs from 41 ^{15}N labeling studies investigating the labeled N redistribution between roots and microorganisms, Kuzyakov and Xu (2013) calculated Michaelis–Menten kinetics from higher stresses on uptake values of nitrate, ammonium, and amino acids by roots and microorganisms, showing that after pulses of these compounds came into the soil from SOM and litter, microorganisms took up most of the nitrogen. However, over the long term, most of the nitrogen moves from soils to roots. This temporal niche differentiation protects the ecosystem from N losses by leaching during periods of little or no root uptake, i.e., during periods of winter dormancy of the vegetation. However, there is continual turnover of microbial biomass, often enhanced by faunal grazing, thus providing roots with available N according to plant demand. Kuzyakov and Xu (2013) suggest that this supports the case for mutualistic interactions between roots and symbiotic microorganisms, such as mycorrhiza, but we suggest, following from the pioneering studies of Ingham et al. (1985), and summarized in Coleman and Wall (2015), that microbivorous micro-, meso-, and macrofauna have a significant role to play in this nutrient cycling process.

Knowledge of the prokaryotes has increased greatly in the past 20 years, with numerous accounts of their phylogeny published (Wu et al., 2009; Bergey's manual at http://www.cme.msu.edu/bergeys/; Torsvik and Øvreås, 2002). The principal concern of bacterial phylogeny is to trace both the extent of species of Bacteria, as well as the Archaea. Until recently, Archaea were considered to be inhabitants of extreme environments, including deep sea trenches and vents and hot springs, but they have been found also in numerous other habitats, including fresh water lakes and forest and agricultural soils (Bintrim et al., 1997; Jurgens et al., 1997; Pace, 1997). Bates et al. (2011) present an evaluation of the global distribution of Archaea in soils, with the principal finding that archaeal gene sequences, although not ubiquitous in their 146 sample soils, were generally quite common (Archaea detected in 87% of samples), and that their relative abundance was related to soil *C:N* (more archaeal gene sequences in soils with lower *C:N*). For more information on soil prokaryote interactions in soils and rhizospheres, see the review by Kent and Triplett (2002).

A method often used to analyze bacterial populations is to amplify DNA extracted from environmental samples by polymerase chain reaction (PCR), using primers universal to the 16S rRNA genes of bacteria and archaea (Lane, 1991; Prosser, 2002). In both tropical (Borneman and Triplett, 1997) and arid southwestern US soils (Kuske et al., 1997), over 50% of the prokaryotic DNA sequences of soil prokaryotes belonged to groups with no representatives in laboratory culture. This has marked implications for identifying prokaryotes involved in biogeochemical cycling and other environmental processes (see Chapter 7: Soil Biodiversity and Aboveground-Belowground Linkages). Either DNA or RNA can be extracted from soils, but a majority of the studies have been based on DNA extraction, which is easier to accomplish efficiently due to the higher lability and turnover of RNA. Ribosomal RNA content in active cells is higher than in inactive ones, thus rRNA-based analyses are a better approach for characterizing active microbial populations in soils (Ogram and Sharma, 2002). Techniques are now available to analyze microbial community structure and function, by analyzing microbial rRNA and mRNA, respectively. Both types of RNA can be extracted from soils and converted to cDNA (complementary DNA) by the enzyme reverse transcriptase for subsequent PCR amplification. Standard PCR analyses using "universal primers for rRNA genes" are not quantitative but do provide very useful qualitative information on dominant microbial populations. As long as suitable primers are available, microbial rRNA copies or mRNA copies can be quantified using quantitative, or "real time" PCR. These latter approaches provide an important means for linking soil microbial community structure and function. As these technologies have advanced at incredible pace, the extraction of DNA and RNA from soils has become more and more feasible, and several "kits" are commercially available to facilitate the extraction of the pertinent portions of bacterial genes for sequencing. The most commonly used technique for analyzing soil microbial populations, and communities, in terms of composition and diversity is massively parallel 454-pyrosequencing which allows previously unimaginable throughput and resolution with regard to DNA in soil (e.g., Lauber et al., 2009; and citing papers).

By using some techniques to "resuscitate" rare bacterial species in soils, Aanderud et al. (2015) used enrichment of soil incubations with heavy water $H_2{}^{18}O$ stable isotope probing to identify bacteria associated with pulses of trace gases (CO_2, CH_4, and N_2O). By using soils from grassland, deciduous forest, and coniferous forests after a wetting event, and sampling a few days later, rare bacteria became up to 60% of the total microbial biomass recovered. This rapid turnover of the bacterial community corresponded with a 5–20-fold increase in net CO_2 production and up to a 150% reduction in the net flux of CH_4 from rewetted soils. Using this technique is thus a way to determine the rare bacterial species in a soil "seed bank" (Aanderud et al., 2015).

Perhaps the greatest difference between bacteria and fungi is to be found in their mode of growth. Fungi have long strands (hyphae) which can grow into and explore many small microhabitats, secreting any of a considerable array of enzymes, decomposing material there, imbibing the decomposed monomers, translocating the carbon and other nutrients back into the hyphal network (Fig. 3.2). In contrast to bacteria, fungi can remain active in soils at very low water potential (−7200 kPa) and are better suited than bacteria to exist in

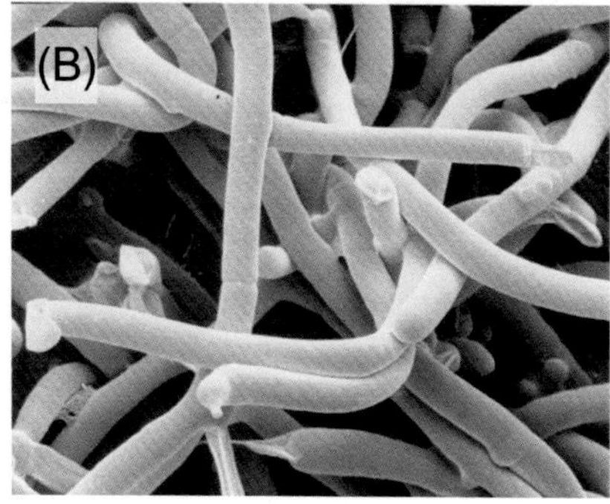

FIGURE 3.2 (A) Fungal hyphae growing across the soil surface. (B) Scanning electron microscopy (SEM) of fungal hyphae. *Source: Photos courtesy of K. Ignatyev and Dr. C. Sanchez. (B) From Dr. Carmen Sánchez, Universidad Autónoma de Tlaxcala, México.*

interpore spaces (Shipton, 1986). Many genera of fungi are closely associated with plants (see Chapter 2: Primary Production Processes in Soils: Roots and Rhizosphere Associates) and animals. Studies of coevolution of fungi with other eukaryotes have been summarized by Pirozynski and Hawksworth (1988). For an extensive overview of the roles that fungi play in terrestrial ecosystems, see Bue'e et al. (2009), Dighton (2003), and Taylor et al. (2014).

While not covered in detail in our book, there is a rapidly growing area of interest in effects of plant pathogens in ecosystems. Because plant pathogens play important roles in mediating plant competition and succession, and the maintenance of plant species diversity, they can have important feedback effects on soil communities and ecosystem processes as well. For a good review of these processes, see Gilbert (2002). We discuss effects of microarthropod grazing on fungal plant pathogens in Chapter 4, Secondary Production: Activities and Functions of Heterotrophic Organisms—The Soil Fauna.

In contrast, bacteria are usually unicellular, or in clustered colonies, occupying discrete patches of soil measuring only a few μm^3 in volume. Bacteria depend on many episodic events for passive movement, such as rainfall and root growth or ingestion by various soil fauna, to enable them to move about. When flagella are present, directed motility in the water film is also possible.

Viruses may play significant roles in the microbial ecologies of soil environments, as they can be a nonfaunal source of mortality, particularly for bacteria. Farrah and Bitton (1990) noted that lytic phages (viruses attacking bacteria) could act so as to restrict the growth of susceptible bacteria, and other phages could transmit genetic information between bacteria. The information on viral numbers and activities in soil in general is quite limited. Temperate phages (as distinct from virulent ones) in desert systems were inactivated on soil particles at acid pH (4.5–6). These phages had virtually no effect on populations of soil bacteria in Arizona soils, but persisted at low densities in their hosts (Pantastico-Caldas et al., 1992). This contrasts markedly with the often-cited deleterious impacts of virulent phages on *Escherichia coli* in liquid cultures in the laboratory. For more information on bacteriophages and interactions with bacteria under starvation conditions, see Schrader et al. (1997). Again, the advent of molecular techniques (such as pyrosequencing and metagenomics) to evaluate the abundance and diversity of viruses in soil are leading to increased appreciation and understanding of their role in horizontal gene transfer and the structure and function of microbial communities (Reavy et al., 2014). This field of inquiry is just in its infancy, and should provide much interesting material in coming years.

Unfortunately for the soil ecologist, the distribution and abundance of microorganisms is so patchy that it is very difficult to determine their mean abundances with accuracy without dealing with a very high variance about that mean, when viewed on a macroscale. Part of this variation is due to the close "tracking" of organic matter "patches" by the microbes. This variability occurs both in space and in time, and is perhaps best captured by the concept of "hot spots" and "hot moments" described in Kuzyakov and Blagodatskaya (2015). There are aggregations of microbes around roots (the oft-cited "rhizosphere") (Lynch, 1990), around fecal pellets and other patches of organic matter (Foster, 1994) and in pore necks (Fig. 3.3) (Foster and Dormaar, 1991). In addition, microorganisms concentrate in the mucus secretions that line the burrows of earthworms (the

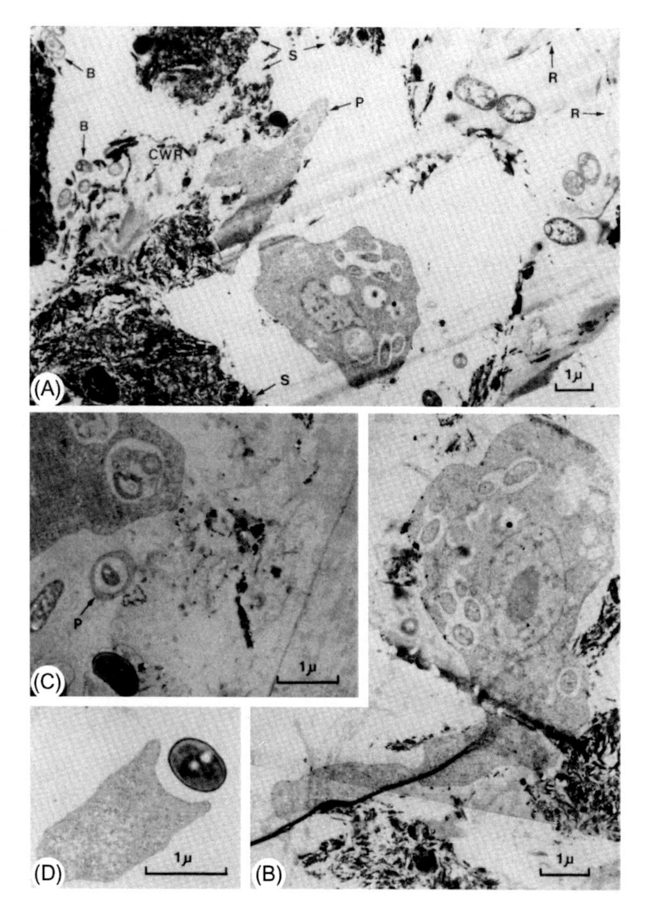

FIGURE 3.3 (A) An amoeba probing a soil microaggregate containing cell wall remnants (CWR) and a micro-colony of bacteria (B); P, pseudopodium; R, root; S, soil minerals; bar, 1 μm. (B) An amoeba with an elongated pseudopodium reaching into a soil pore. The amoeba contains intact ingested bacteria in its food vacuoles. (C) An amoeba with partly digested bacteria in food vacuoles; note bacterium enclosed by a pseudopodium (P). (D) A pseudopodium associated with a Gram-positive microorganism. *Source: From Foster, R.C., Dormaar, J.F., 1991. Bacteria-grazing amoebae in situ in the rhizosphere. Biol. Fertil. Soils 11, 83–87.*

"drilosphere," as defined by Bouché, 1975 and reviewed by Lee, 1985). The phenomenon of "patches" is discussed more in Chapter 6, Soil Foodwebs: Detritivory and Microbivory in Soils.

3.4 ROLES OF SOIL PROTISTS IN SOILS

All too often, the influences of protists (unicellular eukaryotes) are overlooked in many soil ecology studies. These small fauna, covered in detail in Chapter 4, Secondary Production: Activities and Functions of Heterotrophic Organisms—The Soil

Fauna, serve to promote the turnover of bacterial and fungal communities in soils, particularly in the rhizosphere. Recent studies have noted that the genetic divergences between and within major protist groups are much greater than that found in each of the three multicellular kingdoms of multicellular animals, plants, and fungi (Pawlowski et al., 2012).

A large proportion of soil ecology studies have focused on processes occurring in the O and upper A horizons, because so much of the short-term dynamics occurs there. With tools of microbial community analysis, Fierer et al. (2003) used phospholipid fatty acid (PLFA) analysis to examine the vertical distribution of specific microbial groups and their diversity in two soil profiles down to a depth of 2 m. The number of individual PLFAs decreased by c. one-third from the soil surface down to 2 m. Changes in certain ratios of fatty acid precursors and ratios of total saturated/total monounsaturated fatty acids increased with soil depth, indicating that microbes in the lower horizons were more carbon limited at greater depths. Interestingly, approximately 35% of the total amount of microbial biomass was found in soil below a depth of 25 cm Gram-positive bacteria and actinomycetes tended to increase in proportional abundance with depth, whereas Gram-negative bacteria, fungi, and protozoa were highest at the soil surface.

3.5 STOCKS OF CARBON IN DEEP SOIL, FAUNAL INFLUENCES

There has been renewed interest in the role of deep soil stocks of carbon in the terrestrial carbon cycle. Despite their low C content, compared to surface horizons, up to 50% of total soil C stocks reside at from 30 to 100 cm depths. Organic C inputs into subsoils occurs in dissolved form (DOC) following preferential flow pathways as aboveground or root litter and exudates along root channels and via bioturbation. With few exceptions, *C/N* ratio decreases with soil depth, while stable *C* and *N* isotope ratios of SOM are increased, indicating that OM in deep soil horizons is highly processed (Rumpel and Koegel-Knabner, 2011). We note that many macrofaunal groups, particularly earthworms and termites, play significant roles in making the macroporous flow pathways in soils, at times at depths greater than 1 m depth. So, although soil ecologists, biogeochemists, and microbial ecologists recognize that deeper soil accounts for a large fraction of the total SOM, and that the deep SOM is the foundation of biotic webs containing microbes and fauna, the deep soil remains a mostly unexplored frontier in need of much more careful examination. Fortunately, the advent of molecular techniques in microbial ecology and the increased attention to soils below 30 cm depth will insure rapid progress in our understanding of deep soil processes. For example, Uksa et al. (2015) conducted a study in German agricultural fields and found distinct separation of microbial community components, particularly among bacterial phyla, between surface soil and subsoil (in this case 30–75 cm), with oligotrophic bacterial phyla dominating the subsoil layers and copiotrophs dominating the surficial soils associated with rhizosphere and drilosphere soils.

3.6 MICROBIAL HETEROGENEITY AND SPATIAL PATTERNS IN SOILS

Soil is an impressively heterogeneous matrix of minerals and organic matter. Ways in which this heterogeneity in organic matter and texture can influence microbial populations have been widely studied for more than a century. A number of studies using transmission electron microscopy and scanning electron microscopy have revealed the intimate associations of bacteria and fungi with soil aggregates (Figs. 3.4 and 3.5; V. V. S. R. Gupta, Pers. Comm.).

With the development of more sophisticated imaging tools and statistical analyses of data, there have been several studies of microbial spatial patterns at the field or plot scale. Unfortunately, these studies have not spanned the range of spatial variability, which may exist at levels well below the millimeter scale (Nunan et al., 2002). Taking a large volume of soil, both topsoil and subsoil ($3 \times 3 \times 0.9$ m) from an arable field, Nunan et al. (2002) prepared subsampled cores and biological thin sections in which the in situ distribution of bacteria could be quantified. They acquired spatially referenced RGB digital images, using epifluorescence microscopy at 630× magnification. Average bacterial numbers per thin section were calculated using nine replicate images captured from each thin section (Fig. 3.6). Analysis of spatial dependence or continuity of soil bacterial density was performed using geostatistical tools at three scales: (1) centimeter to meter, (2) millimeter to centimeter, and (3) micrometer to millimeter scale, using appropriate semivariogram formulas (for more information on use of semivariograms, see Robertson and Gross, 1994). Spatial structure was found only at the micrometer scale in the topsoil, whereas evidence

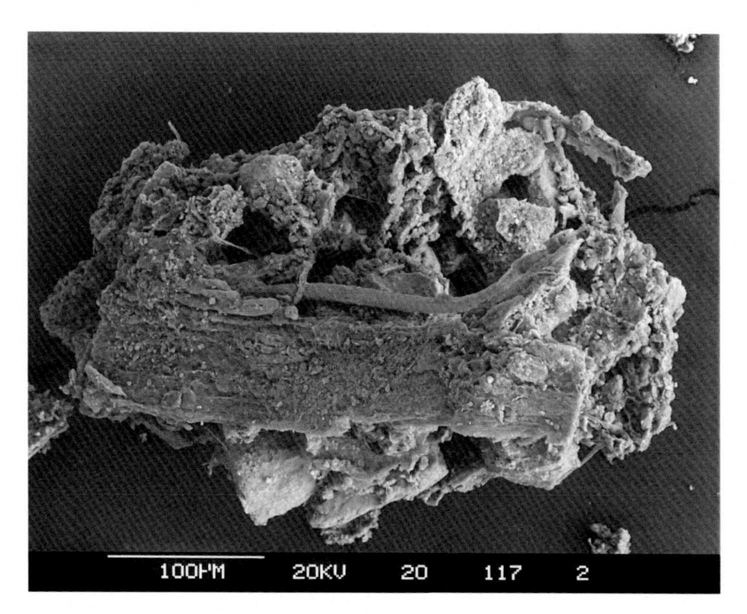

FIGURE 3.4 SEM picture of a macroaggregate (250- to 500-mm diameter) with particulate organic matter and hyphae. *Source: V. V. S. R. Gupta, with permission.*

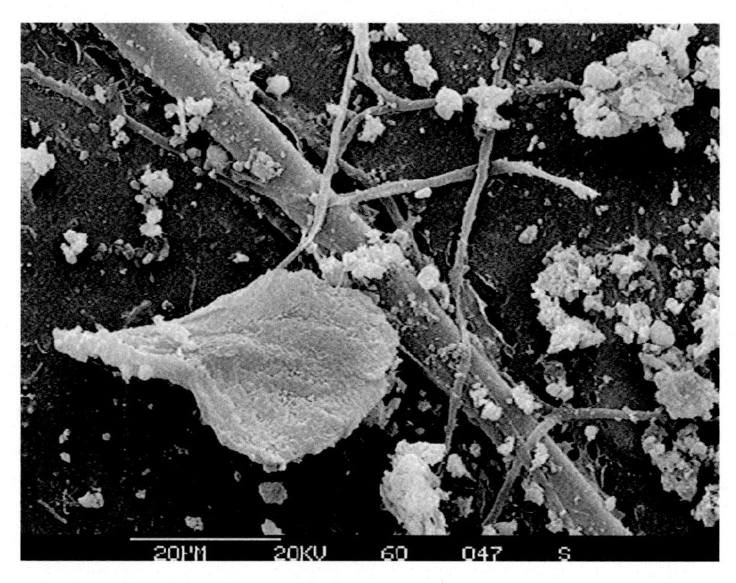

FIGURE 3.5 Amoebae feeding on fungi. *Source: V. V. S. R. Gupta, with permission.*

for nested scales of spatial structure was found in the subsoil at both the micrometer scale and at the centimeter to meter scales. Evidence for spatial aggregation in bacteria was stronger in the topsoil and decreased with depth in the subsoil. Nunan et al. (2002) suggest that factors that regulate the distribution of bacteria in the subsoil operate at two scales, in contrast to one scale in the topsoil, and that bacterial patches are larger and more prevalent in the topsoil.

3.7 APPROACHES FOR ESTIMATING MICROBIAL NUMBERS AND TURNOVER

Textbooks, such as those by Paul (2015), Paul and Clark (1996), and Swift et al. (1979) cover a number of methodological approaches for estimating microbial numbers and turnover in considerable detail. In this book, we present a few principal techniques for measuring numbers and identifying members of the microbial communities. We then relate them to studies of nutrient immobilization and mineralization, covered later in Chapter 5, Decomposition and Nutrient Cycling, on decomposition processes.

3.8 TECHNIQUES FOR MEASURING MICROBIAL COMMUNITIES

Techniques for measuring populations and biomass of microorganisms are either direct, by counting, or by inference from chemical and physical measurements. The following are

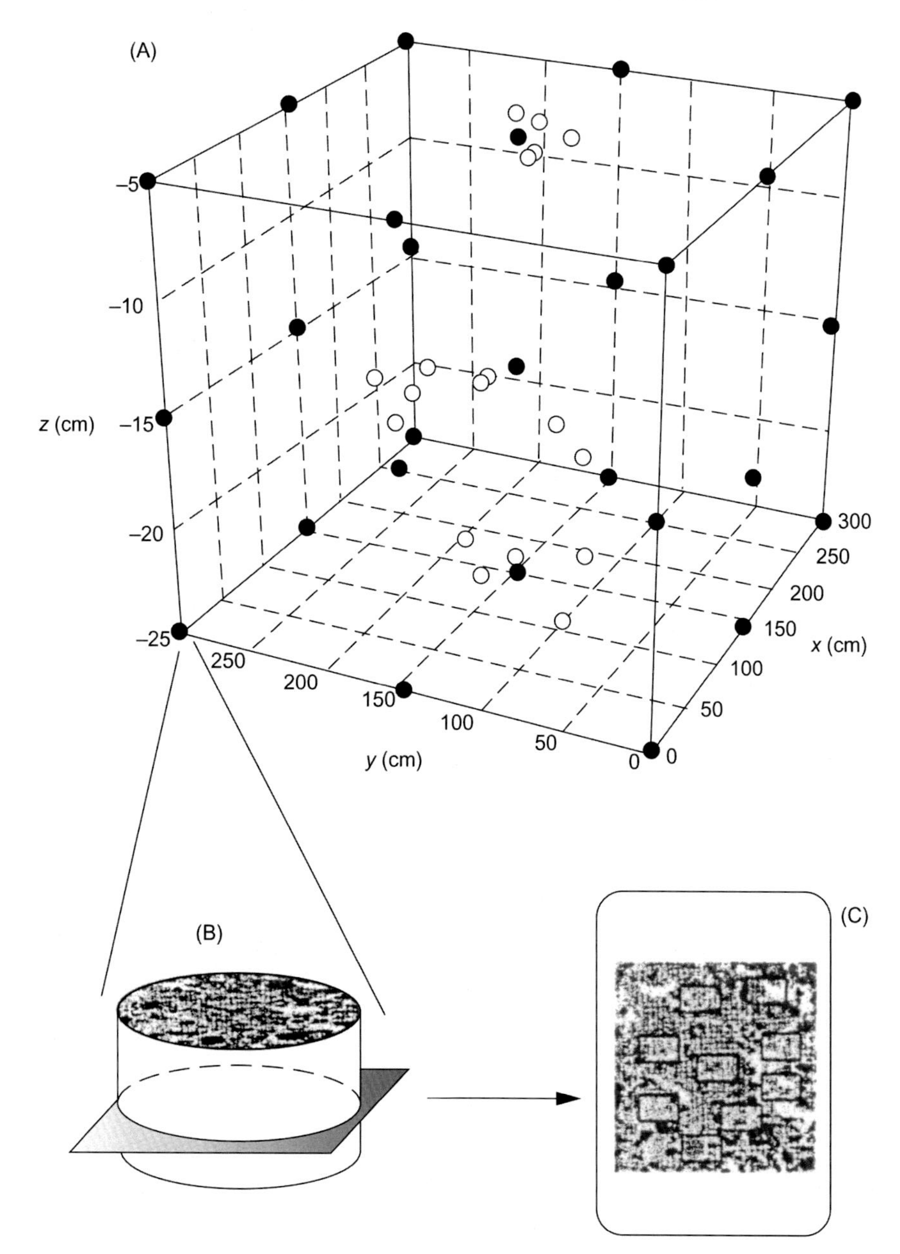

FIGURE 3.6 Spatial distribution of sampling points in topsoil (A). Solid circles form systematic random lattice and open circles form a biased random cluster. An undisturbed core (B) was sampled at each point and a thin section (C) cut from the horizontal plane. Nine spatially referenced images, in which bacteria were mapped, were acquired from each thin section. Average bacterial density per thin section was calculated and the values used to study large-scale variability. Bacterial maps were divided into 100 quadrats and bacterial density in each quadrat calculated. There were 900 quadrats per thin section and these bacterial density values were used to study micro-scale spatial variability. *Source: From Nunan, N., Wu, K., Young, I.M., Crawford, J.W., Ritz, K., 2002. In situ spatial patterns of soil bacterial populations, mapped at multiple scales, in an arable soil. Microb. Ecol. 44, 296–305.*

a few of the more commonly used techniques for studies of microbial standing crops (biomass at the time of sampling) and activity in a community and ecosystem context.

3.8.1 Direct Measures of Numbers and Biomass

Total counts of microbes are made by preparation of soil (c. 10 mg) suspensions spread in thin agar films on microscope slides (Jones and Mollison, 1948). The films are then stained, often with fluorescent dyes, and scanned. More recently, there have been improvements to the direct count technique, such as the membrane filtration technique, which enables one to quickly count fungal hyphae against the filter or a stained background. This approach is generally much faster and easier than the more laborious agar film technique. Other more classical techniques such as viable counts on nutrient-containing agar media are discussed by Parkinson et al. (1971) and Parkinson and Coleman (1991). The viable culture techniques usually recover only 1% or less of the total viable cells, so are useful only for comparative purposes, when one is focusing on a few readily culturable species of bacteria.

Other direct measures include sampling for extractable DNA (Torsvik et al., 1990a, b; Torsvik et al., 1994; Torsvik and Øvreås, 2002), and using the PCR, to multiply specific genes (such as the 16S rRNA) to determine the identities of the organisms of interest. Another approach to microbial community analysis uses signature lipid biomarkers. This technique, pioneered by Dr. David White and colleagues at the University of Tennessee (Tunlid and White, 1992), measures ester-linked polar lipid fatty acids and steroids to determine microbial biomass and community structure. Further comments on these techniques are given in Paul (2015). For microbial community characterizations, various biomarkers are used. Prominent among these are the PLFAs. Phospholipids are found in the membranes of all living cells but not in the storage products of microorganisms. They may be extracted and characterized using gas chromatography/mass spectrometry (Zelles and Alef, 1995).

A variety of techniques have been developed for the isolation and identification of DNA from soil. Techniques include the cell extraction method, and direct lysis method (Saano and Lindström, 1995; Zhou et al., 1996). In studies of microbial community makeup in agroecosystem field soils, Furlong et al. (2002) compared the microbial community composition of earthworm and nonearthworm influenced soils at the Horseshoe Bend agroecosystem field site in Athens, GA. The objective was to compare microbial communities from worm casts and open soil, via the creation of clone "libraries" of the 16S rRNA genes, prepared from DNA isolated directly from the soil and earthworm casts. In the cast soils, representatives of the genus *Pseudomonas* as well as the Actinobacteria and Firmicutes increased in number (Furlong et al., 2002). The results were consistent with a model where a large portion of the microbial population in soils passed through the gastrointestinal tract of the earthworm unchanged while representatives of some bacterial phyla increased in abundance. In Chapter 4, Secondary Production: Activities and Functions of Heterotrophic Organisms—The Soil Fauna, we consider the various faunal groups in soil, and their life-history attributes that have impact on the microbial community makeup and its turnover.

As noted in the introduction to prokaryotes above, we are only now becoming aware of the phylogenetic richness of archaea in soil communities. PCR amplification using primers specific for archaeal 16S rRNA genes allows detection of archaea in diverse habitats (Bomberg et al., 2003). The abundance of crenarchaeal (one of the two kingdoms comprising the archaeal domain) 16S rRNA in both cultivated and native field soils has been estimated to be from 1% to 2% of the total 16S rRNA in these soils (Buckley et al., 1998). A more recent survey of 146 soils confirms this figure, with Archaea being usually 2% of the total unicellular prokaryotic abundance (Bates et al., 2011).

The above procedures are primarily used in determining bacterial community structure. For fungi, several recent studies have made use of the fact that ergosterols are specific to fungi, and the amounts of ergosterols can be quantified to determine the amount of fungal tissues (biomass) present in soils (Eash et al., 1994;Newell and Fallon, 1991; Zelles and Alef, 1995). Molecular phylogeny has been an equally powerful tool for describing fungal communities (see, e.g., Bue'e et al., 2009; Husband et al., 2002), with high-throughput pyrosequencing being the most commonly used technique. Less expensive techniques are also used (and constantly under development), and a sequencing system known as "ion-torrent sequencing" has been successfully applied to questions about fungal community composition (Brown et al., 2013).

3.8.2 Indirect Measures of Biomass: Chemical Methods

Jenkinson (1966) cited an earlier suggestion of Störmer (1908) that a flush of CO_2, evolved after fumigation, was due to the decomposition of organisms killed during fumigation by the surviving microorganisms remaining after fumigation. This relates to the extensive work done on "partial sterilization" of soils, in Great Britain and elsewhere (Powlson, 1975; Russell and Hutchinson, 1909), under the misguided assumption that most soil microorganisms were somehow deleterious to subsequent plant growth, particularly in agricultural fields. Many of the soil heterotrophs are now considered generally beneficial, particularly when viewed in a whole-system nutrient cycling context.

3.8.2.1 *The Chloroform Fumigation and Incubation Technique*

Using a fumigant such as chloroform, and incubating the soil for 10 or 20 days, the size of the flush of CO_2 output can be related to the size of the microbial biomass by the expression: $B = F/k_c$; where B = soil biomass C (in μg C*g^{-1} soil); F = carbon dioxide carbon (CO_2—C) evolved by fumigated soil minus CO_2 evolved by unfumigated soil over the same time period; and k_c = fraction of biomass mineralized to CO_2 during the incubation (Jenkinson and Powlson, 1976). The k_c value, calculated from a range of microorganisms in controlled experiments, is assigned a general value of 0.45 (Jenkinson, 1988).

Jenkinson and Powlson (1976) relied on laboratory measurements of microbial cells added to soil. Voroney and Paul (1984) extended this work to include labile nitrogen, and measured both k_c and k_n (fraction of biomass N mineralized to inorganic N). A review of usage of Carbon-14 to measure microbial biomass and turnover is given by Voroney et al. (1991), with step-by-step procedures for this research. They introduced carbon by labeling

TABLE 3.1 Comparison of Biomass Carbon as Calculated from Direct Microscopy and the Fumigation Incubation Method

Soil	Organic C (%)	pH	Biomass C calculated from biovolume[a] (μg C/g soil)	Biomass C calculated from FI method[b] (μg C/g soil)	Ratio of biomass C from biovolume to biomass C from FI
Arable[c]	2.81	7.6	550	547	1.01
Arable[c]	0.93	8.0	190	220	0.86
Deciduous woodland[c]	4.30	7.5	1540	1231	1.25
Arable[c]	2.73	6.4	390	360	1.08
Grassland[c]	9.91	6.3	3200	3711	0.86
Deciduous woodland[c]	2.95	3.9	330	51	6.47
Secondary rainforest[d]	1.46	7.1	430	540	0.80
Cleared forest[d,e]	1.23	6.2	260	282	0.92

[a] *See Jenkinson* et al. *(1976) for method of calculation.*
[b] *Calculated using* K_c *of 0.45, not 0.5 as in the original paper.*
[c] *Temperate soils from the United Kingdom.*
[d] *Subhumid tropics, Nigeria.*
[e] *Arable cropping for 2 years after clearing secondary forest, Nigeria.*
From Powlson, D.S., 1994. The soil microbial biomass: before, beyond and back. In: Ritz, K., Dighton, J., Giller, K.E. (Eds.), Beyond the Biomass. Wiley/Sayce, Chichester, pp. 3–20. Adapted from Jenkinson, D.S., Powlson, D.S., 1976. The effects of biocidal treatments on metabolism in soil. V. A method for measuring soil biomass. Soil Biol. Biochem. 8, 209–213.

plants via photosynthetic pathways, and followed the carbon into the microbial biomass via root exudates and turnover, and in turn into the SOM.

A wide range of soils has been compared for biomass C calculated from biovolume (the measured volume of the cell), from the chloroform fumigation and incubation (CFI) method, and a ratio of biomass C from biovolume to biomass C from CFI (Powlson, 1994) (Table 3.1). These ratios range from 0.86 to 1.25 from arable lands and up to 6.47 from deciduous woodland. Forest soils, including those with low pH, have proven more difficult to analyze for microbial biomass, and are considered next.

3.8.2.2 *Chloroform Fumigation-Extraction Procedure*

Vance et al. (1987) noted that low pH soils, particularly those in the range below pH 5.0, including many forest soils, were not well characterized for microbial biomass using the CFI procedure. They modified the CFI procedure (Jenkinson and Powlson, 1976) to the chloroform fumigation and extraction (CFE) procedure as follows (Vance et al., 1987): soil samples are fumigated with chloroform for 48 hours, the fumigated and nonfumigated control samples are extracted with 0.5 M K_2SO_4, and the resulting organic extracts are measured for carbon, nitrogen, and other elements. The difference between the total organic carbon from the chloroform-fumigated soils minus the nonfumigated controls, multiplied by the k_{ec} factor (see Chapter 9: Laboratory and Field Exercises in Soil Ecology, for details) is the microbial biomass carbon. For soils with pH values <4, the k_{ec} values are usually lower, from 0.2 to 0.35 (Jenkinson, 1988). The CFE method has proven quite successful, and enables one to obtain microbial biomass values for carbon, nitrogen, phosphorus (Hedley and Stewart, 1982), and sulfur (Gupta and Germida, 1988).

A few authors have expressed concern about the extent of faunal contributions to the fumigation "flush." Protozoan biomass may be a significant contributor in some soils (Ingham and Horton, 1987), but usually constitutes less than 2% of total microbial carbon.

Some general comments on the methodology of the microbial biomass method are necessary. The CFI and CFE methods should be employed within the context or intent of the methods originally described. Because they are bioassays, and not general chemical assays, they are not as robust as the latter. They can be misused, particularly if a great deal of organic matter substrate, waterlogging, or very low pH conditions are encountered (Powlson, 1994). However, the microbial biomass values are useful in the development and exercising of simulation models of labile carbon and nutrient turnover in a wide range of ecosystems (e.g., Jenkinson and Parry, 1989; Parton et al., 1987, 1989a, b). Jenkinson et al. (2004) provide a helpful review of the chloroform fumigation techniques. Interestingly, they consider the fumigation and incubation (FI) technique to be obsolete, and urge caution in the usage of the k values, as had been noted by several investigators.

For reviews of biochemical methods to estimate microbial biomass, see Alef and Nannipieri (1995a) and Sparling and Ross (1993). For more specific details, see Chapter 9, Laboratory and Field Exercises in Soil Ecology, for comments on details of the microbial biomass estimation procedure. The ultimate "take home message" in studies of microbial biomass is the necessity to use more than one method to have some confidence in the numbers and hence biomass of the microorganisms measured. Although more time intensive, it is advisable to compare biomass of microorganisms to direct counts, made microscopically (Parkinson and Coleman, 1991).

3.8.2.3 Physiological Methods

Additional methods for measuring microbial biomass include the substrate-induced-respiration (SIR) technique, first developed by Anderson and Domsch (1978). The SIR technique involves adding a substrate such as glucose to soil, and measuring the respiration resulting from the stimulated metabolic activity in the experimental, versus control treatments, which received no carbon substrate. It is possible to measure the relative contributions of bacteria and fungi by using inhibitors, for example, cycloheximide to inhibit fungal activity, or streptomycin to inhibit bacterial activity. The assumption is that one measures only bacterial activity when fungi are inhibited, and vice versa. The technique requires some care, as texture may affect the apparent "resistance" to biocides. Further details of the technique are given by Alphei et al. (1995), Beare et al. (1990, 1991), Insam (1990), and Kjøller and Struwe (1994).

3.8.2.4 Additional Physiological Methods of Measuring Microbial Activity

There is a large body of literature dealing with the indirectly measured signs of metabolic activity, namely CO_2 output or oxygen uptake. The ratio of the two gases, in terms

TABLE 3.2 Examples of Studies in Soil Microbiology in Which Metabolic Quotients Have Been Applied

Field of Study	Metabolic Quotient[a]
Maintenance carbon requirement	m, qCO_2
Carbon turnover	qCO_2, μ, K_mGLUCOSE, Y, m, qD, C_{mic}/C_{org}
Soil management	qCO_2, qD, V_{max}, C_{mic}/C_{org}
Impact of climate and temperature	qCO_2, C_{mic}/C_{org}, qD
Impact of soil texture and soil compaction	qCO_2, qD
Impact of heavy metals	qCO_2
Ecosystems, ecosystem theory	qCO_2, qD, C_{mic}/C_{org}
Impact of soil animals	qCO_2

[a]*m = maintenance coefficient; qCO_2 = metabolic quotient or specific respiration rate; μ = specific growth rate; K_m = Michaelis–Menten constant; V_{max} = maximum specific uptake rate; Y = growth yield; qD = specific death rate; C_{mic}/C_{org} = microbial carbon to organic carbon ratio expressed as a percentage of microbial carbon to total organic carbon.*

See Anderson (1994) for specific references pertaining to usage of particular metabolic quotients.

Modified from Anderson, T.-H., 1994. Physiological analysis of microbial communities in soil: applications and limitations. In: Ritz, K., Dighton, J., Giller, K.E. (Eds.), Beyond the Biomass. Wiley/Sayce, Chichester, pp. 67–76.

of either uptake or output, is very informative about the principal sources of carbonaceous compounds being metabolized. The ratio of CO_2 evolved to oxygen taken up, or RQ, is lowest for carbohydrates, intermediate for proteins, and highest when lipids are the principal substrate being metabolized (Battley, 1987). In several studies the microbial respiration per unit microbial biomass ($qCO_2 = \mu g\ CO_2\text{-C/mg}C_{mic}/h$) (Anderson, 1994; Anderson and Domsch, 1978; Anderson and Domsch, 1993; Insam and Domsch, 1988) was measured, and found useful as an indicator of the overall metabolic status of a given microbial community. Additional metabolic quotients have been used to study influences of climate and temperature, soil management, impact of heavy metals, and soil animals in ecosystems, notably the microbial carbon/organic carbon ratio expressed as a percentage of microbial carbon to total organic carbon, or C_{mic}/C_{org} (Table 3.2) (Anderson, 1994; Joergensen et al., 1995). This follows from the assumption that terrestrial ecosystems in a near steady state are characterized by a constant flow of nutrients and energy into and out of the ecosystem on a yearly basis, and entering and leaving the microbial biomass pool as well (Fig. 3.7; Anderson, 1994). All of the foregoing is based on aerobic conditions. The extent of anaerobicity can be important at certain times, and needs to be carefully measured.

Enough data sets on microbial biomass C and N have accumulated by now that an extensive synthesis of temporal and latitudinal variation was carried out on data from over 58 studies worldwide. For the entire data set, temporal variability was best predicted by a three-component model incorporating pH, soil C, and latitude (Wardle, 1998). The increasing latitude reflected higher interseasonal variations in temperature, causing greater interseasonal flux of the biomass. A majority of these studies provided data showing <1 turnover of the entire microbial biomass per year, reflecting the extreme scarcity of food

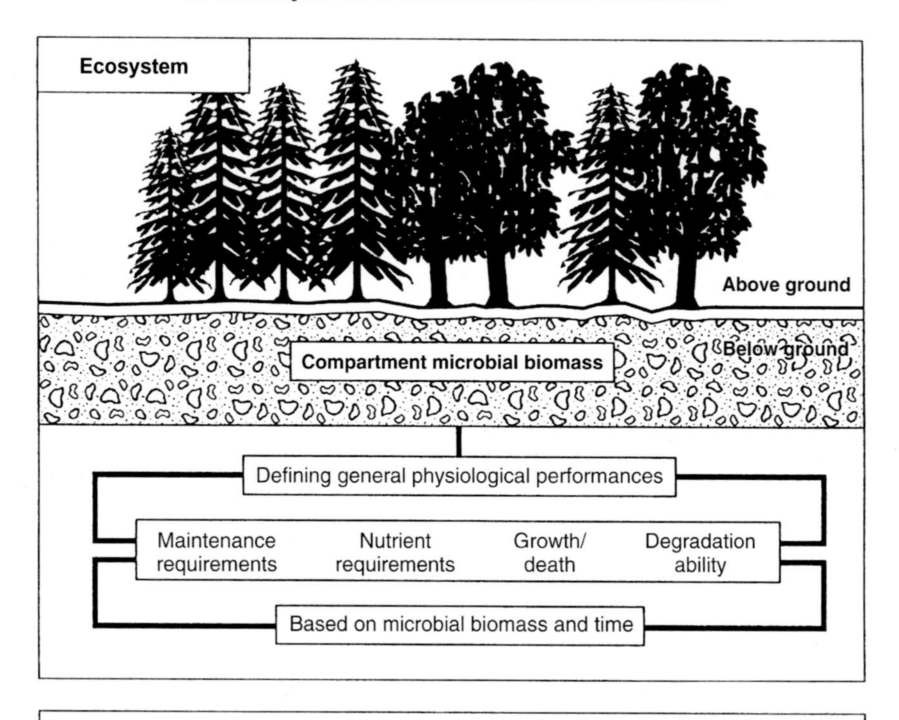

The prevailing assumption is that terrestrial ecosystems in a quasi–steady state are characterised by a constant flow of nutrients and energy, entering and leaving the system on a yearly basis, a well as entering and leaving the microbial biomass compartment. Microbial biomass communities adapt to the flow rate specific to the system.
C_{mic}/C_{org} ratio = percentage of microbial carbon to total soil carbon
Metabolic quotient for CO_2 (qCO_2) = μg CO_2 – C/mg C_{mic}/h

FIGURE 3.7 Working hypothesis for the application of metabolic quotients in ecosystem development at the synecological level. *Source: From Anderson, T.-H., 1994. Physiological analysis of microbial communities in soil: applications and limitations. In: Ritz, K., Dighton, J., Giller, K.E. (Eds.), Beyond the Biomass. Wiley/Sayce, Chichester, pp. 67–76.*

for most of the microbial populations much of the time, for reasons discussed earlier in this chapter.

3.8.2.5 Enzyme Assays and Measures of Biological Activities in Soils

Numerous soil biologists/ecologists have used enzyme assays to measure soil biological activity (Alef and Nannipieri, 1995b; Coleman and Sasson, 1980; Nannipieri, 1994). Oxidoreductases, transferases, and hydrolases have been most studied. There are concerns about using enzyme assays in soil and these are very well expressed by Nannipieri et al. (2002), who note that enzymes in soils can be in many different locations: (1) active and present intracellularly in living cells, (2) in resting or dead cells, (3) in cell debris, (4) extracellularly free in the soil solution, (5) adsorbed by inorganic colloids, or (6) associated in

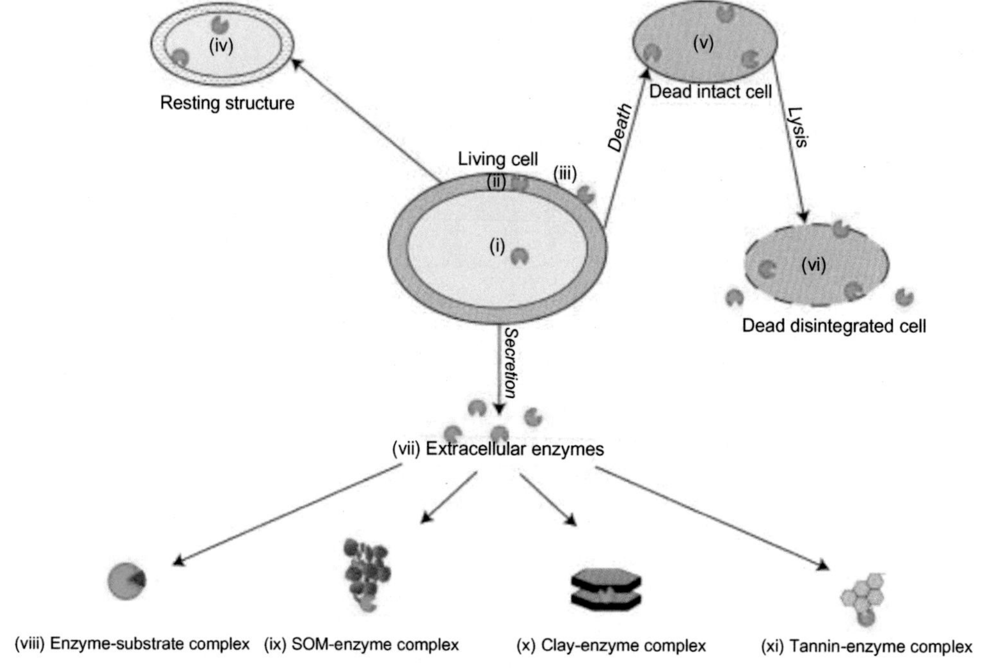

FIGURE 3.8 Locations of enzymes in soil: (i) functioning within the cytoplasm of microbial cells; (ii) restricted to the periplasmatic space of Gram-negative bacteria; (iii) located on the outer surface of cells with their active sites extending into the soil environment, contained within polysomes or retained by biofilms; (iv) situated within resting cells including fungal spores, protozoal cysts and bacterial endospores; (v) attached to entire dead cells and cell debris; (vi) leaking from intact cells or released from lysed cells; (vii) free in soil water; (viii) associated with enzyme–substrate complexes; (ix) complexed with soil organic matter by absorption, entrapment, or copolymerization; (x) sorbed to the external and internal surfaces of clay minerals; (xi) bound to condensed tannins. *Source: From Burns, R.G., DeForest, J.L., Marxsen, J., Sinsabaugh, R.L., Stromberger, M.E., Wallenstein, M.D., Weintraub, M.N., Zoppino, A., 2013. Soil enzymes in a changing environment: current knowledge and future directions. Soil Biol. Biochem. 58, 216–234.*

various ways with humic molecules. However, enzyme-based studies have proceeded apace in the last 15 year, with the focus being placed on extra-cellular enzymes in soil. The array of extracellular and intracellular distributions in the soil environment is expressed in Fig. 3.8 (Burns et al., 2013), which depicts various aspects of overall enzyme diversity related to microbial functional diversity in soil. The utility of soil enzyme assays to measure the influence of global climate change and environmental perturbations in general should be considered by soil ecologists in the future (Burns et al., 2013). Indeed, recent work examining the responses of enzyme activity to changes in soil temperature has shown that experimental warming of soil did not result in consistent changes in extracellular enzyme activity (Machmuller et al., 2016), but these authors warn that the strong seasonal influence on exoenzyme activity observed in their study is driven by climate variables that are also likely to be altered in future climates.

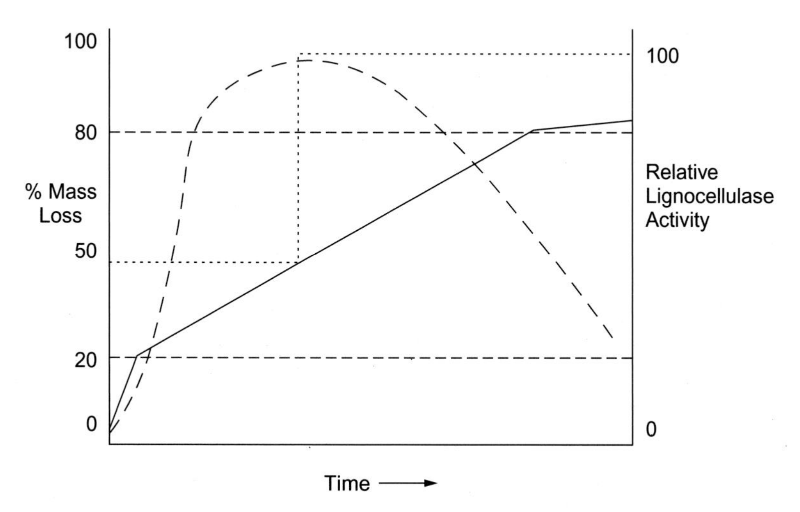

FIGURE 3.9 Idealized plot of lignocellulase activity (heavy dashed line) in relation to litter mass loss (solid line) through time. Lignocellulase activity traces a "bell-shaped" pattern over the course of litter decomposition, peaking (in this example) when cumulative mass loss reaches 45. The dashed horizontal lines at 20% and 80% mass loss highlight breakpoints in the mass loss curve. During the early stages of litter decomposition rapid mass loss is often largely attributable to leaching and mineralization of soluble litter constituents. In the middle stage, lignocellulose degradation predominates. Throughout the late stages, the accumulation of humic condensates depresses microbial activity, stabilizing the remaining material. *Source: From Sinsabaugh, R.L., Moorhead, D.L., Linkins, A.E., 1994. The enzymic basis of plant litter decomposition: emergence of an ecological process. Appl. Soil Ecol. 1, 97–111.*

It should be noted that enzymes related to particular target substrates, such as lignocellulases in leaf litter, may be relatively good predictors of mass loss. After early stages of mass loss, due to leaching and mineralization, the "middle stage" is often strongly correlated with enzyme activity. In the final stages, c. <25% of initial mass remaining, the accumulation of humic condensates depresses microbial activity, stabilizing the remaining material (Fig. 3.9) (Sinsabaugh et al., 1994).

Many models of SOM decomposition are based on first-order kinetics that assume that decomposition rate of a particular C pool is proportional to pool size and a simple decomposition constant ($dC/dt = kC$). In reality, SOM decomposition is catalyzed by extracellular enzymes that are produced by the microorganisms. A theoretical model to explore the behavior of a decomposition-microbial growth system that operates by exoenzyme catalysis used the following relationship: $D_C = K^*_d Enz_C$, where D_C = decomposition of polymeric material to produce available C; K^*_d is a single decomposition constant, and Enz_C = exoenzyme pool (Schimel and Weintraub, 2003). An enzyme kinetics analysis showed there must be some mechanisms to produce a nonlinear response of decomposition rates to enzyme concentration. This nonlinearity induces C limitation, regardless of the potential C supply. In a linked C and N version of the model, adding a pulse of C to an N-limited system increases respiration, while adding N in fact decreases respiration (with C redirected from waste respiration to microbial growth). Previous assumptions in the literature have assumed that the lack of a respiratory response by soil microbes to added N indicates that they are not N limited. This model of Schimel and Weintraub (2003) suggests that, while total C flow may be limited by the functioning of the exoenzyme system, in fact microbial growth may be N limited. This important finding should be the subject of several laboratory and field studies over the next few years. It is notable that the mechanistic-level fine-tuning of exoenzyme kinetics are still being discovered, but better understanding of these basic properties of enzyme activities in soil will improve the predictive power of SOM models in the future (Billings et al., 2016).

3.8.2.6 Direct Methods of Determining Biological Activity

Direct measurements of the activity of soil microorganisms have been a goal of soil biologists for a long time (Newman and Norman, 1943). This results from the basic thermodynamic fact that as organisms undergo metabolic activity, they emit heat from the enthalpy of reactions occurring in net catabolism (Battley, 1987). With the continuing trend toward miniaturization of circuitry and much better, more sensitive thermocouples, it is now possible to obtain direct measures of metabolic activities of organisms in small samples of soils with only a few milligrams of biomass (Alef, 1995; Battley, 1987; Sparling, 1981). Flow microcalorimeters are now available which allow for the simultaneous measurements of CO_2 and N_2O production in soils (Albers et al., 1995).

Another approach to direct measurement of microbial metabolic activity in a more fine-grained fashion is to measure microbial metabolic processes and identify the microorganisms responsible for particular biochemical reactions under field conditions. Radajewski et al. (2000) pioneered stable isotope probing of community-extracted DNA as a laboratory-based means of identifying microbial populations involved in ^{13}C-substrate metabolism. Padmanabhan et al. (2003) combined the approach of Radajewski et al. (2000) with an assay of a range of labile and recalcitrant organic compounds, linking the ^{13}C field release assay of respired $^{13}CO_2$ to DNA extraction analyses of the active microbial populations. Transient peaks of $^{13}CO_2$ released in excess of background were found in glucose- and phenol-treated soil within 8 hours of application. Across the 30-hour time span of the experiment, neither naphthalene nor caffeine additions stimulated $^{13}CO_2$ release above background. A total of 29 full sequences revealed that active populations included relatives of *Arthrobacter, Pseudomonas, Acinetobacter, Massilia, Flavobacterium,* and *Pedobacter* spp. for glucose; *Pseudomonas, Pantoea, Acinetobacter, Enterobacter, Stenotrophomonas,* and *Alcaligenes* spp. for phenol; *Pseudomonas, Acinetobacter* and *Variovorax* spp. for naphthalene; and *Acinetobacter, Enterobacter, Stenotrophomonas,* and *Pantoea* spp. for caffeine. All these genera belong to bacterial divisions or subdivisions that were recovered from soils in over 25% of the studies surveyed in the review. This approach is a useful first step in taking powerful analytical tools to the field. However, Padmanabhan et al. (2003) note that the amendment-based approach used in this study would not be likely to identify less responsive, slow-growing (K-selected) members of the soil microbial community. See Aanderud et al. (2015) for a strong counter to this assertion.

The diversity and biogeography of soil bacterial communities are fundamentally different from that of animals and plants, which show marked gradients of increased diversity from polar regions to tropical ones (Fierer and Jackson, 2006). Microbial biogeography is controlled, they note, primarily by edaphic variables (notably pH). Thus this pattern differs fundamentally from the "macro" organisms' biogeography.

3.9 SOIL STERILIZATION AND PARTIAL STERILIZATION TECHNIQUES

A number of the techniques mentioned earlier involve drastic perturbations to soils, such as fumigation, for the purpose of determining numbers or biomasses of organisms residing within them. Huhta et al. (1989) examined the influence of microwave radiation on soil processes, noting that it seems to have a less drastic impact compared to autoclaving, gamma irradiation, or chloroform fumigation. Microwaving is particularly useful for removing various mesofaunal groups, leaving the microbial communities reasonably intact under moderate thermal energy inputs, of 380 W for 3 minutes (Huhta et al., 1989; Wright et al., 1989). Unfortunately, some unwanted side effects were introduced with microwaving, principally a decreased water-holding capacity. Monz et al. (1991) found that microwaving was of limited use in their agricultural soils. We discuss methods to manipulate fauna, including group-specific chemical inhibitors or repellents, in Chapter 4, Secondary Production: Activities and Functions of Heterotrophic Organisms—The Soil Fauna.

3.10 CONCEPTUAL MODELS OF MICROBES IN SOIL SYSTEMS

3.10.1 Root-Rhizosphere Microbe Models and Experiments

As was noted in the discussion of primary production processes (see Chapter 2: Primary Production Processes in Soils: Roots and Rhizosphere Associates), there are "hot spots" of activity, particularly of microbes in relationship to root surfaces and rhizospheres. When viewed as a transect through the rhizosphere, for example, 2 mm from the root surface or less, there are arrays of rapidly growing bacteria and fungi, which have been called "fast" flora (Trofymow and Coleman, 1982) (Fig. 3.10). Moving up the root toward the shoot into older regions, one finds root hairs and then root cortical cells, which may be sloughing off into the surrounding soil. There are accompanying microbial and root grazers, such as protozoa and nematodes, which are discussed in detail in Chapter 4, Secondary Production: Activities and Functions of Heterotrophic Organisms—The Soil Fauna. Out in the bulk soil, away from the rhizosphere (>4 mm from the root surface), occur some of the slower growing, or "slow" bacteria and fungi, organic matter fragments, and some of the hyphae of either arbuscular mycorrhizal (AM) or ectotrophic mycorrhiza.

The field of mycorrhizosphere (Andrade et al., 1998; Garbaye, 1991) research has taken a quantum leap forward with elegant microscopic methods, in conjunction with molecular tools to pinpoint organisms that are coassociates. Artursson and Jansson (2003) used bromodeoxyuridine (BrdU), as a thymidine analog, to identify active bacteria associated with AM hyphae. After adding BrdU to the soil and incubating for 2 days, DNA was extracted, and the newly synthesized DNA was isolated by immunocapture of the BrdU-containing DNA. The active bacteria in the community were identified by 16S rRNA gene PCR

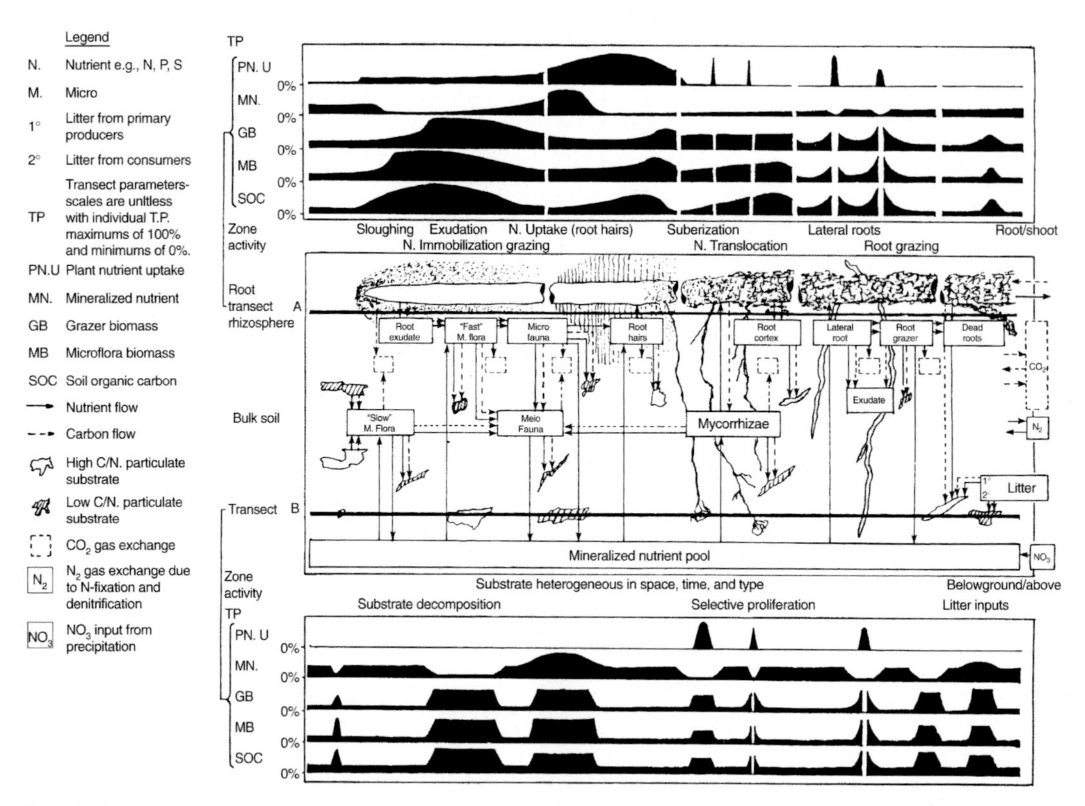

FIGURE 3.10 Conceptual diagram of a root–rhizosphere–soil system. *Source: From Trofymow, J.A., Coleman, D.C., 1982. The role of bacterivorous and fungivorous nematodes in cellulose and chitin decomposition in the context of a root/rhizosphere/ soil conceptual model. In: Freckman, D.W. (Ed.), Nematodes in Soil Systems. University of Texas Press, Austin, Texas, pp. 117–137.*

amplification and DNA sequence analysis. Based on gene sequence information, a selective medium was used to isolate the corresponding active bacteria. *Bacillus cereus* strain VA1, one of the bacteria identified by the BrdU method, was isolated from the soil and tagged with green fluorescent protein. By using confocal microscopy, this bacterium was shown to clearly attach to AM hyphae. This study by Artursson and Jansson (2003) is a pioneering attempt, using molecular and traditional approaches to isolate, identify, and visualize (Fig. 3.11) a specific bacterium that is active in fallow soil and associates with AM hyphae.

Viewed more generally, the entire root–soil–organisms system may be viewed as highly dispersed and episodic at numerous microsites that vary over time, sometimes over hours to days. This has been encapsulated in the phrase: "microbial hot spots and hot moments in soils" (Kuzyakov and Blagodatskaya, 2015). These authors comment that the fraction of active microorganisms in hotspots is 2–20 times higher than in the bulk soil, and their specific activities (i.e., respiration, microbial growth, mineralization potential, enzyme activities, RNA/DNA ratio) may also be much higher. The duration of hot

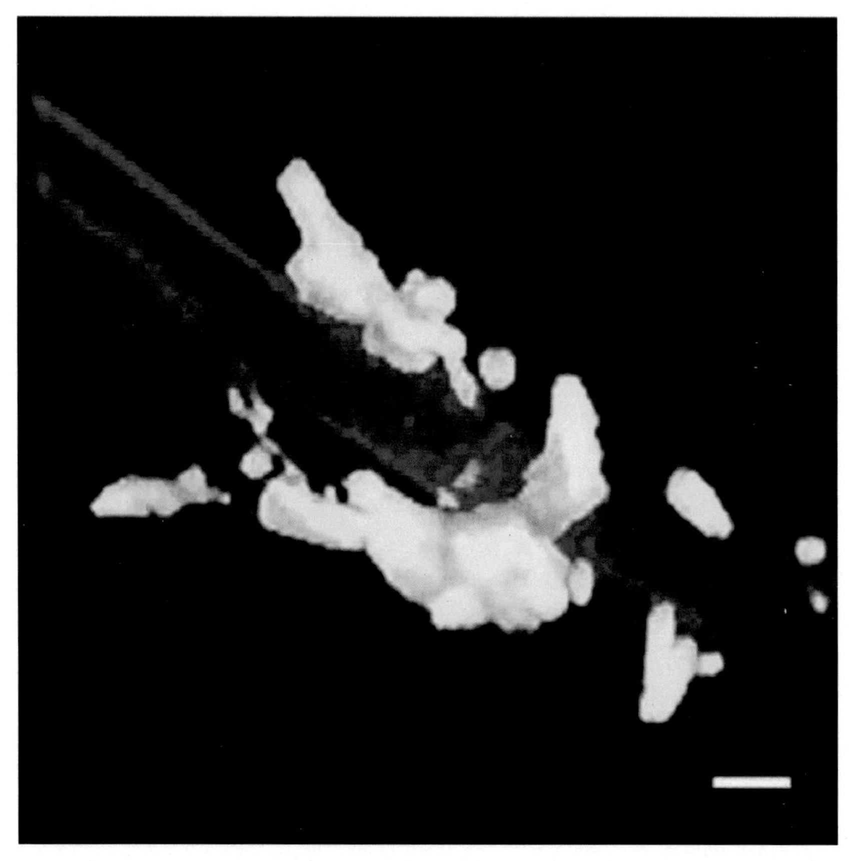

FIGURE 3.11 *Bacillus cereus* strain VA1 (*pnf8*) on a hyphal fragment from field soil. *Source: From Artursson, V., and Jansson, J.K., 2003. Use of bromodeoxyuridine immunocapture to identify active bacteria associated with arbuscular mycorrhizal hyphae. Appl. Environ. Microbiol. 69, 6208–6215.*

moments in the rhizosphere is limited and is controlled by the length of the input of labile organics. It can last a few hours up to a few days. In the detritusphere, however, the duration of hot moments is regulated by the output, by decomposition rates of litter, and lasts for weeks and months. Hot moments induce succession in microbial communities and intense intra- and interspecific competition affecting C use efficiency, microbial growth and turnover. The faster turnover and lower C use efficiency in hotspots counterbalances the high C inputs, leading to the absence of strong increases in C stocks. Consequently, the intensification of fluxes is much stronger than the increase of pools. Maintenance of stoichiometric ratios by accelerated microbial growth in hotspots requires additional nutrients (e.g., N and P), causing their microbial mining from SOM, that is, priming effects. Consequently, priming effects are localized in microbial hotspots and are consequences of hot moments. Kuzyakov and Blagodatskaya (2015) suggested that, irrespective of their volume, the hotspots are mainly responsible for the ecologically relevant processes in soil.

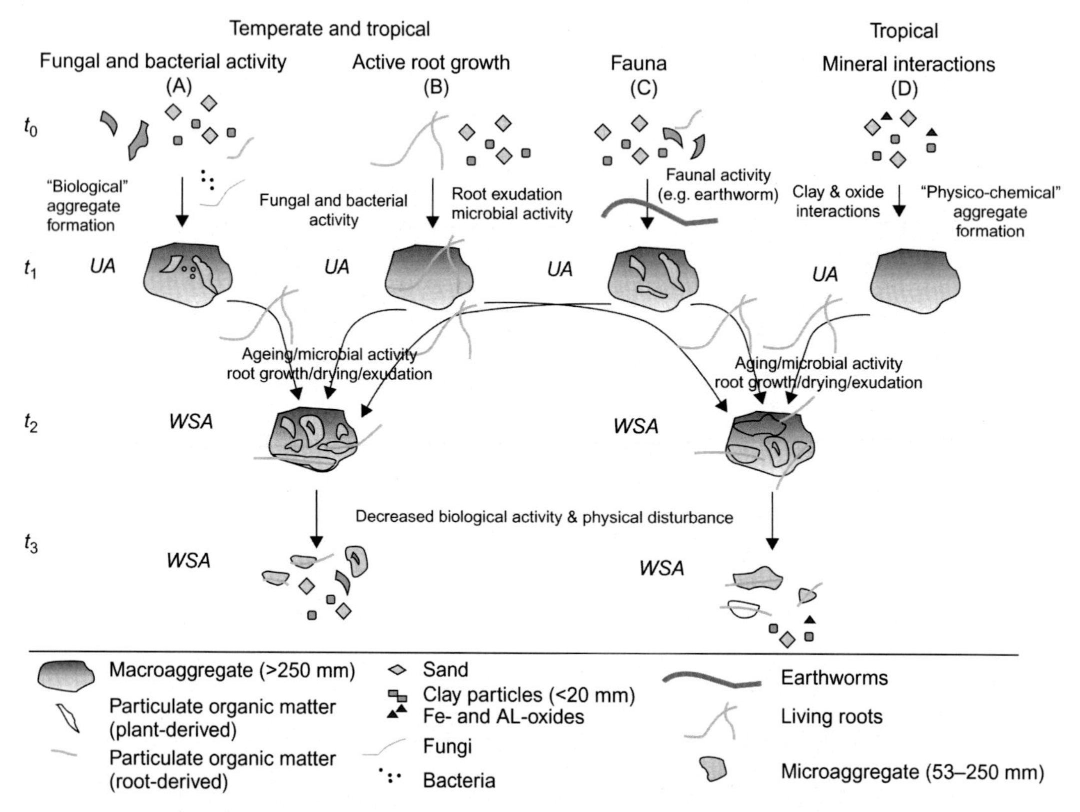

FIGURE 3.12 Aggregate formation and degradation mechanisms in temperate and tropical soils. Fungal and bacterial activity, active root growth, and earthworm activity are the biological aggregate formation agents in both temperate and tropical soils, whereas the mineral–mineral interactions in tropical soils are the physicochemical aggregate formation agents. UA = unstable aggregates; WSA = water-stable aggregates. *Source: From Six, J., Feller, C., Denef, K., Ogle, S.M., de Moraes Sa, J.C., Albrecht, A., 2002b. Soil organic matter, biota and aggregation in temperate and tropical soils—effects of no-tillage. Agronomie 22(7–8), 755–775.*

3.10.2 Soil Aggregation Models

Soil aggregates, as noted in the section on soil structure in Chapter 1, Historical Overview of Soil Science and the Intersection of Soil and Ecology, play a central role in protecting pools of C and N, and are derived from a variety of sources. A mechanism particularly prevalent in many tropical soils is the physical aggregation process, which occurs abiotically as a physicochemical process (Oades and Waters, 1991). In both temperate and tropical soils, there are several biological processes that result in the formation of "biological macroaggregates" (Fig. 3.12, Six et al., 2002b). These include the following: (1) fresh plant- and root-derived residues forming the nucleation sites for the growth of fungi and bacteria. Macroaggregate formation is initiated by fungal hyphae enmeshing fine particles into macroaggregates. Exudates from both bacteria and fungi, produced as a consequence of decomposition of fresh residues, form binding agents that further stabilize macroaggregates ($t_{1,A}$). (2) Biological macroaggregates also form around growing roots in soils, with roots and their exudates enmeshing soil particles, thereby stimulating microbial activity ($t_{O,B}$ to $t_{1,B}$). (3) A third principal mechanism of biological macroaggregate formation in soils in all climates is via the action of soil fauna, particularly earthworms, termites and ants. For example, earthworms often produce casts that are rich in organic matter ($t_{1,C}$) that are not stable when freshly formed and wet. During gut passage, the soil and organic materials are kneaded thoroughly and copious amounts of watery mucus are added as well. This molding process breaks bonds between soil particles, but can lead to casts that are quite stable upon drying. It is also worth noting that soil mesofauna, that is, collembola and mites are important in the SOM formation process through their production of copious amounts of fecal pellets. Effects of meso- and macrofauna on soil structure are discussed further in Chapter 4, Secondary Production: Activities and Functions of Heterotrophic Organisms—The Soil Fauna, on Soil Fauna.

The subsequent fate of macroaggregates follows a fascinating process over time, as noted by Six et al. (2002b) summarizing from several literature sources. At time t_1, the young, freshly formed unstable macroaggregates are only stable when treated in very gentle fashion (i.e., when the aggregates are taken from the field, brought to field capacity, subsequently immersed in water and retained when gently sieved). The formation of water-stable aggregates that can resist slaking (air dying and quick submersion in water before sieving) occurs by three processes (t_1 to t_2): (1) Under moist conditions aging may increase stability by binding through microbial activity. Microbial activity is stimulated inside the biological macroaggregates, including worm casts because of their high organic matter content. In this process, substantial amounts of polysaccharides and other organics are deposited, serving to further stabilize the macroaggregates. (2) Dry−wet cycles can result in closer arrangements of primary particles, leading to stronger bonding and increased aggregate stability. (3) Biological and physicochemical macroaggregates, in the presence of active root growth, can become more stabilized by penetration of the aggregates by roots. This includes the roles of root exudates as cementing agents, and the effects of the accompanying stimulated microbial activity. In addition, roots influence aggregation physically by (1) exerting lateral pressures inducing compaction, and (2) continually removing water during plant transpiration, leading to drying of the soil and cohesion of soil particles around the roots. Note that this process is likely to be enhanced or intensified by mycorrhizal hyphae associated with the plant roots.

During macroaggregate stabilization (t_1 to t_2), the intraaggregate POM is further decomposed by microorganisms into finer POM (Six et al., 1998) (Fig. 3.13). This fine POM is increasingly encapsulated with minerals and microbial products, forming new microaggregates (53−250 μm) within the macroaggregates (Six et al., 1999). Similar processes may arise by stimulation from root exudation and mycorrhizal products, causing further encrustation of microbial products and mineral particles, forming microaggregates around the root-derived POM. Note that this microaggregate formation within macroaggregates is crucial for the long-term sequestration of C because microaggregates have a greater protective capacity to shield C against decomposition compared with macroaggregates.

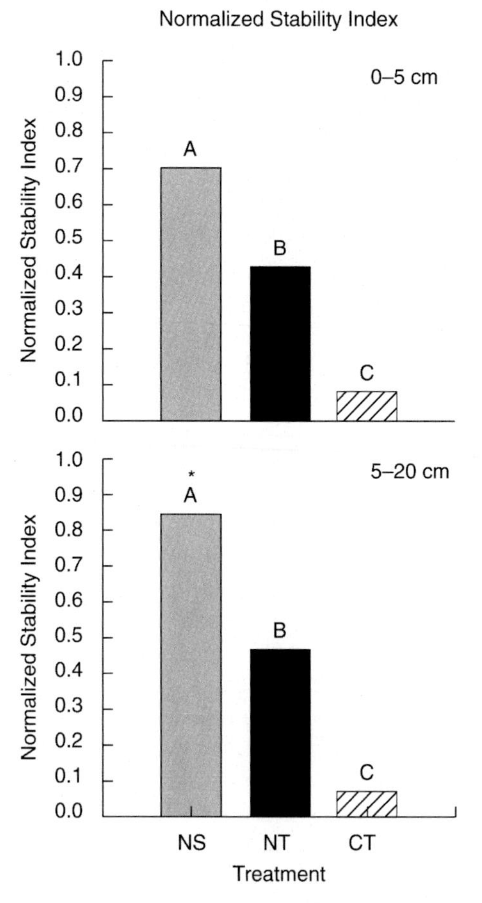

FIGURE 3.13 Effect of management on the normalized stability index (NSI) at the 0-, 5-, and 5- to 20-cm depths. NS = native sod; NT = no-tillage; CT = conventional tillage. Values followed by a different uppercase letter within a depth are significantly different. Values followed by * for the 5- to 20-cm depth are significantly different from corresponding values in the 0- to 5-cm depth. Statistical significance determined at $P>.05$ according to Tukey's HSD mean separation test. *Source: From Six, J., Elliott, E.T., Paustian, K., Doran, J.W., 1998. Aggregation and soil organic matter accumulation in cultivated and native grassland soils. Soil Sci. Soc. Am. J. 62, 1367–1377.*

The final phase of the aggregate turnover cycle (t_2–t_3) occurs when the macroaggregates break down, releasing microaggregates and microbially processed SOM particles. The macroaggregates are more liable to break up over time, as the labile constituents of the coarse sized SOM are consumed, microbial production of binding agents decreases, and the degree of association between the soil matrix and SOM decreases. Fortunately, microaggregates are still stable enough and not as sensitive to disruptive forces as the macroaggregates, and therefore survive (Fig. 3.13). This is borne out by a table of mean residence time (MRT) (in years) of macro- and microaggregate associated C (Table 3.3, Six et al., 2002b). Note the essentially fivefold greater MRT for microaggregates (m) compared with macroaggregates (M).

TABLE 3.3 Mean Residence Time (in Years) of Macroaggregate- and Microaggregate-Associated Carbon

Ecosystem	Aggregate Size Class[a]	(μm)	MRT
Tropical pasture	M	>200	60
	m	<200	75
Temperate pasture grasses	M	212–9500	140
	m	35–212	412
Soybean	M	250–2000	1.3
	m	100–250	7
Corn	M	>250	14
	m	50–250	61
Corn	M	>250	42
	m	50–250	691
Wheat-fallow, no-tillage	M	250–2000	27
	m	53–250	137
Wheat-fallow, conventional tillage	M	250–2000	8
	m	53–250	79
Average±stder[b]	M		42±18
	m		209±95

[a]*M = macroaggregate; m = microaggregate.*
[b]*stder = standard error.*
From Six, J., Feller, C., Denef, K., Ogle, S.M., de Moraes Sa, J.C., Albrecht, A., 2002b. Soil organic matter, biota and aggregation in temperate and tropical soils—effects of no-tillage. Agronomie 22(7–8), 755–775.

Cotrufo et al. (2013) have developed a Microbial Efficiency Matrix Stabilization (MEMS) framework that integrates plant litter decomposition with SOM stabilization. They ask the question: "do labile plant inputs form stable SOM?" A combination of microbial substrate use efficiencies and carbon and nitrogen allocation controls the proportion of plant-derived C and N that is incorporated into SOM, and soil matrix interactions involved in SOM stabilization. Using the MEMS framework, this allows Cotrufo et al. (2013) to consider labile plant constituents being dominant sources of microbial products, relative to input rates, due to greater microbial carbon use efficiency. These microbial products would become major precursors of stable SOM by promoting aggregation and through strong chemical bonding with the soil matrix.

The combined influences of physical, chemical, and biological factors in soil aggregate formation reach a peak when one includes the effects of arbuscular mycorrhizal fungi directly by physical binding. For an extensive account of the many physical, chemical, and biological interactions involved in the dynamics of creation and dissolution of soil structure, refer to the masterful review by Baldock (2002).

3.10.3 Models: Organism- and Process-Oriented

A recurrent theme that resonates throughout the field of soil ecology is the focus on organisms and population dynamics models used by community ecologists, and the use of process models at the ecosystem scale (Moore et al., 1996; Smith et al., 1998). Many ecosystem level models have included dynamics of the soil biota only implicitly, yet the intraseason dynamics of microbes and fauna, as is demonstrated in Chapters 4, Secondary Production: Activities and Functions of Heterotrophic Organisms—The Soil Fauna, and 5, Decomposition and Nutrient Cycling, often have a significant effect on nutrient availability and turnover. One of the more successful combinations of the organismal and process modeling approaches was by Paustian et al. (1990), who used four cropping systems, both annual and perennial crops, that varied in inorganic inputs and organic production in the growing season: (1) barley without fertilizer, (2) barley fertilized with 120 kg N ha^{-1} $year^{-1}$, (3) a meadow fescue field with 200 kg N ha^{-1} $year^{-1}$, and (4) a lucerne field with indigenous nitrogen fixation. The conceptual models of Carbon and Nitrogen flows in the four fields are presented in Figs. 3.14 and 3.15 (Paustian et al., 1990). Although there were no large differences in microbial biomass between treatments, the estimated microbial

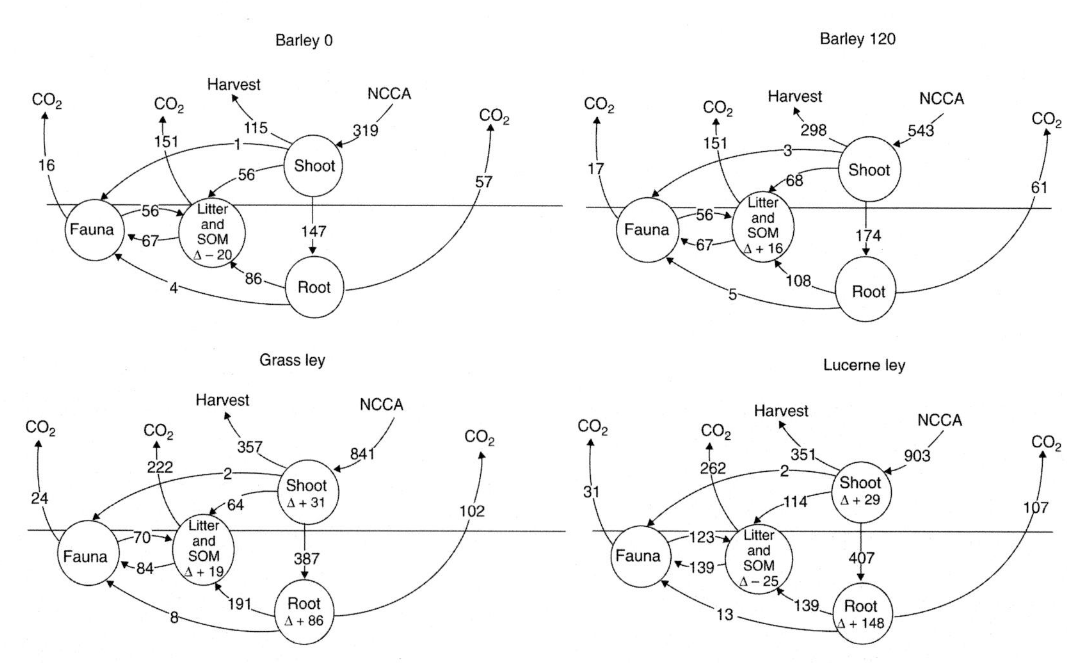

FIGURE 3.14 Budgets of annual carbon flows (g C m^{-2} $year^{-1}$) for barley receiving no nitrogen fertilizer (B0), barley receiving 120 kg N ha^{-1} (B120), a grass ley receiving 200 kg N ha^{-1} (GL) and a N_2-fixing lucerne ley (LL). Budgets are based on data from 1982 to 1983 for the topsoil (0–27 cm). Compartment changes on an annual basis are denoted by delta (Δ) symbols; for the aboveground plant compartment this includes biomass, standing dead litter, and surface litter. Soil organic matter (SOM), including microbial biomass and soil litter, has been combined into a single compartment. NCCA, net canopy carbon assimilation. *Source: From Paustian, K., Andrén, O., Clarholm, M., Hansson, A-C., Johansson, G., Lagerlöf, J., Lindberg, T., Pettersson, R., Sohlenius, B., 1990. Carbon and nitrogen budgets of four agro-ecosystems with annual and perennial crops, with and without N fertilization. J. Appl. Ecol. 27, 60–84.*

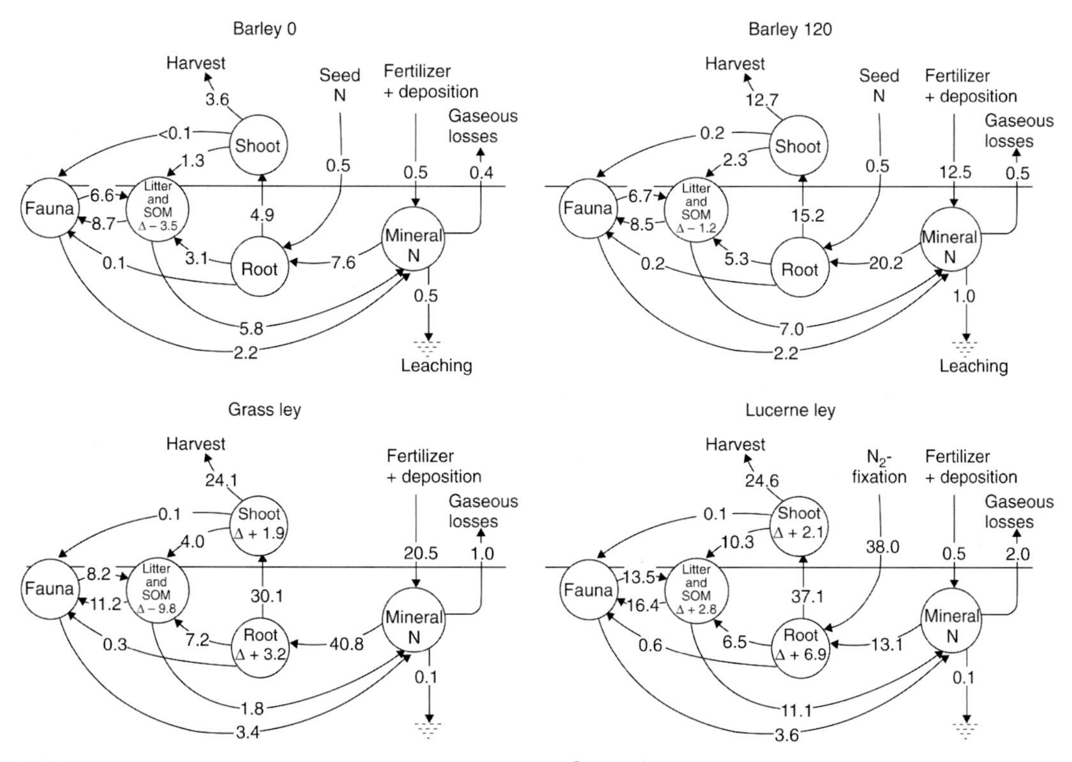

FIGURE 3.15 Budgets of annual nitrogen flows (g Nm^{-2} $year^{-1}$) in the four cropping systems. See Fig. 3.14 for further explanation of symbols. *Source: From Paustian, K., Andrén, O., Clarholm, M., Hansson, A-C., Johansson, G., Lagerlöf, J., Lindberg, T., Pettersson, R., Sohlenius, B., 1990. Carbon and nitrogen budgets of four agro-ecosystems with annual and perennial crops, with and without N fertilization. J. Appl. Ecol. 27, 60–84.*

production was 50% greater in the perennial ley fields than in the barley treatments, with soil meso- and macrofauna biomass, consumption and respiration being significantly higher as well. The main take-home message from this impressive long-term (10-year) study is that microbial and faunal (biotic) interactions do occur, and have a significant effect on nutrient turnover. As we note in Chapter 8, Future Developments in Soil Ecology, with current considerable concerns about the impacts of land use change and global change phenomena on key processes in soils, it is imperative to measure explicit changes in numbers and taxa of soil organisms that carry out these processes.

3.11 SUMMARY

The processes of consumption and decomposition are considered ecologically as system-level catabolism. The primary agents of decomposition are bacteria and fungi, often considered as "microbial biomass."

Microbial production and turnover is determined in a number of indirect ways, both chemical and physiological, as well as in direct fashion, using high-magnification microscopy, or via energetic approaches such as calorimetry.

The microbial biomass, while relatively small (c. 200–400 g m^{-2} in the surface 15 cm) relative to the total SOM pool, has a rapid turnover time and serves as a principal food source for microbivorous fauna. It is also the source of labile nutrients, available for plant roots and other microbes. Hence, the microbial community is indeed the "eye of the needle," through which virtually all of the decomposition carbon and nutrients must pass. In the course of microbial growth and turnover, there is a dynamic process of buildup of macroaggregates, which age and decay into more extensively protected microsites within microaggregates. Roles of fauna in the process of aggregate formation and alteration are also discussed further in Chapter 4, Secondary Production: Activities and Functions of Heterotrophic Organisms—The Soil Fauna.

Some very large strides forward in soil ecology have been taken by investigators who are linking microbial community structure and function. We discuss several examples in Chapter 8, Future Developments in Soil Ecology, regarding the effects of invasive plant species on microbial communities. It is apparent that a combined or synthetic approach, using aspects of functional measures, such as enzyme activities, when combined with the qualitative measures such as PLFA and the more quantitative measures of using 16S rRNA gene probes and cDNA analyses, along with the immunofluorescent approaches presented by Artursson and Jansson (2003) will yield considerable dividends in future studies.

CHAPTER

4

Secondary Production: Activities of Heterotrophic Organisms—The Soil Fauna

4.1 INTRODUCTION

Animals are the other major heterotrophs (cf microbes—Chapter 3: Secondary Production: Activities of Heterotrophic Organisms—Microbes) in soil systems. Soil animals exist in elaborate food webs sometimes containing several trophic levels. Some soil animals are true herbivores, but most subsist on dead plant matter and/or the associated microbes. Still others are carnivores, parasites, or top predators. Actual heterotrophic production by the soil fauna is poorly known because turnover of the faunal biomass, feeding rates, and assimilation efficiencies are difficult to assess. Estimates of biomass of soil animals are not common, and knowledge of material or energy transfer in food webs is fragmentary (Moore and deRuiter, 1991, 2000, 2012). Analyses of soil food webs have emphasized numbers of the various organisms and their trophic resources. Analysis of the structure of these food webs reveals complexity and many "missing links" (Scheu and Setälä, 2002; Walter et al., 1991). Communities of soil fauna offer opportunities for studies of phenomena such as species interactions, resource utilization, or temporal and spatial distributions.

Soil animals are numerous and diverse. The array of species is very large, including representatives of all terrestrial phyla. Many groups are poorly understood taxonomically, and details of their natural history and biology are virtually unknown. For example, it is estimated that only 10% of species of the microarthropods are described, and only 10% of the possible communities have been sampled. Protection of biodiversity in ecosystems clearly must explicitly include the species living in soil, and thus soil must be viewed as critical habitat for biodiversity.

A single soil ecologist cannot hope to become expert in all soil animal groups. When research focuses at the level of the soil ecosystem, two things are required: the cooperation of zoologists and the lumping of animals into functional groups. These groups are often taxonomic, but species with similar biology can also be grouped together for purposes of integration (Coleman et al., 1983, 1993; Coleman and Wall, 2015; Hendrix et al., 1986).

Fundamentals of Soil Ecology
DOI: http://dx.doi.org/10.1016/B978-0-12-805251-8.00004-1

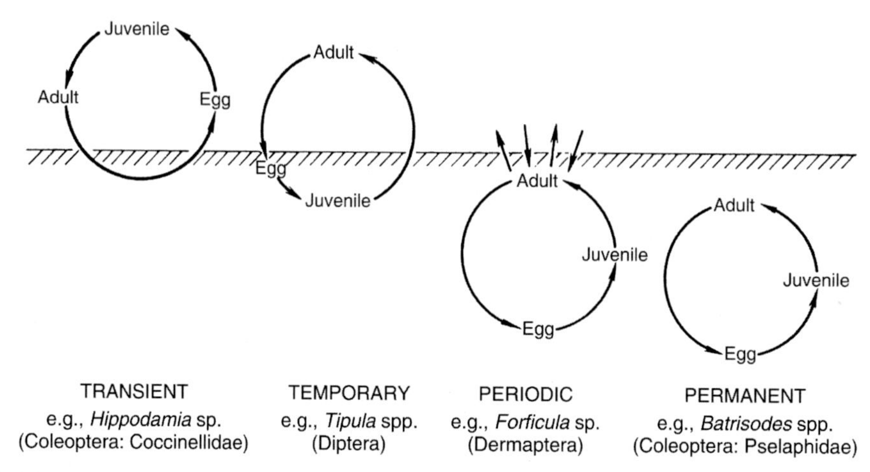

FIGURE 4.1 Categories of soil animals defined according to degree of presence in soil, as illustrated by some insect groups. *Source: From Wallwork, J.A., 1970. Ecology of Soil Animals. McGraw-Hill, London.*

The soil fauna may also be characterized by the degree of presence in the soil (Fig. 4.1) or microhabitat utilization by different life forms. There are transient species exemplified by the ladybird beetle which hibernates in the soil but otherwise lives in the plant stratum of the garden. Gnats (Diptera) represent temporary residents of the soil, because the adult stages live aboveground. Their eggs are laid in the soil and the larvae feed on decomposing organic debris. In some soils dipteran larvae are important scavengers. Cutworms are also temporary soil residents, with larvae feeding on plant seedlings at night, and hiding in soil by day. Periodic residents spend their lives belowground, with adults emerging (such as with cicadas) aboveground only to reproduce. From this perspective, soil food webs are linked to aboveground systems, making trophic analysis more complicated. Even permanent residents of the soil may be adapted to life at various depths in the soil.

Among the microarthropods, collembolans are examples of permanent soil residents (see Fig. 4.1). The morphology of collembolans reveals their adaptations for life in different soil strata. Species that dwell on the soil surface or in the litter layer may be large, pigmented, and equipped with long antennae and well-developed furcula (the springtail that allows them to jump long distances), whereas those dwelling in the mineral soil layers are smaller, unpigmented, and have elongate body forms and reduced furculae—they have no place to jump!

Numerous researchers have marveled at the many and varied body plans and size differences of the soil fauna. A generalized classification based on body length (Fig. 4.2) illustrates convenient device for separating the soil fauna into size classes: microfauna, mesofauna, macrofauna, and megafauna. This classification comprises the entire range from smallest (1–2 μm, e.g., flagellates) to largest (several meters, e.g., the largest known earthworms).

Body width of the soil fauna is related to their microhabitats (Fig. 4.3), and this measure offers another basis for classification. The microfauna (protozoa and small nematodes) exclusively inhabit water films. The mesofauna inhabit air-filled pore spaces and are largely restricted to existing pores (i.e., they do not have the capacity to produce or

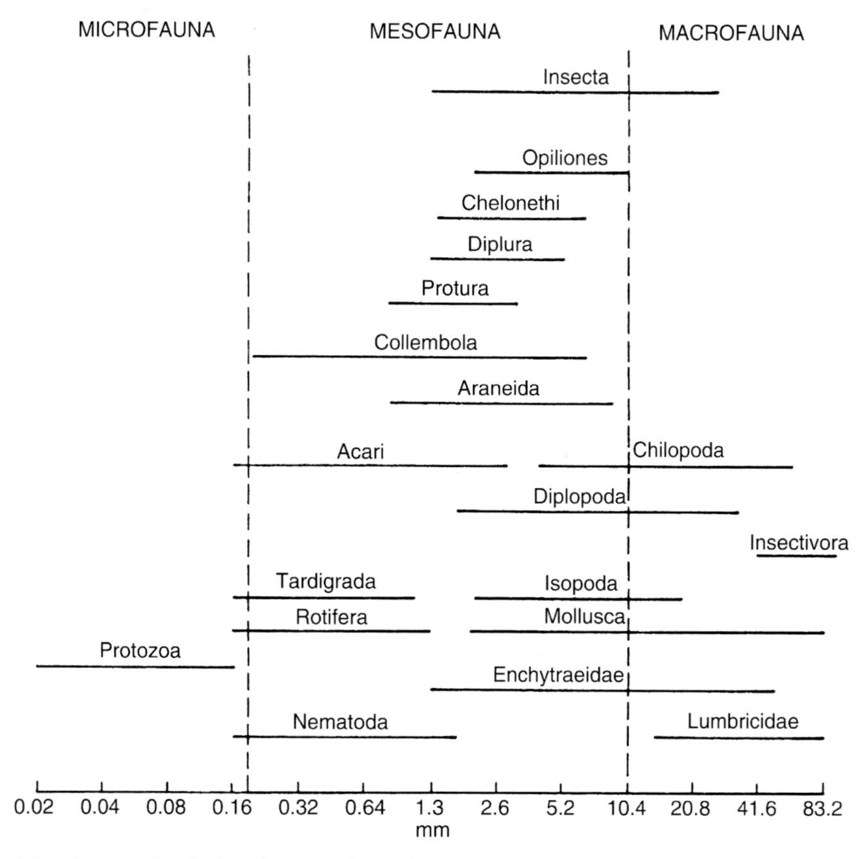

FIGURE 4.2 A generalized classification of soil fauna by body length. *Source: From Wallwork, J.A., 1970. Ecology of Soil Animals. McGraw-Hill, London.*

construct their own living spaces). The macrofauna, in contrast, are often termed ecosystem engineers (*sensu* Jones et al., 1994), and can physically modify the soil environment through the feeding and burrowing activities (Lavelle and Spain, 2001; Lavelle et al., 2016; Lee, 1985; van Vliet and Hendrix, 2007). For example, methods for studying these faunal groups are in large part size-dependent. Methods for studying the microfauna rely on techniques ranging from traditional microbiology (e.g., serial dilution and laboratory culture (see exercise from Chapter 9)) to cutting-edge molecular techniques (e.g., Wu et al., 2011). Mesofauna must be extracted from the soil using specialized techniques and then examined with microscopy. Macrofauna can be sampled by careful sorting known areas or volumes of soil and litter in the field.

There is, of course, considerable gradation in the classification based on body width. The smaller mesofauna exhibit characteristics of the larger microfauna, and so forth. Nevertheless, the classification continues to have utility. Finally, the vast range of body sizes among the soil fauna suggests that their effects on soil processes take place at a range of spatial scales. Three levels of participation have been suggested (Coleman and Wall, 2015; Lavelle et al., 1995, 2016; Wardle, 2002). "Ecosystem engineers," such as earthworms,

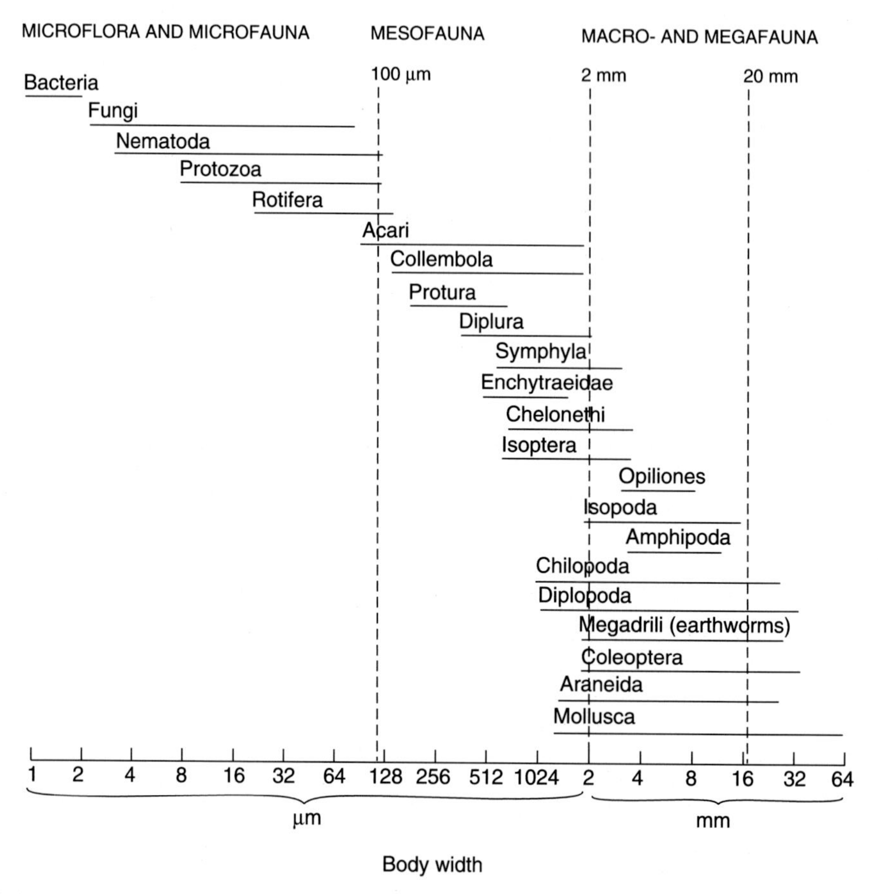

FIGURE 4.3 Size classification of organisms in decomposer food webs by body width. *Source: From Swift, M.J., Heal, O.W., Anderson, J.M., 1979. Decomposition in Terrestrial Ecosystems. University of California Press, Berkeley, CA.*

ants, and termites, produce biogenic structures that alter soil physical and chemical characteristics including aggregate structure, water, and air movement, and soil organic matter distribution. "Litter transformers," the microarthropods, fragment decomposing litter and improve its availability to microbes. "Micro-food webs" include the microbial groups and their direct microfaunal predators (nematodes and protozoans). These three levels operate on different size, spatial, and timescales (Fig. 4.4) (Lavelle et al., 1995; Wardle, 2002).

4.2 THE MICROFAUNA

The free-living protozoa of litter and soils are extremely diverse and are classified into at least seven different "kingdom"-level lineages (Pawlowski et al., 2012). This understanding of protistan diversity and phylogeny is relatively recent and belies the apparent simplicity of these single-celled organisms. For practical purposes, we will consider them in four ecological groups which are loosely based on traditional taxonomic classifications: the

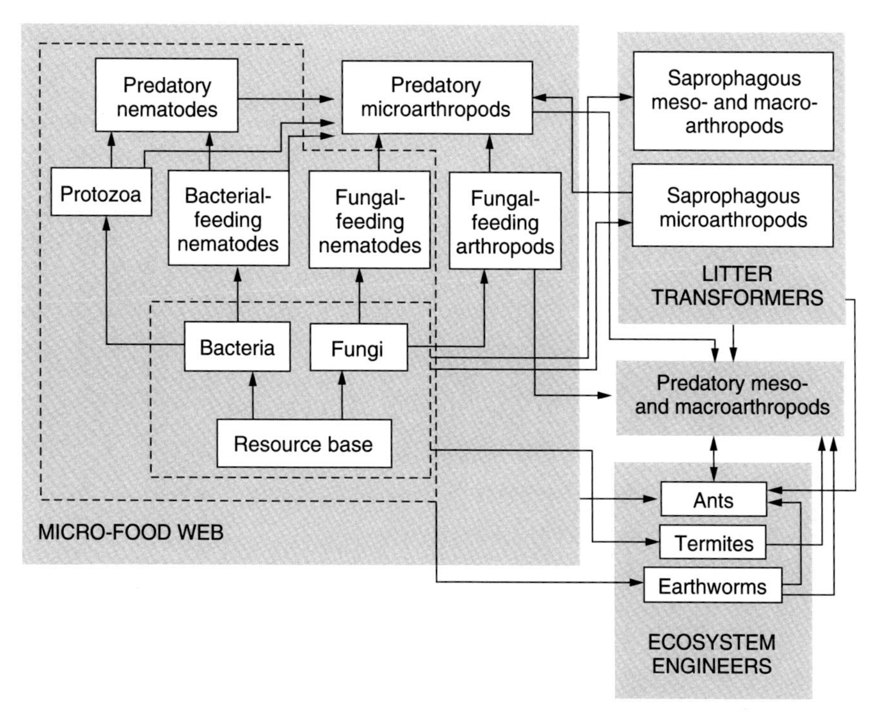

FIGURE 4.4 Organization of the soil food web into three categories—ecosystem engineers, litter transformers, and micro-food webs *Source: After Wardle, D.A., 2002. Communities and Ecosystems: Linking the Aboveground and Belowground Components. Princeton University Press, Princeton, NJ, and Lavelle, P., Lattaud, D.T., Barois, I., 1995. Mutualism and biodiversity in soils. Plant Soil, 170, 23–33.*

flagellates, the naked amoebae, testacea, and ciliates (Lousier and Bamforth, 1990). A general comparison of body plans is given in Fig. 4.5, showing representatives of the four major types. After a brief overview of the groups, we will consider aspects of their enumeration and identification.

1. *Flagellates.* Named for the whip-like propulsive organ, the flagellum (of which there can be one or more per organism), these are among the more numerous and active of the protozoa. They play a significant role in nutrient turnover by their often intensive feeding activities, with bacteria as their principal prey items (Kuikman and Van Veen, 1989; Zwart and Darbyshire, 1992). Numbers have been reported to vary from 100 per gram in desert soils to more than 10^5 per gram in forest soils (Bamforth, 1980).
2. *Naked amoebae.* These are among the more voracious of the soil protozoa, and are very numerous and active in a wide range of agricultural, grassland, and forested soils (Clarholm, 1981, 1985; Elliot and Coleman, 1977; Gupta and Germida, 1989). The dominant mode of feeding for the amoebae, as for the larger forms such as ciliates, is phagotrophic (i.e., engulfing food items), with bacteria, fungi, algae, and other fine particulate organic matter being the majority of the ingested material (Bamforth, 1980; Bryant et al., 1982; Koller et al., 2013). The highly plastic mode of existence of the naked amoebae is impressive; they have the ability to explore very small cavities or pores in

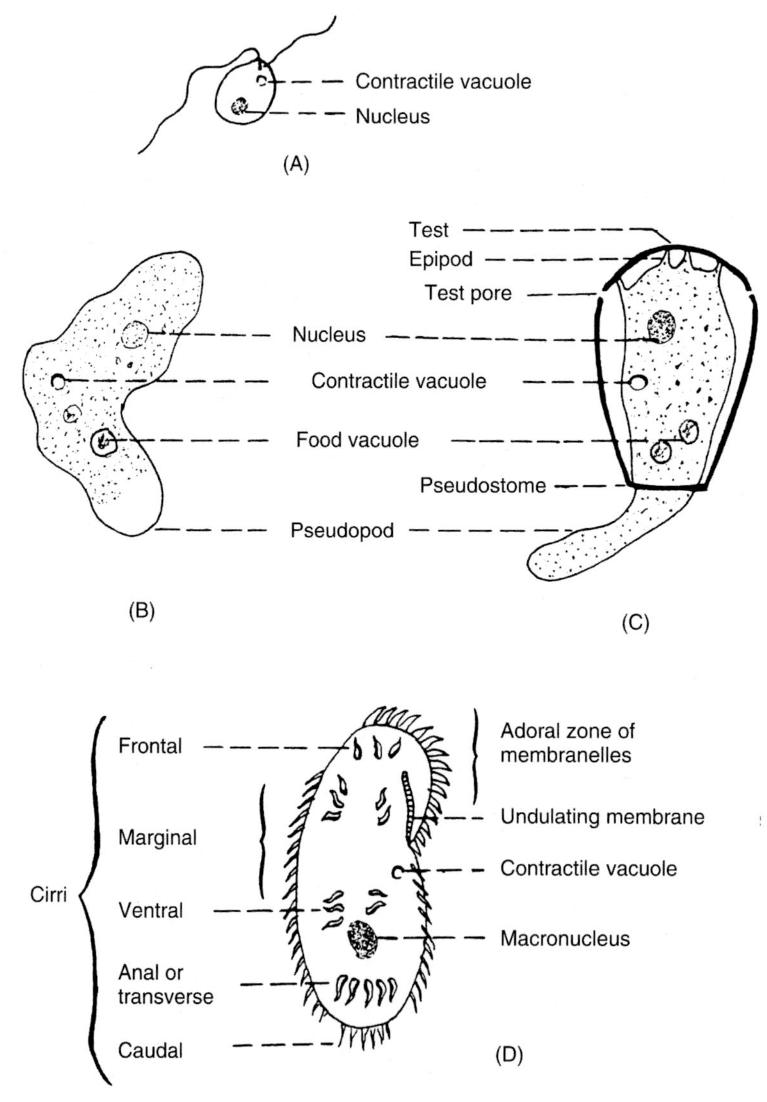

FIGURE 4.5 Morphology of four types of soil protozoa: (A) flagellate (Bodo); (B) naked ameba (Naegleria); (C) testacean (Hyalosphenia); and (D) ciliate (Oxytricha). *Source: From Lousier, J.D., Bamforth, S.S., 1990. Soil Protozoa. In: Dindal, D.L. (Ed.), Soil Biology Guide. Wiley, New York, NY, pp. 97–136.*

soil aggregates and feed on bacteria that would otherwise be considered inaccessible to predators (Foster and Dormaar, 1991).

3. *Testate amoebae.* When compared with the naked amoebae, testate amoebae are often less numerous, except in moist, forested systems where they thrive. However, they are more easily censused by a range of direct filtration and staining procedures (Lousier and Parkinson, 1981). Detailed community and biomass production studies of testate amoebae have been carried out in forested French sites by Coûteaux (1972, 1985) and in

Canadian aspen forest lands (Lousier and Parkinson, 1984). For example, Lousier and Parkinson (1984) noted a mean annual biomass of 0.07 g dry weight m^{-2} of aspen woodland soil, much smaller than the annual mass for bacteria or fungi, of 23 and 40 g m^{-2}, respectively. However, the annual secondary production of testate amoebae (new tissue per year) was 21 g dry weight m^{-2}, or essentially the entire average standing crop of the bacteria in that site.

4. *Ciliates.* These protozoa, which have their own unusual life cycles and complex reproductive patterns, tend to be restricted to very moist or seasonally moist habitats. Their numbers are lower than other groups, with a general range of 10–500 per g of litter/soil. Ciliates can be very active in entering soil cavities and pores and exploiting bacterial food sources inside these spaces (Foissner, 1987a). As with other protozoa, ciliates have resistant or encysted forms from which they can emerge when conditions become favorable for growth and reproduction, with the presence of suitable food resources (Foissner, 1987a). Ciliates, along with flagellates, and naked and testate amoebae, can quickly reproduce asexually by fission. The flagellates, naked and testate amoebae can also reproduce by syngamy, or fusion of two cells. For the ciliates, sexual reproduction occurs by conjugation, with the micronucleus undergoing meiosis in two individuals, and the two cells joining at the region of the cytosome and exchanging haploid gametic nuclei. Each cell then undergoes fission to produce individuals, which are genetically different from the preconjugant parents (Lousier and Bamforth, 1990).

Much progress has been made in delineating the major groups of protists, and their incredibly large biodiversity. A universal DNA barcode for protists has been developed along with a series of group-specific barcodes, which will enable a new burst of taxonomic research on this fascinating large group (Pawlowski et al., 2012). An example of the extensive number of protist groups is given in Fig. 4.6 and Table 4.1 (Pawlowski et al., 2012).

As noted above for the testate amoebae, some genera of ciliates are considered indicative of acid humus, and other more typical of higher-pH or "mild" humus. See Table 4.2 for species characteristic of mull and mor soils.

There is an exercise in Chapter 9, Laboratory and Field Exercises in Soil Ecology, where methods of extraction for protozoa are described along with a laboratory protocol for extraction and enumeration of these organisms from field fresh soil samples.

4.2.1 Distribution of Protozoa in Soil Profiles

Although protozoa are considered to be distributed principally in the upper few centimeters of a soil profile, they are also found at depths of more than 200 meters (m) in groundwater environments (Sinclair and Ghiorse, 1989). Small (2–3 μm cell size) microflagellates were found to decrease 10-fold in numbers during movement through 1 m in a sandy matrix under a trickling-filter facility (in dilute sewage), as compared to 10-fold reduction in bacterial transport over a 10 m distance (Harvey et al., 1995).

4.2.2 Impacts of Protozoa on Ecosystem Function

Several investigators have noted the obvious parallels between the protozoan microbe interaction in water films in soil and on root surfaces and in open water aquatic systems

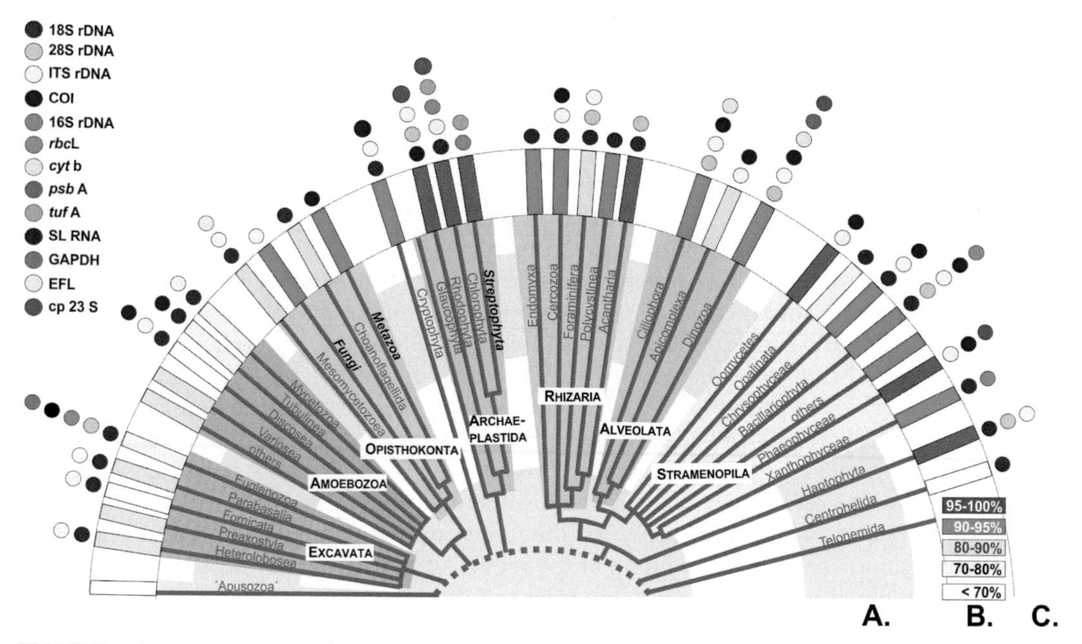

FIGURE 4.6 Current state-of-the art phylogeny and barcode markers for the main protistan lineages. (A) A recent phylogeny of eukaryotic life. (B) Mean V4 18S rDNA genetic similarity between all congeneric species within each lineage, available in GenBank. (C) Currently used group-specific barcodes. The dashed line indicates the incertitude concerning the position of the root in the tree of eukaryotic life. The unresolved relationships between eukaryotic groups are indicated by polytomies. The names of the three multicellular classical "kingdoms" are highlighted. *Source: From Pawlowski, J., Audic, S., Adl, S., Bass, D., Belbahri, L., Berney, C., 2012. CBOL Protist Working Group: Barcoding Eukaryotic Richness Beyond the Animal, Plant and Fungal Kingdoms. PLos Biol. 10 (11), 1–5.*

(Clarholm 1994; Coleman 1976; Coleman 1994a; Stout 1963). The so-called microbial loop of Pomeroy (1974) has proven to be a powerful conceptual tool; rapidly feeding protozoa may consume several standing crops of bacteria in soil (see Chapter 6: Soil Foodwebs: Detritivory and Microbivory in Soils) every year (Clarholm 1985; Coleman 1994b). Darbyshire and Greaves (1967) noted that this tendency is particularly marked in the rhizosphere which provides a ready food source for microbial prey. This was demonstrated impressively for protozoa in arable fields (Cutler et al., 1923), and more recently for bacteria, naked amoebae and flagellates in the humus layer of a pine forest in Sweden after a rain (Clarholm, 1994). Bacteria and flagellates began increasing immediately after a rainfall event and rose to a peak in 2–3 days; naked amoebae rose more slowly and peaked at days 4–5, and then tracked the bacterial decrease downward as did the flagellates (Fig. 4.7). Further information on protozoan feeding activities and their impacts on other organisms and ecosystem functions are given in the Griffiths (1994), Zwart et al. (1994), and Bamforth (1997). Bonkowski et al. (2000) suggest that protozoa, and the bacteria that they feed upon the rhizosphere, produce plant-growth-promoting compounds that stimulate plant growth above and beyond the amounts of nitrogen mineralized in the rhizosphere. The range and scope of feeding carried out by soil protists is proving to be more

TABLE 4.1 Number of Cataloged Morphospecies and V4 18S rDNA OTU-97% Among the 60 Main Eukaryotic Lineages

Supergroup	Division/Phylum	Cataloged Morpho-species	OTU (V4 rDNA, 97%)[a]
Alveolata	Apicomplexa	6,000	626
	Ciliophora	800	1,157
	Dinophyceae	2,280	426
	Syndiniales	0	391
Amoebozoa	Tubulinea + Arcellinida	1,100	58
	Discosea	180	77
	Dictyostelia + Myxogastria	1,062	107
Archaeplastida	Chlorophyta (−Streptophyta)	9,000	740
	Streptophyta	350,000	1,447
	Rhodophyta	5,000	640
Excavata	Euglenozoa	1,520	296
	Fornicata	146	45
	Heterolobosea	80	640
	Parabasalia	466	180
Opisthokonta	Choanoflagellida	250	88
	Fungi	377,200	3,694
	Metazoa	1,200,000	8,750
Rhizaria	Cercozoa	600	749
	Foraminifera	12,000	36
	Acantharea	160	36
	Polycystinea	850	91
Stramenopiles	Bacillariophyta	20,000	603
	Chrysophyceae-Synurophyceae	1,200	273
	Labyrinthulea	40	186
	Oomycota	676	66
	Opalinata	400	53
	Phaeophyceae	1,750	49
	Xanthophyceae	600	28
Incertae sedis	Centroheliozoa	150	71
	Haptophyta	350	127

[a]*OTU-97% found in the Genbank nt and env_nt divisions of Genbank (Jan 2012). Total of 86,587 sequences.*
Modified from Pawlowski, J., Audic, S., Adl, S., Bass, D., Belbahri, L., Berney, C., 2012. CBOL Protist Working Group: Barcoding eukaryotic richness beyond the animal, plant and fungal kingdoms. PLOS Biol. 10 (11), 1–5.

TABLE 4.2 Representative Species of the Testacean and Ciliate Communities in Mull and Mor Soils

	Testaceans		Ciliophora
Type of Humus	**Characteristic Species**	**Ratio of Full and Empty Shells**	**Characteristic Species**
Mull	*Centropyxis plagiostoma*	< 1:2–5	*Urosomioda agilis*
	Centropyxis constricta		*Urosoma* spp.
	Centropyxis elongata		*Hemisincirra filiformis*
	Plagiopyxis minuta		*Engelmanniella mobilis*
	Geopyxella sylvicola		*Grossglockneria hyalina*
	Paraquadrula spp.		*Colpoda elliotti*
Moder and mor	*Trigonopyxis arcula*	> 1:2–5	*Frontonia depressa*
	Plagiopyxis labiata		*Bryometopus sphagni*
	Assulina spp.		*Dimacrocaryon amphileptoides*
	Corythion spp.		
	Nebela spp.		

From Foissner, W., 1987a. Soil protozoa: fundamental problems, ecological significance, adaptations in ciliates and testaceans, bioindicators, and guide to the literature. Progr. Protistol. 2, 69–212.

complex than formerly thought. Thus facultative feeding of various protist taxa on bacteria and fungi, the source of both major energy channels, strongly implies that a clear split of the energy channels at lower trophic levels does not exist and that more complex energy flows prevail in soil food webs (Geisen, 2016).

Protozoa and other microfauna are quite sensitive to environmental insults, and changes in the distribution and activities are diagnostic of changes in soil health (Gupta and Yeates, 1997). We address the issue of soil health in Chapter 8, Future Developments in Soil Ecology.

4.3 THE MESOFAUNA

4.3.1 Rotifera

Among the small fauna, rotifera are often found where a significant proportion of water films exist in soils. They are usually considered to be aquatic organisms and may not be listed in major compendia of soil biota (Dindal, 1990); they are a genuine, albeit secondary, component of the soil fauna (Wallwork, 1976). While sampling for nematodes in the surface layers of agricultural fields near La Selva, in the Atlantic coastal forest of Costa Rica, one of the authors (Coleman) found virtually no nematodes, but a large number of rotifers (tens of thousands per square meter), despite the soils being far from water saturation. The field was being maintained in a "bare fallow" regime with frequent weeding or denudation of vegetation, to deliberately reduce organic inputs. However, there seemed to be ample cyanobacteria and perhaps other unicellular primary producers, which would have

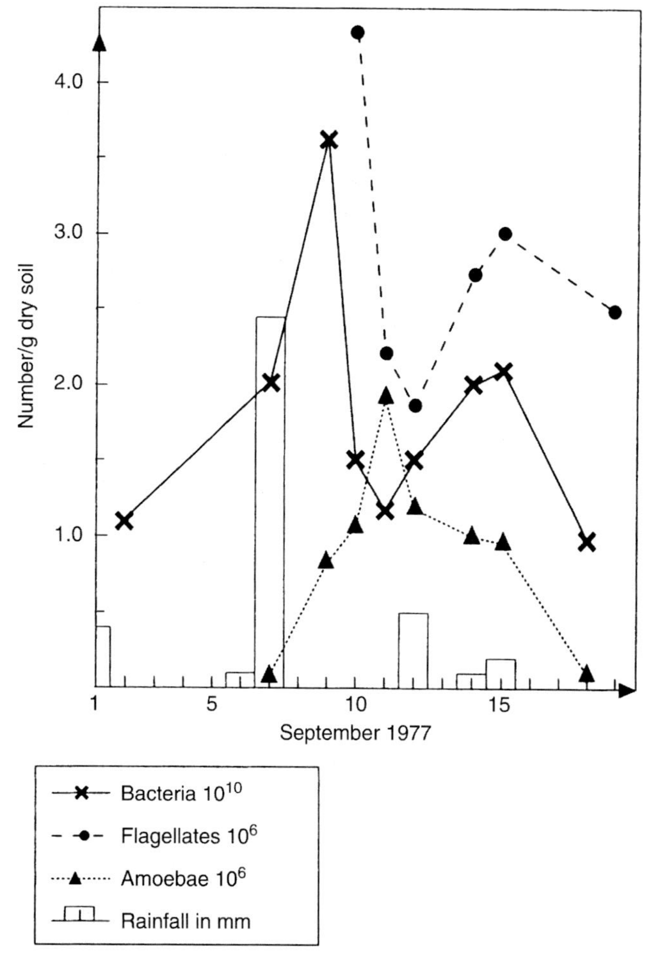

FIGURE 4.7 Daily estimates of numbers of bacteria, amoebae, and flagellates in the humus layer of a pine forest after rain. *Source: From Clarholm, M., 1994. The microbial loop in soil. In: Ritz, K., Dighton, J., Giller, K.E. (Eds.), Beyond the Biomass. Wiley/Sayce, Chichester, pp. 221–230.*

provided food for the rotifers. Some rotifers have been found in bagged leaf litter on forest floors in the Southern Appalachian Mountains.

4.3.1.1 Features of Body Plans and General Ecology

More than 90% of soil rotifers are in the order Bdelloidea, or worm-like rotifers. In these creeping forms, the suctorial rostral cilia and the adhesive disk are employed for locomotion (Donner, 1966). Rotifers also form cysts to endure times of stress or lack of resources. Additional life history features of interest include the construction of shell from a body secretion, which may have particles of debris and/or fecal material adhering to it. Some rotifers will use the empty shells of testate amoebae. The Bdelloidea are vortex feeders, creating currents of water that conduct food particles to the mouth for ingestion (Wallwork, 1970). The importance of these organisms is largely unknown, although they may reach numbers exceeding 10^5 per square meter in moist organic soils (Wallwork, 1970).

Rotifers are extracted from soil samples and enumerated using methods similar to those used for nematodes (see next section).

4.3.2 Nematoda

Nematodes, or roundworms, are among the most numerous of the multicellular organisms found in any ecosystem. As with the protozoa, they primarily inhabit water films, or water-filled pore spaces in soils. Nematodes have a very early phylogenetic origin, but as with many other invertebrate groups, the fossil record is fragmentary. They are classified among the triploblastic pseudocoelomates (three body layers: ectoderm, mesoderm, and endoderm). In other words, nematodes have a body cavity for the gastrointestinal tract, but it is less well differentiated than that for the true coelomates, such as annelids and arthropods.

The overall body shape is cylindrical, tapering at the ends (Fig. 4.8). In general, nematode body plans are characterized as consisting of "a tube within a tube." The outer tube consists of a cuticle and muscular body wall, and the inner tube consists of the alimentary canal (collectively the stylet, pharynx/esophagus, intestine, and rectum which opens externally at the anus. The reproductive structures are quite complex, as shown in Fig. 4.8. Some species are parthenogenetic, reproducing without sex. It is possible to view the internal structures of most nematodes because they have virtually transparent cuticles. The nematodes can be keyed out fairly readily to family or genus under a moderate magnification ($\sim 100\times$) binocular microscope or in a Sedgewick-Rafter chamber on an inverted microscope (Wright, 1988), but species-specific characteristics must be determined under high magnification, using compound microscopes.

4.3.2.1 Nematode Feeding Habits

Nematodes feed on a wide range of foods. A general trophic grouping is bacterial feeders, fungal feeders, plant feeders, and predators and omnivores. For the purposes of our overview, one can use anterior (stomal or mouth) structures to differentiate feeding, or trophic, groups (Fig. 4.9; Moore and de Ruiter, 2012; Yeates and Coleman, 1982; Yeates et al., 1993), although tools such as stable isotopes and metagenomics are revealing more precise information on food sources and affiliation with trophic groups (Crotty et al., 2012a; Porazinska et al., 2012). Plant-feeding nematodes have a hollow stylet that pierces cell walls of higher plants. Some species are facultative, feeding occasionally on plant roots or root hairs. Others, recognized for their damage to agricultural crops and forest plantations, are obligate parasites of plants and feed internally or externally on plant roots. The effect nematodes have on plants is generally species-specific and can include alterations in root architecture, water transport, plant metabolism, or all of these. Recently, plant parasites in the Tylenchida and some fungal feeding Aphelenchida were found to have cell wall degrading enzymes. Some of the genes for secretion of endogluconases (cellulases) and pectase lyase genes appear to play direct roles in the nematode parasitic process (Karim

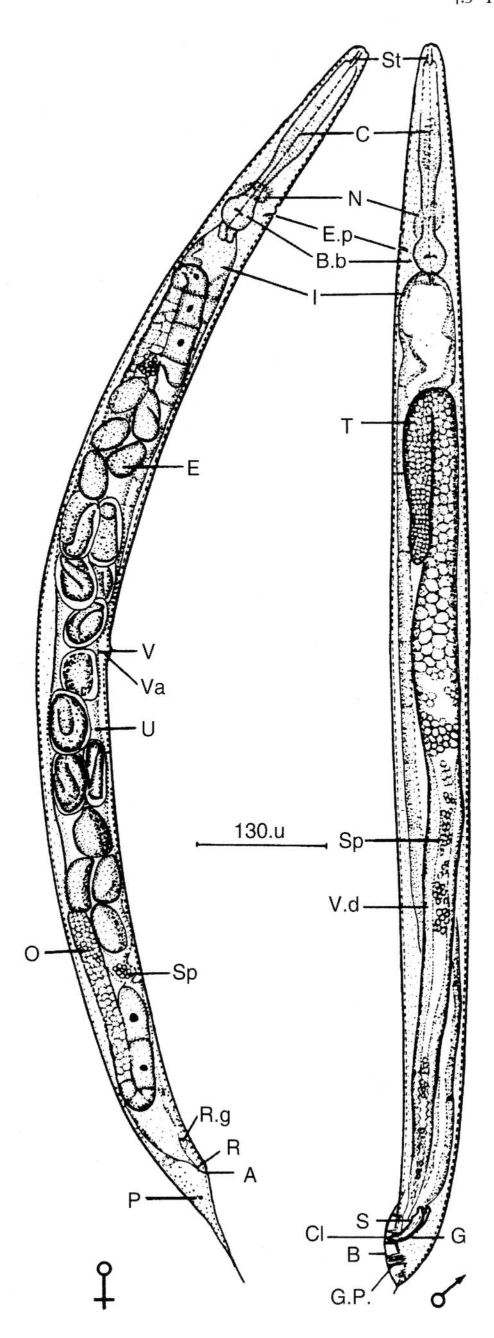

FIGURE 4.8 Structures of a *Rhabditis* sp., a secernentean microbotrophic nematode of the order Rhabditida. (Left) Female. (Right) Male. *ST*, Stoma; *C*, corpus area of the pharynx; *N*, nerve ring; *E.p*, excretory pore; *B.b*, basal bulb of the pharynx; *I*, intestine; *T*, testis; *E*, eggs; *V*, vulva; *Va*, vagina; *U*, uterus; *O*, ovary; *SP*, sperm; *V.d*, vas deferens; *R.g*, rectal glands; *R*, rectum; *A*, anus; *S*, spicules; *G*, gubernaculum; *B*, bursa; *P*, phasmids; *G.P*, genital papillae; *CL*, cloaca. *Source: Courtesy of Proceedings of the Helminthological Society of Washington. From Poinar (1983).*

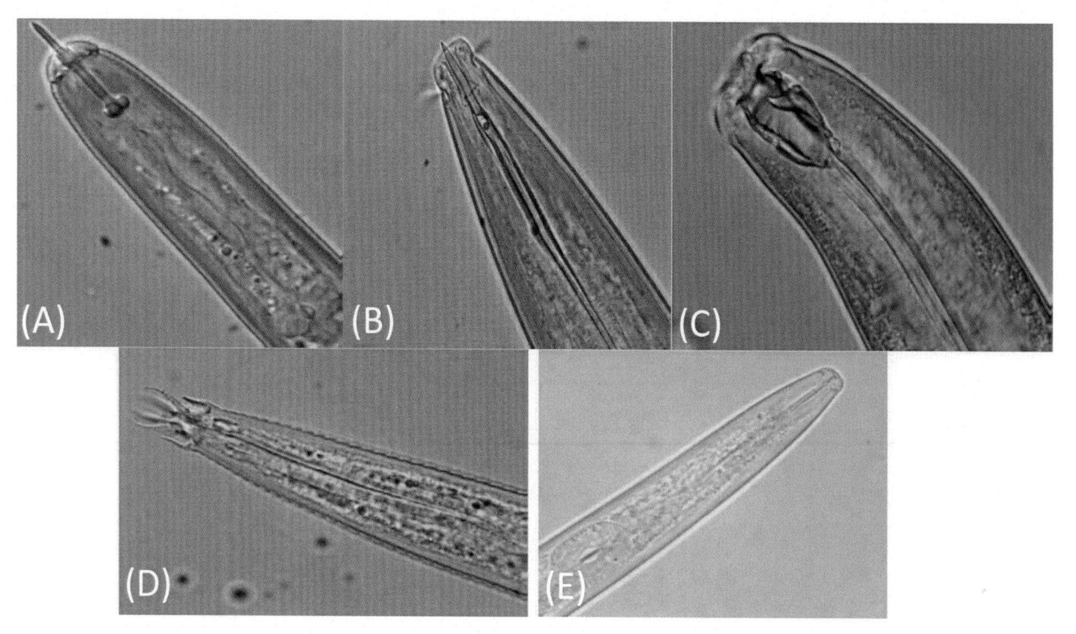

FIGURE 4.9 Nematode functional group head appearances, used by permission of European Union. (A) Plant feeding nematode: *Globodera pallida* with an extendable spear used to penetrate roots of host plant species. Note the knobs on the base of the spear that anchor muscles extending forward to the head. When these muscles contract, the spear juts forward. (B) Omnivorous nematode: *Prodorylaimus filarum* has a spear without knobs. Omnivorous species can feed on algae, protists, other nematodes and then, when these primary food sources are unavailable, switch to feeding on fungi and bacteria. (C) Predaceous nematode: This predator with a large dorsal tooth (*Mylonchulus sigmaturus*) eats other nematodes. (D) Bacterial-feeding nematode: *Acrobeles mariannae* has ornate head appendages (probolae). (E) Fungal-feeding nematode: *Aphelenchus* sp. have a tiny spear to pierce fungal hyphae. *Source: Modified from Orgiazzi, A., Bardgett, R.D., Barrios, E., Behan-Pelletier, V., Briones, M.J.I., Chotte, J.-L., 2016. Global Soil Diodiversity Atlas. Publications Office of the European Union, Luxembourg; photos by H. Helder, H. van Megen, and J.G. Baldwin.*

et al., 2009). Their enzyme products modify plant cell walls and cell metabolism (Davis et al., 2000, 2004), information that, with genome sequencing of plant parasitic nematodes such as Meloidogyne incognita (Abad et al., 2008), has increased development of antiplant parasite strategies for improved crop production. These genes have greatest similarity to microbial genes for cellulases, and it appears that horizontal gene transfer has occurred from microbes to these plant parasites (Sommer and Streit, 2011).

The feeding categories are a good introduction, but the feeding habits of many genera are either complex or poorly known. Thus, immature forms of certain nematodes may be bacterial feeders, then become predators on other fauna once they have matured (Allen-Morley and Coleman, 1989). Some of the stylet-bearing nematodes (e.g., the family Neotylenchidae) may feed on roots, root hairs, and fungal hyphae (Yeates and Coleman, 1982). Some bacterial feeders (e.g., *Alaimus*) may ingest 10 μm width cyanobacterial cells (Oscillatoria) despite the mouth of the nematode being only 1 μm wide indicating that the cyanobacterial cells can be compressed markedly by the nematode (Yeates, 1998). Recent laboratory studies (Venette and Ferris, 1998) have confirmed that population growth of bacterial feeding nematodes is strongly dependent on the species of bacteria ingested. The six nematode species used in the study of Venette and Ferris (1998) reached maximal population growth rates when ingesting from 10^4 to 10^5 colony-forming units per nematode. Population growth rates (λ) under these controlled conditions ranged from 1.1 to greater than $12\,d^{-1}$, making these organisms ideal for detecting rapid changes in the environment.

Although specialized in nature, the feeding habits and impacts of entomopathogenic nematodes are quite marked in several soil environments worldwide (Hominick, 2002). The nonfeeding "infective juveniles," or third instar dauer larvae of nematodes in the family Heterorhabditidae, live in the soil and search for hosts and disperse. An infective juvenile enters the insect host (which it senses along a CO_2 gradient) through a spiracle or other opening, punctures a membrane, then regurgitates the symbiotic bacterium *Photorhabdus luminescens*, which kills the host within 48 hours. A rapidly growing bacterial population then digests the insect cadaver and provides food for the exponentially growing adult nematode population inside (Strong et al., 1996). The symbiotic bacteria produce antibiotics and other antimicrobial substances that protect the host cadaver and adult nematodes inside from invasion by alien bacteria and fungi from the soil (Strong et al., 1999). When the cadaver is exhausted of resources, reproduction shunts to infective juveniles, which break through the host integument and disperse into the soil. For example, as many as 410,000 *Heterorhabditis hepialus* infective juveniles are produced in a large ghost moth caterpillar (Strong et al., 1996). In pot experiments, Strong et al. (1999) found that *Lupinus arboreus* seedlings, whose seedling survival decreased exponentially with increasing densities of root feeding caterpillars, had virtually the entire negative effect of the herbivore canceled upon the introduction of the entomopathogenic nematode into the system. For more information on dynamics of entomopathogenic nematodes in soil food webs, see Strong (2002).

For identifying fungal-based food chains, Ruess et al. (2002) have shown that the measurement of fatty acids specific to fungi can be traced to the body tissues of fungal-feeding nematodes. Although still in early stages of development, this technique shows considerable promise for more detailed biochemical delineation of food sources of specific feeding groups of nematodes.

Because of the wide range of feeding types and the fact that they seem to reflect ages of the systems in which they occur (i.e., annual vs perennial crops, old fields, pastures, and more mature forests), nematodes have been used as indicators of overall ecological condition (Bongers, 1990; Ferris et al., 2001; Ferris and Bongers, 2006; Vervoort et al., 2012). This is a growing area of research in land management and soil ecology, one in which the intersection between nematode and other soil faunal community analysis and ecosystem function could prove to be useful for assessing land damage and restoration (Wilson and Kakouli-Duarte, 2009; Domene et al., 2011).

4.3.2.2 Nematode Zones of Activity in Soil

As noted in Chapter 2, Two Primary Production Processes in Soils: Roots and Rhizosphere Associates, the rhizosphere is a zone of considerable metabolic activity for root-associated microbes. This extends also to the soil fauna, which may be concentrated in the rhizosphere. For example, Ingham et al. (1985) found up to 70% of the bacterial- and fungal-feeding nematodes in the 4%–5% of the total soil that was rhizosphere, namely the amount of soil 1–2 mm from the root surface (the rhizoplane). In comparison, Griffiths and Caul (1993) found that nematodes migrated to packets of decomposing grass residues, with considerable amounts of labile substrates therein, in pot experiments. They concluded that nematodes are seeking out these "hot spots" of concentrated organic matter, and that protozoa, also monitored in the experiment, do not.

Nematodes are very sensitive to available soil water in the soil matrix. Elliott et al. (1980) noted that the limiting factor for nematode survival often hinges on the availability of soil pore necks, which enable movement between soil pores. In other studies, Yeates et al. (2002) measured the movements, growth, and survival of three genera of bacterial-feeding soil nematodes in undisturbed soil cores maintained on soil pressure plates. Interestingly, the nematodes showed significant reproduction even when diameters of water-filled pores were approximately 1 μm. This information should prove useful when determining biological interactions under field conditions, and indicates that soil nematodes may be more active over a wider range of soil moisture tensions than had been thought to be the case previously.

4.3.2.3 *Nematode Extraction Techniques*

Nematodes may be extracted by a variety of techniques, either active or passive in nature. For more accuracy in determination of populations, the passive or flotation techniques are generally preferred. The principal advantage of the oldest, active method, namely Baermann funnel method, is that it is simple, requiring no specialized equipment or electricity. It is based on the fact that nematodes in the soil will move about in the wetted soil and fall into the funnel itself. Thus samples are placed on coarse tissue paper, on a coarse mesh screen, and then placed in the cone of a funnel and immersed in water. Once they crawl through the moist soil and tissue paper, the nematodes fall down into the neck of the funnel. Because nematodes have only circular and not longitudinal muscles, they do not stay in suspension in the water, and fall to the bottom of the funnel stem which is kept closed with a rubber hose and clamp. At the conclusion of the extraction (typically 48 hours), the nematodes in solution are drawn off into a tube and kept preserved for examination later. One drawback to the technique is that it allows dormant nematodes to become active and be extracted, so it may give a slightly inflated estimate of the true, "active" population at a given time. Other methods include filtration, or decanting and sieving, and flotation/centrifugation (Christie and Perry, 1951; Coleman et al., 1999) to remove the nematodes from the soil suspension. Elutriation methods can be employed for handling larger quantities of soil, usually greater than 500 g, or to recover large amounts and a greater diversity of nematodes. Elutriation methods rely on fast mixing of soil and water in funnels. Semiautomatic elutriators, which enhance the number of soil samples to be extracted, are available and used for assessing root mycorrhizae, rhizobia, root mass, and nematode species (Byrd et al., 1976). There are many references comparing methods, including Whitehead and Hemming (1965), Schouten and Arp (1991), and McSorley and Frederick (2004). Anhydrobiotic nematodes can be extracted in a high-molarity solution, such as sucrose, which prevents the nematodes from rehydrating (Freckman et al., 1977) (Fig. 4.10).

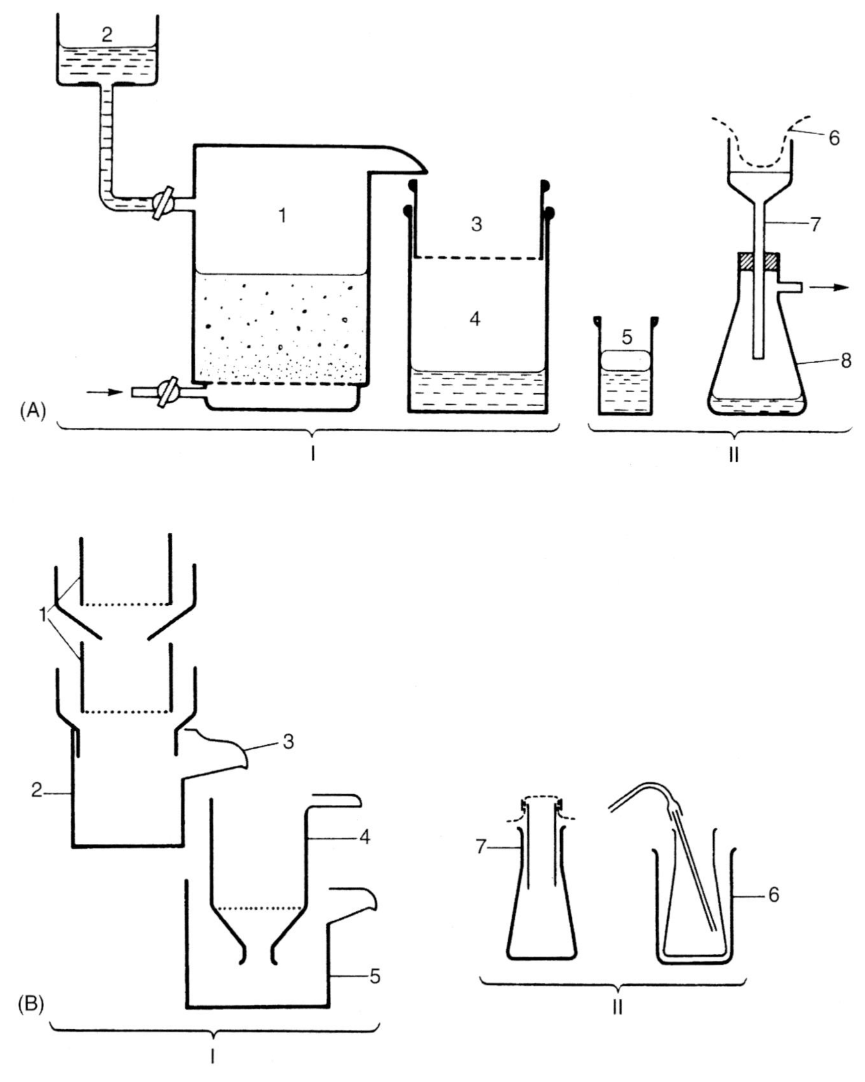

FIGURE 4.10 Flotation apparatus for extraction of soil invertebrates (Ladell's modified flotation apparatus). (A) Edwards and Denis, 1962; (B) Salt, 1953. I: Separation of organic matter from the mineral components of the soil; II: Separation of soil invertebrates and organic residue. The process uses these materials: (A) 1 = flotation vessel with air supply from below, 2 = container with salt solution, 3 = sieve, 4 = sedimentation tank, 5 = glass oil separator, 6 = gauze with the material under study, 7 = glass funnel, 8 = flask connected to suction pump. (B) 1 = vessels with coarse and fine sieves, 2 = container collecting the residue, 3 = outlet, 4 = final collecting sieve, 5 = water tank, 6 = water bath, 7 = glass for oil–water flotation. *Source: From Kaczmarek (1993).*

4.3.3 Tardigrada

These interesting little micrometazoans (ranging from 50 μm, the smallest juvenile, to 1200 μm, the largest adult) (Nelson and Higgins, 1990) are also called "water bears" because of their microursine appearance. They were named "Il Tardigrado," literally slow-stepper, because their slow movements resembled those of a tortoise, by the famous Italian abbot and natural history professor Lazzaró Spallanzani in 1776 (Nelson and Higgins, 1990).

Tardigrades are members of the monophyletic group known as Ecdysozoa, a clade of all molting animals that includes nematodes and arthropods (Garey, 2001). Tardigrades are bilaterally symmetrical with four pairs of legs, equipped with claws on the distal end, of various sizes and forms (Fig. 4.11). The sizes and shapes of the claws are used in keying genera and species. Perhaps their greatest notoriety in recent times has come from the marked recuperative powers that they show after having been kept dry in a state of "suspended animation" for many years or even decades. These studies (Møbjerg et al., 2011; Wright, 2001) have found that tardigrades recover well even after extreme environmental insults such as being plunged into liquid nitrogen. More generally, a series of five types of latency or virtual cessation of metabolism have been described: encystment, anoxybiosis, cryobiosis, osmobiosis, and anhydrobiosis (Crowe, 1975). All these are subsumed under the more general term "cryptobiosis" (Keilin, 1959), or hidden life, which was first described by Antonie van Leeuwenhoek in a Royal Society lecture in 1702 (Wright 2001). Being highly resistant, or resilient, to various environmental insults, tardigrades exemplify a recurrent thread throughout biology in general, and soil biology in particular: the selective advantages to "waiting out" a spell of bad microclimate and being able to reactivate and become active in a given patch in the soil, years or decades later. However, this amazing capacity to tolerate environmental stressors is much more than just a biological curiosity. The tardigrade's evolved resilience may serve as a genetic resource for use in the human health arena, as tardigrade-unique proteins have been shown to reduce radiation-induced damage to DNA in cultured human cells. This implies that there could be potential therapeutic applications for protecting in other animals from radiation derived from the tardigrade genome (Fig. 4.12; Hashimoto et al., 2016). Further high-tech research has involved sending tardigrades into space aboard the space shuttle Endeavor, for use in experiments conducted at the International Space Station. Thanks to the tardigrade's ability to withstand desiccation, vacuum, and temperature extremes, they are ideal model organisms for examining the effects of microgravity (or other space-related stressors) on multicellular organisms (Rizzo et al., 2015).

Tardigrades occur predominantly in the surface 1–3 cm of many grassland soils, but certain genera (e.g., the *Macrobiotus*-group species) are quite numerous at depths up to 10 cm in subalpine coniferous forest (Ito and Abe, 2001). They may serve as "early-warning devices" for environmental stress. Tardigrades were found to be the most sensitive organism measured in a several-year study of the effects of dry-deposition of SO_2 on litter and soil of a mixed-grass prairie ecosystem (Grodzinski and Yorks, 1981; Leetham et al., 1982). They are thought to feed on algal cells and debris in the interstices of moss thalli and probably have a rather broad diet of various microbial-rich bits of organic matter. Tardigrades have also been observed to feed voraciously on nematodes when in culture (G.E. Yeates, personal communication). Tardigrades have been found in large numbers (up to 2000 per 10 m^2 of soil surface), and are particularly associated with lichens, mosses, liverworts, and rosette angiosperms (Nelson and Higgins, 1990; Nelson and Adkins, 2001). They are also found in very cold, dry habitats, such as the Antarctic dry valleys, where

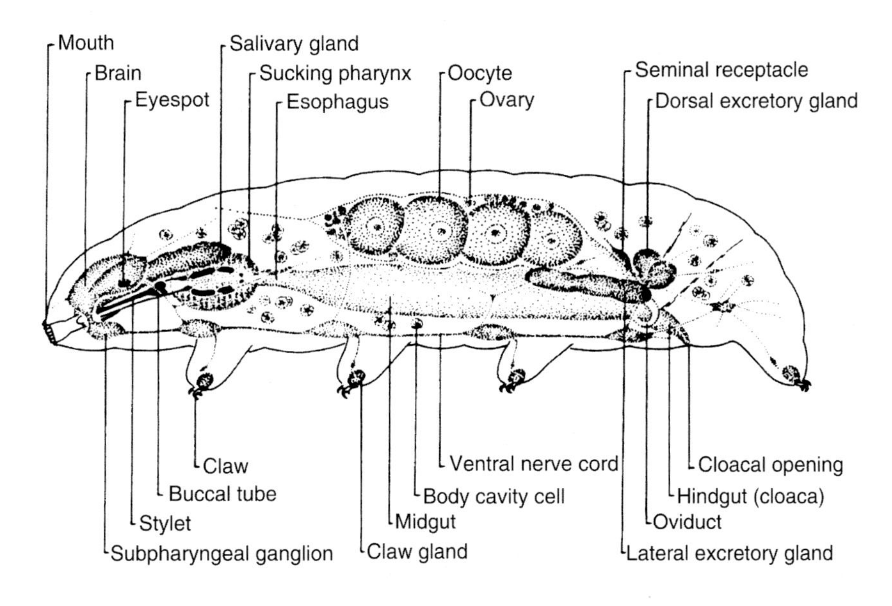

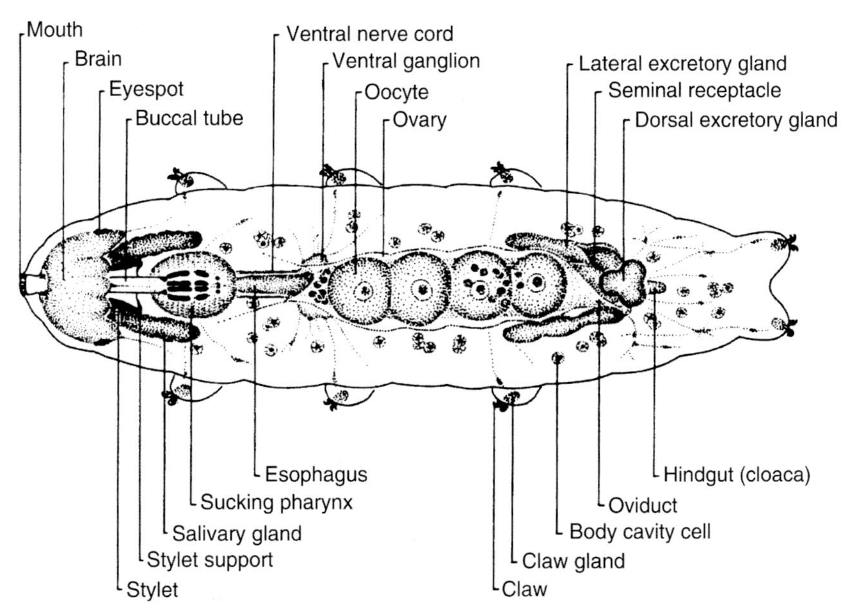

FIGURE 4.11 Eutardigrade internal anatomy (*Macrobiotus hufelandii*, female); lateral view (above) and dorsal view (below). *Source: Dindal, D.L., 1990. Soil Biology Guide. John Wiley & Sons, New York, NY. Reprinted by permission of John Wiley & Sons.*

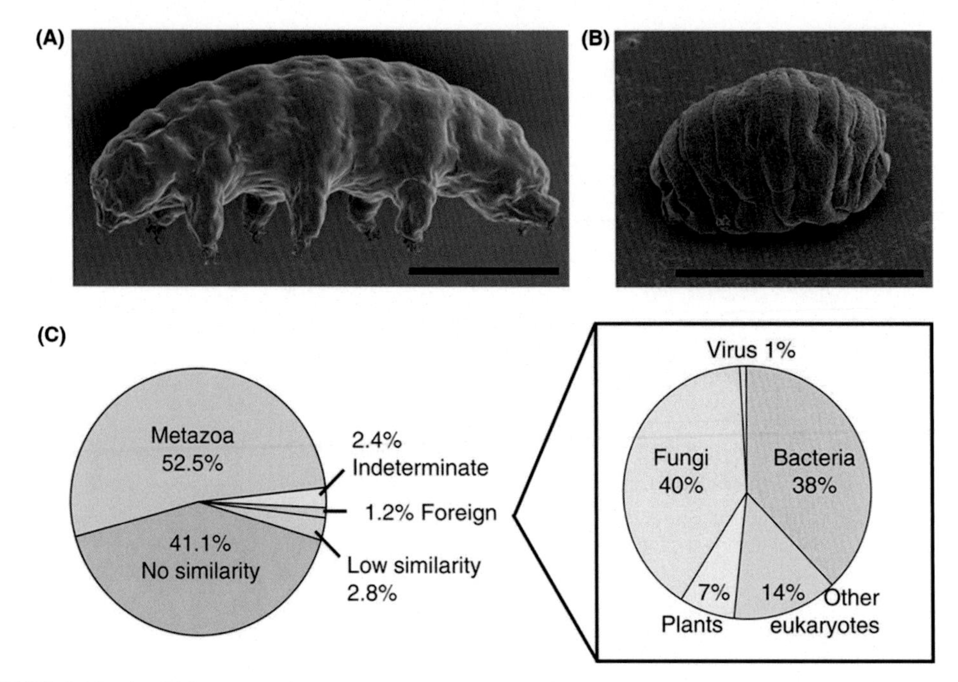

FIGURE 4.12 (A, B) Scanning electron microscopy images of the extremotolerant tardigrade, *Ramazzottius varieornatus*, in the hydrated condition (A) and in the dehydrated state (B), which is resistant to various physical extremes. Scale bars, 100 μm. (C) Classification of the gene repertoire of *R. varieornatus*, according to their putative taxonomic origins and distribution of best-matched taxa in putative HGT genes. *Source: From Hashimoto, T., Horikawa, D.D., Saito, Y., Kuwahara, H., Kozuka-Hata, H., Shin-I, T., 2016. Extremotolerant tardigrade genome and improved radiotolerance of human cultured cells by tardigrade-unique protein. Nat. Commun. 7, 12808.*

they feed on the particulate organic detritus brought in by windward movement of algal cells from lake ice at one end of the long valleys (D. Wall, personal communication).

Tardigrades may be extracted from soils and various substrates by Baermann funnels (Petersen and Luxton, 1982), or flotation and sieving through a 44 μm sieve, or by the sucrose flotation and centrifugation technique used for extracting nematodes (Christie and Perry, 1951).

4.3.4 Microarthropods

Large numbers of the microarthropod group (mainly mites and collembolans) are found in most types of soils. A square meter of forest floor may contain hundreds of thousands of individuals representing thousands of different species (Fig. 4.13). Microarthropods have a significant impact on the decomposition processes in the forest floor (see Chapter 5: Decomposition and Nutrient Cycling) and are important reservoirs of biodiversity in forest ecosystems.

Microarthropods also form an important set of linkages in food webs. Many microarthropods feed on fungi and nematodes, thereby linking the microfauna and microbes with the mesofauna. Microarthropods in turn are prey for macroarthropods such as spiders,

FIGURE 4.13 An example of the thousands of species of soil microarthropods. *Source: Photo by K. Lamoncha.*

beetles, ants, and centipedes, thus bridging a connection to the macrofauna. Even some of the smaller megafauna (toads, salamanders) feed upon microarthropods. We emphasize, again, the need to study soil as an ecosystem. Analysis of one part of the food web, the microarthropods for example, falls short if other components are ignored.

In the size spectrum of soil fauna (see Figs. 4.2 and 4.3), the mites and collembolans are found among the mesofauna. Members of the microarthropod group are unique, not so much by their body size as by the methods used for sampling them. Microarthropods are too small and numerous to be sampled as individuals. Instead, small pieces of habitat (soil, leaf litter, or similar materials) are collected and the microarthropods extracted from them in the laboratory. In this manner they resemble certain of the microfauna such as nematodes, rotifers, or tardigrades. Most of the methods used for microarthropod extraction are either variations of the Tullgren funnel ("Berlese funnel"), which uses heat to desiccate the sample and force the arthropods into a collection fluid, or flotation in solvents or saturated sugar solutions followed by filtration (see Chapter 9: Laboratory and Field Exercises in Soil Ecology). Edwards (1991) gives an extensive review of these procedures. Both approaches to sampling microarthropods have their proponents. Generally, flotation methods work well in low organic, sandy soils, whereas Tullgren funnels perform best in soils with high organic matter content. Flotation procedures are much more laborious than is Tullgren extraction.

Choice of method also depends upon the objectives of the sampling program. If numbers of individuals are to be measured, a large set of small samples may be needed. Estimations of species number may be better served by fewer, larger samples. In any case, extraction methods are never completely efficient and, indeed, efficiency of sampling is seldom estimated (André et al., 2002). Consequently, Walter and Proctor (1999) concluded that enumeration of microarthropods was an "intellectually vacuous" exercise. However,

TABLE 4.3 Abundance of Microarthropods in Soils From Various Ecosystems

Ecosystem	Microarthropods (10^3 per m^2)	Reference
Fallow crop fields, Nigeria	40–68	Adejuyigbe et al. (1999)
Corn tillage plots, Guelph, Canada	16–17	Winter et al. (1990)
No-tillage plots, North Carolina	1–30	House et al. (1987)
Cedar plantation, Nagoya, Japan	48–149	Hijii (1987)
Deciduous forest, Tennessee	36.9	Reichle et al. (1975)
Deciduous forest, North Carolina	88	Lamoncha and Crossley (1998)
Burned tallgrass prairie, Kansas	35–50	Seastedt (1984a)
Unburned tallgrass prairie, Kansas	63–77	Seastedt (1984a)
Mediterranean desert, Negev	1–2	Steinberger and Wallwork (1985)
North American desert, American Southwest	1–8	Steinberger and Wallwork (1985)
Phryganic ecosystem, Greece	20–60	Sgardelis et al. (1981)

valid comparisons of microarthropod abundance in different habitats may be obtained even if extraction efficiencies, though unknown, are similar.

Microarthropod densities vary during seasons within and between different ecosystems (Table 4.3). Generally, temperate forest floors with large accumulations of organic matter support high numbers, whereas tropical forests where the organic layer is thin contain lesser numbers of microarthropods (Seastedt, 1984b). Disturbance or perturbation of soils usually depresses microarthropod numbers. Tillage, fire, and pesticide applications typically reduce populations but recovery may be rapid and microarthropod groups respond differently.

Soil mites usually outnumber collembolans but these become more abundant in some situations. In the springtime, forest leaf litter may develop large populations of "snow fleas" (*Hypogastrura nivicola* and related species). Among the mites themselves the oribatids usually dominate but the delicate Prostigmata may develop large populations in cultivated soils with a surface crust of algae. Immediately following cultivation, numbers of astigmatic mites have been seen to increase dramatically (Perdue, 1987).

Soil microarthropods are significant reservoirs of biodiversity but it is not clear exactly how diverse they may be. Estimation of species richness is a difficult problem for many types of soil organisms (fungi, bacteria and nematodes, for example, as well as microarthropods). In an extensive review, André et al. (2002) report that at most 10% of soil microarthropod populations have been explored and 10% of species described. Thus, according to those authors, the contribution of soil fauna to global biodiversity remains an enigma. Consequently, the mechanisms underlying the large species diversity of the microarthropods continue to elude us. The decline in numbers of taxonomic specialists for these groups has been noted (Behan-Pelletier and Bissett, 1993) as a contributing factor to our inadequate information base for microarthropods.

Unlike the macroarthropods, the mites and collembolans have little or no effect on soil structure. Their dimensions allow them to use existing spaces in soil structure. Even the large, soft-bodied members of the mite group Prostigmata do not seem to create their own passageways. Some litter-feeding species do burrow into substrates such as petioles of decaying leaves and create tunnels, but these have no direct effect on soil structure per se. The microarthropods resemble the microfauna in this characteristic.

4.3.4.1 *Collembola*

Among microarthropods, collembolans are often equal to soil mites in numerical abundance. They are worldwide in distribution and occur in all biomes, from tropic to arctic and from forest to grassland and desert and throughout the soil profile. Collembolans (Fig. 4.14A–D) have the common name of "springtails" from the fact that many of the species are able to jump by means of a lever attached to the bottom of the abdomen. They also have a unique ventral tube (*collophore*), which seems to function in osmoregulation, and a springing apparatus (*furcula* and *tenaculum*) ventrally on the abdomen (absent in some groups). Most species are small, at most a few millimeters long, but may be brightly colored. They are ubiquitous members of the soil fauna, often reaching abundances of 100,000 or more per square meter. They occur throughout the upper soil profile, where their major diet appears to be fungi associated with decaying vegetation. In the rhizosphere, they are often the most numerous of the microarthropods. Surface-dwelling forms,

FIGURE 4.14 (A) An entomobryomorph collembolan; (B) A symphypleonan collembolan (*Calvatomina* sp.) from Tasmania; (C) A poduromorph collembolan; (D) A neelipleonan collembolan (*Neelus murinus*) from Mexico. Note the absence of the furcula (jumping apparatus) on the eyeless, soil-dwelling poduromorph, in contrast to the other litter-dwelling forms. *Source: All photos by A. Murray, used by permission.*

inhabiting the litter layer, are usually well equipped with furculas (e.g., Fig. 4.14A). Residents of the deeper soil layers generally have no furcula, or only a rudimentary one, and typically lack eyes (e.g., Fig. 4.14C) pigmentation and eyes (Petersen, 2002).

Recent insect classification has removed the Collembola from the Insecta (Delsuc et al., 2003). They are sometimes lumped together with Protura and Diplura in a class named Entognatha, although they do not appear to be closely related to one another. Indeed, some workers treat Collembola at the level of class and separate from other microarthropods (Janssens and Christiansen, 2011).

Their mouthparts are held in a unique cone-shaped structure. Collembolans lack features such as compound eyes or wings, but do resemble insects by having three body regions. The head bears a pair of antennae. The thorax is three-segmented and bears three pairs of legs. The collembolan abdomen consists of only six segments, less than the insect model. The collembolan ventral tube, the collophore, is not found in other groups of arthropods.

The classification of the Collembola has been relatively stable at the generic level, although many species remain unnamed and the classical taxonomy of the group was based almost entirely on external morphology (Hopkin, 1997). Resolution of the higher taxonomic categories will require a close examination of the fauna of the entire world (Christiansen and Bellinger, 1998), and will necessarily include molecular approaches to develop accurate phylogenies. There have been several attempts to refine the understanding of ordinal-level phylogenetic relationships of the Collembola, but as yet, the group remains plagued by uncertainties deriving from evidence of paraphyletic relationships within several previously accepted orders and families (Carapelli et al., 2014; Xiong et al., 2008). In North America, the taxonomic analysis by Christiansen and Bellinger (1998) has been the standard reference and provides keys to genera and known species, although higher-level relationships have changed somewhat (Janssens and Christiansen, 2011). On a worldwide basis, the literature is scattered over a wide variety of journals and other publications. Hopkin (1997) offers regional checklists of the collembolan fauna. The publications of Gisin (1962, 1963, 1964, and other cited therein) are essential for the study of European Collembola. The online resource, www.collembola.org, represents a clearinghouse of information, with a list of links to published keys to the Collembola across the globe, hundreds of photographs, and a continuously updated checklist of species.

Identification of collembolans requires use of a microscope and magnifications as high as 400 ×. Preliminary sorting of samples to family levels can be performed with a dissecting microscope, once some familiarity with the group has been gained. Recognition of genera and species will require slide mounts (see Chapter 9: Laboratory and Field Exercises in Soil Ecology). Collembolans will float on the surface of many collection fluids, due to their very hydrophobic cuticle, and special collection fluids are recommended (see Chapter 9: Laboratory and Field Exercises in Soil Ecology).

FAMILIES OF COLLEMBOLA

In the current system of classification, 33 families of collembolans are arranged in four major groups (Janssens and Christiansen, 2011). We adopt the conventions of these authors here, with the understanding that the relationships within the Collembola are likely to change in future iterations of their taxonomy and systematics. Importantly, Collembola is treated as a class in this system, and the four major groupings are therefore orders.

The so-called linear collembolans (formerly classified as "Arthropleona"), representing the great majority of species, occur in two orders: the Entomobryomorpha (Fig. 4.14A) and the Poduromorpha (Fig. 4.14C).

Poduromorpha collembolans include the important families Hypogastruridae and Onychiuridae, whose species are dwellers in mineral soil layers, and the family Poduridae which consists of a single darkly pigmented species, *Podura aquatica*, whose natural habitat is standing water.

Onychiurid collembolans almost always have no furcula; eyes are reduced or absent and, if present, are unpigmented. They possess pseudocelli, cuticular organs which have nothing to do with vision but which can extrude a defensive oil when disturbed, an alarm pheromone (Usher and Balogun, 1966). Onychiurids feed in the rhizosphere. Curl and Truelove (1986) argue persuasively that these collembolans are attracted to plant roots and are important in rhizosphere dynamics. In experiments, collembolans protected cotton plants from the root pathogen *Rhizoctonia solani* by selectively grazing that fungus from the plant roots. These rhizosphere inhabitants may prove to be effective biological control agents (Fig. 4.15; Curl and Truelove, 1986). Onychiurids are not well sampled with Tullgren funnels; they do not appear to respond to the heating and drying process in Tullgren extractions. Estimates of numbers of Onychiurids are best made with flotation methods (see Chapter 9: Laboratory and Field Exercises in Soil Ecology) (Edwards, 1991).

The family Hypogastruridae includes several common species whose populations may build up to huge numbers. These include the "snow flea," *H. nivicola* (Fig. 4.16). That

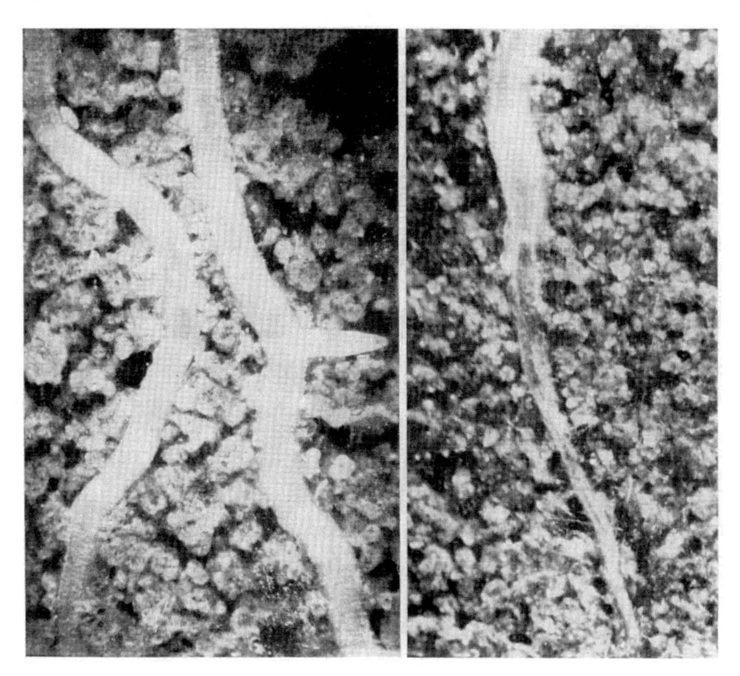

FIGURE 4.15 Collembolan protection of roots from infection by *R. solani*. (Left) Roots from pathogen-infested soil with mycophagous collembola; (right) diseased root from pathogen-infested soil without collembolan. *Source: From Curl, E.A., Truelove, B., 1986. The Rhizosphere. Springer-Verlag, Berlin, New York, NY.*

FIGURE 4.16 *Hypogastrura nivicola*, the "snow flea." Common in wintertime in Appalachian forest floors in western North Carolina. *Source: Photo by S. Justis. Photo courtesy of collembola.org.*

FIGURE 4.17 *Tomocerus longicornis* with curled antennae. *Source: Photo by S. Hopkin. Photo courtesy of collembola.org.*

species multiplies under winter snows and, on warm days, appears to boil out onto the surface (Christiansen, 1992). Another related species, *Hypogastrura armata*, is common in the litter layer of hardwood forests during the winter months (Snider, 1967).

The Entomobryomorpha (Fig. 4.14B) includes the large family Entomobryidae in which the furcula is well developed. The collembolans are primarily dwellers of surficial soil layers, in forest canopies or on tree trunks. Laboratory cultures of one species, *Sinella curvisetta*, have found a valuable role as prey for cultures of spiders (Draney, 1997). Species in the family Tomoceridae (Fig. 4.17) include large forms with long antennae, found in upper litter layers of forest floors throughout the Holarctic region.

Members of the family Isotomidae are a highly variable set of species; the group is in need of serious taxonomic revision (Christiansen and Bellinger, 1998). This family includes *Folsomia candida* (Fig. 4.18), a species widely used in laboratory experiments and in the assessment of the effect of toxic substances. Its reproductive biology has been thoroughly explored and culture methods well developed (Snider, 1973). In fact, the

FIGURE 4.18 *Folsomia candida* adults and juveniles. The largest individual is 2 mm in length. *Source: From Hopkin, S.P., 1997. The Biology of Springtails (Insecta: Collembola). Oxford University Press, Oxford.*

microbial gut flora of *F. candida* has been extensively explored using DNA probing methods. It was found to be a frequently changeable but selective habitat, possibly indicating that soil microarthropods could modify the species makeup of soil microbial communities (Thimm et al., 1998). The European "glacier flea," *Isotoma saltans*, is active on ice at temperatures below freezing and feeds on pollen grains trapped on the glacier surface (Christiansen, 1992).

The other two major groups of Collembola are the order Symphypleona (Fig. 4.14B), and the order Neelipleona (Fig. 4.14D), and include the spherical or globular collembolans. These are smaller groups than the Poduromorpha and Entomobryomorpha and are much more uniform in habits (Christiansen and Bellinger, 1998). The Symphypleonan family Sminthuridae is a large and cosmopolitan one, being active jumpers, dwellers in surficial litter layers, on vegetation, and in the canopies of tropical humid forests. Often brightly colored, these collembolans are readily collected with Tullgren funnels or pitfall traps (see Chapter 9: Laboratory and Field Exercises in Soil Ecology), but many also be collected by sweeping through grassy vegetation with a white enamel pan. The family Neelidae (the only family within the Neelipleona) consists of tiny globular forms lacking eyes and with short antennae. The family is cosmopolitan but poorly studied relative to other groups (Hopkin, 1997).

POPULATION GROWTH AND REPRODUCTION

Many collembolans are opportunistic species, capable of rapid population growth under suitable conditions. They often respond to disturbances of the soil environment. In agricultural systems, spurts of growth may follow plowing or cultivation (Hopkin, 1997). In forests, fire may stimulate collembolan abundances, as may the broadscale application of pesticides (Butcher et al., 1971). Collembolans occur in aggregations. In samples of soils, they are not found at random, but occur in groups. Aside from the statistical problems of assessment of population size, aggregations pose ecological questions as well. In laboratory investigations, Christiansen (1970) and Barra and Christiansen (1975) analyzed collembolan responses to habitat variables (i.e., moisture and substrate) and food resources.

Although these were important, the major variable seemed to be a behavioral one. Collembolans possess aggregation pheromones (Krool and Bauer, 1987), which probably function in bringing the sexes together for reproductive purposes. Earlier, reproductive pheromones were identified by Waldorf (1974). Many collembolan species are capable of rapid, even explosive, population growth under ideal conditions. Gist et al. (1974) analyzed life tables for *Sinella curviseta* under laboratory conditions for 170 days; they found an intrinsic rate of increase of 0.036 per day and a replacement rate (R_0) of 515 per female.

Sperm transfer is by means of spermatophores, either actively passed or deposited on pedicels and located by the females. Eggs are laid in groups. Development is continuous; the number of instars ranges between 2 and 50 or more (Christiansen and Bellinger, 1998). Collembolans become sexually active with the fifth or sixth instar but continue to molt throughout life, in contrast with the Insecta, which do not molt after reproduction.

Parthenogenic reproduction is common in many collembolan species, including the commonly cultured *F. candida*. Many species are bisexual, especially those in the Entomobryidae (e.g., *S. curviseta*).

COLLEMBOLAN FEEDING HABITS

Collembolans are generally considered to be fungivores, with occasional ingestion of other animals, decomposing plant or animal residue, or fecal material. As noted previously, they are often considered to be nonspecific feeders but this conclusion is controversial (Petersen, 2002). Gut content analysis of field-collected specimens or field observation in rhizotrons (Gunn and Cherrett, 1993) often reveals a wide variety of materials, including fungi, plant debris, and animal remains. Laboratory choice studies, in contrast, have found that collembolans have specific food preferences, choosing one fungal species over others (similar discrepancies in feeding analysis have been noted for oribatid mites; see later section on mites). Bengtsson et al. (1994), in laboratory experiments, reported that the collembolan *Protaphorura armata* showed an increased dispersal rate if a favored fungus was present as far away as 40 cm (cited in Petersen, 2002).

In any case, the particular feeding of Collembola is likely to be a species-specific phenomenon, with a wide range of possible behaviors. The advent of molecular techniques for investigating the ingested material holds promise for working out the details of detrital food webs. The work by Heidemann et al. (2014) is an example of how careful preparation of DNA primers can allow identification of prey species in gut contents of consumers. These authors thus demonstrated that certain species of field-collected Collembola were indeed ingesting nematodes, although other collembolans in laboratory feeding trials would not eat nematodes even when they had no other food choices (Heidemann et al., 2014).

Models of soil food webs usually place collembolans as fungivores (e.g., Coleman, 1985; Hunt et al., 1987; Moore and de Ruiter, 2000). However, like many of the soil fauna, collembolans in general defy such exact placement into trophic groups, as mentioned earlier (Heidemann et al., 2014). Living plant tissue may be consumed and even dead animal material or feces in cultures. Some species are known herbivores (e.g., *Sminthurus viridis*) and have been documented to reach pest status under certain circumstances (Crotty et al. 2014). Many collembolan species will eat nematodes when those are abundant (Gilmore and Potter, 1993). Some species may be significant in the biological control for nematode populations (Gilmore, 1970). Feeding on nematodes does not seem to be selective;

collembolans do not distinguish between saprophytic and plant parasitic nematodes. In the words of Hopkin (1997), "Indeed the opportunistic nature of the feeding behavior of many species of Collembola may be one reason for their success." This earlier view is supported by stable isotopic, and gut lipid analyses suggesting that collembolan communities had members occupying multiple trophic levels (spanning herbivore–detritivore–predator levels) in European beech forests (Fiera, 2014).

COLLEMBOLAN IMPACTS ON SOIL ECOSYSTEMS

The direct effect of collembolans on ecosystem processes such as energy flow appears to be quite small. Their biomass is relatively tiny, their respiration rates are but a fraction of total soil CO_2 efflux, and their feeding rates account for only a small amount of microbial activity. They share these characteristics with other soil microarthropods (Gjelstrup and Petersen, 1987). These conclusions have led Andrén et al. (1999) to a sardonic statement, to wit: "Soil animals exist. I like soil animals. They respire too little. Ergo, they must CONTROL something!" Those authors caution us to avoid an overly enthusiastic appraisal of the importance of microarthropods in soil ecosystems. Nevertheless, manipulation experiments have shown important impacts of collembola on nitrogen mineralization, soil respiration, leaching of dissolved organic carbon, and plant growth (Filser, 2002). These system responses may be viewed as indirect effects. Assessing the importance of Collembola in soil ecosystems needs to be done in the context of the intact system and may be expected to vary with temperature, moisture, season, and interactions with other soil biota.

Grazing upon fungal hyphae appears to be the major contribution of Collembola in the decomposition process. Such grazing on fungal hyphae may be selective, thus influencing the fungal community (Table 4.4). Indirectly, such direct effects on the fungal community may have indirect effects on nutrient cycling (Moore et al., 1987). Selective grazing by the collembolan *Onychiurus latus* changed the outcome of competition between two basidiomycete decomposer fungi (Newell, 1984a,b), allowing an inferior competitor to prosper. Grazing upon fungi may actually increase general fungal activity in soils and stimulate

TABLE 4.4 Compensatory Growth of Fungi in Response to Collembolan Grazing

Fungal Species	Collembolan Species	Growth Relative to Controls
Botrytis cinerea	*Folsomia fimetaria*	–[a]
Coriolus versicolor	*Folsomia candida*	–/+[b]
Mortierella isabellina	*Onychiurus armatus*	–[a]
Verticillium bulbillosum	*Onychiurus armatus*	+
Penicillium spinulosum	*Onychiurus armatus*	+
Field soil dilution	*Hypogastrura tullbergi*	+[a]
	Folsomia regularis	+[a]

[a]*Bacteria were present.*
[b]*Increase with fungi grown on high nutrient medium, otherwise decrease.*
After Lussenhop, J., 1992. Mechanisms of microarthropod–microbial interactions in soil. Adv. Ecol. Res. 23, 1–33.

fungal growth. The relationship between fungal and collembolan population dynamics is not straightforward, however, because some collembolan species may reproduce more successfully on least favored foods (Walsh and Bolger, 1990). Collembola have been demonstrated to have complex interactions with several fungal species simultaneously. Cotton was grown in a greenhouse with four fungal species, the pathogen *R. solani* and three known biocontrol fungi (including two sporulating Hyphomycetes), and the rhizosphere-inhabiting collembolan *Proisotoma minuta*. The collembolan preferentially fed on the pathogenic fungus and avoided the biocontrol fungi (Lartey et al., 1994).

4.3.5 Acari (Mites)

The soil mites, Acari, chelicerate arthropods related to the spiders, are the most abundant microarthropods in many types of soils. In rich forest soil, a 100 g sample extracted on a Tullgren funnel may contain as many as 500 mites representing almost 100 genera (Table 4.5). This much diversity includes participants in three or more trophic levels and varied strategies for feeding, reproduction, and dispersal. Often, ecologists analyze samples by a preliminary sorting of mites into suborders. Identification of mites to the family level is a skill readily learned under the tutelage of an acarologist. Expert assistance is necessary for identification of soil mites to genus or species. By combining slide mounts with examination of specimens in alcohol, reasonably accurate sorting of samples can be performed.

Four groups of mites occur frequently in soils, all historically considered to be suborders. However, these have now been reassigned to differing taxonomic ranks. It should be said that the taxonomic and phylogenetic relationships of the Acari are still not well known, and will be likely to remain in flux well into the future, but here we report the currently accepted names and ranks of the major soil groups. The four predominant groups of soil mites are represented by one group within the superorder Parasitiformes, the order Mesostigmata; and three groups within the superorder Acariformes, the suborders Prostigmata and Oribatida, and a group within the Oribatida, cohort Astigmatina. Specimens in alcohol can usually be assigned to one of these groups. Slide mounts are required for placement of dubious specimens. Techniques for slide preparations are given in Chapter 9, Laboratory and Field Exercises in Soil Ecology. Krantz and Walter (2009) provide keys to all families of the Acari. Keys to families or superfamilies of soil mites

TABLE 4.5 Densities (number per $m^{2)}$ of Soil Microarthropods in Five Land-Use Types

Environment	Collembola	Oribatida	Prostigmata	Mesostigmata
Secondary shrub forest	4,878	4,380	559	1,148
Deciduous forest	6,479	5,241	765	917
Coniferous plantation	11,157	3,528	446	954
Mixed forest	7,192	5,019	864	963
Subalpine meadow	6,352	4,411	2,324	417

Modified from Wu, P., Liu, X., Liu, S., Wang, J., Wang, Y., 2014. Composition and spatio-temporal variation of soil microarthropods in the biodiversity hotspot of northern Hengduan Mountains, China. Eur. J. Soil Biol. 62, 30–38.

may be found in Dindal (1990). As is the case for the collembolans, taxonomic aids may be found in various Internet-based platforms.

The soil mites are a subset of the Acari. Occasionally, mites from other habitats are extracted from soil samples. During autumnal leaf drop in deciduous forests, foliage-inhabiting mites may enter the soil food web and may be found in Tullgren extractions. This group includes, for example, plant feeding mites (red spider and false spider mites) and their predators. Occasionally, mites parasitic upon vertebrates may be seen in soils from the vicinity of mammal nests, or host-seeking larval forms may occur in Tullgren samples. These stragglers doubtless enter the soil food web as food items, but the more numerous species are the true soil mites.

Among the four mite groups, the oribatids are the characteristic mites of the soil and are usually fungivorous, detritivorous, or both, and sometimes even predatory on soil nematodes (Heidemann et al., 2014). Mesostigmatid mites are nearly all predators on other small fauna, although some few species are fungivores and these may become numerous in some situations. Astigamtine mites are found associated with rich, decomposing nitrogen sources and are seldom abundant except in agricultural soils or stored products. The Prostigmata contains a broad diversity of mites with a variety of feeding habits and strategies. Very little is known of the niches or ecological requirements of most soil mite species, but some tantalizing information is emerging from field research (Hansen, 2000; Klarner et al., 2013; Walter et al., 1987). As noted previously for Collembola, some mites may play significant roles as consumers of plant pathogenic fungi. In ecosystems where most primary productivity occurs belowground (e.g., grasslands), where nematode biomass is high in root rhizospheres, nematophagous arthropods (including mites) could be significant predators of plant-feeding nematodes (Gerson et al., 2003).

4.3.5.1 *Oribatid Mites*

Oribatids (Figs. 4.19–4.22) are an ancient group. They have the richest fossil record of any mite group, dating back to the Devonian period, or 350–400 million years ago (Labandeira et al., 1997). Specimens from a Devonian site near Gilboa, NY, contain organic matter visible in their guts, attesting to a long relationship between oribatid mites and decomposing vegetable matter (Norton et al., 1987). Some oribatid species are Holarctic in distribution, being widely distributed in forest floors of Europe and North America. A general catalog of oribatids in North America was published by Marshall et al. (1987). Oribatid mites occur in all terrestrial ecosystems, from arctic (Behan-Pelletier and Norton, 1983) to tropics (Balogh and Balogh, 1988, 1990; González et al., 2001) and from deciduous forest (Hansen, 2000; Lamoncha and Crossley, 1998) to desert (Santos and Whitford, 1981).

A combination of three factors makes oribatid mites unique among the soil fauna. First, their sheer numbers are impressive. They are the most numerous of the microarthropods (Travé et al., 1996). Second, they often possess juvenile polymorphism. Many immature stadia do not resemble their adults. Unlike most other mites, the immatures of some of the Oribatei (the "higher" oribatids) (Norton, 1984) are morphologically so dissimilar to the adult stadia that it is frequently impossible to correlate the two based on morphology alone (compare Figs. 4.19 and 4.20). Despite the differences in morphology, immature stages and adults can usually be cultured on the same food source. Third, oribatids reproduce relatively slowly, in contrast to other microarthropods. One or two generations per year are usual, and

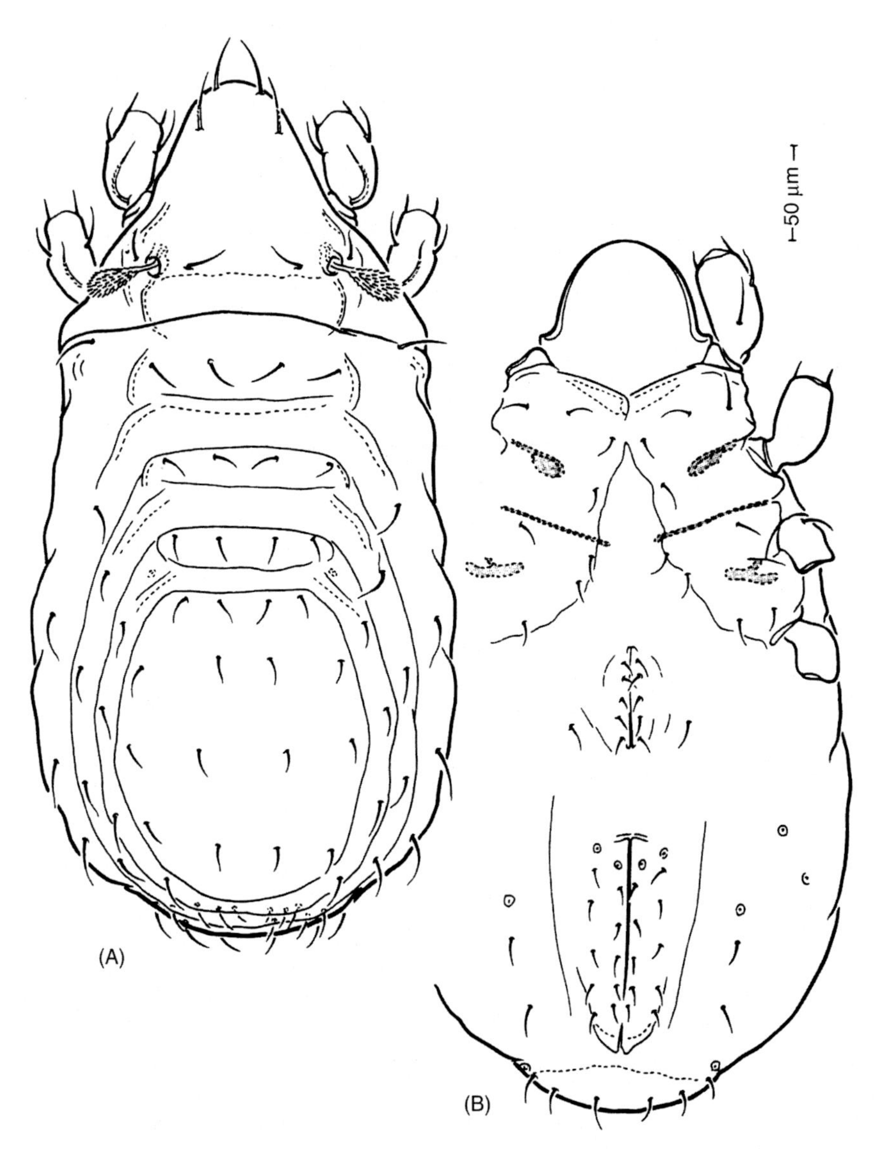

FIGURE 4.19 *Eueremaeus columbianus* (Berlese), tritonymph: (A) dorsal aspect and (B) ventral aspect. *Source: From Behan-Pelletier, V.M., Bissett, B., 1993. Biodiversity of nearctic soil arthropods. Can. Biodivers. 2, 5–14.*

females do not lay many eggs. Indeed, some common species are parthenogenic. Oribatids are sometimes considered to be "K" specialists, and, in contrast, collembolans would be "r" specialists, or opportunistic species (MacArthur, 1972). However, oribatids' K-style traits may be a constraint of low secondary production, not an adaptation of specialization.

Oribatids differ from other microarthropods by having a sclerotized, often calcareous, exoskeleton. In this they resemble the millipedes, snails, and isopods. Most oribatids are brown to tan in color, although some primitive species are nearly colorless.

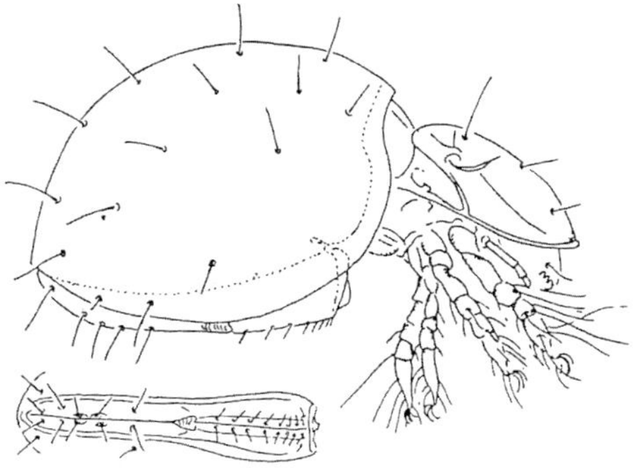

FIGURE 4.20 A "box" mite of the oribatid family Phthiracaridae. For protection, the legs can be withdrawn beneath the hinged prodorsum. *Source: After Baker et al., 1958.*

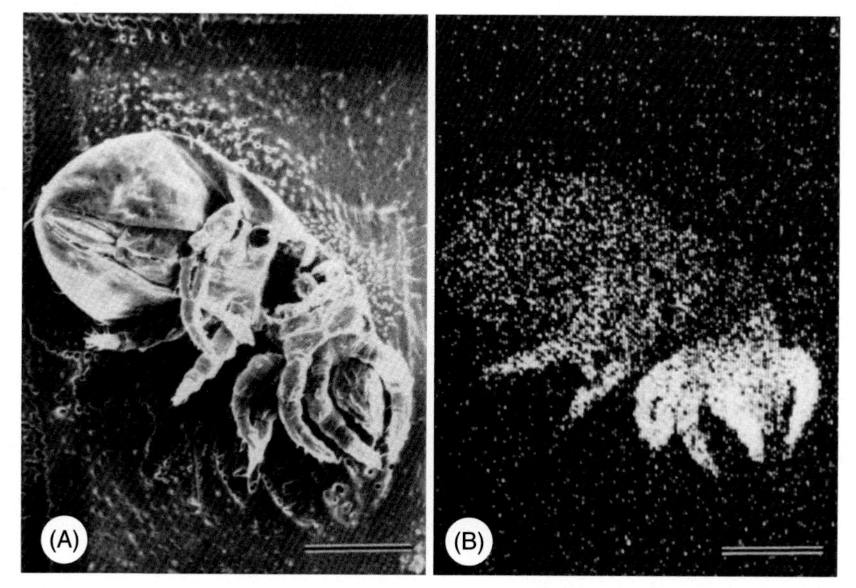

FIGURE 4.21 Oribatid scanning electron micrograph with calcium in exoskeleton. (A) and (B) are micrographs of an oribatid mite (*Hypochthonius* sp.). A 100 μm marker is indicated at lower right. (A) Image made by secondary electrons. (B) Map of calcium distribution made from X-ray image. *Source: From Todd, R.L., Crossley Jr., D.A., Stormer Jr., J.A., 1974. Chemical composition of microarthropods by electron microprobe analysis: a preliminary report. Proceedings of the 32nd Annual Proceedings of the Electron Microscopy Society of America, St. Louis, MI.*

The exoskeleton contains high calcium levels even in the primitive, lightly colored species (Fig. 4.21) (Todd et al., 1974). When present, the chemical form of the mineral deposits is principally calcium carbonate, calcium oxalate, or calcium phosphate. Presumably, oribatids are able to sequester calcium by feeding on fungi. Senescent fungal hyphae contain crystals of calcium oxalate, which may be metabolized by the mites (Norton and Behan-Pelletier, 1991). Oribatids often possess a cerotegument, a secreted layer of the integument that may be highly sculptured (Norton et al., 1997).

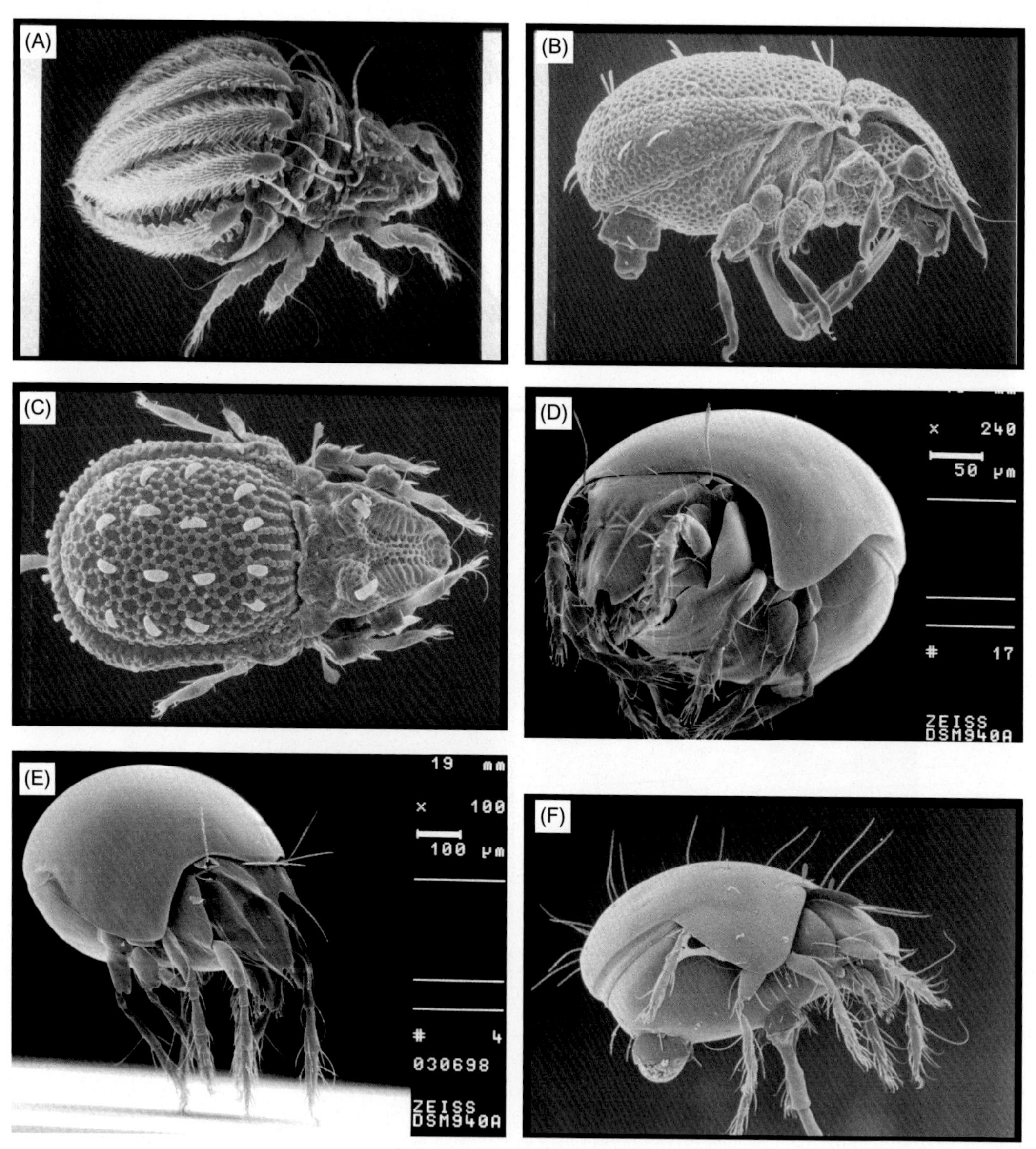

FIGURE 4.22 Scanning electron micrographs of oribatid mites, showing modified setae (A–F). *Source: From Valerie Behan-Pelletier, personal communication.*

ABUNDANCE AND DIVERSITY OF ORIBATID MITES

Oribatid densities in forest soils are in the range of 50,000–500,000 individuals per square meter (Table 4.5). Coniferous forests typically support high numbers of oribatid mites, followed by deciduous hardwood forests, grassland, desert, and tundra. In arctic tundra and in some grassland or savanna habitats, oribatids may be outnumbered by prostigmatid mites (see later section). Cultivation of agricultural soils reduces oribatid numbers to an average of around 25,000 individuals per square meter. Population cycles in agroecosystems are often influenced by harvest and cultivation procedures, which change patterns of residue input into soils.

The diversity of oribatid mite species is large in soils from many different localities. In a southern Appalachian hardwood forest, Hansen (2000) reported 170 species of adult oribatid mites from litter and soil samples, with as many as 40 species in a single 20 cm^3 sample core through the soil profile. Many oribatid mite species are widely distributed across a variety of habitats in Europe and North America, including common species such as *Oppiella nova*, *Tectocepheus velatus*, and *Scheloribates laevigatus*. Tropical soils also contain a diverse community of oribatid mites. Noti et al. (2003) examined three ecosystems in a sere in the Democratic Republic of Congo and reported a total of 149 species of oribatids. They found that the number of species dropped regularly from forest to savanna, where sampling revealed 105 oribatid species. This high species diversity has caused some authors (i.e., Anderson, 1975b; André et al., 2001) to consider "the enigma of the oribatids" (as a comparison to the "the paradox of the plankton") in view of the seeming uniformity of forest floor habitats. Indeed, explaining this prodigious diversity of apparently similar species with similar requirements is a long-standing question (Bolger, 2001; Hansen, 2000).

Tropical and temperate forest canopies may support large numbers of oribatid mites, in such abundance that canopy-inhabiting mites have been dubbed "arboreal plankton" (Walter and Proctor, 1999). For more information on tropical oribatids, see Franklin et al. (2007).

Tree canopies have long been known to support phytophagous spider mites and their mite predators (Phytoseiidae and relatives). Oribatid mites in canopies had been largely ignored until work by Winchester (1997), and Behan-Pelletier and Walter (2000). There has been considerable work since that time examining the ecology of canopy-dwelling oribatids in the temperate rainforests of British Columbia. Lindo and Winchester (2006) investigated patterns of diversity in tree canopies and compared these to diversity patterns in forest floor and found that although the species richness of canopy oribatid communities was less than that of the forest floor, the canopy communities were distinct and had representatives (20% of total species) that were found exclusively in the canopy. For Antarctic oribatids, see Stary and Block (1998).

POPULATION GROWTH

Populations of oribatid mites in forest floors show peaks of activity beginning in spring and continuing through the summer months, and again in mid-autumn (for those species producing two generations per year). Peaks of abundances of immatures show a gradual progression during the summer (Fig. 4.23) (Reeves, 1967). Population densities for many species are markedly higher during these months. Is there a sequence of oribatid species—a succession—in decomposing forest leaf litter, corresponding to a succession of fungal species? Crossley and Hoglund (1962) found such a general relationship. In a detailed study, Anderson (1975a) concluded that the dominant oribatid species rapidly colonized litterbags containing beech and chestnut leaf disks. The mites fed upon the succession of fungi, but no succession of the mites themselves was demonstrated.

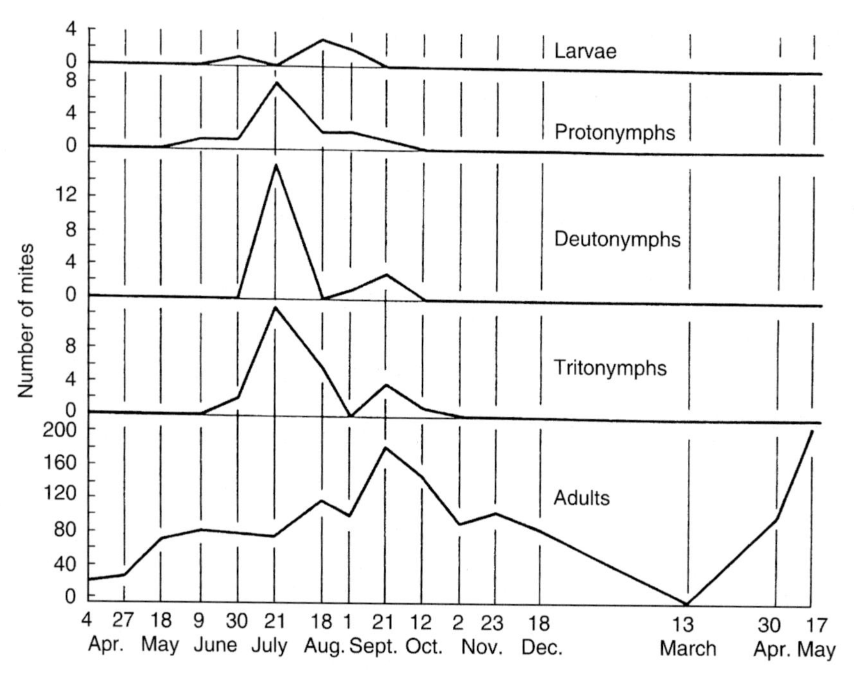

FIGURE 4.23 Oribatid immatures and adult abundances by season (larvae, nymphs, and adults of Oppia subpectinata (Oudemans)). *Source: From Reeves, R.M., 1967. Seasonal distribution of some forest soil Oribatei. In: Evans, G.O. (Ed.), Proceedings of the 2nd International Congress of Acarology. Akademiai Kiado, Budapest, pp. 23–30.*

ORIBATID FEEDING HABITS

Information about feeding habits and nutrition of oribatid mites had been elusive, despite numerous detailed studies (e.g., Luxton, 1972, 1975, 1979; Mueller et al., 1990; Walter and Proctor, 1999), but stable isotope and molecular techniques may allow closer examination of these relationships than had been previously possible. Classical techniques for evaluating feeding behavior involved laboratory feeding trials, and oribatids in culture have been demonstrated to eat a variety of substrates (but these results have always been presented with the caveat that mites may feed differently under field conditions). In the simplest classification, oribatids are separated into feeders on fungi (microphytophages), on decomposing vegetable matter (macrophytophages), or on both (panphytophages) (Schuster, 1956). This classification has some utility, but more specific feeding habits can be identified. Some species, the phthiracarids or box mites, are largely macrophytophages. Many oribatids appear to be indiscriminant fungal feeders, ingesting fungal hyphae or fruiting bodies of a variety of species (Mitchell and Parkinson, 1976; Siepel and de Ruiter-Dijkman, 1993). Others may be selective. Adults of *Liacarus cidarus*, when offered a variety of foods, preferred the mold *Cladosporidium* (Arlian and Woolley, 1970) (Table 4.6). Anderson (1975a) studied competition between two generalist feeders, *Hermanniella granulata* and *Nothrus sylvestris*. When isolated, these two species used similar food sources, but when kept together in soil–litter microcosms, the two species changed their feeding and

TABLE 4.6 Feeding Behavior of Adult *L. cidarus* When Offered Different Resources in Laboratory Cultures

Resource	Heavy Feeding	Light Feeding	Intermittent Feeding	Failed to Feed
Lichens	–	X	–	–
Yeast and sugar	–	–	X	–
Pine cone scales	–	X	–	–
Cladiosporium	X	–	–	–
Aspergillus	–	–	–	X
"Mushrooms"	–	–	X	–
Potato dextrose agar and *Cladiosporium*	X	–	–	–
Potato dextrose agar	–	X	–	–
Pine litter	–	–	X	–
Yeast	–	–	–	X
Trichoderma	–	–	–	X

After Arlian, L.G., Woolley, T.A., 1970. Observations on the biology of Liacarus cidarus (Acari: Cryptostigmata, Liacaridae). J. Kansas Entomol. Soc. 43, 297–301.

their utilization of habitat space. *Hermanniella* moved into the litter layer (Oi) while the *Nothrus* population increased in fermentation layer (Oe).

Several lines of evidence—gut analyses, feeding trials in cultures, chemical considerations—suggest that oribatid mites, as a group, are fungal feeders. Some exceptions exist, such as the phthiracarid group, which may tunnel into coniferous needles or twigs and ingest the spongy mesophyll. Others bore into the pedicels of oak leaves (Hansen, 1999). These species may contain a gut flora of bacteria that allow them to digest decomposing woody substrates. Possibly, their nutrition is derived from bacteria or fungi embedded in the ingested material. Examination of gut contents for most other oribatids shows that they feed primarily on fungi. A survey of 25 species from the North American arctic found that more than 50% were panphytophagous (i.e., feeding on microbial and plant sources), but nearly all contained fungal hyphae or spores. Similarly, a study of Irish species showed that 15 of 16 species were generalist feeders, having both fungi and plant remains in their guts (Behan and Hill, 1978; Behan-Pelletier and Hill, 1983). Occasional fragments of collembolans were discovered in some of the guts. Morphology of the chelicerae seems related to feeding type (Kaneko, 1988). Xylophagous (wood-feeding) species have large, robust chelicerae. Fragment feeders are generally smaller and have smaller chelicerae.

Further refinement of guild designations for oribatid mites has been made, based on their digestive capabilities as evidenced by their cellulase, chitinase, and trehalase activity in field populations (Siepel and de Ruiter-Dijkman, 1993). Using these three enzymes, it was possible to recognize five major feeding guilds: herbivorous grazers, fungivorous grazers, herbo-fungivorous grazers, fungivorous browsers, and opportunistic herbo-fungivores. In this classification, grazers are species which can digest both cell walls and cell contents; browsers are those which can digest only cell contents. Siepel and de Ruiter-Dijkman also recognized two minor guilds: herbivorous browsers and omnivores. Further work with other groups of gut enzymes may provide additional information concerning resource utilization, and fungal feeding in particular, by guilds of oribatid mites. This is a most rewarding area for research in biodiversity, considering the multitude of species and niche dimensions for oribatids in forest floor habitats (Anderson, 1975b; Hansen, 2000).

Considerable progress has been made in determining the details of trophic (feeding) interactions in microarthropods, particularly for the oribatid mites. Stable isotope techniques have provided powerful new information on the diet of mites over time in the field. The relative positions of mite gut contents and tissues in the amount of ^{13}C and ^{15}N stable isotope signatures have enabled the assignment of oribatid mites into feeding guilds (Pollierer et al., 2009; Schneider et al., 2004). Unfortunately, this requires a minimum mass of mites to enable the analyses to be performed. An alternative approach uses an analysis of mite chelicerae (the chewing mouth parts) along with measurement, where possible, of the stable isotope ratios of oribatid mites present in moss communities in the forest floor. Using this approach, cheliceral morphology was found to be an inexpensive and quick filter for estimation of the mites' feeding preferences. In cases of ambiguous trophic relationships, the dietary preferences were then resolved through isotope analyses (Perdomo et al., 2012). Chamberlain et al. (2006) used a combination of fatty acid gut analysis with compound-specific C isotope analysis to show that two collembolan species consumed nematodes, confirming earlier observations (Ruess et al., 2004). Studying the fates of DNA from two prevalent species of soil nematodes, Heidemann et al. (2011) found that eight species of oribatid mites and one mesostigmatid would act as either predators or scavengers on soil nematodes, thus unraveling the complexity of microarthropod food webs.

The oribatids themselves are prey for some insects such as scydmaenid beetles (Molleman and Walter, 2001), pselaphid beetles (Park, 1947), or ants (Matsuko, 1994). Some vertebrates such as salamanders have been found to contain oribatids in their guts. In general, the hard exoskeleton of oribatid mites protects them from smaller predators such as prostigmatic and mesostigmatic mites (see later sections).

ORIBATID IMPACTS ON SOIL ECOSYSTEMS

Oribatid mites can affect organic litter decomposition and nutrient dynamics in forest floors, but they appear to do so indirectly by grazing on microbial populations or fragmenting plant detritus, and thus influencing the decomposition process (Petersen and Luxton, 1982; Seastedt, 1984b). Calcium dynamics may be an exception, because oribatid mites can store and process a significant portion of the calcium input in litterfall (Gist and Crossley, 1975). Several mechanisms of microarthropod–microfloral interaction have been proposed (Lussenhop, 1992), including consequences of selective grazing, dispersal of fungal spores or inocula, and stimulation of fungal growth or bacterial activity. It is difficult to interpret the importance of oribatid mites relative to soil processes outside the context of the entire suite of soil biota, as these organisms work in concert through myriad interactions, and the impacts to soil processes represent the *net* effect of all these interactions. The milieu of decomposer organisms, vegetable matter in various stages of decomposition, and localized "hot spots" of activity requires careful analysis and has been a focus of much research (Bolger et al., 2000; Edwards, 2000; Hansen, 2000; Moore and de Ruiter, 2000). In spite of the complexity of microarthropod decomposer communities, there has been one laboratory incubation demonstrating that a single species of oribatid mite (*Scheloribates moestus*) had measurable positive effects on extracellular enzyme concentrations and microbial respiration, along with extractable C and N (Wickings and Grandy, 2011). This "proof-of-concept," reductionist approach allows soil ecologists to assign attribution of at least one small part of the net effect observed in the decomposition process.

4.3.5.2 Prostigmatic Mites

The Prostigmata contains a large array of soil species (Fig. 4.24). Many of these species are predators, but some families contain fungal-feeding mites and these may become numerous. Like the oribatids, prostigmatic mites are an ancient group with fossil representatives from the Devonian era. Keys to families of Prostigmata (also known as Actinedida) are provided by Krantz and Walter (2009) and Kethley (1990).

Some families of the prostigmatic mites include species which are predators, microbial feeders, plant feeders, or parasites (Kethley, 1990). The fungal feeding species (such as members of the family Eupodidae) are opportunistic, able to reproduce rapidly following a disturbance or a sudden shift in resources. Small species of Prostigmata are the common mites of Antarctic soil surfaces, of drained lake beds with algal blooms, of plowed and fertilized agricultural fields, of tidal marshlands, or burned prairie soils, and so forth (Lussenhop, 1976; Luxton, 1967, 1981b; Perdue and Crossley, 1989; Seastedt, 1984a; Strandtmann, 1967; Tevis and Newell, 1962). Species of the families Eupodidae, Tarsonemidae, Nanorchestidae, and some of their relatives feed on algae or fungi; their populations may grow rapidly to large sizes. In these situations, the Prostigmata may become more numerous than the oribatid mites. In general, they are more numerous in temperate than in tropical or subtropical habitats (Luxton, 1981b).

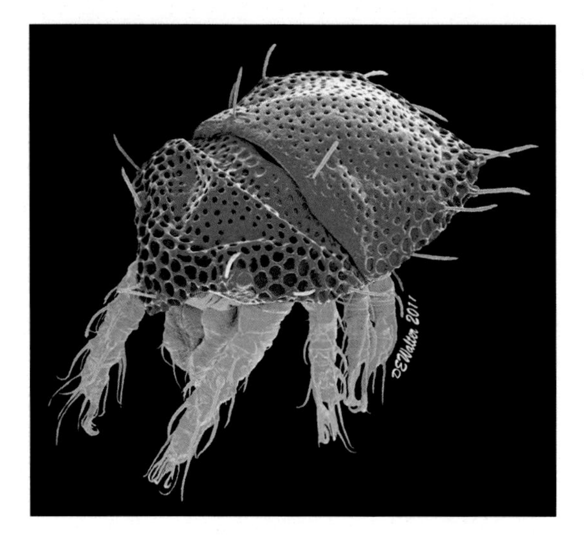

FIGURE 4.24 Prostigmatid mite: *Eustigmaeus frigida*. *Source: Photo by D. Walter. From macromite.wordpress.com.*

Nematodes are an important part of the diet of the smaller Prostigmata. Whitford and Santos (1980) found that mites in the family Tydeidae were important regulators of nematode populations in desert soils of the southwestern United States. Some of the smaller predatory species may utilize fungi on occasion. Walter (1988) observed predation on nematodes by prostigmatic mites in the families Bimichaelidae and Alicorhagiidae; members of the latter family also ingested fungal hyphae but with lower reproductive success. Many types of fungal-feeding Prostigmata have small, stylet chelicerae and may simply pierce fungal hyphae (members of the Nanorchestidae and Nematalychidae families) (Walter, 1988). That author also examined gut contents in slide mounts of more than 500 small prostigmatic mites collected in various locations in North America and elsewhere. Many of these specimens contained a fungal food bolus (Table 4.7). Both fungal feeding and predation on nematodes appear to be widespread among the tiny Prostigmata. Kethley (1990) lists the known feeding habits of 30 families or superfamilies of soil-inhabiting Prostigmata, and summarizes their biology, ecology, and type of habitat.

In general, the larger predaceous Prostigmata feed upon other arthropods or their eggs; the smaller species are nematophagous. Some Prostigmata have well-defined patterns of predation. The "grasshopper mite," *Allothrombium trigonum*, feeds exclusively upon grasshopper eggs; the larval stages of the mite are parasitic on grasshoppers. The large "red velvet mites" (*Dolicothrombium* species), which erupt in numbers following desert rains, are predaceous on termites. The pestiferous "chiggers" are the larval stages of mites in the family Trombiculidae; the adults are predaceous on collembolans and their eggs. Collembolans may be an important diet item for the larger Prostigmata such as members of the families Bdellidae, Cunaxidae, and the trombidioid families. Eggs of Collembola were used successfully to culture pest chiggers (*Eutrombicula alfreddugesi* and relatives) (Crossley, 1960).

It is difficult to assess the importance of the soil-dwelling Prostigmata, even in those cases when their numbers escalate. Those species predaceous on nematodes may have an impact (Whitford and Santos, 1980), but it is seldom quantified. The biomass of prostigmatic mites is generally small, only a fraction of the total acarine mass (Kethley, 1990), and their total respiration is comparatively small (Luxton, 1981b). When populations of fungal or algal feeders reach high population sizes, we suspect that they may have some impact on their food base, but the magnitude of the effect is unknown.

4.3.5.3 Mesostigmatic Mites

The Mesostigmata (Fig. 4.25) contains fewer soil-inhabiting species than do Oribatida or Prostigmata. Krantz and Ainscough (1990) include keys to families and genera of the soil-inhabiting species of Mesostigmata. Many of the Mesostigmata are parasitic on vertebrates or invertebrates (Krantz, 1978), and some of these may be captured in soil samples. The true soil species are almost all predators. A few species (in the Uropodidae, for example) are polyphagous, feeding on fungi, nematodes, and juvenile insects (Gerson et al., 2003), and may become somewhat numerous in agroecosystems (Mueller et al., 1990). Mesostigmatic mites are not as numerous as oribatids or prostigmatid mites, but are universally present in soils and may be important predators. As with the Prostigmata, the larger species tend to feed on small arthropods or their eggs; the smaller species are mainly nematophagous. Some of the species in the family Laelapidae are voracious

TABLE 4.7 Percentages of Field-Collected Prostigmatid Mites with Boluses or a Central Mass of Particulate Fungal Hyphae in their Guts

Taxon	**Individuals Examined (no.)**	**With Fungal Bolus (no.)**	**Other Inclusions**
Terpnacaridae	38	26	—
Alicorhagiidae			
Alicorhagia	98	23	Nematode stylets (3) tardigrade claws (1)
Stigmalycus	32	10	Nematode (1)
Grandjeanicidae	20	17	Diatoms
Oehserchestidae	38	10	–
Lordalychidae	49	0	Fungal mass
Bimichaelidae			
Alycus	52	0	–
Bimichaelia	36	0	–
Pachygnathus	1	0	–
Petralycus	13	0	–
Nanorchestidae			
Nanorchestes	24	0	Spore-like bodies (1)
Speleorchestes	25	0	Cecal masses (11)
Nematalycidae			
Cunliffia	24	0	Crystals? (21)
Gordialycus	31	0	–
Psammolycus	13	0	–
Sphaeroluchidae	18	0	–

After Walter, D.E., 1988. Predation and mycophagy by endeostigmatid mites (Acariformes: Prostigmata). Exp. Appl. Acarol. 4, 159–166.

predators on red spider mites and false spider mites feeding on aboveground vegetation. In the soil itself, members of the genus *Hypoaspis* (Fig. 4.26) are important predators of small insect larvae.

Walter and Ikonen (1989) found that mesostigmatid mites were the most important predators of nematodes in grasslands of the western United States. In contrast, they found that the larger Prostigmata were predators on arthropods or their eggs. Of 63 species of mesostigmatic mites tested, only 6 did not readily feed on nematode prey. The mites each consumed 3–8 nematodes per day. Western grassland soils have little surface plant litter.

Forest floors, with abundant surface litter, contain a larger spectrum of mesostigmatic mites. The forest litter inhabitants (families Veigaiidae and Macrochelidae, for example) are bigger

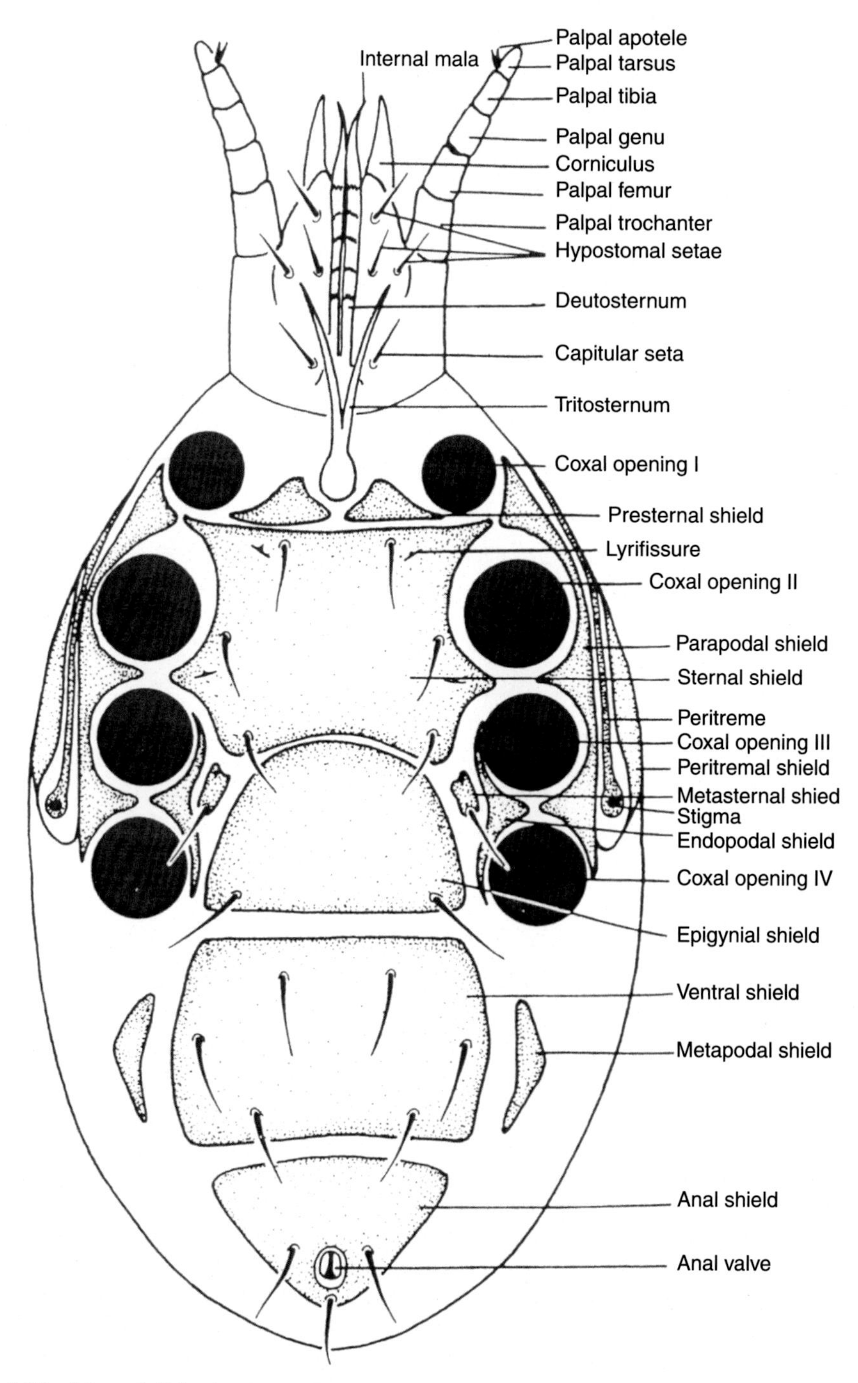

FIGURE 4.25 Macrochelid mite. *Source: From Krantz, G.W., 1978. A Manual of Acarology, second ed. Oregon State University Book Stores, Corvallis, OR.*

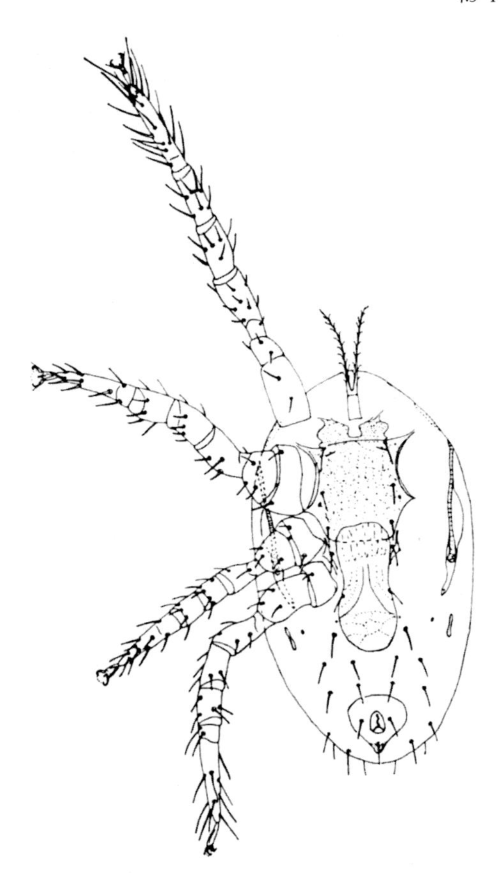

FIGURE 4.26 *Hypoaspis marksi. Source: From Strandtmann and Crossley, 1962.*

species and are predaceous on arthropods or their eggs. Mesostigmatic mites of the mineral soil layer are the smaller, colorless Rhodacaridae and relatives, and are nematophagous. The size of soil Mesostigmata diminishes with increasing depth in the soil (Coineau, 1974). In forest floor habitats, members of the genus *Veigaia* are inhabitants of litter and humus layers; smaller species (*Dendrolaelaps*) occur in the humus–soil interface, and the minute *Rhodacarellus* is found in mineral soil (Krantz and Ainscough, 1990). Many species of Mesostigmata have a close relationship with various insect species, a relationship that often includes the soil environment (Hunter and Rosario, 1988). Several genera in the cohort Gamasina are also considered useful as bioindicators of habitat and soil conditions (Karg, 1982).

In one of the most detailed examinations of a mesostigmatid mite community to date, Klarner et al. (2013) examined the stable isotope composition of all the species of mesostigmatids they found in a European beech forest ecosystem. They used stable isotope analysis of body tissues to assign mites to trophic categories, and found that the mites in this community had ^{15}N signatures indicative of species that occupied four trophic levels ranging from primary decomposer to tertiary predators. Klarner et al. (2013) further showed that stable isotopes could be used to evaluate feeding niche space, and demonstrated that congeneric species frequently occupied different niche space relative to their ^{15}N and ^{13}C signatures.

4.3.5.4 Astigmatine Mites

The Astigmatina (Fig. 4.27) are the least common of the soil mites, although they may become abundant in some habitats (Luxton, 1981a). The free-living Astigmatina favor moist environments high in organic matter. The Astigmatina contain some important pests of stored grain. They become abundant in some agroecosystems following harvest, or after application of rich manures. In agroecosystems of the Piedmont region of the southeastern United States, Perdue and Crossley (1989) found a marked increase in Astigmatina following autumnal harvest and tillage (Fig. 4.28). Incorporation of residues and winter rains produced moist, organic residues suitable for the mites. The springtime plowing, under drier conditions, did not lead to increases of astigmatic mites. Tomlin (1977) described a buildup of astigmatic mites following pipeline construction in Ontario, Canada. The mites were associated with accumulations of residue under moist conditions. Philips (1990) provided keys to families and genera of soil-inhabiting Astigmatina.

Most of the soil Astigmatina are microbial feeders. Those with chelate chelicerae are able to chew vegetable material, fungi, and algae (Philips, 1990). Members of the Anoetidae have reduced chelae; their palpi are highly modified strainers for filter-feeding on microbial colonies (Philips, 1990).

Occasionally, species in the family Acaridae become pests in microbiology laboratories, where they reproduce rapidly on agar plates. They are readily cultured on baker's yeast; a few grams of yeast left unattended in a collembolan culture will soon become infested with acarids. We have found astigmatine mites as contaminants in some Tullgren funnel extractions. When fresh agricultural products are stored in the laboratory, large populations of Astigmatina may develop, and may wander into Tullgren funnels during extractions. Similar population excursions of prostigmatid mites (family Cheyletidae) have also yielded extensive contamination of Tullgren samples. It is good practice to operate some empty "control" funnels to check for the possibility of wandering microarthropods in the funnel room.

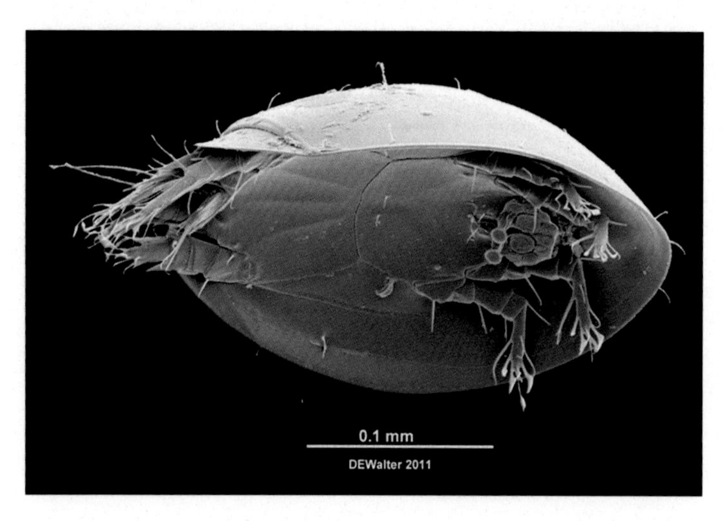

FIGURE 4.27 An astigmatine mite. *Source: Photo by D. Walter. From macromite.wordpress.com.*

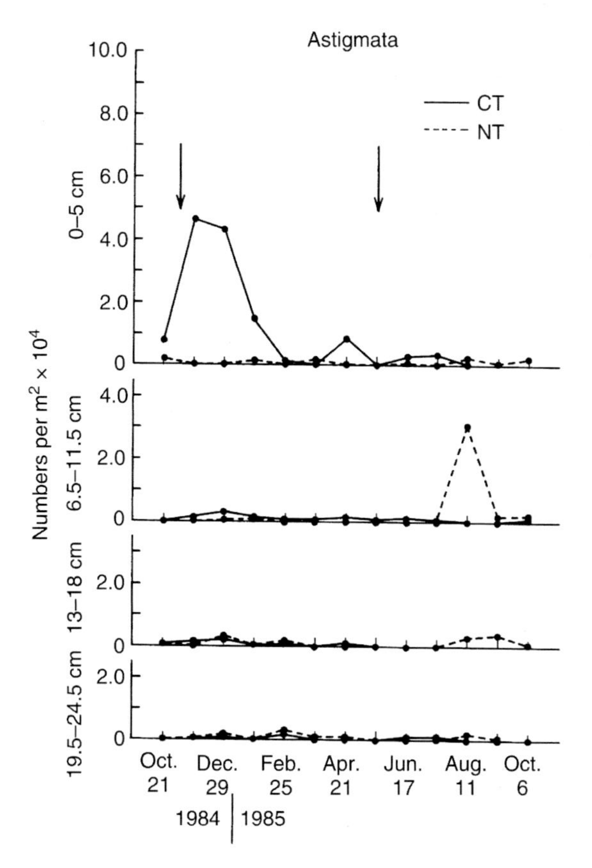

FIGURE 4.28 Vertical distribution of astigmatine mites in conventional and no-tillage agroecosystems. Arrows indicate autumn and spring dates for mowing, tillage, and planting. Numbers increased under conventional tillage following autumn tillage, but not following spring tillage. *Source: From Perdue, J.C., Crossley Jr., D.A., 1989. Seasonal abundance of soil mites (Acari) in experimental agroecosystems: effects of drought in no-tillage and conventional tillage. Soil Tillage Res. 15, 117–124.*

4.3.6 Other Microarthropods

In addition to mites and collembolans, Tullgren extractions contain a diverse group of other small arthropods. Although not numerous in comparison to mites and collembolans, Tullgren extractions may have abundances of several thousand "other" microarthropods per square meter. Collectively, these microarthropods have relatively small biomass, and probably have no major impact on soil ecology, but such a judgment may be premature, given the general lack of information about their ecology.

Small spiders and centipedes, occasional small millipedes, insect larvae, and adult insects occur in soil cores extracted on Tullgren funnels. Most of these are better sampled as macroarthropods using hand-sorting or trapping methods. Some insects (small larvae of carabid and elaterid beetles, thrips, pselaphid beetles, and tiny wasps, for example) are sometimes numerous enough to be effectively sampled from soil cores. Social insects, such as ants and termites, require special sampling considerations.

4.3.6.1 *Protura*

Proturans (Fig. 4.29) are small, wingless, primitive *arthropods* readily recognized by their lack of antennae and eyes (Bernard, 1985). Seldom as numerous as other microarthropods, proturans occur in a variety of soils worldwide, often associated with plant roots and litter. Keys to families and genera of the Protura were published by Copeland and Imadaté (1990) and by Nosek (1973).

Numbers reported in the literature range between 1000 and 7000 per square meter at best (Petersen and Luxton, 1982). They penetrate the soil to surprising depths (25 cm), considering that they do not appear to be adapted to burrowing (Copeland and Imadaté, 1990; Price, 1975). Their feeding habits remain unknown. Observations that they feed on mycorrhizae (Sturm, 1959) have not been verified, but their occurrence in the rhizosphere of trees with mycorrhizae would support Sturm's observations.

4.3.6.2 *Diplura*

Diplurans are small, elongate, delicate, primitive arthropods. They have long antennae and two abdominal cerci. Most diplurans are euedaphic, but some are nocturnal cryptozoans, hiding under stones or tree bark during the day. They occur in tropical and temperate soils in low densities. In Georgia Piedmont agroecosystems, the authors have sampled dipluran populations with Tullgren extractions, finding populations of approximately 50 per square meter.

Two common families of diplurans are found in soils, readily separated by their abdominal cerci. Campodeidae (Fig. 4.30) have filiform cerci; Japygidae (Fig. 4.31) have sclerotized cerci modified into pincers. Keys to families and subfamilies were provided by Ferguson (1990a). The japygids are predators on small arthropods (such as collembolans), nematodes, and enchytraeids. The cerci are used in capturing prey. Campodeids are predators on mites and other small arthropods, but also ingest fungal mycelia and detritus (Ferguson, 1990a). Also see Allen (2003) for information on two new species of epigean Litocampa (Insecta: Diplura: Campodeidae) from the Southeastern Appalachians. These animals are adapted for life in the soil by their elongate narrow form, sensory antennae, and sensory cerci.

FIGURE 4.29 Proturan. *Source: Photo by A. Murray.*

FIGURE 4.30 Campodeidae. *Source: Photo by A. Murray.*

FIGURE 4.31 Japygidae. *Source: Photo by A. Murray.*

4.3.6.3 Microcoryphia

The jumping bristletails (family Machilidae) were formerly included in the order Thysanura with silverfish and relatives, but now are placed in the separate order Microcoryphia. They are closely related to one another (Ferguson, 1990b). Machilids, when disturbed, can leap a distance of 10 cm (Denis, 1949). They feed on a variety of primitive plant materials such as lichens and algae. We have observed machilids (presumably *Machilis* sp.) on rocky cliff faces in Georgia. They emerge at dusk from cracks in the rock surface on Stone Mountain and other granite domes on the Georgia Piedmont, and may reach densities of 50 per square meter. They return to their crevices at dawn. They fall prey to spiders (*Pardosa lapidocina*) on these outcrops (Nabholz et al., 1977).

4.3.6.4 Pseudoscorpionida

Pseudoscorpions (Fig. 4.32) are minute copies of their more familiar relatives, the scorpions, except that they lack tails and stingers. They occur throughout the terrestrial world except for Arctic and Antarctic regions. Pseudoscorpions are small cryptozoans, hiding under rocks and tree bark, but they are occasionally extracted from leaf litter samples in

FIGURE 4.32 Pseudoscorpion. *Source: Photo by A. Murray.*

FIGURE 4.33 Symphylid. *Source: Photo by A. Murray.*

Tullgren funnels. They are predaceous on small arthropods, nematods, and enchytraeids. Keys to families and genera were provided by Muchmore (1990).

False scorpions are found in a number of habitats but not in large numbers. They can move readily through small spaces and crevices. Wallwork (1976) notes that two important habitat features for pseudoscorpions are high humidity and the availability of small crevices. Forest litter provides both of these features, as do bark of decomposing logs, caves, nests of small mammals and similar habitats.

Hand collecting is successful, but Tullgren extraction is the usual means of sampling pseudoscorpions (Hoff, 1949; Muchmore, 1990).

4.3.6.5 *Symphyla*

Symphylids (Fig. 4.33) are small, white, eyeless, elongate, many-legged invertebrates that resemble tiny centipedes. They differ from centipedes in several characteristics, but superficially symphylids have only 12 body segments and 12 pairs of legs, whereas centipedes have at least 15 pairs of legs (with the first pair modified into fangs). Edwards (1990) provides a partial key to genera, and notes that the North American fauna of symphylids is badly in need of further revision. Symphylids are part of the true euedaphic fauna, occurring in forest, grassland, and cultivated soils. They are omnivorous and can feed on the soft tissues of plants or animals (Edwards, 1959). Some species reach pest status in greenhouse soils where they feed on roots of seedlings (Edwards, 1990). Symphylids have silk glands near the end of the abdomen. The function of the silk strands for these soil dwellers is obscure.

4.3.6.6 Pauropoda

Pauropods (Fig. 4.34) are tiny (1.0–1.5 mm long) terrestrial myriapods with 8–11 pairs of legs and a distinctive morphological feature—branched antennae (Scheller, 1988). They are white to colorless and blind; these characteristics make them members of the true euedaphic fauna. Pauropods occur in soils worldwide but are not well known. They are commonly collected in Tullgren extractions but are seldom numerous, usually fewer than 100 per square meter. In forests, they inhabit the lower litter layers, F-layers, and mineral soil; they also occur in agricultural soils. It is generally assumed that pauropods are fungus feeders, but they may also be predaceous. Little information has been accumulated about their biology or ecology (Scheller, 1990). The taxonomy of the group is in need of revision. Although not considered particularly diverse, probably less than 20% of extant species have been described. For example, Scheller (2002) working in the Great Smoky Mountains National Park has found more than 30 species of pauropods previously undescribed in that region, with seven or eight of these entirely new to science.

4.3.7 Enchytraeidae

In addition to earthworms (discussed in the next section), another important family of terrestrial oligochaete is the Enchytraeidae. This group of small, unpigmented worms (Fig. 4.35), also known as "pot-worms," is classified within the "microdrile" oligochaetes and consists of some 710 valid species in 33 genera (Schmelz and Collado, 2015). Species from 19 of these genera are found in soil, the remainder occurring primarily in marine and freshwater habitats (Table 4.8) (Schmelz and Collado, 2015; van Vliet et al., 2012). The Enchytraeidae are thought to have arisen in cool temperate climates where they are commonly found in moist forest soils rich in organic matter; interestingly, Tynen (1972) described the occurrence of "ice worms," which were enchytraeids that emerged onto

FIGURE 4.34 Pauropoda. *Source: Photo by A. Murray.*

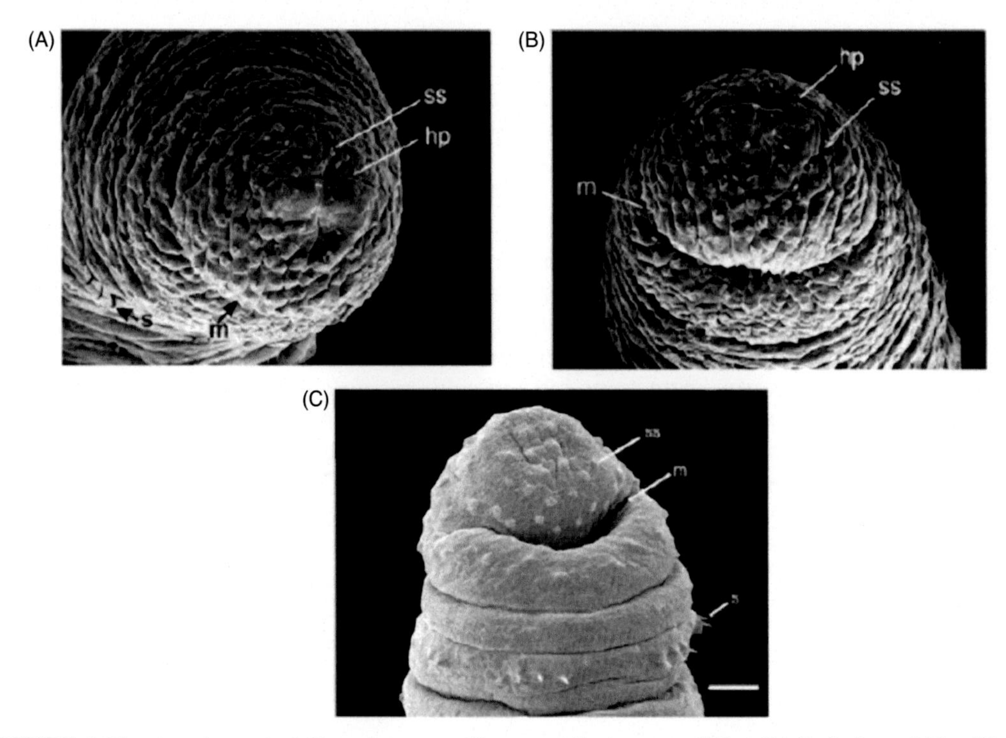

FIGURE 4.35 Anterior end of *Mesenchytraeus solifugus* and *Enchytraeus albidus*. (A) End view of *M. solifugus* displaying head pore (hp), and sensory structures (ss) at tip of prostomium. The mouth (m) and one cluster of setae (s) are visible in the lower region of the image. (B) Ventral view of *M. solifugus* displaying mouth, head pore, and sensory structures. (C) Ventral view of *E. albidus* displaying mouth, sensory structures, and four setal clusters. Scale bar = 50 μm. *Source: From Shain et al., 2000.*

snow- and ice-covered ground in British Columbia. Various species of enchytraeids are now distributed globally from subarctic to tropical regions.

Taxonomic organization of the European Enchytraeidae was definitively treated by Nielsen and Christensen (1959a, 1959b, 1961, and 1963); and there has been a growing body of work describing new species and developing phylogenies and taxonomies for Enchytraeidae around the world (Schmelz and Collado, 2012, 2015). From this work it is clear that there is much more discovery to be made in terms of describing the true extent of enchytraeid diversity globally. Identification of adult enchytraeids is difficult, but genera may be identified by observing internal structures through the transparent body wall of specimens mounted on slides (Fig. 4.36).

The Enchytraeidae are typically 10–20 mm in length and they are anatomically similar to the earthworms, except for the miniaturization and rearrangement of features overall. They possess setae (with the exception of one genus), and a clitellum in segments XII and XIII, which contains both male and female pores. Sexual reproduction in enchytraeids is hermaphroditic and functions similarly to that of earthworms. Cocoons may contain one or more eggs and maturation of newly hatched individuals ranges from 65 to 120 days, depending on species and environmental temperatures. Enchytraeids also display asexual

TABLE 4.8 Enchytraeid Genera and Their Occurrence in Various Environments

Genera Occurring in Soil	Other Genera	Environment
Achaeta	Aspidodrilus	Epizoic on earthworms
Bryodrilus	Barbidrilus	Freshwater
Buchholzia	Enchylea	Only found in Enchytraeid culture
Cernosvitoviella	Enchytraeina	Marine
Chamaedrilus	Grania	Marine
Enchytraeus	Pelmatodrilus	Epizoic on earthworms
Enchytronia	Propappus	Freshwater
Euenchytraeus	Randidrilus	Marine
Fridericia	Stephensoniella	Marine
Guaranidrilus		
Hemienchytraeus		
Hemifridericia		
Henlea		
Isosetosa		
Lumbricillus		
Marionina		
Mesenchytraeus		
Oconnorella		
Stercutus		
Tupidrilus		

From van Vliet, P.C.J., 2000. Enchytraeids. In: Sumner, M. (Ed.), Handbook of Soil Science. CRC Press, Boca Raton, FL, pp. C70–C77. Taxonomy per Schmelz, R.M., Collado, R., 2015. Checklist of taxa of Enchytraeidae (Oligochaeta): an update. Soil Organ. 87, 149–152.

strategies of parthenogenesis and fragmentation, which enhance their probability of colonization of new habitats (Dósza-Farkas, 1996).

Enchytraeids ingest both mineral and organic particles in the soil, although typically of smaller size ranges than those ingested by earthworms. Numerous investigators have noted that finely divided plant materials, often enriched with fungal hyphae and bacteria, are a principal portion of the diet of enchytraeids; microbial tissues are probably the fraction most readily assimilated, because enchytraeids lack the gut enzymes to digest more recalcitrant soil organic matter (Brockmeyer et al., 1990; van Vliet et al., 2012). Didden (1990, 1993) suggested that enchytraeids feed predominantly on fungi, at least in arable soils, and classified a community as 80% microbivorous and 20% saprovorous. As with several other members of the soil mesofauna, the mixed microbiota that occur on decaying organic matter, either litter or roots, are probably an important part of the diet of these

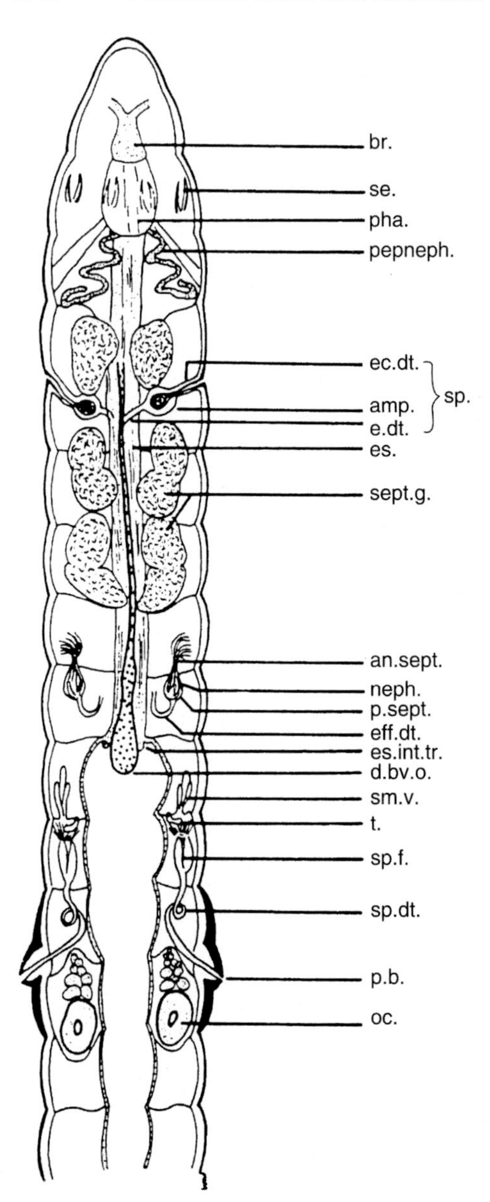

FIGURE 4.36 Morphological characters of an enchytraeid worm. *amp.*, Ampulla, *an. sept.*, ante-septal; *br.*, brain; *d.bv.o.*, dorsal blood vessel origin; *ec.g.*, ectal gland; *eff.dt.*, efferent duct; *e.op.*, ental opening; *es.*, esophagus; *es.int.tr.*, esophageal intestinal transition; *m.pha.*, muscular pharynx; *neph.*, nephridia; *oc.*, oocyte; *pha.*, pharynx; *p.b.*, penial bulb; *pepneph.*, peptonephridia; *p.sept.*, postseptal; *se.*, setae; *sept.g.*, septal gland; *sm.v.*, seminal vesicle; *sp.*, spermatheca; *sp.dt.*, sperm duct; *sp.f.*, sperm funnel; *t.*, testes. *Source: Dindal, D.L., 1990. Soil Biology Guide. John Wiley & Sons, New York, NY. Reprinted by permission of John Wiley & Sons.*

creatures. The remaining portions of the soil organic matter, after the processes of ingestion, digestion, and assimilation, enter the slow-turnover pool of soil organic matter. Zachariae (1963, 1964) studied the nature of enchytraeid feces and found that they had no identifiable cellulose residues. In addition, Zachariae suggested that the so-called collembolan soil, which is said to be dominated by collembolan feces (particularly low pH mor soils) were really formed by the action of enchytraeids. Mycorrhizal hyphae have been found in the fecal pellets of enchytraeids from pine litter (Fig. 4.37). There is also the strong likelihood

FIGURE 4.37 An enchytraeid wom: *Euenchytraeus clarae*. *Source: Photo by* Damiano Zanocco.

that enchytraeids consume and further process larger fecal pellets and castings of soil fauna such as collembolans and earthworms (Rusek, 1985; Zachariae, 1964).

Enchytraeid densities range from less than 1000 to more than 140,000 individuals per square meter in intensively cultivated agricultural soil in Japan and a peat moor in the United Kingdom, respectively (Table 4.9). In a subtropical climate, Coleman et al. (1994a) reported enchytraeid densities of 4000–14,000 per square meter in agricultural plots on the Piedmont of Georgia, whereas van Vliet et al. (1993) found higher densities (20,000–40,000 per square meter) in surface layers of deciduous forest soils in the southern Appalachian Mountains of North Carolina. Although enchytraeid densities are typically highest in acid soils with high organic content, Didden (1995) found no statistical relationship between average density and annual precipitation, annual temperature, of soil pH over a broad range of data; local variability may be at least as great as variation on a wider scale. Enchytraeid densities show both spatial and seasonal variations. Vertical distributions of enchytraeids in soil are related to organic matter horizonation; up to 90% of populations may occur in the upper layers in forest and no-tillage agricultural soils, but densities may be higher in the Ah horizon of grasslands (Davidson et al., 2002). Seasonal trends in enchytraeid abundance appear to be associated with moisture and temperature regimes (van Vliet et al., 2012).

Enchytraeids have been shown to have significant effects on organic matter dynamics in soil and on soil physical structure. Litter decomposition and nutrient mineralization are influenced primarily by interactions with soil microbial communities. Enchytraeid feeding on fungi and bacteria can increase microbial metabolic activity and turnover, accelerate release of nutrients from microbial biomass, and change species composition of the microbial community through selective grazing. However, Wolters (1988) found that enchytraeids decreased mineralization rates by reducing microbial populations and possibly by occluding organic substrates in their feces. Thus the influence of enchytraeids on soil organic matter dynamics is the net result of both enhancement and inhibition of microbial activity, depending on soil texture and population densities of the animals (van Vliet et al., 2012; Wolters, 1988).

Enchytraeids affect soil structure by producing fecal pellets which, depending on the animal size distribution, may enhance aggregate stability in the 600–1000 μm fraction (Didden, 1990). In forest floors, these pellets are composed mainly of fine humus particles, but in

TABLE 4.9 Enchytraeid Abundances (annual average number/m^2) in Different Ecosystems and Locations

Ecosystem	Location	Density (no. m^{-2})
FOREST		
Douglas fir	Wales	134,300
Pinus radiata 50 stems/ha	New Zealand	64,002
Pacific silver fir, mature stand	WA, United States	49,400
Pinus radiata 200 stems/ha	New Zealand	39,270
Spruce	Norway	34,700
Rhododendron-oak 1160 m altitude	NC, United States	32,630
Rhododendron-oak 750 m altitude	NC, United States	26,811
Pine	Norway	22,900
Pinus radiata 100 stems/ha	New Zealand	21,391
Scots pine forest	Sweden	16,200
Deciduous forest	United Kingdom	14,590
Spruce	South Finland	13,400
Pacific silver fir, young stand	WA, United States	11,400
Pinus radiata 0 stems/ha	New Zealand	10,647
Spruce	South Finland	8,200
Spruce	North Finland	4,000
ARABLE LAND		
Sugarbeet	The Netherlands	30,000
Winterwheat	The Netherlands	19,437
NT corn-clover	GA, United States	16,830
CT corn-clover	GA, United States	15,270
Potato field	Poland	13,200
Barley, no N	Sweden	10,000
Rye field	Poland	9,800
Barley, 120 kgN	Sweden	8,100
Rice/wheat/barley (organic)	Japan	4,940
Rice/wheat/barley (conven.)	Japan	525
MOOR		
Juncus peat	United Kingdom	145,000
Nardus	United Kingdom	71,000

(*Continued*)

TABLE 4.9 (Continued)

Ecosystem	Location	Density (no. m^{-2})
Blanket bog	United Kingdom	40,000
Fen	Canada	5,600
GRASSLAND		
Grassland soil	Sweden	24,000
Lucerne ley	Sweden	9,900
Grassland 10 sheep/ha	Australia (NSW)	6,000
Grassland 30 sheep/ha	Australia (NSW)	2,300

Modified from van Vliet, P.C.J., 2000. Enchytraeids. In: Sumner, M. (Ed.), Handbook of Soil Science. CRC Press, Boca Raton, FL, pp. C70–C77.

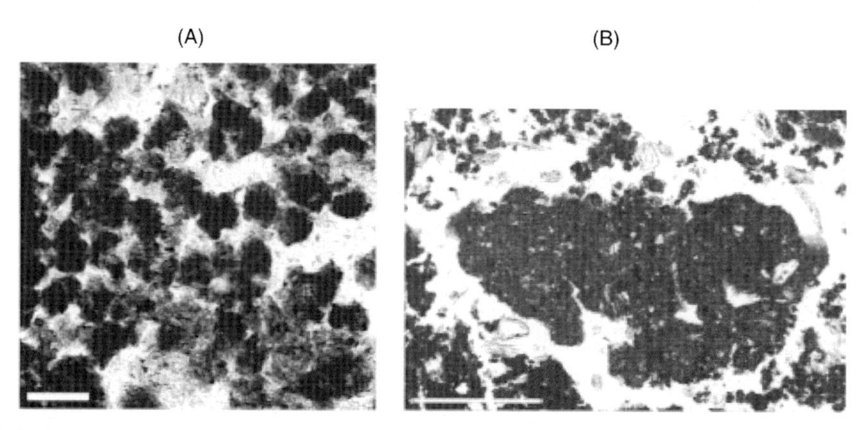

FIGURE 4.38 Thin section micrographs of fecal pellets in a grassland soil. (A) Derived from enchytraeids (scale bar = 0.5 mm) and (B) derived from earthworms (scale bar = 1.0 mm). *Source: From Davidson, D.A., Bruneau, P.M.C., Grieve, I.C., Young, I.M., 2002. Impacts of fauna on an upland grassland soil as determined by micromorphological analysis. Appl. Soil Ecol. 20, 133–143.*

mineral soils, organic matter and mineral particles may be mixed into fecal pellets with a loamy structure (Kasprzak, 1982). Davison et al. (2002) estimated that enchytraeid fecal pellets constituted nearly 30% of the volume of the Ah horizon in a Scottish grassland soil (Fig. 4.38). Encapsulation or occlusion of organic matter into these structures may reduce decomposition rates. Burrowing activities of enchytraeids have not been well studied, but there is evidence that soil porosity and pore continuity can increase in proportion to body size (50–200 μm in diameter) (Didden, 1990; Rusek, 1985). Van Vliet et al. (1993) observed that enchytraeids in small microcosms increased soil porosity and hydraulic conductivity, depending on the distribution of organic matter and enchytraeid population density.

Enchytraeids are typically sampled in the field using cylindrical soil cores of 5–7.5 cm diameter; large numbers of replicates may be needed for a sufficient sampling due to the clustered distribution of enchytraeid populations (van Vliet et al., 2012). Extractions are often done with a wet-funnel technique (O'Connor, 1955), similar to the Baermann funnel extraction method used for nematodes. In this case, soil cores are submerged in water on the funnel and exposed for several hours to a heat and light source from above; enchytraeids move downward and are collected in the water below (see van Vliet et al., 2012, for a comparison of modifications of this technique).

4.4 THE MACROFAUNA

4.4.1 Macroarthropods

Larger insects, spiders, myriapods, and others are considered together under the appellation "macroarthropods." Typical body lengths range from about 10 mm to as much as 15 cm or more (the largest scolopendromorph centipedes) (Shelley, 2002). The group includes an artificial mix of various arthropod classes, orders, and families. Like the microarthopods, the macroarthropods are defined more by the methods used to sample them rather than by measurements in body size.

Large soil cores (10 cm diameter or greater) may be appropriate for euedaphic species; arthropods can be recovered from them by using flotation techniques (Edwards, 1991). Mechanical or hand-sorting of soils and litter is more time-consuming, but yields better estimates of population size. In rare instances, capture-mark-recapture methods have been used to estimate population sizes of selected macroarthropod species, but the assumptions for this procedure are violated more often than not (Southwood, 1978).

Pitfall traps have been widely used to sample litter- and surface-dwelling macroarthopods (Banerjee, 1970; Greenslade, 1964; Mikhail, 1993) (see Chapter 9: Laboratory and Field Exercises in Soil Ecology). This method catches arthropods that blunder into cups filled with preservative. Absolute population estimates are difficult to obtain with pitfall traps (Gist and Crossley, 1973), but the method yields comparative estimates when used with caution.

Many of the macroarthropods are members of the group termed "cryptozoa," a group consisting of animals that dwell beneath stones, under bark, or in cracks and crevices (Cole, 1946). Cryptozoans typically emerge at night to forage; some are attracted to artificial lights. The cryptozoan fauna is poorly defined and is not an ecological community in the usual sense of the term. The concept remains useful, however, for identifying a group of invertebrate species with similar patterns of habitat utilization.

4.4.1.1 Importance of the Macroarthopods

However they are sampled, the macroarthropods are a significant component of soil ecosystems and their food webs. Macroarthropods differ from their smaller relatives in that they may have direct effects on soil structure. Termites and ants in particular are important movers of soil, depositing parts of lower strata on top of the litter layer (Fig. 4.39). Emerging nymphal stages of cicadas may be numerous enough to influence soil structure. Larval stages of soil-dwelling scarabaeid beetles sometimes churn the soil in grasslands. These and other macroarthropods are part of the complex that has been termed "ecosystem engineers" (Jones et al., 1994).

FIGURE 4.39 A fire ant (*Solenopsis invicta*) mound. *Source: Photo by M. Taylor.*

Some macroarthropods participate in both above- and belowground parts of terrestrial ecosystems. Many macroarthropods are transient or temporary soil residents (see Fig. 4.1), and thus form a connection between food chains in the "green world" of foliage and the "brown world" of the soil. Caterpillars descending to the soil to pupate or migrating armyworm caterpillars are prey to ground-dwelling spiders and beetles. A ground beetle species *Calosoma sycophanta* ("the searcher") was imported from Europe for biological control of the gypsy moth (Kulman, 1974).

Macroarthropods may have a major influence on the microarthropod portion of belowground food webs. Collembola, among other microarthropods, are important food items for spiders, especially immature stadia, thus providing a macro- to microconnection. Other macroarthropods such as cicadas emerging from soil may serve as prey for some vertebrate animals (Lloyd and Dybas, 1966), thus providing a link to the larger megafauna.

Among the macroarthropods there are many litter feeding species, such as the millipedes, that are important consumers of leaf, grass, and wood litter. These arthropods have major influences on the decomposition process, thereby impacting rates of nutrient cycling in soil systems. Further, the reduction, consumption, and even burial of vertebrate carrion are largely accomplished through the actions of soil-dwelling insects (Payne, 1965).

The vast array of macroarthropod species in soil systems constitutes a major reservoir of biological diversity. As with the mites and collembolans, the functional significance of this diversity is not immediately evident, and its effects on process-level phenomena are complex and often indirect. However, it is becoming clear that on a global scale, loss of biodiversity (and functional diversity) from decomposer communities results in changes to the decomposition function in both aquatic and terrestrial ecosystems (Handa et al., 2014).

4.4.1.2 Isopoda

Terrestrial isopods (Fig. 4.40) are crustaceans, but are typical cryptozoans, occurring under rocks, logs, and in similar habitats. Although they are distributed in a variety of habitats, including deserts, they are susceptible to desiccation. Adaptations to resist desiccation include nocturnal habits, the ability to roll up into a ball, low basal respiration rates, and restriction of respiratory surfaces to specialized areas. Considered to be general saprovores, isopods can feed upon roots or foliage of seedlings. Isopods possess heavy, sclerotized mandibles, and are capable of considerable fragmentation of

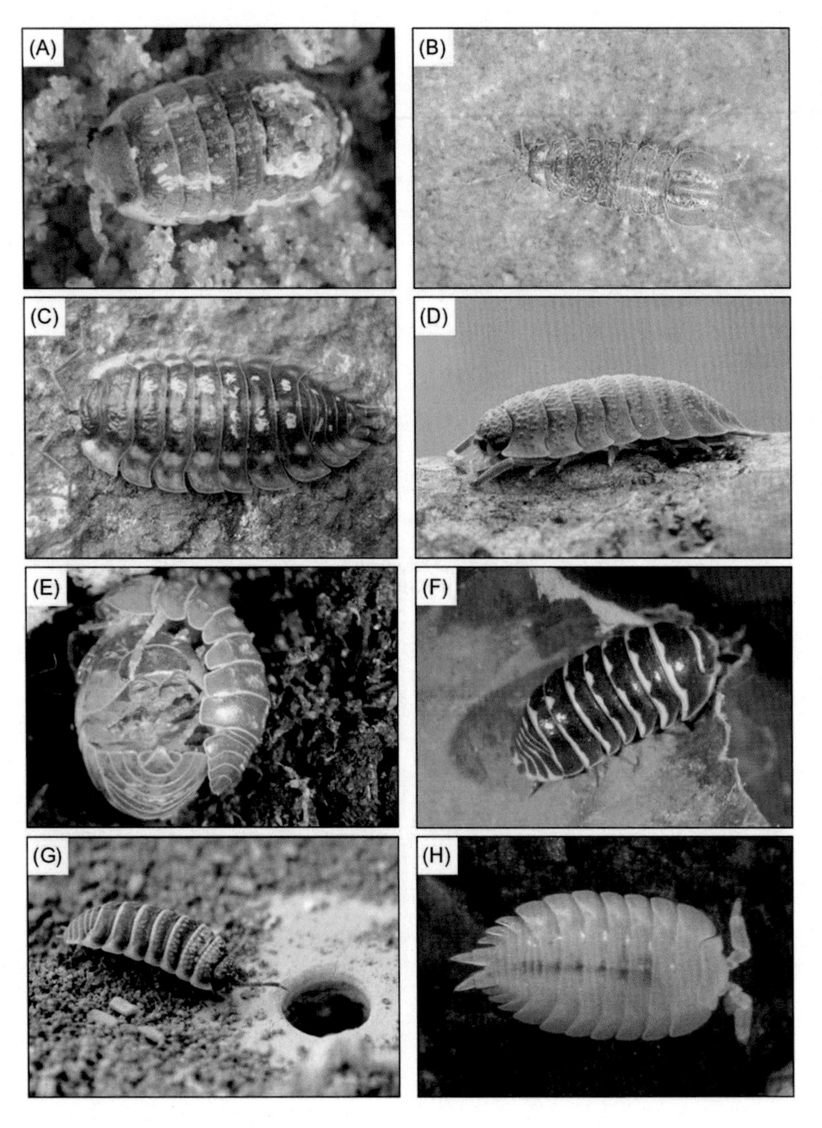

FIGURE 4.40 Diversity of terrestrial isopods from littoral to desert habitats. (A) The semi-terrestrial *Tylos europaeus*, inhabiting the littoral zone; (B) the freshwater isopod *Asellus aquaticus*; (C) the common woodlouse *Oniscus asellus*; (D) the rough woodlouse *Porcellio scaber*; (E) the pill-bug *Armadillidum vulgare* during mating; (F) the "zebra pillbug" *Armadillidium maculatum*; (G) the desert isopod *Hemilepistus reaumuri* near the entrance of a burrow in Tunisia; and (H) the ant woodlouse *Platyarthrus hoffmannseggii*. This species is blind and lives in ant nests. *Source: Picture credit: Martin Zimmer (B), Sören Franzenburg (D), Giuseppe Montesanto (G), UMR CNRS 7267 (A, C, E, F, H). From Boushon et al. (2016).*

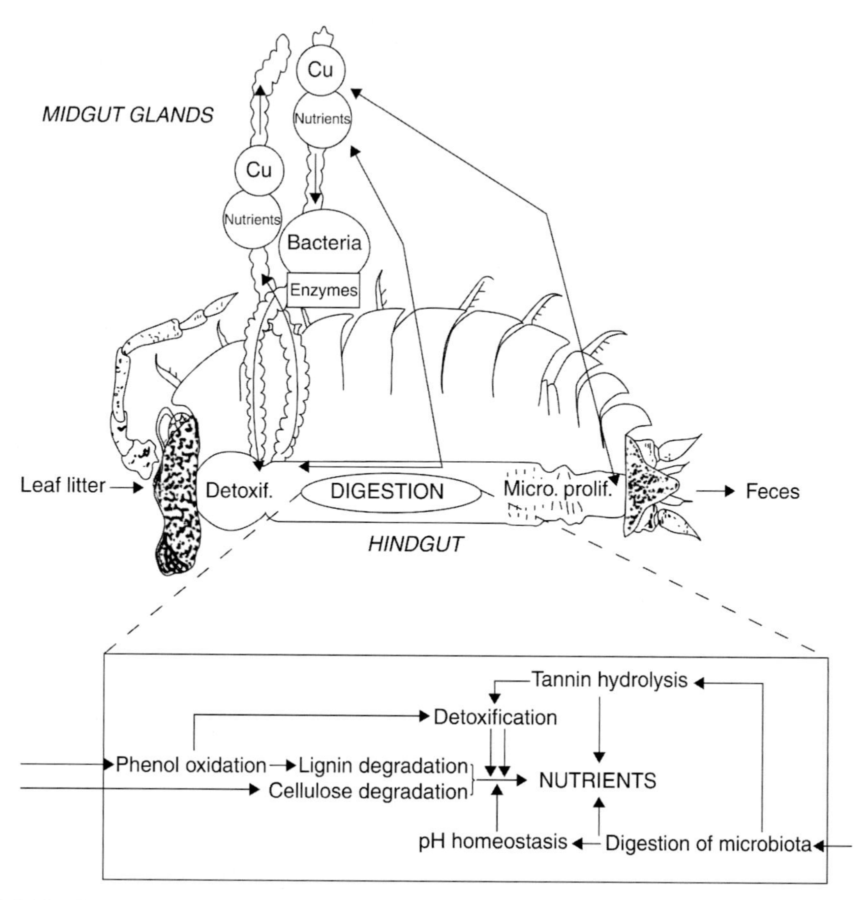

FIGURE 4.41 Digestive processes in the hindgut of *Porcellio scaber* (Porcellionidae), including the ingestion of leaf litter, detoxification of ingestion phenolics (Detoxif.) in the foregut, digestion in the anterior hindgut through the activity of endogenous and bacterial enzymes, adsorption of nutrients and copper, microbial proliferation (Microb. Prolif.) in the posterior hindgut, and egestion of feces. *Source: From Zimmer, M., 2002. Nutrition in terrestrial isopods (Isopoda: Oniscidea): an evolutionary-ecological approach. Biol. Rev. 77, 455–493.*

decaying vegetable matter. They display some selectivity in preferences for different leaf species. Digestive processes in the terrestrial isopods encompass a wide extent of biochemical complexity, with detoxification of ingested phenolics in the foregut, digestion by endogenous and bacterial enzymes in the anterior hindgut, absorption of nutrients, and microbial proliferation in the posterior hindgut (Fig. 4.41) (Zimmer, 2002). In the laboratory, terrestrial isopods feed upon fecal pellets dropped by themselves or by any other isopod (Zimmer, 2002). There is some doubt about how common this trait is expressed in the field due to the difficulty of finding feces beneath the litter layer. However, microbially inoculated feces represent a microbial "hot spot" generating microbial metabolites that might allow the isopod to "home in" on a desirable food source (Zimmer et al., 1996).

The isopod *Porcellio* has an excretory system that exposes its products to the external environment. Urine from the nephridia is channeled into a water-conducting system on the ventral surface. Ammonia is lost to the atmosphere and oxygen is absorbed during this flow. The ammonia-free water is then reabsorbed in the rectum (Eisenbeis and Wichard, 1987).

4.4.1.3 Diplopoda

Millipedes (Diplopoda) (Fig. 4.42) are a group of widely distributed saprophages. They are major consumers of organic debris in temperate and tropical hardwood forests, where they feed on dead vegetable matter. Millipedes are also inhabitants of arid and semiarid regions, despite their dependence on moisture. Millipedes lack a waxy layer on their epicuticle and are subject to rapid desiccation in environments with low relative humidity. Some are true soil forms, others seem restricted to leaf litter or to cryptozoan habitats. They can be loosely grouped into (1) tubular, round-backed forms such as the familiar *Narceus*; (2) flat-backed forms (many polydesmid millipedes); and (3) pillbug types, which roll into a ball. Millipedes range widely in length. Typical North American forms are 5–6 cm in length; tropical ones may reach nearly 20 cm in length. Hoffman (1999) has published a checklist of millipedes of North and Central America. Keys to North American families of millipedes were published by Hoffman (1990). For an account of millipede biology and ecology, see Hopkin and Read (1992).

Millipedes become abundant in calcium-rich, high rainfall areas in tropical and temperate zones. The Southern Appalachian Mountains of the eastern United States support a large millipede population. Millipedes can be important in calcium cycling. They have a calcareous exoskeleton, and because of their high densities they can be a significant sink for calcium. Millipedes are major consumers of fallen leaf litter, and may process some 15%–20% of calcium input into hardwood forest floors. In desert areas, millipedes are active following rains, especially in desert shrub communities. They avoid hot, dry

FIGURE 4.42 (A) *Narceus americanus* (photo R. Carerra-Martínez) and (B) *Polydesmus* sp.; and (C) Spirobolidae. *Source: Photos B.A. Snyder.*

conditions by concealment under vegetation or debris (Crawford, 1981). Millipedes are vulnerable to desiccation, because their cuticle generally lacks a water-proof layer, their gas exchange system is not closed, and they lose a considerable amount of water through the mouth, in defecation, and during reproduction (Wolter and Eckschmitt, 1997).

Millipedes appear to be selective feeders, avoiding leaf litter high in polyphenols and favoring litter with high calcium content (Neuhauser and Hartenstein, 1978). Freshly fallen leaves are generally avoided, even though assimilation efficiency is much higher from that source (David and Gillon, 2002). Some millipedes are obligate coprophages. When McBrayer (1973) cultured millipedes in containers, which excluded their feces, the millipedes lost weight. When a small tray containing feces was added to the cultures, the millipedes consumed it and prospered. Such obligate coprophagy indicates a close relationship with bacteria necessary for digestion of vegetable material. It is not known whether millipedes possess a unique gut flora of microbes.

4.4.1.4 Chilopoda

Centipedes (Chilopoda) are common predators in soil, litter, and cryptozoan habitats (Fig. 4.43). They are all elongate, flattened, and highly vagile forms. Centipedes occur in biomes ranging from forest to desert. The large desert centipedes (Scolopendromorpha) are some 15 cm in length; tropical centipedes may exceed 30 cm (Shelley, 2002). Lithobiids are the common brown, flat centipedes of litter in hardwood forests. The elongate, slim geophilomorph centipedes are euedaphic in forest habitats where they prey on earthworms, enchytraeids, and Diptera larvae (Lock and Dekonink, 2001). Like the millipedes, centipedes lose water through their cuticles at low relative humidities. They avoid desiccation by seeking moist habitats, and by adjusting their diurnal activities to humid periods in desert and sand dune habitats.

Centipedes are distinguished from superficially similar organisms by the presence of forcipules, the modified first segment upon which the head rests (Edgecombe and Giribet, 2007; Mundel, 1990). This segment bears the pincerlike fangs, which have venom ducts opening at their tips. Five orders of centipedes are recognized, and Mundel (1990) provides keys to orders and families of centipedes of the world.

All centipedes are predators but may ingest some leaf litter on occasion—it can sometimes be seen in their guts. Centipedes are fast runners and actively pursue and capture small prey such as collembolans.

4.4.1.5 Scorpionida

The scorpion, the archetypal generalized arachnid with its long, segmented, stinger-bearing abdomen and chelate palpi, needs no description. It was obvious to the ancients—the only zodiacal sign bearing the name of a soil organism. Scorpions (Fig. 4.44) are inhabitants of warm, dry, tropical, and temperate regions, but reach their highest diversity in deserts. They are highly mobile predators of other arthropods and occasionally even of small vertebrates. The selection of prey items utilized is large; for one species of scorpion more than 100 different prey were recorded (Crawford, 1990). Different species demonstrate different patterns of predation, some being "sit and wait" predators and others acting as mobile hunters (Crawford, 1990). Scorpions are also cannibalistic to an unusual extent (Williams, 1987).

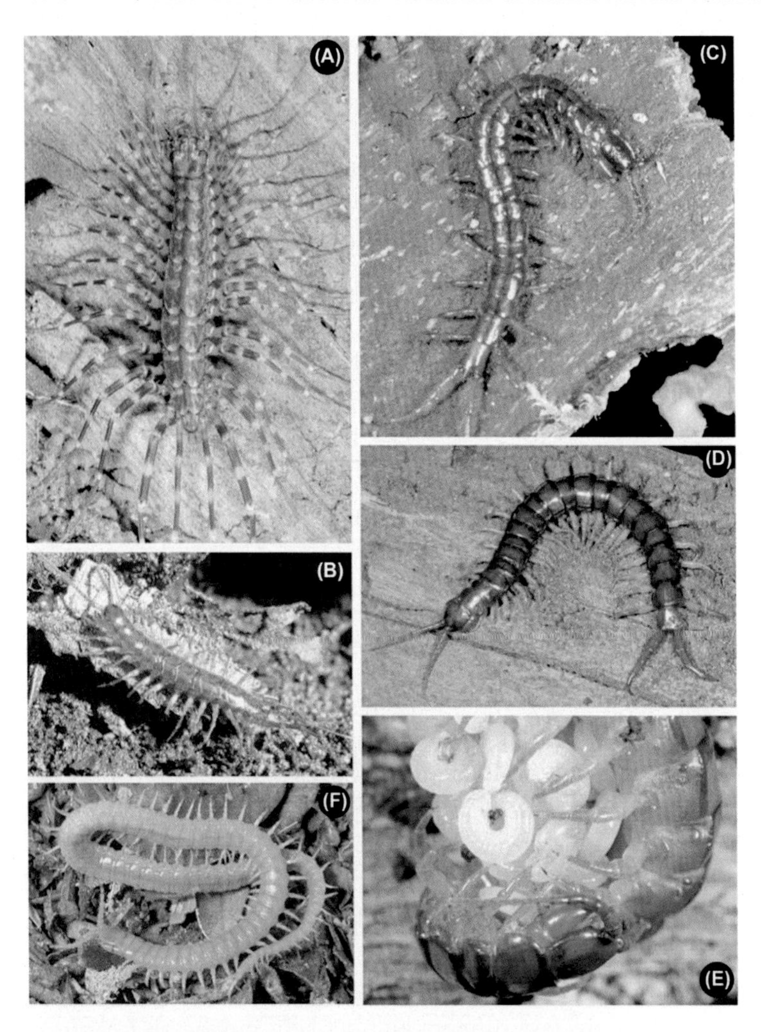

FIGURE 4.43 Life images of the five extant centipede orders. (A) Scutigeromorpha: *Parascutigera* sp., a scutigerid from Lane Poole Reserve (Western Australia). (B) Lithobiomorpha: *Cermatobius japonicus*, a henicopid from Mount Takao (Japan). (C) Craterostigmomorpha: *Craterostigmus* cf. tasmanianus from Mount Cook/Aoraki National Park (New Zealand). (D) Scolopendromorpha: *Cormocephalus aurantiipes*, a scolopendrid from Lane Poole Reserve (Western Australia). (E) Scolopendromorpha: *Cormocephalus hartmeyeri* from Porongurup National Park (Western Australia) brooding peripatoid stage hatchlings. (F) Geophilomorpha: *Zelanophilus provocator*, a geophilid from Hinewai Scenic Reserve (New Zealand). *Source: From Edgecombe, G. D., Giribet, G., 2007. Evolutionary biology of centipedes (Myriapoda: Chiolopoda). Annu. Rev. Entomol., 52, 151–170.*

FIGURE 4.44 A scorpion, *Vaejavis carolinensis*, showing its sting and chelate pedipalps. *Source: From Orgiazzi, A., Bardgett, R.D., Barrios, E., Behan-Pelletier, V., Briones, M.J.I., Chotte, J.-L., 2016. Global Soil Biodiversity Atlas. Publications Office of the European Union, Luxembourg ; Photo by M. Hedin, T. Iwane, and/or F. Krohn.*

Typical cryptozoans, scorpions hide under rocks or logs, or in crevices, during the day and emerge at night to feed. In the southeastern United States, scorpions may be trapped by placing wet cloth on the ground at night; the dampness will attract them. The scorpion cuticle will fluoresce under black light, offering a nighttime survey procedure.

Scorpions' stings are painful, about the same as a honeybee sting, but of shorter duration. In the southeastern United States, the tiny scorpion *Vejovis carolinus*, commonly found under the bark of pine stumps, invades houses on occasion. A species with similar habits, *Centruroides vittatus*, occurs from east Texas to the southwestern United States (Shelley and Sisson, 1995). Only a few species of scorpions are deadly. Of the 1500 species worldwide, only about 20–25 are dangerous, all in the family Buthidae. Where dangerous species occur, antivenom is usually available (Jackman, 1997).

The impact of scorpions on their ecosystems is unknown. They are not numerous, but in desert ecosystems they may be the dominant predators (Polis, 1991).

4.4.1.6 *Araneae*

Spiders (Araneae) (Fig. 4.45A–D) are another familiar group of carnivores. They are solitary hunters, exhibiting a range of strategies from "sit and wait" with silken webs to active pursuit of prey. They are found in all terrestrial environments except truly polar (Arctic/Antarctic) regions. Many species are found in aboveground habitats, but some are cryptozoans in litter and on the soil surface. Some small spiders are euedaphic. Some of the small litter-inhabiting spiders could be considered microarthropods. Wolf spiders (Lycosidae) (Fig. 4.45C) are common wandering predators in leaf litter and on soil surfaces, and are often captured in pitfall traps. They are conspicuous ground-dwelling predators in agroecosystems (Draney, 1997).

Like many other arthropod groups, the taxonomy and phylogeny of spiders is a dynamic discipline. There are 114 families and 3958 genera in the order, and like other groups, the higher-level relationships within the order are currently being untangled with molecular approaches (Garrison et al., 2016). Kaston's (1978) guide, *How to Know the Spiders*, is an excellent introduction for the novice arachnologist. Ubick et al. (2005) is invaluable for workers in the United States. Regional works such as Jackman (1997) should not be overlooked, nor should information available through various Internet outlets (e.g., the World Spider Catolog (http://wsc.nmbe.ch/)). Bradley (2013) offers keys to families of spiders together with color illustrations of common species.

Sampling methods for spiders run the gamut from Tullgren extraction to hand collection, sorting of litter samples, and pitfall trapping. Spiders have complex behavioral patterns including mating rituals and defense of territories. Wolf spiders may wander some distance between forest, meadow, and agroecosystems, so that assessment of population size may be complicated (Draney, 1997).

Little is known about the ecology of most of the smaller soil- and litter-dwelling spiders (Linyphiidae, Fig. 4.45 and relatives). Most information about spiders comes from studies of web-spinning species or the jumping spiders (Salticidae, Fig. 4.45B) in vegetation (Foelix, 1996). There are large numbers of small spiders in deciduous leaf litter (about 100 per square meter). Their habitat usage and prey selection are not well known. Spiders in laboratory microcosms show some prey selectivity, but leaf litter spiders doubtless feed more opportunistically.

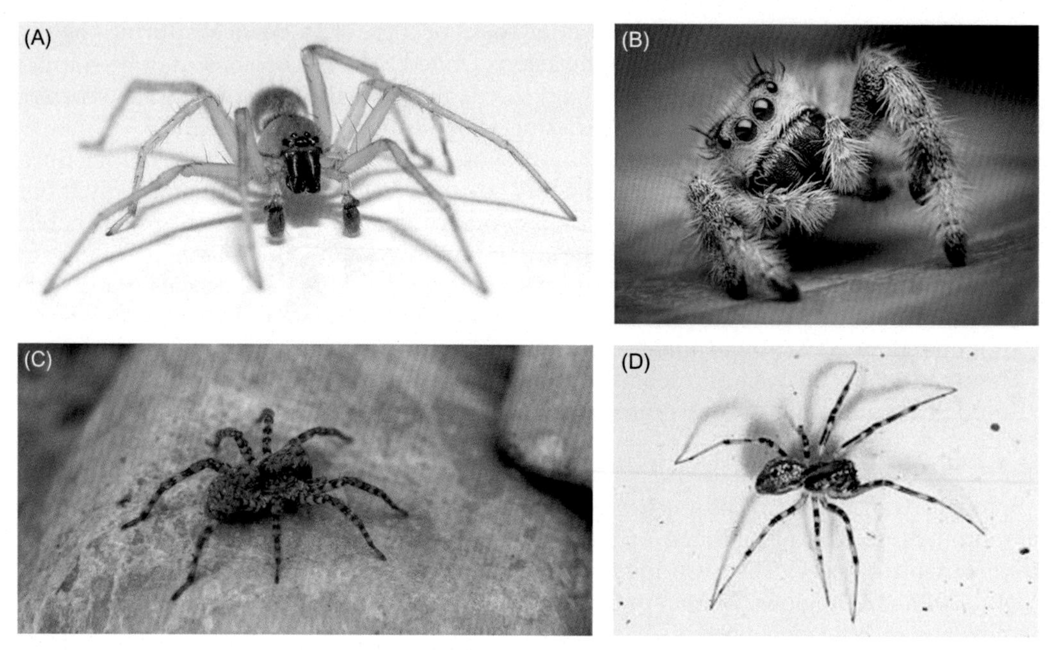

FIGURE 4.45 Spiders. (A) A Clubionid spider; (B) female jumping spider (Salticidae); (C) a wolf spider (Lycosidae); (D) a sheet web weaver (Linyphiidae). Source: *(A) André Karwat, published under a creative commons license and available at https://commons.wikimedia.org/wiki/File:Clubiona_sp._-_front_(aka).jpg. (B) Thomas Shahan, published uner a creative commons license and available at https://commons.wikimedia.org/wiki/File:Female_Jumping_Spider_-_Phidippus_regius_-_Florida.jpg. (C) OhWeh, published under a creative commons license and available at https://en.wikipedia.org/wiki/Pardosa_amentata#/media/File:Wolfsspinne_Arctosa_maculata_Weibchen_Mangfalltal-002.jpg. (D) Judy Gallagher, published under a creative commons license and avaiable at https://commons.wikimedia.org/wiki/File:Sheetweb_Spider_-_Tapinopa_bilineata,_Woodbridge,_Virginia.jpg*

A number of spider species are ant mimics (Foelix, 1996). These species copy the body shapes and colorings of ants, and move in an ant-like manner. The spider's front legs are elongate and thin, and mimic the antennae of the ant. The significance of this mimicry is not known, but it is suspected to confer some protection from predation by birds. Spiders that live together with ants seldom prey upon the ants (Jackson and Willey, 1994).

The impact of spiders on their ecosystems has not been thoroughly investigated, but some progress has been made. Their effectiveness as biological control agents has been discounted because of their slow reproduction rates. As noted earlier, spiders are strongly territorial, with complicated mating behaviors—adaptations that tend to hold down population sizes even when prey is abundant (Wise, 1993). In woodland forest floors, the large number of spiders argues that they must have an impact on insect populations there. In contrast, numbers of spiders in agricultural systems seems lower (Draney, 1997). In an experimental study, Lawrence and Wise (2000) found a "top-down" effect of spiders on litter decomposition rates. When spiders were removed from the experimental areas of a forest floor, collembolan populations increased. Subsequently, straw in litterbags decomposed more rapidly in those areas. These results suggest that spider predation may reduce collembolan populations enough to lower rates of litter disappearance on the forest floor.

OPILIONES

Harvestmen (Opiliones) are delicate, shy forms that are among the largest arachnids in woodlands. Their bodies are small but their legs may be unusually long, suggesting that their habitat is litter surface or exposed areas. Smaller, shorter-legged forms inhabit loose leaf litter or small spaces (Edgar, 1990). Others are inhabitants of caves, and have reduced eyes and reduced pigmentation (Goodnight and Goodnight, 1960). Opilionids have no venom glands, yet are considered to be largely predaceous. Some species occur high in foliage, others in subcanopy, some on the soil surface, and some (smaller forms) within the litter layer. Opilionids are slow reproducers, usually one generation per year. They are active predators in the daylight but seem to be primarily crepuscular (active at dawn and dusk). They possess repugnatorial glands, the secretion of which is offensive to predators (Blum and Edgar, 1971).

Opilionids (Opilio means "shepherd" in Latin) resemble the mites in that the cephalothorax (prosoma) and abdomen (opisthosoma) are broadly fused, so that the body is oval. They differ superficially from the mites in that opilionids have 6–10 segments in their abdomen; mites have none. Of course, they are larger than mites as well. Edgar (1990) provides a key to genera and species of North American Opiliones (exclusive of Mexico).

SOLIFUGAE

Solifugae, or solpugids, are desert arachnids with large distinctive curved chelicerae, often as long as the cephalothorax (Fig. 4.46). They are ferocious predators capable of rapid movement. Common names include sun-spiders, false-spiders, wind-spiders, or wind-scorpions (in reference to their rapid movement), or camel-spiders, among other names. The term "camel spider" refers to a prominent arch-shaped plate on the prosoma (Punzo, 1998). Solifugae occur in tropical and temperate deserts worldwide.

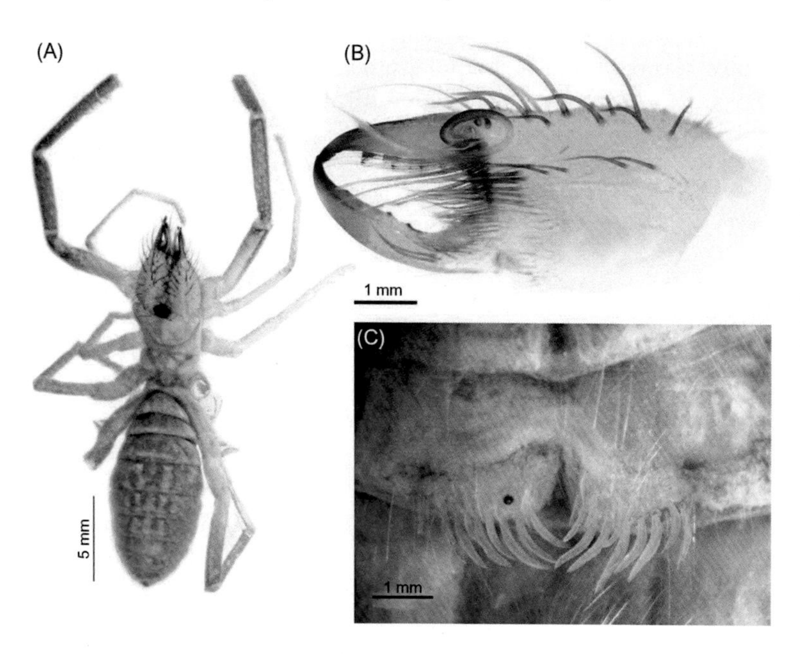

FIGURE 4.46 A solpugid, *Karschia mastigofera*: (A) habitus, dorsal view; (B) chelicerae, internal view showing flagellum; and (C) ventral view of opisthosoma. *Source: From Khazanehdari, M., Mirshamsi, O., Aliabadian, M., 2016. Contribution to the solpugid (Arachnida: Solifugae) fauna of Iran. Turkish J. Zool. 40 (4), 608–614.*

Most species of Solifugae are nocturnal predators, emerging from relatively permanent burrows to feed upon a variety of arthropods. They do not have poison glands. As generalist predators, they have also been known to feed upon small lizards, birds and mammals (Punzo, 1998). In North American deserts, immature stages of solpugids feed extensively on termites (Muma, 1966).

UROPYGI

This order of arachnids contains some of the largest species, up to 10 cm in length. The North American form, *Mastigoproctus giganteus* (Fig. 4.47), occurs from Florida to Arizona (Jackman, 1997). The distinctive long, whip-like tail has no stinger. When disturbed, this arachnid emits acetic acid from a gland at the base of the tail, giving rise to the common name, "vinegaroon." Uropygids are nocturnal predators, utilizing natural retreats or burrowing into sand. They have poor vision and depend up on vibrations to locate their prey.

4.4.1.7 *The Pteyrgote Insects*

Many of the higher, winged insects (Pterygota) are residents of soils and participate in food webs there. Some are permanent soil inhabitants, and all stages of the life history are found in soil or on the soil. Immature stages of other species are true soil dwellers—white grubs, wire worms, cutworms, for example—whereas their flying adult forms live in vegetation and feed in aboveground food chains. All of the major winged insect orders—the Coleoptera (beetles), Lepidoptera (moths and butterflies), Hymenoptera (ant, bees, and wasps), and Diptera (flies)—include soil-dwelling species. The termites (Order Blattodea, Epifamily Termitoidae) are essentially soil insects and are saprophages. The Homoptera (aphids, cicadas), Orthoptera (grasshoppers and crickets), and minor orders such as the Dermaptera (earwigs) contain soil-dwelling species or life history stages. Space does not permit us to review thoroughly these extensive and important groups, or to discuss all species groups that impact soils or soil food webs. We refer readers to general entomology texts such as Rosomer and Stoffolano (1994), or Tripplehorn and Johnson (2004), and to field guides such as that of White (1983) for aids in identification and basic biology of

FIGURE 4.47 The giant whip scorpion or "vinegaroon," *M. giganteus giganteus* (Lucas). *Source: Photo by R. Mitchell, University of Florida.*

these groups. We can offer only a very superficial treatment of the higher insects. Nevertheless, this group includes important species that are root feeders, predators, and modifiers of soil structure—animals that the soil ecologist can hardly ignore.

COLEOPTERA

Beetles, the largest order of insects, have soil species that are predatory, phytophagous, or saprovores. Some are permanent residents, others are temporary, and many are transient members of soil food webs. Beetles are particularly abundant in tropical ecosystems, where many species remain to be scientifically described and named. For identification of beetles to family, see White's (1983) guide. Evans (2014) provides excellent color illustrations of common ground beetles, tiger beetles, scarab beetles, and others.

There are many other online guides available for beetle identification, and these can sometimes by very useful (e.g., insectidentification.org and BugGuide.net).

The Carabidae (Fig. 4.48), the ground beetles, are among the more familiar insects caught in pitfall traps or active on the soil surface of agroecosystems (Purvis and Fadel, 2002). *Harpalus pennsylvanicus* is frequently caught in pitfall traps. Some members of the genus ingest seeds but most are predators. Larval stages are euedaphic and may be sampled with Tullgren extraction or in hand-sorted quantitative soil pits. Adults and larval stages of *Calosoma sycophonta*, the searcher, climb trees in search of prey. Bell (1990) provides taxonomic keys to adults and larvae of Carabidae.

Darkling beetles, family Tenebrionidae (Fig. 4.49), are abundant in desert ecosystems; their habits are similar to those of the carabid beetles. Most of them are scavengers or saprophytic on decaying vegetation (White, 1983). Adults are surface active whereas larvae are euedaphic. Pitfall traps catch large numbers of desert tenebrionids in the springtime, when adults emerge from the pupal stage. The larvae resemble wireworms (larval elaterid beetles) and are called "false wireworms." They are considered to be saprovores (Crawford, 1990).

Rove beetles (Staphylinidae) are a large family of common, distinctive species (Fig. 4.50), often caught in pitfall traps. Most species appear to be predaceous (both larvae and adult) but a few are saprophagous. The adults are agile runners on the soil surface. Frequently, the tip of the abdomen is turned up as they scurry along the ground. They are attracted to decaying vegetation and carrion (Dillon and Dillon, 1961). Most species have well-developed wings and can fly, but wing reduction is the general rule for euedaphic species. Keys to adult and larval stages of soil-inhabiting genera of Staphylinidae are provided by Newton (1990).

Scarab beetles (Scarabaeidae) (Fig. 4.51) are members of a large family of beetles, some quite colorful (e.g., the familiar metallic green Japanese beetle, *Popillia japonica*), with others multicolored, and often brown and black. Males may have horns on the head or pronotum.

Scarab beetles may be separated into two groups based on their feeding habits (Dillon and Dillon, 1961). One group contains species that feed upon leaves, flowers, and pollen as adults, and on plant roots or decaying wood as larvae; the other group feeds on dung and carrion. Some scarab beetles excavate burrows under dungpats, and provision these burrows with dung for their larvae to feed upon during development; other species live on the surface of dungpats (Curry, 1994). "Tumble bugs" chew off a piece of dung, work it into a ball, roll this ball to a burial site, and deposit an egg inside the dungball. The sacred scarab of ancient Egypt is a member of this group (Tashiro, 1990). Some scarabs are

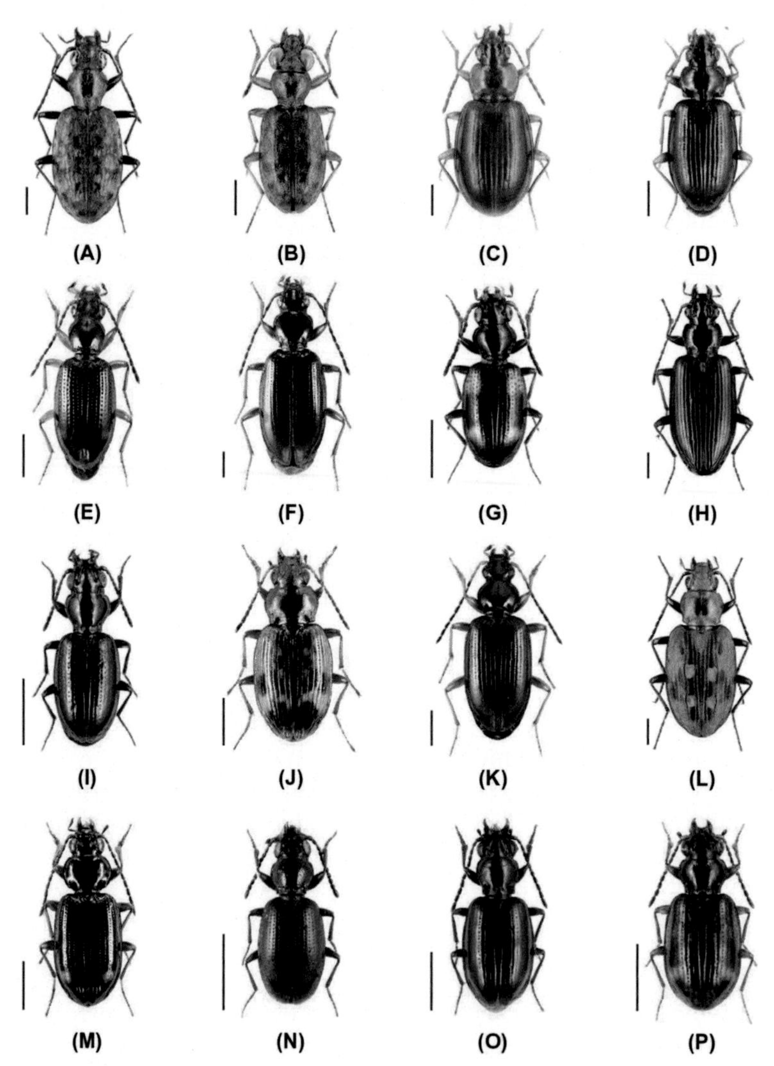

FIGURE 4.48 Representative images of carabid beetle species. (A) *Asaphidion caraboides* (Schrank, 1781); (B) *Asaphidion flavipes* (Linnaeus, 1761); (C) *Ocys harpaloides* (Audinet-Serville, 1821); (D) *Ocys quinquestriatus* (Gyllenhal, 1810); (E) *Sinechostictus elongatus* (Dejean, 1831); (F) *Sinechostictus ruficornis* (Sturm, 1825); (G) *Bembidion (Bembidion) quadrimaculatum* (Linnaeus, 1761); (H) *Bembidion (Bembidionetolitzkya) fasciolatum* (Duftschmid, 1812); (I) *Bembidion (Emphanes) azurescens* (Dalla Torre, 1877); (J) *Bembidion (Notaphus) semipunctatum* (Donovan, 1806); (K) *Bembidion (Ocydromus) testaceum* (Duftschmid, 1812); (L) *Bembidion (Bracteon) litorale* (Olivier, 1790); (M) *Bembidion (Philochthus) biguttatum* (Fabricius, 1779); (N) *Bembidion (Talanes) aspericolle* (Germar, 1829); (O) *Bembidion (Trepanedoris) doris* (Panzer, 1796); and (P) *Bembidion (Trepanes) octomaculatum* (Goeze, 1777). Scale bars = 1 mm. *Source: From Raupach, M.J., Hannig, K., Moriniere, J., Hendrich, L., 2016. A DNA barcode library for ground beetles (Insecta, Coleoptera, Carabidae) of Germany: The genus Bembidion Latreille, 1802 and allied taxa. ZooKeys, 592, 121.*

FIGURE 4.49 A beetle (Tenebrionidae). Source: *Image by Didier Descouens, published under a creative commons license and available at https://commons.wikimedia.org/wiki/File:Tenebrio_molitor_MHNT.jpg.*

FIGURE 4.50 *Staphylinus caesareus*: A predaceous staphylinid beetle. Source: *Image by Tiia Monto, published under a creative commons license and available at https://commons.wikimedia.org/wiki/File:Busteni-insect_3.jpg.*

important due to their role in hastening the decay of dung from large animals. In Australia, where the native coprophagous fauna were ill-adapted to the dung of domesticated animals, this dung accumulated, fouling pastures and immobilizing nutrients (Gillard, 1967).

Tiger beetles (Cicindellidae) are predators whose larvae dig pits where they sit and wait for prey (Fig. 4.52). Adults are rapid runners and fliers, often suddenly pouncing on prey. Conspicuous on the soil surface in open, sunlit areas, adult tiger beetles are usually iridescent green or blue.

Wireworms (larvae of the family Elateridae) are significant root feeders in forests and agroecosystems, where they can be destructive to certain crops. Tan to brown, wireworms are slender and have hard covering on their bodies. Adult elaterids are called "click

FIGURE 4.51 Some scarabaeid beetles: (A) the Flightless Dung Beetle (*Circellium bachuss*) at Addo Elephant National Park, South Africa. (B) *Hoplia coerulea* (C) *Cetonia aurata*, a rose chafer; (D) a scarab larva. Source:*(A) Kay-Africa, published under a creative commons license and available at https://commons.wikimedia.org/wiki/File: Flightless_Dung_Beetle_Circellium_Bachuss,_Addo_Elephant_National_Park,_South_Africa.JPG. (B) Siga, published under a creative commons license and available at https://commons.wikimedia.org/wiki/File:Hoplia_coerulea_front.jpg. (C) Chrumps, published under a creative commons license and available at https://en.wikipedia.org/wiki/Cetonia_aurata#/media/ File:Cetonia-aurata.jpg. (D) David Cappaert, published under a creative commons license, https://commons.wikimedia.org/ wiki/File:Scarab%C3%A9e_Japonais_3_Stades_larvaires.jpg.*

beetles" because of their ability to snap the hinge between the plates of the sternum. If the beetle is on its back, or otherwise disturbed, the beetle can right itself (or give a predator a surprise) by snapping and propelling itself into the air, turning over repeatedly in the process (Dillon and Dillon, 1961). Adult elaterids are captured occasionally in pitfall traps.

Beetles, as a group, bridge the gap between mesofauna and macrofauna. In general, they are of more economic importance for their phytophagous activities aboveground than for their participation in soil food webs. However, there are some important root feeding pests among the beetles, including some weevils (Curculionidae), which can be pests in agricultural and forestry situations (e.g., Campos-Herrera et al., 2015; Coyle et al., 2011). Likewise, some scarab species can be important pests in turfgrass (Gyawaly et al., 2016). On the other hand, some predatory beetle species can be beneficial in agricultural systems as they prey upon and reduce populations of other pest arthropods. Likewise, beetles serve the important function of processing and reducing dung and carrion in ecosystems,

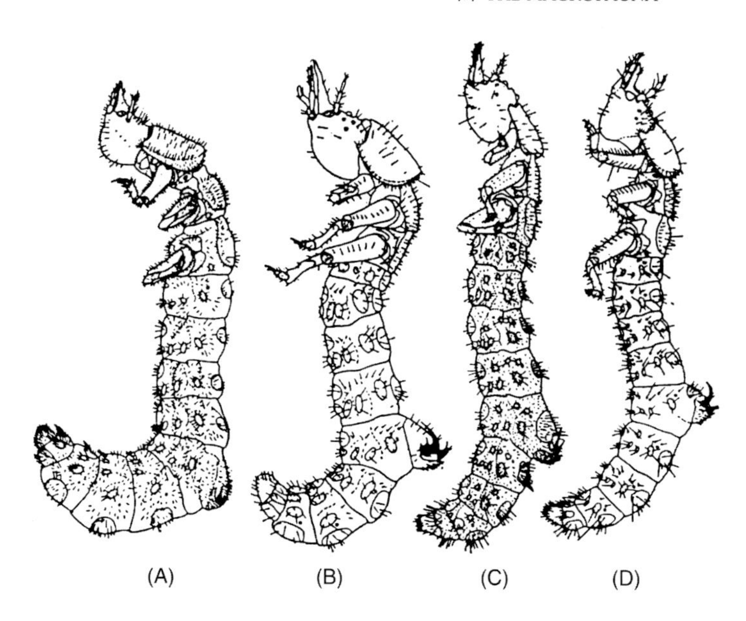

FIGURE 4.52 Larvae of tiger beetles, family Cicindelidae. (A) *Amblychila cylindriformis*; (B) *Omus californicus*; (C) *Tetracha carolina*; and (D) *Cicindela limbalis*. The predaceous larvae lie in wait in vertical burrows with their heads flush with the soil surface, and held in place by hooks on the hump protruding from the fifth abdominal segment. *Source: From Frost (1942).*

and can be important in the early stages of wood decomposition on the forest floor (Hanula, 1995; Wallwork, 1982).

HYMENOPTERA

The order Hymenoptera is one of the largest orders of insects. It contains two groups of soil insects of large importance: the ants (Fig. 4.53), and the ground-dwelling wasps (Fig. 4.54). The Formicidae (ants) are probably the most significant family of soil insects, due to the very large influence they have on soil structure. Bees and wasps, other hymenopteran insects, also impact soils because they may nest there. The ants are in a category by themselves.

Ants are widely distributed (from arctic to tropics), numerous, and diverse. Ant communities contain many species, even in desert areas (Whitford, 2000). Local species diversity is large, especially in tropical areas (Kemmpf, 1964). Populations of ants can also be quite large. About one-third of the total animal biomass in Amazonian forests is composed entirely of ants and termites, with each hectare containing in excess of 8 million ants and 1 million termites (Hölldobler and Wilson, 1990). Furthermore, ants are social insects, living in colonies with several castes (Fig. 4.53).

Ants have large impact on their ecosystems. They are major predators of small invertebrates (including oribatid mites) (E.O. Wilson, personal communication). Their activities reduce the abundance of other predators, such as spiders and carabid beetles (Wilson, 1987). Ants are "ecosystem engineers," moving large quantities of soil—as much as earthworms do in some ecosystems (Hölldobler and Wilson, 1990). Ant influences on soil structure are particularly important in deserts (Table 4.10) (Whitford, 2000), where earthworm densities are low.

Given the large diversity of ants, identification to species is problematic for any but the taxonomist skilled in the group. Wheeler and Wheeler (1990) offer keys to subfamilies and genera of the Nearctic ant fauna. Bolton's (1994) identification guide to the ant genera of

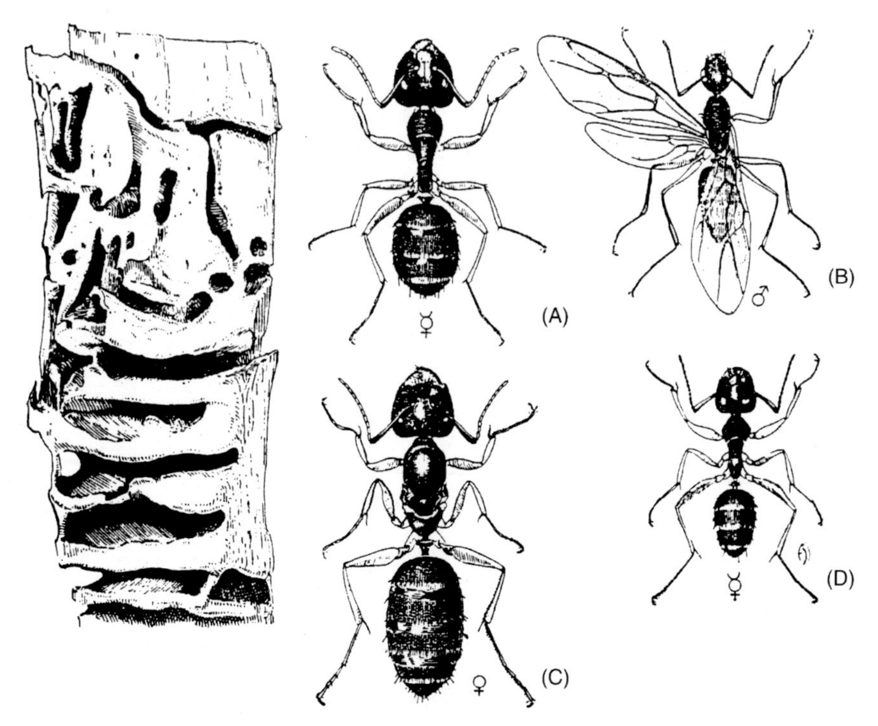

FIGURE 4.53 Carpenter ants and their galleries in deadwood. Shown are a large neuter worker (A), a winged male (B), a wingless female (C), and a small neuter worker (D). *Source: From Henderson (1952).*

FIGURE 4.54 A digger-wasp (family Sphecidae). Source: *Image by Judy Gallagher, published under a creative commons license and available at https://commons.wikimedia.org/wiki/File:Great_Golden_Digger_Wasp_-_Sphex_Ichneumoneus,_Juliette,_Ga.jpg.*

the world is very well illustrated. For a review of the ants, their biology, ecology, and social structure, the work by Hölldobler and Wilson (1990) is unsurpassed.

Most of the solitary wasps in the superfamily Vespoidea construct nests in the soil (Fig. 4.54). The adult female wasp first constructs a small nest cavity. Then a suitable prey item (another insect or a spider) is located, which the wasp then stings (paralyzing the prey), and hauls to the nest. An egg is laid on the paralyzed victim, and

TABLE 4.10 Estimated Quantities of Soil[a] Brought to the Surface by Desert Ants and Termites During the Construction of Feeding Galleries and Nest Chambers, and in Nest Repair

Ant Species	Location	Turnover Rate
Ant Community	***Atriplex vesicaria* shrub steppe**	**350–420**
Aphaenogaster barbigula (funnel ants)	*Cailtris-Eucalyptus* open woodland	3360
Ant community	Heath, Western Australia	310
Ant community	Wandoo woodland, Western Australia	200
Ant communities	Variety of Chihuahuan Desert shrublands and grasslands	21.3–85.8
TERMITE SPECIES		
Heterotermes aureus	Sonoran Desert, Arizona, United States	750
Gnathamitermes perplexus		
Macrotermes subhyalinus	Senegal	675–950
Gnathamitermes tubiformans	Chihuahuan Desert, United States	
	Mixed grassland–shrubland	4095
	Creosotebush shrubland	801
	Black-grama grassland	981
	Watershed	2600

[a]*kg* ha^{-1} $year^{-1}$.

From Whitford, W.G., 2000. Keystone arthropods as webmasters in desert ecosystems. In: Coleman, D.C., Hendrix, P.F. (Eds.), Invertebrates as Webmasters in Ecosystems. CAB International, Wallingford, UK, pp. 25 – 41.

it is then entombed. Some of the social wasps, especially *Vespula* spp., nest in the ground. Often natural cavities such as abandoned rodent burrows are used as nesting sites. Vespids are carnivorous, feeding their larvae on captured prey, although adult wasps generally feed on nectar, sap, or similar juices (Michener and Michener, 1951). Members of the wasp families Sphecidae and Crabronidae are also known to be ground nesters, with the cicada killer wasp (*Sphecius speciosus*) being among the largest and most visible member. These loosely colonial wasps capture cicadas that are sometimes larger and heavier than themselves in a noisy aerial battle, ending with the cicada being pulled into a burrow and provisioning the young of the wasps (Hastings et al., 2010).

DIPTERA

Many of the true flies can be considered soil insects, at least in some stage of their life histories. Nearly three-quarters of the dipteran families in North America have some contact with soil ecosystems (McAlpine, 1990). This listing excludes strictly aquatic families, aboveground herbivores, and some parasitic species. Many species that live in aboveground habitats pupate in the soil, thus participating, involuntarily, in soil food webs. McAlpine provides a well-illustrated key to families of Diptera that have relations with soil ecosystems.

Many species of fly larvae are important saprovores in soils. They are restricted to moist situations rich in organic matter. Some larvae are predatory and these have adaptations to reduce moisture loss; they occur in drier situations (Teskey, 1990). Fly larvae have a major impact on decomposition rates of carrion. Together with some beetle species, maggots of various types hasten the decomposition rate significantly. When Payne (1965) used window screen to exclude insects from decomposing remains of baby pigs, the bodies became mummified and decomposed slowly compared to those exposed to insect attack. Fly larvae are also important in forensic entomology, where their identification has been helpful in determining the time of death for human corpses (Catts and Haskell, 1990). Some flies are considered pests in agricultural situations with the "leatherjacket" larvae of the family Tipulidae (crane flies) being an important example. The soil-dwelling larvae of these flies are root herbivores of permanent pasture systems in the United Kingdom, and can reduce the yield of forage available for grazing livestock significantly when they reach high densities (Blackshaw and Hicks, 2013).

ISOPTERA

The blattodean infraorder Isoptera (Fig. 4.55), the termites, are among the most important soil fauna, in terms of their impact on soil structure and on decomposition processes

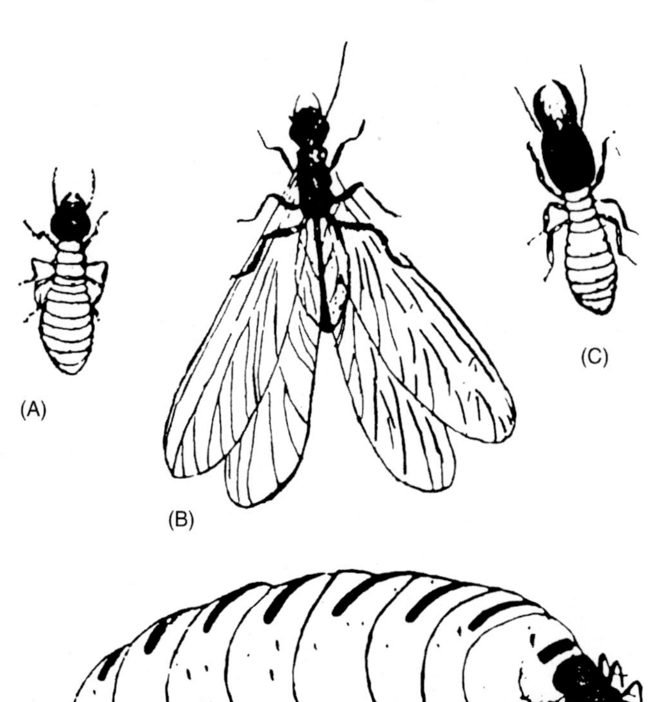

FIGURE 4.55 Isoptera (castes of termites: (A) worker, (B) winged reproductive, (C) soldier, and (D) queen. *Source: Courtesy of Banks and Snyder and the US National Museum. From Borror et al. (1981).*

worldwide. Termites are social insects with a well-developed caste system. Through their intimate association with their gut symbionts, termites have the ability to digest wood, and thus have an enormous economic impact in some parts of the world (Bignell and Eggleton, 2000; Lee and Wood, 1971). Termites are highly successful, constituting up to 75% of the insect biomass and 10% of all terrestrial animal biomass in the tropics (Bignell, 2000; Wilson, 1992).

Termites in the primitive families, such as Kalotermitidae, possess a gut flora of protozoans, which enables them to digest cellulose. Their normal food is wood that has come into contact with the soil. Most species of termites construct galleries of soil and some are builders of spectacular mounds (Fig. 4.56). Members of the phylogenetically advanced family Termitidae do not have protozoan symbionts, but possess a formidable array of microbial symbionts (bacteria and fungi) that enable them to process and digest the recalcitrant organic matter in tropical soils, and to grow and thrive on such a diet (Bignell, 1984; Breznak, 1984; Pearce, 1997). Interestingly, no single adaptive feature or mechanism appears to distinguish the guts of soil-feeding termites. As a result, approximately 67% of the genera in the family Termitidae now consist of these forms. A speculative and generalized sequence of events in a typical Termitidae soil-feeder is given in Fig. 4.57 (Brauman et al., 2000).

FIGURE 4.56 Thermal imagery of a termite mound: (A) visible spectrum image and (B) thermal image indicating contrasting temperature profiles for night (left half) and day (right half). *Source: Photos by Hunter King and Sam Ocko.*

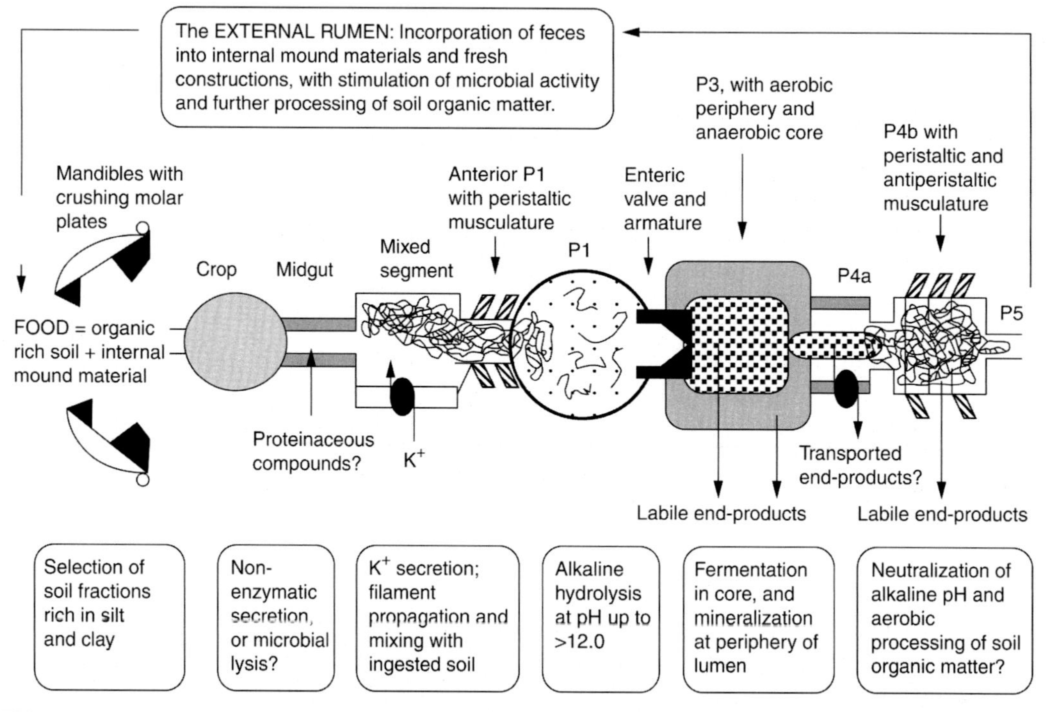

FIGURE 4.57 Hypothesis of gut organization and sequential processing in soil-feeding Cubitermes-clade termites. The model emphasizes the role of filamentous prokaryotes, the extremely high pH reached in the P1, and the existence of both aerobic and anaerobic zones within the hindgut. Major uncertainties have question marks. Not to scale. *Source: From Brauman, A., Bignell, D.E., Tayasu, I., 2000. Soil-feeding termites: biology, microbial associations and digestive mechanisms. In: Abe, T., Bignell, D.E., Higashi, M. (Eds.), Termites: Evolution, Sociality, Symbioses, Ecology. Kluwer Academic, Dordrecht, The Netherlands, pp. 233–259.*

A number of inquilines (organisms existing in, and sharing, common space) occur in termite nests—ants, collembolans, mites, centipedes, and beetles that have become morphologically specialized for that habitat.

Although termites are mainly tropical in distribution, they occur in temperate zones and deserts as well. Termites have been called the tropical analogs of earthworms, because they reach a large abundance in the tropics and process large amounts of leaf litter. Three nutritional categories include wood-feeding species, plant- and humus-feeding species, and fungus growers. This latter group lacks intestinal symbionts and depends on cultured fungus for nutrition. Termites have an abundance of unique microbes living in their guts. Using the criterion of 97% sequence similarity to define OTUs (operational taxonomic units), one study of bacterial microbiota in the gut of a wood-feeding termite (*Reticulitermes speratus*) found 268 OTUs of bacteria (based on 16s rRNA, including 100 clostridial, 61 spirochaetal, and 31 bacteroides-related taxa (Hongoh et al., 2003). More than 90% of the OTUs identified in this study had not been previously sampled. Others were monophyletic clusters that had been recovered from other termite species. It should be noted that cellulose digestion in termites, which was once considered to be solely due to the activities of fungi, protists, and occasionally bacteria, has now been convincingly demonstrated to be endogenous to termites. Endogenous cellulose-degrading enzymes occur in the midguts of two species of higher termites in the genus *Nasutitermes*, and in the Macrotermitinae (which cultivate basidiomycete fungi in elaborately constructed gardens) as well (Bignell, 2000).

In contrast to the carbon degradation situation, only prokaryotes are capable of producing nitrogenase to "fix" N_2, or dinitrogen. This process occurs in the organic matter rich, microaerophilic milieu of termite guts. Some genera have bacteria that fix relatively small amounts of nitrogen, but others, including *Mastotermes* and *Nasutitermes*, have from 0.7 to >21 μg of N fixed per gram of fresh weight. This equals 20–61 μg of N per colony per day, which would double the N content if N_2 fixation was the sole source of nitrogen and the rate per termite remained constant (and the N content of termites is assumed to be around 11% on a dry weight basis) (Breznak, 2000).

Termites are one of three major earth-moving groups of soil invertebrates (along with earthworms and ants). Mound-building termite species have a major impact on the distribution and composition of soil mineral and organic matter. Where there is rich, well-drained grassland (e.g., in the Ivory Coast), humivorous and fungus-growing termites are common. In areas of poor drainage, these species are absent and grass-feeders, such as Trinervitermes and some Macrotermes spp. are common. In some cases, farmers grow crops on mounds. This is advantageous in areas that are flooded, such as paddy fields in Southeast Asia (e.g., Thailand) (Pearce, 1997). In desert regions of North America (e.g., the Chihuahuan Desert of southern New Mexico), termites are considered "keystone species," removing and processing large amounts of dead and dying net primary production every year (see Table 4.10) (Whitford, 2000). For a masterful exposition on the role of termites in ecosystems worldwide, refer to Bignell and Eggleton (2000) and Jouquet et al. (2011).

4.4.1.8 Other Pterygota

As we noted earlier, most terrestrial insect orders have members that participate in soil systems, either by burrowing, pupating, or even feeding there. At times they may be present in some numbers or exert an unusual influence on soil food webs.

The orthopteran, grasshoppers and crickets, lay eggs in soils and some are active on the soil surface. Crickets, family Gryllidae, may be abundant in pitfall traps set in meadows or agroecosystems (Blumberg and Crossley, 1983).

The Psocoptera, psocids, are a small order of insects that occasionally become abundant in leaf litter. They feed on organic detritus, algae, lichens, and fungus (Aldrete, 1990).

The order Homoptera, cicadas, aphids, and others, has members that are important as belowground herbivores and soil movers. Cicadas are noisy, active flyers as adults. The immature stages of cicadas feed on the roots of perennial plants until mature, a period of time that can be as long as 17 years for the periodical cicadas (*Magicicada* spp.). In some ecosystems, massive emergences of periodical cicadas can represent significant fluxes of nutrients and energy from below- to aboveground (Whiles et al., 2001), and these nutrients may have temporary "pulsed resource" benefits to plants growing in the understory of forests where emergences occur (Yang, 2004). Other cicadas emerge on shorter time frames, with a cohort emerging annually. In tallgrass prairie soils of North America, cicadas are abundant insects, and their annual emergence represents a significant redistribution of nutrients from belowground to aboveground (Callaham et al., 2002).

Gastropoda

Terrestrial gastropods (snails and slugs) (Figs. 4.58 and 4.59) (Burch and Pearce, 1990) are major players among herbivores and detritivores in many ecosystems, particularly agroecosystems (Byers et al., 1989). Gastropods have been studied much less than the arthropod fauna in forests. They tend to require moist conditions and the presence of significant amounts of calcium for their metabolic needs, but some gastropods exist successfully in low pH and low calcium environments (Burch and Pearce, 1990). The terrestrial

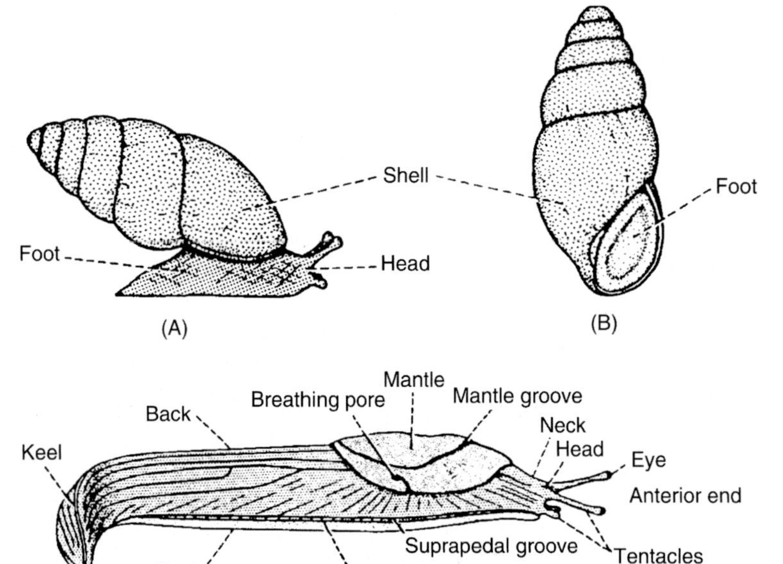

FIGURE 4.58 Terrestrial gastropods (snails and slugs): At top, a nonoperculated (pulmonate) snail. It does not have a protective operculum to seal the shell aperture when the animal has withdrawn into its shell. (A) Active snail; (B) inactive snail withdrawn into its shell, with only the surface of its foot showing. At bottom, slug body terminology. *Source: From Burch, J.B., Pearce, T.A., 1990. Terrestrial Gastropoda. In: Dindal, D. (Ed.), Soil Biology Guide. Wiley, New York, NY, pp. 201–309.*

FIGURE 4.59 A banana slug in the Hoh rainforest, Washington State. Public domain image courtesy of the National Parks Service and available at https://commons.wikimedia.org/wiki/File:Banana_slug_in_the_Hoh_Rainforest.jpg.

gastropod faunas have become rich and diversified as they have invaded many habitats. More than 1200 species occur in North America north of Mexico, but this is viewed to be a surprisingly small degree of richness compared to similarly sized land masses (Nekola, 2014), although the number of species is similar to that found in Europe (South, 1992). Land gastropods are ubiquitous in the eastern United States, with at least 500 species, including both snails and slugs, being described from this region (Hubricht, 1985; cited in Hotopp, 2002). Snails seem to key in on structural attributes of their environment. In an extensive study of community patterns of 108 snail species of the Great Lakes region, Nekola (2003) found that soil surface architecture—organic horizon soils (deeper than 4 cm of "duff"), versus thin O-horizon soils—accounted for 43% of the variability of snail distributions. Unlike some other members of the macroinvertebrate fauna (e.g., earthworms), the terrestrial gastropod fauna in North America is largely composed of native species with only 6% of the total number of species listed as naturalized (Nekola, 2014).

In Danish beech forests, Petersen and Luxton (1982) measured 822 mg m^{-2} gastropod biomass, an amount that was exceeded only by millipedes and earthworms in the fauna of that forest floor. Terrestrial gastropods feed primarily on plants, but may prefer decaying or senescent tissues. Numerous basidiomycetes are consumed as well, including some that highly toxic to mammals. Only a few gastropods feed on animals, but several may feed on carrion. The feces of gastropods retrieved from the wild include soil particles, which may be due to humic acids required as a substrate in their diet (this is particularly true for helicid snails grown in culture) (Speiser, 2001). Feeding rates on leaf litter range from 9.3 to 28.1 mg g^{-1} live weight of slugs in European forest floors (Jennings and Barkham, 1975). Assimilation rates of *Agriolimax reticulatus* feeding on fresh *Ranunculus repens* (lotus) leaves were greater than 78%, and snails' assimilation rates ranged between 40% and 70%, feeding on either fresh leaf or leaf litter material (Mason, 1974). Pallant (1974) estimated rates of assimilation for several slug species to average 161.3 J · 100 mg^{-1} dry weight in grasslands, and 141.5 J · 100 mg^{-1} dry weight in woodlands, in Europe. For an extensive review of terrestrial gastropod phylogenies, see Dayrat et al. (2011).

SAMPLING TECHNIQUES FOR GASTROPODS

Two different sampling techniques were employed by Hotopp (2002), working in the forests of the central Appalachian Mountains: (1) timed searching and (2) sieving litter. The former approach constitutes a 10-minute search of leaf litter surface, rocks, woody debris, and live plant stems across a 200 m^2 sample plot. It tends to be more efficient in finding large snails and slugs. The latter method consists of placing litter from Oi and Oe horizons on a 10 mm sieve placed to a depth of about 10 cm of litter material, shaking it 50 times, turning it over, and shaking 50 more times. This method is more efficient in retrieving small specimens including the ecological dominant, *Punctum minutissimum*, which is about 1 mm in width.

4.4.2 Oligochaeta—Earthworms

Earthworms are the most familiar, and with respect to soil processes, often the most important of the soil fauna. As observed by many farmers and gardeners and reported in the popular literature, the importance of earthworms arises from their influence on soil structure (e.g., aggregate or crumb formation, soil pore formation) and on the breakdown of organic matter applied to soils (e.g., fragmentation, burial, and mixing of plant residues). These observations have led to numerous studies on the potential benefits of earthworms in agriculture, waste management, and land remediation (Edwards, 2004).

While the scientific literature on earthworms officially began with Linnaeus' taxonomic description of *Lumbricus terrestris* more than 250 years ago, the modern era of earthworm research began with Charles Darwin's (1881) last book, *The Formation of Vegetable Mold Through the Actions of Worms, With Observations of Their Habits*, which called attention to the beneficial effects of earthworms: "*It may be doubted where there are many other animals which have played so important a part in the history of the world, as have these lowly organized creatures.*" Since then, a vast literature has established the importance of earthworms as biological agents in soil formation, organic litter decomposition, and redistribution of organic matter in the soil (see Edwards and Bohlen, 1996; Lavelle et al., 2016; Lee, 1985; Hendrix, 1995).

Despite the common reference in the popular literature to "the earthworm," there is great diversity and a wide range of adaptations to environmental conditions among earthworm fauna. More than 5000 earthworm species have been described, and it is estimated that considerably more await discovery and description (Fragoso et al., 1999; Lavelle and Lapied, 2003).

Earthworms are classified within the phylum Annelida, class Clitellata, and order Haplotaxida. Although there is not universal agreement taxonomic classification, recent analyses suggest that there are 20 families, with some comprising aquatic or semiaquatic worms (e.g., Almidae, Sparganophilidae, Lutodrilidae), and the rest consisting of the terrestrial forms. Nineteen of the families are classified as Crassiclitella (those with clitella formed from greater than one cell layer thickness), and these are the earthworms proper (James and Davidson, 2012). Species within the families Lumbricidae and Megascolecidae are ecologically the most important in North America, Europe, Australia, and Asia; some of these species have been introduced worldwide by human activities and now dominate the earthworm fauna in many temperate areas (Hendrix et al., 2008). Likewise, several tropical species in the families Glossoscolecidae, Eudrilidae, and Megascolecidae have become pantropical in distribution. Such "peregrine" or "anthropochorous" species are highly successful in many agricultural, or otherwise disturbed, areas and often show significant effects on soil processes (Brown et al., 2013; Fonte et al., 2010; Lavelle et al., 1999). Different localities may be inhabited by all native species, all exotic species, a combination of native and exotic species, or by no earthworms at all. Relative abundance and species composition of local fauna depend greatly on soil, climate, vegetation, topography, land-use history, and especially on past invasions by exotic species (e.g., Callaham et al., 2006; Lobe et al., 2014).

4.4.2.1 Earthworm Distributions and Abundance

As noted previously, earthworms occur worldwide in habitats where soil water and temperature are favorable for at least part of the year; they are most abundant in forests and grasslands of temperate and tropical regions, and least so in arid and frigid environments. Across this range of habitats, earthworms display a wide array of morphological, physiological, and behavioral adaptations to environmental conditions (Lee, 1985). Even in unsuitable regions, earthworms may inhabit local microsites where conditions are suitable (e.g., urban gardens, desert oases), especially if well-adapted species have been introduced (Gates, 1967). During unfavorable conditions, many species are adapted to enter a temporary quiescent state termed estivation, or diapause (Fig. 4.60) or produce resistant cocoons that hatch when conditions improve (Edwards and Bohlen, 1996). Within habitats, earthworms often show patchy spatial distributions corresponding to factors such as vegetation, soil texture, or soil organic matter; feeding preferences dictate vertical distributions of species within the soil profile.

Abundance and biomass of earthworms establish them as major factors in soil biology, leading Blakemore (2002) to remark: "*And while birdwatchers get excited about a few kilograms of birdlife, or the grazier is concerned about a couple of 100's kg per hectare of livestock in a pasture, almost totally ignored is an underground biomass of earthworms often far in excess of those above that may total 2 or 3 tonnes per hectare.*" Earthworm densities in a variety of habitats worldwide ranges from less than 10 to more than 2000 individuals m^{-2}, the highest values occurring in fertilized pastures, and the lowest in acid or arid soils (coniferous or sclerophyll forests) (Table 4.11). Typical densities from temperate deciduous or tropical forests and certain arable systems range from less than 100 to more than 400 individuals m^{-2}. Intensive land management (especially soil tillage and application of toxic chemicals) often reduces the density of earthworms, or may completely eliminate them. Conversely, degraded soils converted to conservation management (e.g., no-tillage) often show increased earthworm densities and associate soil properties after a suitable period of time (Curry et al., 1995; Edwards and Bohlen, 1996). Biomasses of lumbricid species in temperate regions of the world, where they have been spread by human activities, often exceed that of other animal groups. In the Piedmont region of Georgia in the United States, for example, Hendrix et al. (1987) reported an earthworm

FIGURE 4.60 An earthworm in an estivation chamber in dry soil on the South Carolina Piedmont during summer. Earthworms reduce their surface area to volume ratio by forming into a ball, and waiting out very dry periods. *Source: Photo by Zachary Brecheisen, used with permission.*

TABLE 4.11 Typical Ranges of Earthworm Density and Biomass in Various Habitats

Habitat	Earthworms (m^{-a})	Fresh Weight (gm^{-m})
Temperate hardwood forest	100–200	20–100
Temperate coniferous forest	10–100	30–35
Temperate pastures	300–1000	50–100
Temperate grassland	50–200	10–50
Sclerophyll forest	<10–50	<10–30
Taiga	<10–25	≤10
Tropical rainforest	50–200	<10–50
Arable soil	<10–200	<10–50

Summarized from Lee, K.E., 1985. Earthworms: Their Ecology and Relationships With Soils and Land Use. Academic Press, Sydney, Australia, and Edwards, C.A., Bohlen, P.J., 1996. Earthworm Biology and Ecology, third ed. Chapman and Hall, London.

dry-matter biomass of 10 g carbon m^{-2} in no-tillage agricultural plots, a value larger than all other fauna combined.

4.4.2.2 *Biology and Ecology*

Earthworms are soft-bodied, segmented animals ranging in length from a few millimeters to more than 3 m (and with claims of up to 6 m for some South African species). One example of such large earthworms is those currently being discovered in the interior reaches of the Amazon rainforests of Brazil (Fig. 4.60). Morphological details vary greatly among earthworm groups and many such details (e.g., position of the reproductive organs) are used in taxonomic distinctions among species (Schwert, 1990). Nonetheless, a number of features are common to most earthworms (Fig. 4.61). In general, earthworms consist of a simple, tube-within-a-tube body plan, the outer tube constituting the body proper, and the internal tube comprising the alimentary canal. Ingested material (e.g., mineral soil, particulate organic matter) is drawn through the mouth into a muscular buccal cavity and then through the pharynx into the esophagus. Many species have a muscular esophageal gizzard that grinds and mixes food material as it passes through. The esophagus in many species also contains a calciferous gland that functions in calcium metabolism and regulation of CO_2 levels in the blood (Briones et al., 2008; Versteegh et al., 2014). The remainder of the gut consists of the intestine that, in many endogeic species, has an infolding of the gut wall known as the typhlosole, which greatly increases the absorptive surface area of the intestine. The overall length of the gut and the configuration of the typhlosole vary with species, probably as a function of diet.

Earthworms are hermaphroditic, each individual possessing both male and female reproductive organs (testes, ovaries, and associated structures) (Edwards and Bohlen, 1996). During sexual reproduction, sperm is exchanged between two individuals, and stored in sperm sacs or spermathecae. This sperm is then later released, along with the eggs, into cocoons secreted by the glandular clitellum, which is the characteristic

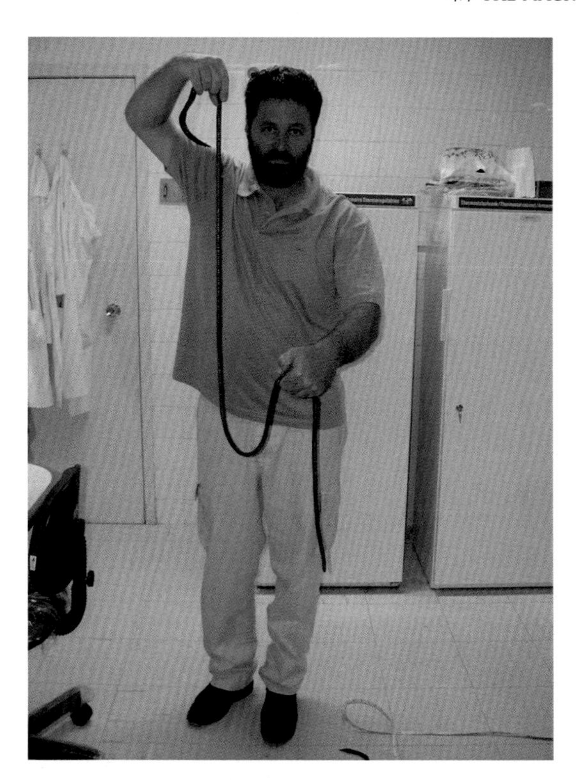

FIGURE 4.61 A giant Amazonian earthworm, *Rhinodrilus priollii*. *Source: Photo by George Brown.*

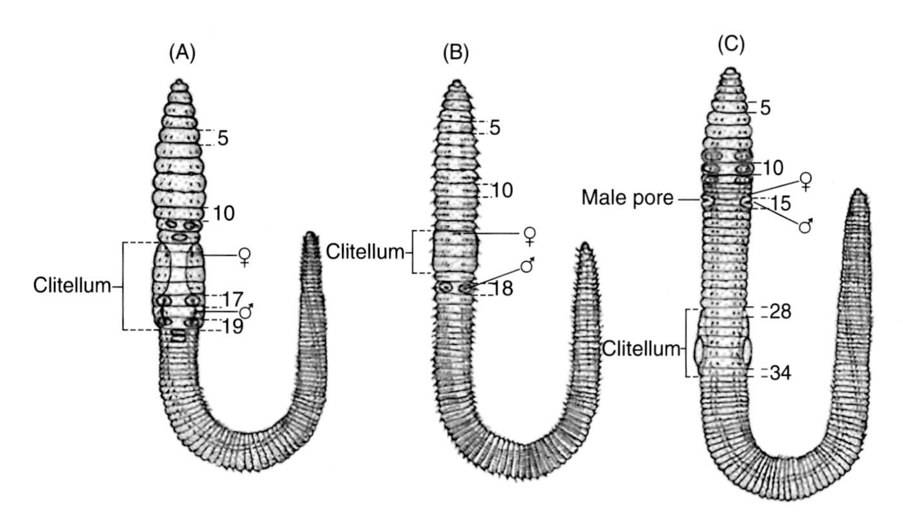

FIGURE 4.62 General external characteristics of representatives from three earthworm families: (A) Acanthodrilidae, (B) Megascolecidae, and (C) Lumbricidae. *Source: From Blakemore, R.J., 2002. Cosmopolitan earthworms: an eco-taxonomic guide to the peregrine species of the world. VermEcology, Kippax, Australia.*

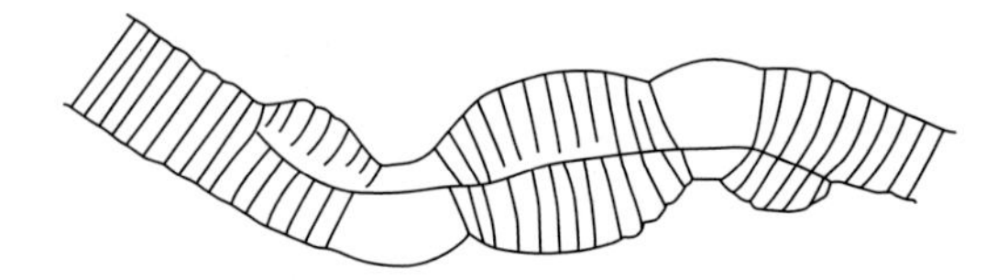

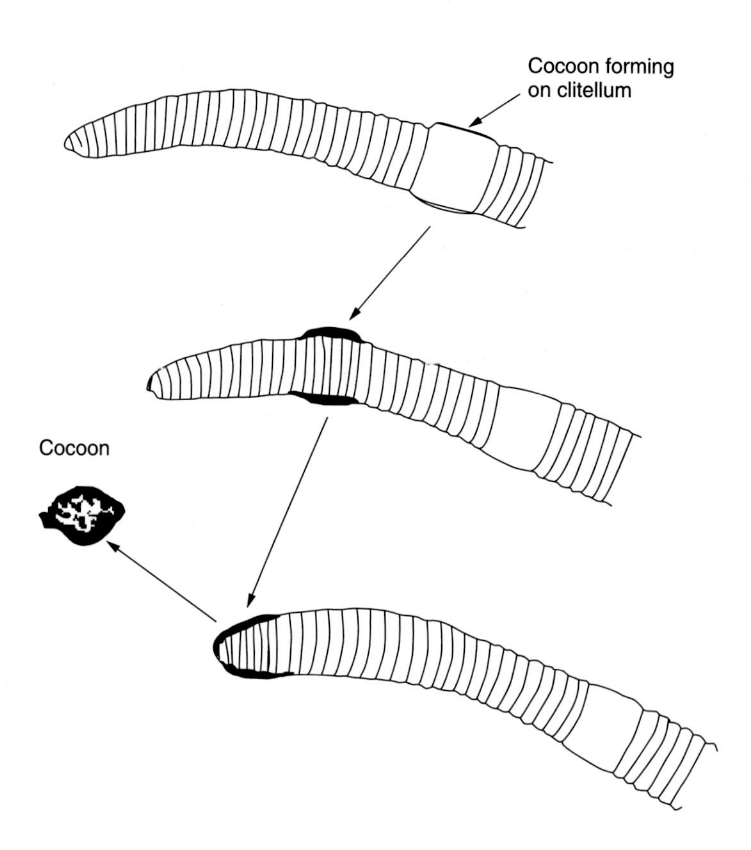

FIGURE 4.63 Copulation and subsequent cocoon formation during sexual reproduction in lumbricid earthworms. *Source: From Edwards, C.A., Bohlen, P.J., 1996. Earthworm Biology and Ecology. third ed. Chapman and Hall, London.*

thickening or saddle-shaped structure often seen near the anterior end of sexually mature individuals (Figs. 4.62 and 4.63). One to several embryos may from within each cocoon, depending on earthworm species. Some earthworms reproduce parthenogenetically, whereby an ovum develops without fertilization by a sperm. Parthenogenesis provides an effective means by which certain species can establish new populations in new habitats; such species are often the successful peregrines and anthropochores discussed previously.

Earthworms are often grouped into functional categories based on their morphology, their behavior and feeding ecology, and their microhabitats within the soil (Bouché, 1977, 1983; Lavelle, 1983; Lee, 1959, 1985). These categories describe the ways by which different earthworm species utilize resources within a soil volume (Table 4.12) (Fig. 4.64). Epigeic

TABLE 4.12 Ecological Categories, Habitat, Feeding, and Morphological Characteristics of Earthworms

Category	Subcategory	Habitat	Food	Size and Pigmentation
Epigeic	Epigeic	Litter	Leaf litter, microbes	<10 cm, highly pigmented
	Epi-endogeic/epi-anecic	Surface soil	Leaf litter, microbes	10–15 cm, partially pigmented
Endogeic	Polyhumic	Surface soil or root zone	Soil with high organic content	<15 cm, filiform unpigmented
	Mesohumic	Upper 0–20 cm soil	Soil from 0–10 cm strata	10–20 cm, unpigmented
	Endo-anecic	0–50 cm soil, some make burrows	Soil from 0–10 cm strata	>20 cm, unpigmented
	Oligohumic	15–80 cm soil	Soil from 20–40 cm strata	>20 cm, unpigmented
Anecic	Anecic	Lives in burrows in soil	Litter and soil	>15 cm, anterodorsal pigmentation

Modified from Barois, I., 1999. Ecology of earthworm species with large environmental tolerance and/or extended distributions. In: Lavelle, P., Brussaard, L., Hendrix, P. (Eds.), Earthworm Management in Tropical Agroecosystems. CABI Publishing, New York, NY, pp. 57–85.

FIGURE 4.64 A pair of *Lumbricus terrestris* earthworms mating. Source: *Image by Beentree, published under a creative commons license and available at https://commons.wikimedia.org/wiki/File:Earthworm_klitellum_copulation_beentree.jpg.*

and epi-endogeic species are often polyhumic (prefer organically enriched substrates) and utilize plant litter on the soil surface and carbon-rich upper layers of mineral soil; poly-, meso-, and oligohumic endogeic species inhabit mineral soil with high (e.g., in the rhizosphere), moderate, and low organic matter content, respectively; and anecic species exploit both the surface litter as a source of food and the mineral soil as a refuge in which they make permanent burrows. The familiar *L. terrestris* is an example of an anecic species, constructing burrows and pulling leaf litter down into them. In contrast, *Bimastos parvus* (the American log worm) exploits leaf litter and decaying logs, with little involvement with the soil—an epigeic species. The ubiquitous European lumbricid, *Aporrectodea caliginosa*, and several megascolecids (e.g., *Diplocardia* spp., native to eastern North America) are endogeic in life habits. Some earthworm species appear to be intermediate between two categories; for example the epi-endogeic *Lumbricus rubellus*, which can inhabit litter layers and/or form shallow horizontal burrows in mineral soil. Even though some species might not exactly fit, these categories have become a very popular, and useful, means for segregating earthworm communities into functional groups of species.

Within a particular soil, less than half a dozen earthworm species are typically found, and the species within such an association often effectively partition the soil volume according to their functional categories. Further, the activities of earthworms influence soil processes in various ways according to these functional categories. For example, epigeic species promote the breakdown and mineralization of surface litter, whereas anecic species incorporate organic matter deeper into the soil profile, and facilitate aeration and water infiltration through their formation of burrows.

4.4.2.3 *Influence of Earthworms on Soil Processes*

Earthworms, as ecosystem engineers (Lavelle et al., 1998), have pronounced effects on soil structure as a consequence of their burrowing activities as well as their ingestion of soil and production of castings (Lavelle and Spain, 2002; van Vliet and Hendrix, 2007). Casts are produced after earthworms ingest mineral soil and/or particulate organic matter, mix them together, and enrich them with organic secretions in the gut, and then egest the material as a slurry or as discrete fecal pellets within or upon the soil, depending on earthworm species. Darwin (1881) observed that surface castings by earthworms buried chalk to considerable depths in soil over a 20-year period, and a similar study which recreated the original conditions at the Down House estate in southern England confirmed that the descendants of Darwin's worms were still actively burrowing and burying objects on the soil surface (Butt et al., 2016). Turnover rates of soil through earthworm casting range from 40–70 t ha^{-1} $year^{-1}$ in temperate grasslands (Bouché, 1983) to 500–1000 t ha^{-1} $year^{-1}$ in tropical savannas (Lavelle, 1978). In an extreme case of this ecosystem engineering, earthworms in the Llanos of Colombia and Venezuela are responsible for the construction of a whole ecosystem type, the "surales" which are vegetated mounds consisting entirely of soils cast up by earthworms above the level of the water in what would otherwise be submerged and saturated soils (Zangerlé et al., 2016)

During passage through the earthworm gut, casts are colonized by microbes that begin to break down soil organic matter. As casts are deposited into the soil, microbial colonization and activity continue until readily decomposable compounds are depleted. Eventually, casts may harden into stable soil aggregates. Mechanisms of cast stabilization include organic bonding of particles by polymers secreted by earthworms and microbes, mechanical stabilization by plant fibers and fungal hyphae, and stabilization due to wetting and drying cycles and age-hardening effects (Tomlin et al., 1995). Earthworm casts are usually enriched with plant-available nutrients and thus may enhance soil fertility; plant-growth–promoting substances have also been suggested as constituents of earthworm casts. Castings from vermicomposting operations are sold commercially as soil amendments that purportedly enhance plant growth (Edwards, 1998).

Earthworm burrowing in soil creates macropores of various sizes, depths, and orientations, depending on species and soil type (Fig. 4.65). Burrows tend to be similar in diameter to that of the earthworms that produced them ranging from about 1 mm to larger than 10 mm in diameter and constituting among the largest of soil pores (Edwards and Shipitalo, 1998). Burrows of epigeic earthworms (e.g., *Dendrobaena octaedra*) are often small and limited to upper layers of soil; they may be horizontal to vertical in orientation. Endogeic species (e.g., *Diplocardia mississippiensis* or *Pontoscolex corethrurus*) may form networks of variously oriented burrows, as the earthworms ingest soil and cast behind them as they burrow. These networks may form continuous pores over some depth, but castings within the burrows may impede free water movement. Anecic earthworms (e.g., *L. terrestris*) may create deep vertical burrows that form continuous macropores to depths of 1 m or more (van Vliet and Hendrix, 2007). These burrows tend to be very stable because their walls are lined with organic matter drawn in or secreted by earthworms, and they often have higher bulk density than that of surrounding soil (Lee, 1985). Continuous macropores resulting from earthworm burrowing may enhance water infiltration by functioning as bypass flow pathways through saturated soils. These pores may or may not be important in solute transport, depending on soil water content, nature of the solute, and chemical exchange properties of the burrow linings (Edwards and Shipitalo, 1998).

The influence of earthworms on organic matter and nutrient cycling in soils is closely related to the density and feeding ecology of resident populations, as described previously (Barois, 1999; Lee, 1985). Epigeic species typically inhabit the surface litter and the O and

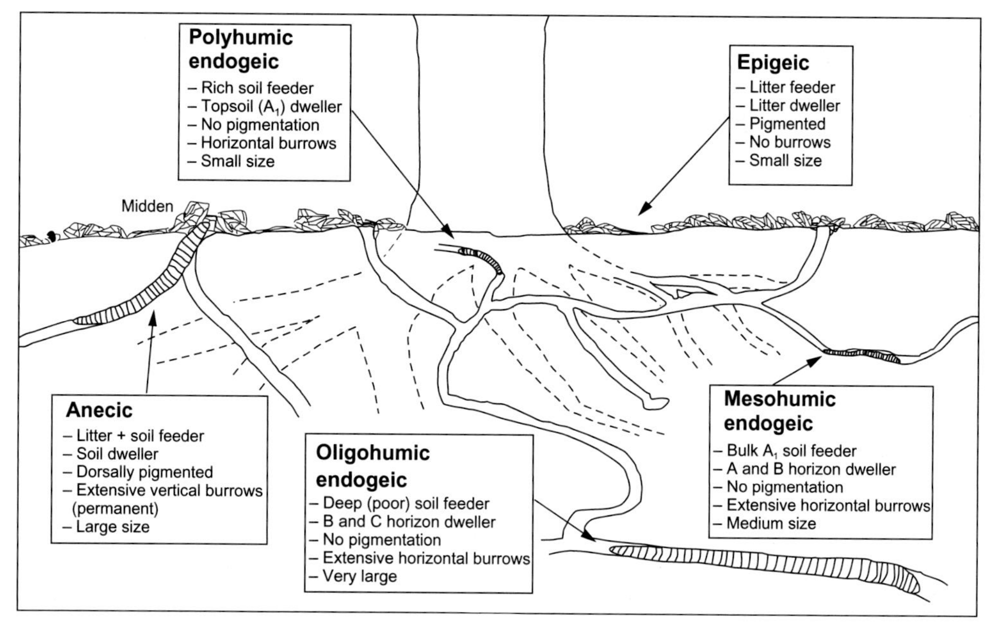

FIGURE 4.65 Pictorial representation of some of the characteristics of earthworm ecological strategies (categories) as proposed by Bouché (1977), Lavelle (1981), and Lavelle et al. (1989). *Source: From Brown, G.G., 1995. How do earthworms affect microfloral and faunal community diversity? Plant Soil 170, 247–269.*

upper A horizons of soil where they mix mineral soil and plant residues, fragment organic particles, inoculate them with microbes, and thus increase organic matter decomposition rates. Anecic earthworms pull surface litter into their burrows, thus transporting organic material deeper into the soil profile. They cast on the soil surface, mixing organic and mineral particles in the litter layer. The activities of both epigeic and anecic earthworms produce "mull" soil horizons, in which organic matter is intimately incorporated into the upper mineral soil of a well-developed A horizon overlain with a recently deposited litter layer. The extreme case is termed "vermimull," in which the Ah horizon is granular and characterized by organic–mineral complexes consisting of earthworm casts (Green et al., 1993). Endogeic earthworms feed within the soil on organic matter and microbes associated with plant roots or mineral soil. As mentioned previously, they are termed oligo-, meso-, or polyhumic, depending on the level of organic matter decomposition (Brown, 1995). Mineralization of organic matter in earthworm casts and burrow linings produces zones of nutrient enrichment that are different from those in bulk soil. These zones are referred to as the "drilosphere" and are often sites of enhanced activity of plant roots and other soil biota (Lavelle et al., 1998).

Despite the many beneficial effects of earthworms on soil processes, some aspects of earthworm activities may be undesirable (Edwards and Bohlen, 1996; Lavelle et al., 1998; Parmelee et al., 1998). Detrimental effects include removing and burying of surface residues that would otherwise protect soil surfaces from erosion; producing fresh casts that increase erosion and surface sealing; increasing compaction of surface soils; depositing castings on the surface of lawns and golf greens where they are a nuisance; dispersing weed seeds in gardens and agricultural fields; transmitting plant or animal pathogens; riddling irrigation ditches; making them less able to carry water; increasing losses of soil nitrogen through leaching and denitrification; and increasing soil carbon loss through enhanced microbial respiration. Furthermore, there have been reports of earthworms transmitting pathogens, either as passive carriers or as intermediate hosts, raising concerns that some earthworm species could provide a mechanism for the spread of certain plant and animal diseases.

Thus it is the net result of positive and negative effects of earthworms, or any other soil biota, that determines whether they have detrimental impacts on ecosystems (Lavelle et al., 1998). An effect, such as mixing of O- and A-horizons, may be considered beneficial in one setting (e.g., urban gardens) and detrimental in another (e.g., native forests).

4.4.2.4 Earthworm Management

There is interest in managing earthworms to utilize their beneficial effects in organic waste reduction, in land reclamation, and in reduced-intensity agriculture (Edwards and Bohlen, 1996; Lee, 1995). Because of their effects on organic matter decay, earthworms are increasingly being used to accelerate decomposition of organic waste materials. Vermicomposting involves culturing of earthworms outdoors in beds or in confined chambers in the presence of waste materials, which are reduced in volume and carbon–nitrogen ration as they are processed by earthworms and decomposed by enhanced microbial activity within the earthworms and their castings (Edwards, 1998). A variety of approaches and designs has been developed for vermicomposting systems, but the basic principle is the feeding of acceptable organic materials to earthworms in continuous or batch culture, and the collection of processed wastes that ultimately consist of stabilized castings. Earthworm biomass is also harvested from vermicomposting systems for a variety of uses, including further composting operations, animal protein, and fishing bait. Organic wastes that have been used successfully in vermicomposting include animal manures, sewage sludge, food production wastes, and horticultural residues. Small-scale vermicomposting is becoming popular for reduction of household wastes such as kitchen scraps and yard trimmings. Earthworm species typically used for vermicomposting included the European lumbricids, *Eisenia fetida* and *L. rubellus*, often called "red worms"; the African "nightcrawler," *Eudrilus eugeniae*; and the Asian "blue worm," *Perionyx excavatus*. The latter two species are tropical and best suited to composting under warm conditions (Edwards, 1998). A number of vermicomposting publications have appeared in the popular literature in the last decade including the periodical *Worm Digest* and the ever-popular *Worms Eat Our Garbage* (Appelhoff et al., 1993).

The potential for earthworms to ameliorate soils during land reclamation or in degraded agricultural sites is also of increasing interest (Baker, 1998; Lee, 1995). In many situations, it may be desirable to introduce earthworms. Techniques have been developed for large-scale inoculation of areas devoid of earthworms (e.g., reclaimed polders) and for introduction of species that may perform desired functions (e.g., epiendogeic species for thatch removal from pastures). It is usually necessary that favorable soil conditions (e.g., adequate water and organic matter, appropriate temperatures) exist at the time of inoculation and/or that refugia (e.g., blocks of native sod or containers of native soil) are provided from which earthworms may disperse (Butt, 2008). Introductions of earthworms into unfavorable environments often fail.

Mixed-species assemblages of earthworms may influence a wider array of soil processes, such as organic matter turnover as well as soil structural properties, than a single species can (Lee, 1995). Introductions of such assemblages might include one or more anecic species that make deep vertical burrows and that cast on the surface and bury residues, and one or more endogeic species that feed belowground on dead roots and organic matter and that make horizontal burrows. Inclusion of epigeic species might accelerate decomposition of plant residues on the soil surface.

Whether introduced earthworms displace native species or occupy areas devoid of native species as a result of disturbance is a subject of some debate, but it is clear that at the continental scale, and over geologic time frames, the exotic species clearly take advantage of profoundly disturbed soils where glaciers were present 10,000–12,000 years ago and no native fauna exist (earthworm fauna in glaciated soils exclusively consist of introduced species (Hendrix et al., 2008; James, 2004)). The establishment of earthworms in the glaciated soils of North America is primarily through human-mediated transport—the dumping of ship ballast, movement of horticultural materials, and use of nonnative earthworms as fish bait have all been implicated as being important vectors of the invasive species (Callaham et al., 2006; Hendrix and Bohlen, 2002). More recently, however, it has become clear that invasive earthworms are now actively colonizing less disturbed, or even pristine forested habitats of the glaciated portions of the continent (Cameron et al., 2007; Frelich et al., 2006; Hale et al., 2005). In one study the age of roads constructed in northern boreal forests was a strong predictor of the extent of earthworm invasion (Cameron and Bayne, 2009). These authors also showed that the distance of a particular forested habitat from active agriculture was related to earthworm invasion, and they further developed a predictive model showing the relationship between road density and the likely extent of new earthworm invasions. Another study found that road building and harvesting in a small portion of a forested watershed was enough disturbance to allow earthworms to become established throughout the watershed, and that the earthworms were passively dispersed along stream corridors into areas otherwise unaffected by the harvesting activity (Costello et al., 2011).

The interaction is not so clear between disturbance, native species, and the establishment of nonnative species in soils where native species still exist (i.e., nonglaciated soils), but this interaction may involve competitive exclusion or other biotic interactions. In such soils, it is extremely rare to find pure assemblages of native earthworm species particularly if there has been any tillage, or other soil profile disruption (Fig. 4.66; Hendrix et al., 2008). However, these profile-disrupting disturbances need not affect vast areas of land for nonnative species to establish. Researchers have found that roads built through otherwise undisturbed habitats can serve as corridors for the introduction of nonnative earthworms into less disturbed habitats (Kalisz and Dotson, 1989; Kalisz and Wood, 1995). In still other nonglaciated soils in North America, workers have observed a relationship between land-use history and the prevalence of nonnative earthworms in soil invertebrate communities, with a clear indication that perturbations to vegetation and soil result in a dominance of the earthworm community by European species (e.g., Callaham et al. 2006, Sanchez de Leon and Johnson-Maynard, 2009, Winsome et al., 2006).

4.4.2.5 *Earthworm Sampling and Identification*

A variety of sampling methods have been used for collecting earthworms, both quantitatively and qualitatively (see Chapter 9: Laboratory and Field Exercises in Soil Ecology). Hand digging and sorting of soil is the most commonly used method for quantitative sampling of earthworms. Pits of known dimensions (e.g., $25 \times 25 \times 25\ cm^3$) are dug with a shovel, often in layers of defined thickness, and the soil broken by hand. More elaborate modifications of this method, which may improve collection of juveniles and cocoons, include dry or wet sieving of soil through screens of known mesh size, and flotation of

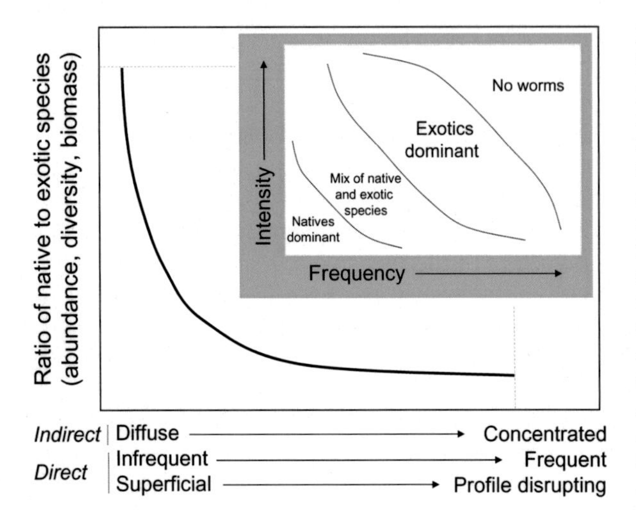

FIGURE 4.66 Hypothesized relationships between types of disturbance and expected proportions of native and exotic species. Direct disturbances are categorized by frequency of occurrence and by intensity or degree of physical perturbation of the soil profile. Frequency and intensity are related, but we envision a scenario where extreme disruption of the soil profile need not be frequent to effect change in earthworm communities (inset). Indirect disturbances include road density, human population density, degree of fragmentation, and diversity of land-use types, among others. *Source: Modified from Hendrix, P.F., Callaham Jr., M.A., Drake, J.M., Huang, C.-Y., James, S.W., Snyder, B.A., 2008. Pandora's box contained bait: the global problem of introduced earthworms. Ann. Rev. Ecol. Evol. Syst., 39, 593–613.*

sieved material in high-density solutions to separate earthworms and other soil fauna. Earthworms may be collected alive or immediately preserved in 70% ethanol or 5% formalin for later counting and identification.

Earthworms are also collected by applying solutions of chemical irritants to the soil, which bring earthworms to the surface where they can be hand collected. A number of chemicals have been used, including $HgCl_2$, $KMnO_4$, formalin, and mustard powder slurry (Iannone et al., 2012; Singh et al., 2016; Zaborski, 2003). The latter has received much attention recently because of its safety and availability. Chemical extraction techniques may be effective on anecic earthworms such as *L. terrestris* but may be less so on other species. Effectiveness varies with earthworm species and activity, soil water content, porosity, and temperature. Comparisons with hand sorting should be done before adopting extraction techniques for quantitative sampling.

Mechanical or electrical stimulation may also bring earthworms to the soil surface. A technique known as "grunting" employs a wooden stake driven into the soil, vibration of the stake with a bow or flat piece of metal, and collection of earthworms that emerge. It is thought that the response of earthworms to these subterranean vibrations is an antipredator behavior, and may be related to their instinctive reaction to moles tunneling in the soil (Catania, 2008) Some megascolecid species have been sampled with this technique (Hendrix et al., 1994; Mitra et al., 2009) but it may not be effective on other earthworm groups.

Electrical extraction of earthworms uses metal rods connected to a source of electrical current and inserted into the soil; the current brings earthworms to the surface. Different voltages and amperages have been used with varying degrees of success; effectiveness of the technique is highly dependent on soil water content, electrolyte concentration, and temperature. As with mechanical vibration, the soil volume sampled with electrical current is not known and therefore these methods may be best suited for qualitative or comparative sampling (Lee, 1985). However Schmidt (2001) used a commercially available "Octet" device to quantitatively sample earthworm communities in arable soils in Ireland; the electrical method gave estimates of species composition and population size comparable to those from hand digging, with the exception of one very small species. The electrical technique is potentially very dangerous and should only be used with extreme caution.

Earthworm populations also may be sampled with trapping techniques. Pitfall traps (described earlier) may give some idea of surface-active earthworm species present in an area. Baited traps consist of porous containers (e.g., clay flower pots) filled with bait such as animal manure and buried in the soil for appropriate periods of time. Trapping techniques are highly selective of certain earthworm species and thus are best suited for qualitative or comparative sampling (Callaham et al., 2003; Lee, 1985).

For earthworm species that cast on the soil surface (e.g., *L. terrestris*), numbers and types of castings may give an indication of population activity. Because casting is dependent on soil water content and temperature, this technique is highly variable and not suitable for quantitative estimates of population density.

Many earthworm species in the family Lumbricidae can be identified from external body characteristics if the specimens are sexually mature. Several taxonomic keys are useful for the common lumbricids found worldwide (Reynolds, 1977; Schwert, 1990; Sims and Gerard, 1985). Most other earthworms require dissection for accurate taxonomic identification; position and characteristics of sexual organs, the gut and associated glands, and other structures are required. The procedures must be done carefully and require a degree of skill and practice. Additional general taxonomic references include Edwards and Bohlen (1996), Fender and McKee-Fender (1990), Fragoso et al. (1999), Jamieson (1988), and James (1990).

4.5 GENERAL ATTRIBUTES OF FAUNA IN SOIL SYSTEMS

In recent years, interest has been shown by soil scientists and ecologists in measuring "soil quality." This elusive concept has been the subject of entire symposia and volumes resulting from them (e.g., Doran et al., 1994). As defined by soil scientists, soil quality can be considered as the degree or extent to which a soil can: (1) promote biological activity (plant, animal, and microbial); (2) mediate water flow through the environment, and (3) maintain environmental quality by acting as a buffer that assimilates organic wastes and ameliorates contaminants (Linden et al., 1994). Many environmental scientists are attempting to use the concept of indicator organisms or indicator communities as a way to determine overall soil "health" (e.g., Bongers, 1990; Ettema and Bongers, 1993; Ferris et al., 2001; Foissner, 1994; Linden et al., 1994). Because of their large size and public awareness of them, earthworms are often considered a sign of soil "health" (Hendrix, 1995; Linden et al., 1994). All of the biota play important roles in affecting and influencing soil processes. As summarized in Table 4.13, each of the biotic groups has significant impacts.

TABLE 4.13 Influences of Soil Biota on Soil Processes in Ecosystems

	Nutrient Cycling	Soil Structure
Microflora	Catabolize organic matter	Produce organic compounds that bind aggregates
	Mineralize and immobilize nutrients	Hyphae entangle particles onto aggregates
Microfauna	Regulate bacterial and fungal populations	May affect aggregate structure through interactions with microflora
	Alter nutrient turnover	
Mesofauna	Regulate fungal and microfaunal populations	Produce fecal pellets
	Alter nutrient turnover	Create biopores
	Fragment plant residues	Promote soil organic matter protection
Macrofauna	Fragment plant residues	Mix organic and mineral particles
	Stimulate microbial activity	Redistribute organic matter and microorganisms
		Create biopores
		Promote SOM protection
		Produce fecal pellets

From Hendrix, P.F., Crossley Jr., D.A., Blair, J.M., Coleman, D.C., 1990. Soil biota as components of sustainable agroecosystems. In: Edwards, C.A., Lal, R., Madden, P., Miller, R.H., House, G. (Eds.), Sustainable Agricultural Systems. Soil and Water Conservation Society, Ankeny, IA, pp. 637–654.

Among the fauna, microfauna have a principal role via interactions with the microflora. The mesofauna and macrofauna create fecal pellets, and produce biopores of various sizes, which affect water movement and storage as well as root growth and proliferation. Perhaps, more important, over the longer term, they have marked effects on soil organic matter stabilization processes as well (Wolters, 1991). Based on biological characteristics, there are three general trophic systems: microtrophic (protozoa, nematodes, and some enchytraeids), mesotrophic (the mesofauna), and macrotrophic (the large fauna capable of breaking through physical barriers of soil) (Heal and Dighton, 1985).

Further concerns about fauna as indicators of soil quality led Linden et al. (1994) to erect a hierarchical array of three categories in which fauna and soil quality interact, namely: (1) organisms and populations, relating to behavior, physiology, and numbers; (2) communities, with concerns about functional groups (i.e., guilds of burrower and nonburrowers, trophic groups, and biodiversity); and (3) biological processes, relating to the several processes and properties listed in Table 4.14 (Curry and Good, 1992; Linden et al., 1994). These processes are considered in greater detail in later chapters on decomposition and nutrient cycling processes. In general, the concept of using soil biota as indicators of soil quality or ecosystem function has advanced the most in European

TABLE 4.14 Properties of Soil Fauna for Use as Indicators of Soil Quality

1. Organisms and populations
Individuals
Behavior, morphology, physiology
Populations
Numbers and biomass
Rates of growth, mortality, and reproduction
Age distribution
2. Communities
Functional groups
Guilds (e.g., burrowers vs nonburrowers, litter vs soil dwellers, etc.)
Trophic groups
Food chains and food webs (microbivores, predators, etc.)
Biodiversity
Species richness, dominance, evenness
Keystone species
3. Biological processes
Bioaccumulation
Heavy metals and organic pollutants
Decomposition
Fragmentation of organic matter
Mineralization of C and nutrients
Soil structure modification
Burrowing and biopore formation
Fecal deposition and soil aggregation
Mixing and redistribution of organic matter

From Curry, J.P., Good, J.A., 1992. Soil Fauna Degradation and Restoration. Adv. Soil Sci. 17, 171–215 and Linden, D.R., Hendrix, P.F., Coleman, D.C., van Vliet, P.C.J., 1994. Faunal indicators of soil quality. In: Doran, J.W., Coleman, D.C., Bezdicek, D.F., Stewart, B.A. (Eds.), Defining Soil Quality for a Sustainable Environment. SSSA Special Publication No. 35. American Society of Agronomy, Madison, WI, pp. 91–106.

contexts where earthworms have been identified as one potentially effective indicator (Griffiths et al., 2016).

4.6 FAUNAL FEEDBACKS ON MICROBIAL COMMUNITY COMPOSITION AND DIVERSITY

Since the time of Darwin's epochal book on soil biology (1881), there has been considerable interest in the effects of fauna on microbial communities. Satchell (1983), in a centenary celebration of Darwin's book, stated that, by the culture methods existing up until then, there was little or no indication that earthworms have a qualitatively different flora in their guts or in their castings. Yet there are numerous studies that have indicated an increase in microbial numbers or activity during or after passage through the gut and in the drilosphere (Barois, 1999; Daniel and Anderson, 1992; Drake and Horn, 2007; Kristufek et al., 1992; Schoenholzer et al., 1999). Recent studies using both molecular and culture-based analyses of agricultural soil and burrows and casts of the epigeic earthworm *L. rubellus* have revealed interesting differences between earthworm- and nonearthworm-influenced soils (Furlong et al., 2002). Clone libraries of the 16S rRNA genes were prepared from DNA isolated directly from the soil and earthworm casts. Representatives of the *Pseudomonas* genus as well as the Actinobacteria and Firmicutes increased in number, and one group of unclassified organisms found in the soil library was absent in that of the cast. In fact, Singleton et al. (2003) isolated a new species of bacterium, *Solirubrobacter paulii*, from the intestinal wall of *L. rubellus*. This was not found in the soil library, and may represent the first known instance of a bacterium unique to the gut wall of an earthworm.

Studies of microbial community similarity have been conducted comparing termite mounds and nearby tropical soils. Harry et al. (2001) used RAPD (random amplified polymorphic DNA) molecular markers to estimate the similarity of microbial communities in the mounds of several termite species and surrounding soils. They studied four species of soil-feeding termites and one species of fungal-feeding termite in a tropical rain forest area of the Nyong River basin in Southern Cameroon. They found that microbial communities of the mounds of the soil-feeding termite species were clustered in the same clade, whereas those of the mounds of the fungus-growing species were distinct like those of the control soils. The microbial changes were dependent upon their mound building and the fungal-feeding species using saliva as particle cement in its mounds.

4.7 SUMMARY

Animals in soils are a large, numerous, and diverse group of species, organized into complex food webs. In addition to a formal taxonomic classification, the soil fauna may be classified in several ways: persistence in the soil, distribution through the soil profile, body shape, and body size. The latter classification, body size, has the advantages of separating fauna into groups collected and quantified in similar manners. Methods for study of the microfauna including the protozoa are essentially the methods of microbiology.

Among the mesofauna, the abundant and ubiquitous nematodes have significant impacts on microbial populations and on roots. Another group, the microarthropods, contains mites and collembolans that feed on plant debris rich in fungus, nematodes, and microarthropods, and provides complex food webs, whose connections may vary opportunistically. The macrofauna contains a large group of arthropods, including the familiar isopods and millipedes as detritus feeders, and scorpions, spiders, and other predators. Pterygote (winged) insects are numerous in soils. The termites and ants are important soil movers (bioturbators) in many situations, as are earthworms. The earthworms may be the single most important groups of soil animals, in terms of their feeding upon detritus and their effects on soil structure. But the entire fauna is involved in maintenance of soil health. The microfauna and microfloral interactions, the feeding of the mesofauna on microbial-rich detritus, and the creation of biopores and the bioturbation effects of the larger mesofauna all interact in creating soil quality (see Chapter 8: Future Developments in Soil Ecology).

As noted in the summary to Chapter 3, Secondary Production: Activities of Heterotrophic Organisms—Microbes, the prospects of linking microbes and fauna by meaningful qualitative (structural) and quantitative (functional) techniques are growing rapidly, and the future is bright for synthesis in soil ecology studies.

CHAPTER

5

Decomposition and Nutrient Cycling

5.1 INTRODUCTION

The bulk of terrestrial net primary production (NPP), along with the bodies and excretions of animals, is returned to the soil as dead organic matter. Some 90% of NPP eventually enters the soil system through dead plants in grasslands; leaves, roots, and wood in forests; and organic residue in agricultural fields. Indeed, ecosystems may be viewed as consisting of four functional subsystems: the production subsystem, the consumption subsystem, the decomposition subsystem, and the abiotic subsystem. The decomposition subsystem serves to reduce dead residues to CO_2 and soil organic matter (SOM), and to release nutrient elements for entry into soil food webs, and ultimately for re-accumulation by plants. The decomposition process drives complex belowground food webs, in which chemical forms of nutrient elements become modified. It is responsible for the creation of long- and short-lived organic compounds important in nutrient dynamics, and it fuels the formation of soil structure.

Terrestrial plant growth is highly dependent on the decomposition system, particularly in oligotrophic soils where nutrient stocks are held in litter organic matter and SOM, rather than in mineral soil. Heterotrophic organisms in the soil are ultimately responsible for ensuring the availability of nutrients for primary production (Wardle, 2002). Thus, the two subsystems, primary production and decomposition, are dependent upon each other. We emphasize, again, the necessity for evaluation of entire ecosystems when considering their respective parts. The soil subsystem performs crucial functions within terrestrial ecosystems, however modified they may be. Decomposition processes in highly modified agricultural systems still involve a significant variety of heterotrophic organisms with characteristic abilities (Coleman et al., 2006; Wasylik, 1995).

Decomposition per se is the catabolism of organic compounds in plant litter or other organic detritus. As such, decomposition is mainly the result of microbial activities. Few soil animals have the enzymes that would allow them to digest plant litter. Animal nutrition depends upon the action of microbes, either free living in the soil or specialized in the rhizosphere or in animal guts. However, the term "decomposition" is often used more generally to refer to the breakdown or disappearance of organic litter. In that context, the decomposition of organic residue involves the activities of a variety of soil biota, including both microbes and fauna, which interact together. The term "litter breakdown" has been applied to the interactive process, which results in the disappearance of organic litter.

Fundamentals of Soil Ecology
DOI: http://dx.doi.org/10.1016/B978-0-12-805251-8.00005-3

Continuing interest in decomposition is apparent from the large number of studies of the process that have been published during the past 25 years. A recent search (using Web of Science) of the terms "decomposition" and "soil" as part of the title of publications appearing in refereed journals resulted in more than 1100 titles, and that number was in excess of 19,000 when those terms were searched as topics of the publications (Personal observations, access date March 15, 2017). Improved understanding of the decomposition process has accompanied the refinement of methods and conceptual models. The litterbag technique (Bocock and Gilbert, 1957; Crossley and Hoglund, 1962; Edwards and Heath, 1963; Shanks and Olson, 1961) has become a major tool in these studies, despite its limitations (Heal et al., 1997). Radioactive tracers (Coleman and Fry, 1991) have been replaced by methods using stable isotopes of C and N (Boutton and Yamasaki, 1996; Crotty et al., 2012; Maraun et al., 2011; Nadelhoffer and Raich, 1992). But see Trumbore (2009) and He et al. (2016) for an updated look at the utility of radiocarbon to study the dynamics of soil carbon. Early models of mass loss (Jenny, 1941; Olson, 1963) defining a decomposition constant, *k*, are being supplanted by more sophisticated models which consider different constituents of litter and enzyme kinetics (Sinsabaugh, 2010; Moorhead et al., 2012).

5.2 INTEGRATING VARIABLES

In studies of soil systems, rates of litter breakdown have been used as integrating variables. That is, since litter breakdown rates are the result of the combined activities of the soil biota, breakdown rates may be used to evaluate the effects of disturbance on the entire system. For example, conversion of agricultural systems into conservation tillage regimes will affect soil biology, notably by shifting the composition of microbial communities and increasing earthworm population densities, but with changes in other soil biota as well (Beare et al., 1992; Coleman et al., 2012; Doran, 1980a,b; Parmelee et al., 1990). We can evaluate the consequences of these changes in soil biota by measuring litter breakdown rates (Fig. 5.1) (Crossley et al., 1992).

Other integrating variables include: (1) soil respiration, (2) formation of soil structure, and (3) nutrient dynamics. All of these variables are readily measured, and all are important for ecosystem function. Soil respiration estimates biological activity generally and is dominated by microbes, with an important contribution by roots (Cheng et al., 1993; Freschat et al., 2013; Kuzyakov, 2002). Soil structure is the result of combined actions of biota and climate on mineral substrates. Nutrient dynamics are the most valuable of the integrating variables for predicting primary productivity.

Although microbes are responsible for the biochemical degradation of organic litter, fauna are important in conditioning the litter and aiding in microbial actions. The soil scientist Hans Jenny characterized soil fauna as mechanical blenders: "They break [down] plant material, expose organic surface areas to microbes, move fragments and bacteria-rich excrement around and up and down, and function as homogenizers of soil strata" (Jenny, 1980). Breakdown rates for organic litter integrate the effects of these various activities into a single set of variables. The combination of microbial and faunal activities results in a set

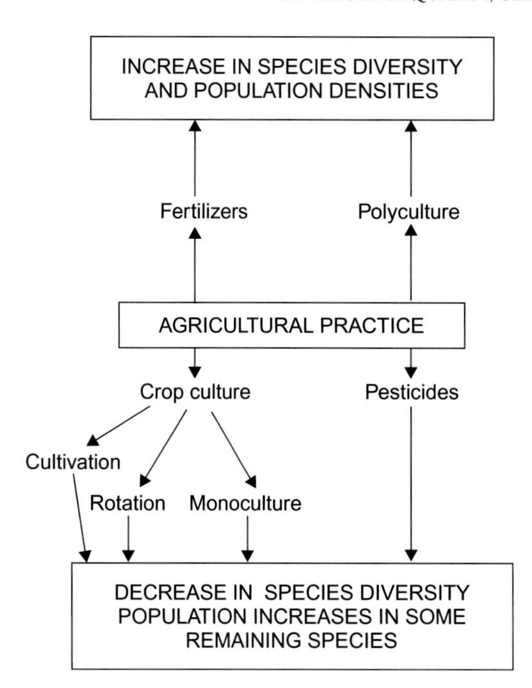

FIGURE 5.1 Changes in species diversity of soil microarthropods as a function of agricultural practice. *Source: From Crossley Jr., D.A., Mueller, B.R., Perdue, J.C., 1992. Biodiversity of microarthropods in agricultural soils: relations to functions. Agric. Ecosyst. Environ., 40, 37–46.*

of positive interactions of the type termed "facilitation" by Bruno et al. (2003). Results of the interactions are likely to be more significant than that by either component, animal, or microbial, acting alone.

5.3 RESOURCE QUALITY, CLIMATE, AND LITTER BREAKDOWN

Litter breakdown rates vary between and among ecosystems on localized and broad geographical scales, as functions of soil biota, substrate quality, microclimate, and ecosystem condition. In general, we view breakdown and decomposition as the result of biota acting on substrates of varying quality within the constraints of climate and soil properties.

Resource quality is defined principally by the chemical composition of organic residues deposited on or in the soil. Sugars and starches (i.e., labile substrates) are easily digested by microbes and other soil biota, whereas tannins, lignins, and other compounds rich in polyphenols (i.e., recalcitrant substrates) can be utilized directly only by certain specialized organisms (e.g., white-rot fungi). Cellulose and hemicelluloses are intermediate in their degradability. Hence, the relative proportions of these classes of compounds in organic materials greatly influence the overall rate of decomposition of those materials (Fig. 5.2) (Berg, 1986, 2014). Organic litter in most terrestrial ecosystems is a mixture of relatively labile and relatively recalcitrant substrates—thin, calcium-rich dogwood (*Cornus florida*)

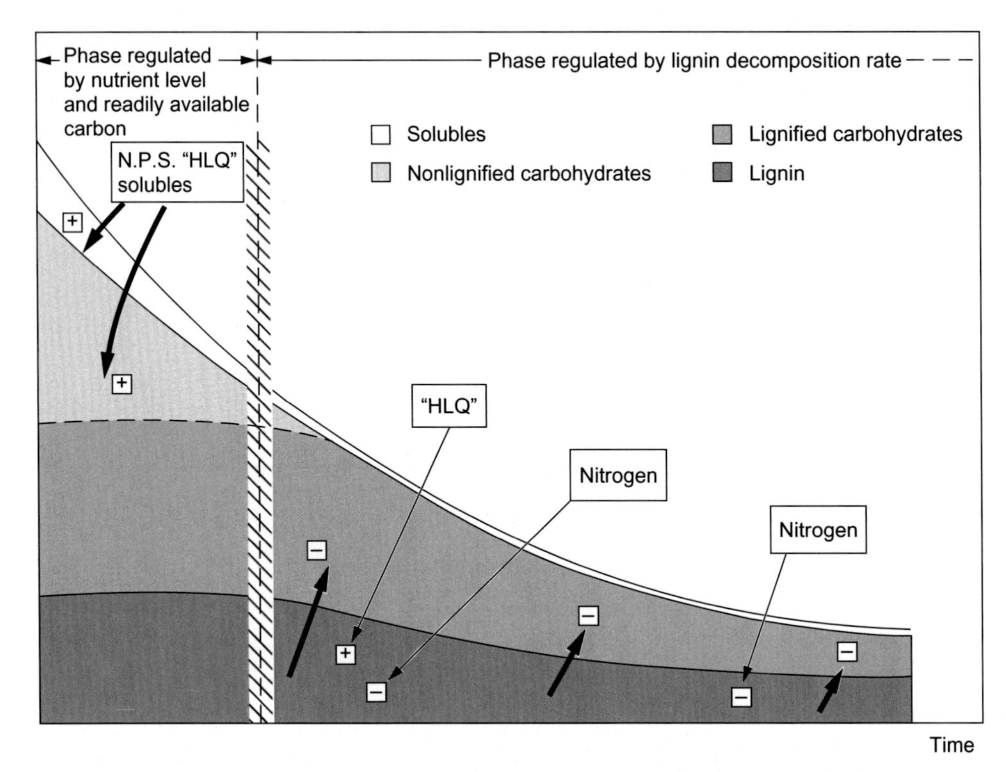

FIGURE 5.2 Model for decomposition of some organic components in Scots pine (*Pinus sylvestris*) needle litter. In the early phase of decomposition, high concentrations of nutrients such as nitrogen (N), phosphorus (P), and sulfur (S) exert a rate-enhancing influence on mass loss of the nonlignified parts of the litter. Also, high concentrations of easily degraded solubles and celluloses influence a high mass loss rate. In the late stage where mainly lignified material remains, lignin mass loss is governing, which in its turn is negatively affected by high nitrogen concentrations and positively by high concentrations of celluloses in the lignified material. The negative effect of lignin on cellulose degradation is indicated by black arrows. A (+) indicates a rate-enhancing influence and (−) a negative one. HLQ designates the quotient between holocellulose and lignin plus holocellulose. *Source: From Berg, B., 1986. Nutrient release from litter and humus in coniferous forest soils—a mini review. Scand. J. Forest Res. 1, 359–369.*

leaves versus thick, highly lignified, oak leaves (*Quercus* spp.) or conifer needles (*Pinus* spp.), for example. Even in agricultural systems, differences between leaves and stalks of corn (Zea *mays*), for example, represent different substrate qualities with different breakdown rates. Woody litter, high in tannins and lignins, may have breakdown rates measured in decades or even centuries for large logs in cool climates (Harmon and Chen, 1991; Zhou et al., 2007). Fine-root turnover may be measured in days, but coarse roots, with highly suberized tissues, turnover in years.

On a broad geographical basis, the change in breakdown rates as a function of latitude is generally predictable (Fig. 5.3) (Meentemeyer, 1978). However, the effect of

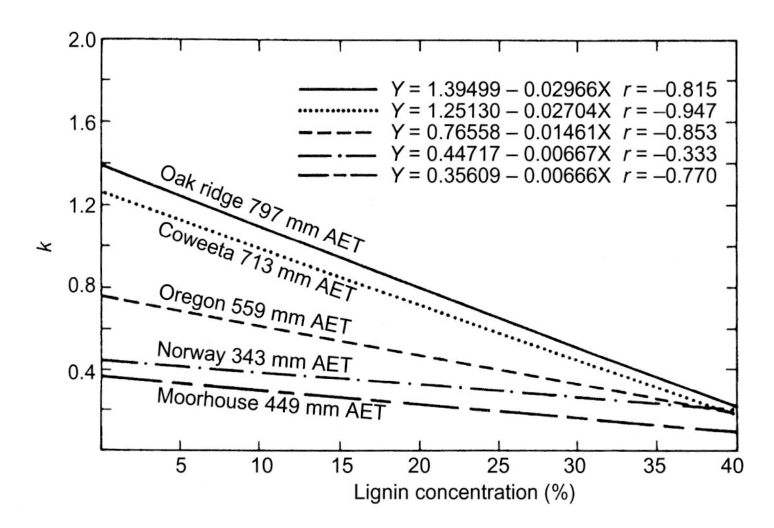

FIGURE 5.3 Simple correlation–regression between initial lignin concentration (%) and annual decomposition rate (k) for five locations ranging in climate from subpolar to warm temperate. *AET*, actual evapotranspiration. *Source: From Meentemeyer, V., 1978. Macroclimate and lignin control of decomposition. Ecology, 59, 465–472.*

latitude is not strictly a direct effect of climate; the abundance of the various soil biotas also changes with latitude (Fig. 5.4) (Swift et al., 1979). Indeed, Bradford et al. (2014) found that local-scale variation in moisture and variables such as macroinvertebrate abundance and fungal colonization were more important than broadscale climate variables in explaining wood decomposition rates. For example, adaptations of the soil biota to desert conditions allow breakdown rates to proceed more rapidly than predicted by temperature–moisture considerations. Members of the desert soil biota are active nocturnally, when temperatures moderate and light dew may accumulate (Whitford, 2000). Litter breakdown in tropical systems may be strongly influenced by seasonality of litterfall as well as faunal abundance. González and Seastedt (2001) found that all three groups of factors (climate, substrate quality, and soil fauna) independently influenced the decomposition rate of leaf litter in tropical dry and subalpine forests. Soil fauna had a disproportionately larger effect on litter decomposition in a tropical wet forest, than in tropical dry or subalpine forests.

Decomposition rates may vary along elevational gradients as well, but not as predictably. In a study reported for Arizona, United States, plant litter decomposition was measured along a gradient from desert to pinyon-juniper woodland and up into a ponderosa pine forest (Murphy et al., 1998). Decomposition was more rapid at the upper, cooler elevation that was also moister. In these systems moisture, not temperature, was the overriding variable. Similarly, in a hardwood forested ecosystem at Coweeta Hydrologic Laboratory, North Carolina, decomposition did not vary predictably along an elevation gradient (Hoover and Crossley, 1995; Hansen, 2000). Decomposition was slowest at a low elevation, very mesic, cove hardwood site, and was most rapid at intermediate elevation sites. Microclimate—temperature and moisture around decomposing substrates—regulates activity rates of the biota. Disturbed ecosystems and successional ones also may have litter breakdown rates that are slower than predicted from broad regional temperature–moisture conditions. Even factors such as differing clay

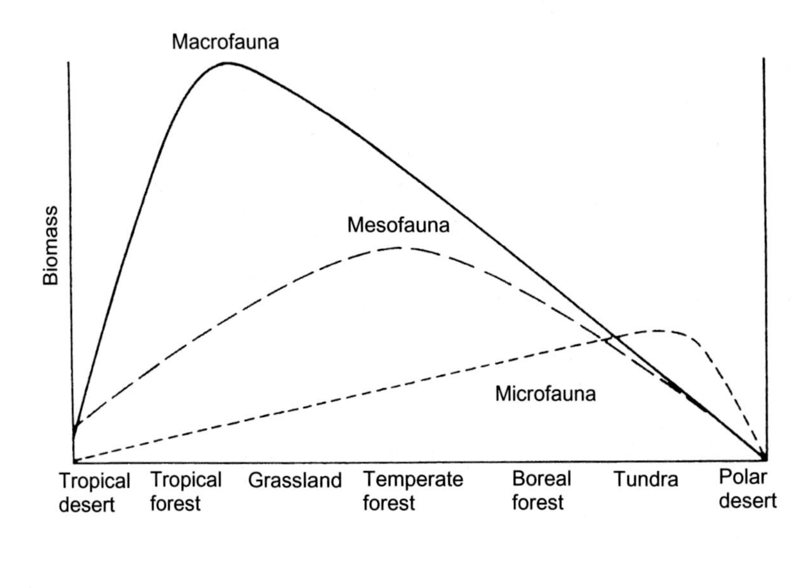

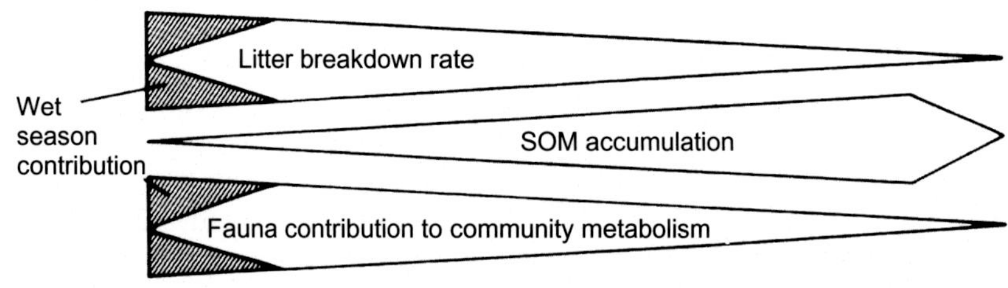

FIGURE 5.4 Hypothetical patterns of latitudinal variation in the contribution of the macro-, meso-, and microfauna to total soil fauna biomass. The effects on litter breakdown rates of changes in the relative importance of the three fauna size groups are represented as a gradient together with the faunal contribution to soil community metabolism. The favorability of the soil environment for microbial decomposition is represented by the cline of SOM accumulation from the poles to the equator; SOM accumulation is promoted by low temperatures and waterlogging where microbial activity is impeded. *Source: From Swift, M.J., Heal, O.W., Anderson, J.M., 1979. Decomposition in Terrestrial Ecosystems. University of California Press, Berkeley, CA.*

mineralogy can influence the rate of decomposition, as demonstrated by Fissore et al. (2016) in a study where they found that mesocosms constructed with kaolinitic clays exhibited faster decay rates of woody material than mesocosms with montmorillonite clays. Alteration of microclimates may reduce faunal activities, and substrate quality of foliage may change during plant succession. Furthermore, edaphic factors, particularly soil texture (i.e., relative proportions of sand, silt, and clay), greatly influence local microclimatic conditions by regulating the availability of surface water films for soil microbes and microfauna, and water-holding capacity of the bulk soil for meso- and macrofauna. Thus, soil–water relations exert indirect control on litter decomposition through their influence on soil biological activity.

5.4 DYNAMICS OF LITTER BREAKDOWN

The disappearance of litter on forest floors follows approximately a simple first-order equation:

$$dX/dt = -kX$$

where X is the standing stock of litter and k is the annual fractional rate of disappearance. Olson (1963) proposed that it was a characteristic of mature forests that rates of litter production and disappearance were equal, so that annual production (L) would be balanced by breakdown ($-kX$). Olson used the symbol X_{ss} to designate the standing stock of litter on the ground at steady state (i.e., when litter production and disappearance are equal). Then, the ratio of input (L) to standing stock (X_{ss}) provides an estimate of breakdown rate k.

$$k = L/X_{ss}$$

Olson (1963) estimated decomposition rates (k) for evergreen forests in various parts of the world (Fig. 5.5). Values for k ranged from 4 for rapid decomposition in tropical regions, through 0.25 for eastern US pine forests, to 0.02 for higher latitude pine forests. Subsequent analyses using more sophisticated, mechanistic models have shown similar trends in litter decomposition across climatic gradients (e.g., Moore and de Ruiter, 2000, 2012; Parton et al., 1989a). However, Bradford et al. (2016) caution that local-scale factors should be accounted for and incorporated into models aiming to evaluate controls on decomposition over broader scales.

What is estimated here is the rate of leaf or needle litter breakdown. Olson's (1963), as most other models, did not consider inputs of organic litter belowground, although he did use the entire mass of carbon per m^2 in estimates of X_{ss}. Current studies of root dynamics (see Chapter 2: Two Primary Production Processes in Soils: Roots and Rhizosphere Associates) are providing estimates of root breakdown rates, but these are more difficult to measure than leaf litter breakdown rates and consequently are less well known. Root death and decay may account for as much as one-half of the annual carbon addition to soils in forests or even more in grasslands; but as is often the case, dynamics within the soil are obscure (Freschet et al., 2013).

The simple exponential model using a single constant, k, to represent decomposition rate continues to be widely used. It is not difficult to estimate k using litterbag techniques (see later). The simple model loses its attractiveness when patterns of litter breakdown are examined more closely. Leaf litter often is a combination of leaf species, each with different breakdown rates. Furthermore, each species contains both labile and recalcitrant fractions. Wieder and Lang (1982) examined several different models, and concluded that the single exponential model shown earlier or double exponential models (including fast and slow components) best describe breakdown rates over time "with an element of biological realism." Jenkinson et al. (1991) and Paustian et al. (1997) considered single-pool, multiple litter pool, and continuous spectrum models of litter decomposition, and provided mathematical representations of decomposition rates. Their results show that litter quality is a key factor for accurately modeling decomposition dynamics.

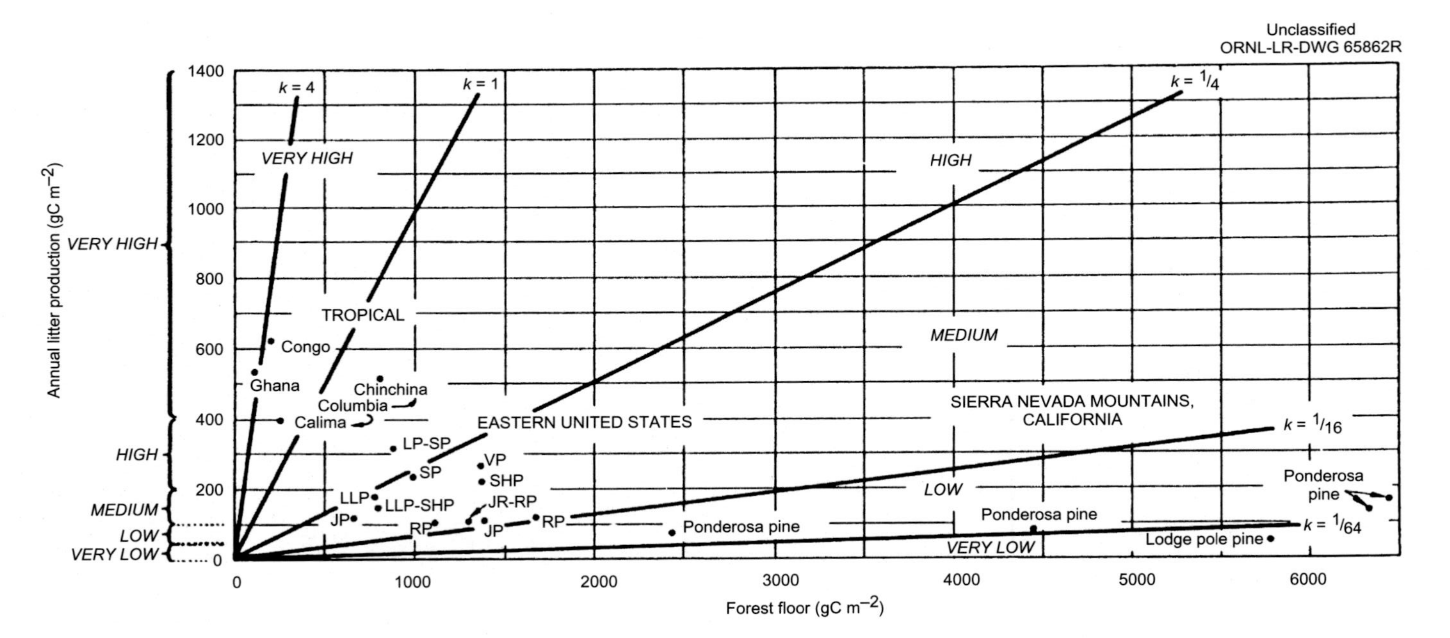

FIGURE 5.5 Estimates of decomposition rate factor k for carbon in evergreen forests, from the ratio of annual litter production (L) to (approximately) steady-state accumulation of forest floor (X_{SS}). *Source: From Olson, J.S., 1963. Energy storage and the balance of producers and decomposers in ecological systems. Ecology, 44, 322–331, with permission.*

5.5 DIRECT MEASUREMENT OF LITTER BREAKDOWN

In deciduous forests with annual pulses of leaf drop, it is possible to measure litter breakdown directly from litter samples taken through time. If combined with estimates of the mass of litterfall, these samples provide a good measure of the dynamics of litter breakdown (Fig. 5.6) (Witkamp and van der Drift, 1961). As the year progresses the litter layer mass, *L*, becomes transformed into F layer (see Chapter 1: Introduction to Soil: Historical Overview, Soil Science Basics, and the Fitness of the Soil Environment). Sampling over several years reveals year-to-year variation in masses of litter input and rates of breakdown (Table 5.1). It should be noted that two systems of organic layer horizonation are used in the soils literature: L, F, and H refer to litter, fermentation (or fragmented), and humus layers, respectively (see Green et al., 1993, for a complete description of this biologically based system); these are equivalent to the O_i, O_e, and O_a layers often used in forest soils literature.

Rates of litter breakdown are measured more easily by using confined leaf litter. Mesh bags (litterbags) containing a known mass of leaf litter are placed on the forest floor at the time of leaf drop. Litterbags are then collected on a time schedule and the remaining mass is measured (Fig. 5.7). Litterbags have been a valuable tool for comparative studies of rates of litter breakdown (Fig. 5.8). Such studies include mass loss rates by different tree species, and have shown the importance of elemental contents, lignin, C/N ratios, and other

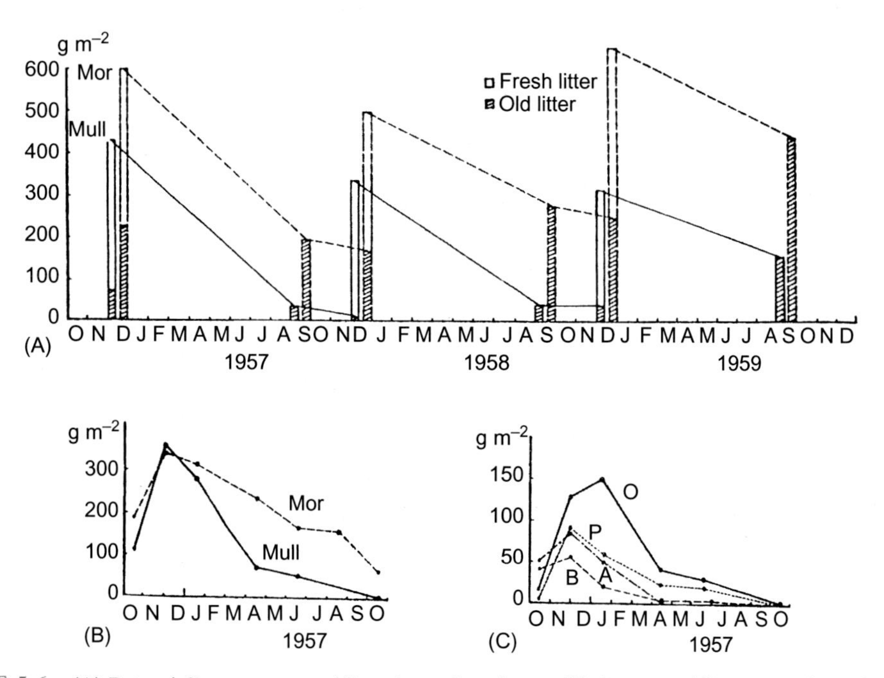

FIGURE 5.6 (A) Rate of disappearance of litter in mull and mor. (B) Amount of litter on a cleaned surface in mull and mor. (C) Amount of fresh oak (O), birch (B), poplar (P), and alder (A) litter in mull. *Source: From Witkamp, M., van der Drift, J., 1961. Breakdown of forest litter in relation to environmental factors. Plant Soil, 15, 295–311.*

TABLE 5.1 Summary of Litter Decomposition Experiments Conducted on Clear-Cut (WS 7) and Control (WS 2) Watersheds, at the Coweeta Hydrologic Laboratory[a]

	74–75		75–77		77–78			
	WS 7, Precut		WS 2, Precut		WS 7, Postcut		WS 2, Postcut	
Species	**Rate**	**% Remaining**	**Rate**	**% Remaining**	**Rate**	**% Remaining**	**Rate**	**% Remaining**
Liriodendron tulipifera	−0.682	49.4	−0.656	49.8	−0.545	60.0	−0.814	47.3
Acer rubrum	−0.529	49.0	−0.477	57.9	−0.324	71.5	−0.368	67.9
Quercus prinus	−0.336	69.3	−0.285	72.2	−0.242	79.3	−0.3000	76.0
Cornus florida	−1.309	27.8	−0.711	47.8	−0.531	59.6	−0.825	43.4
Robinia pseudoacacia	−0.250	72.4	−0.530	49.7	−0.330	69.1	−0.330	70.7

[a]*Data are summarized for a 1-year study on WS 2 and 7 (1974–75), a 2-year study on WS 2 (1975–77), and a 1-year study on WS 2 and 7 (1977–78). Values are shown for decay rate (per year) and % remaining after 1 year. Only first-year decay results are presented here. From Crossley Jr. (unpublished).*

FIGURE 5.7 View of leaf litterbag.

resource quality factors (Table 5.2) (Blair, 1988a) (Berg, 2014; Melillo et al., 1982). Decomposition rates also vary between habitats and forest types, and litterbags have proved to be useful in delineating and analyzing differences (Table 5.3) (Cromack, 1973; Heneghan et al., 1998, 1999).

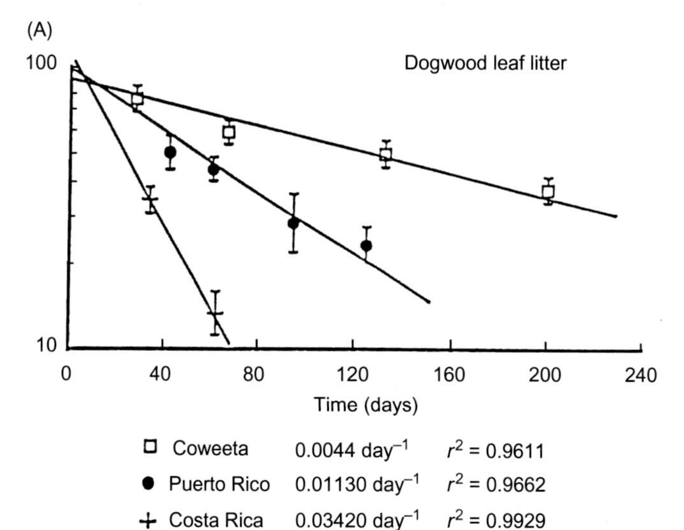

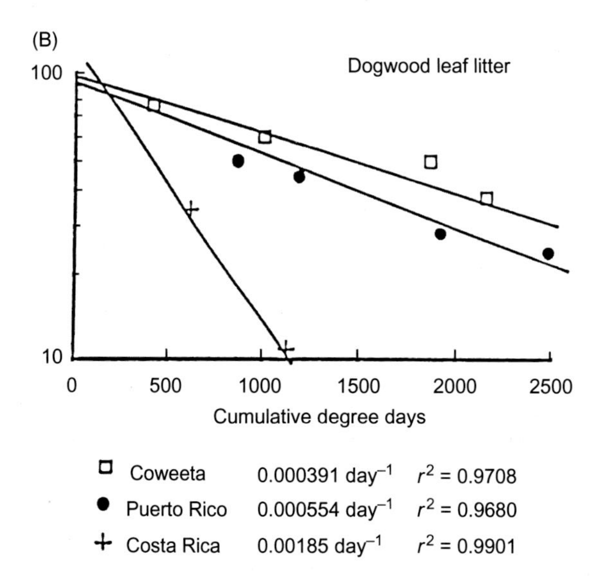

FIGURE 5.8 Litterbags and masses remaining showing days elapsed (A) and degree days (B) at one temperate (Coweeta) and two tropical (Puerto Rico and Costa Rica) sites. *Source: From Crossley and Haines, unpublished (1990).*

Use of litterbags does have its problems. Fine-mesh bags, with openings of 1–2 mm, will exclude most macrofauna and thus underestimate decomposition rates. Larger meshes allow larger fragments to escape the bags, thus overestimating decomposition. The microclimate within litterbags tends to be moister than that of unbagged litter, and thus more favorable for microbial activity (Vossbrinck et al., 1979). In most cases litterbags probably underestimate actual breakdown rates and do not account for the fate of the fine particulate organic matter that falls from the bags and becomes part of the F, H, and mineral SOM pools (Heal et al., 1997). Their usefulness for comparative studies and for nutrient measurements makes them important tools, nevertheless.

TABLE 5.2 Initial Litter Quality Variables as Predictors of First-Year Decay Rates[a]

Initial Litter Quality Variable	r^2	Slope	Y-Intercept
% Nitrogen	0.271	−0.978	−1.29
C:N ratio	0.138	0.007	−0.12
% Lignin	0.987	0.029	−0.96
Lignin:N ratio	0.967	0.027	−1.05
% Water soluble	0.322	−0.015	−0.06
% Ethanol soluble	0.426	−0.040	−0.32

[a]*Coefficients of determination (r^2), slopes, and Y-intercepts of regressions relating first-year decay rate constants (k) to initial litter quality variables for litter of the three species examined.*
From Blair (1988a).

TABLE 5.3 First-Year Litter Breakdown Rates for Single Species[a]

Species	Year	% Weight Remaining	Exponential Loss Rate (k^a)	Correlation Coefficient (r)
White pine	1969–70	59.5	−0.52(12)	−0.888[b]
	1970–71	65.4	−0.42(19)	−0.480[c]
Chestnut oak	1969–70	57.8	−0.55(33)	−0.866[b]
	1970–71	51.4	−0.66(31)	−0.890[b]
White oak	1969–70	47.6	−0.74(36)	−0.887[b]
	1970–71	49.6	−0.70(32)	−0.906[b]
Red maple	1969–70	43.6	−0.83(36)	−0.934[b]
	1970–71	48.6	−0.72(30)	−0.839[b]
Dogwood	1969–70	30.8	−1.18(37)	−0.939[b]
	1970–71	26.0	−1.35(33)	−0.948[b]

[a]Where *the annual exponential loss rate (k) is estimated from a semilogarithmic regression (base e) of monthly weight loss of litterbags.*
[b]*Denotes* $P < 0.01$.
[c]*Denotes* $P < 0.05$.
From Cromack, K., 1973. Litter Production and Litter Decomposition in a Mixed Hardwood Watershed and in a White Pine Watershed at Coweeta Hydrologic Station, North Carolina. Ph.D. Dissertation, University of Georgia, Athens, GA.

Alternatives to litterbags and modifications of the standard approach have been reported. For example, individual leaves tied together by their petioles on a string ("trot-lines") were used by Crossley Jr. (unpublished data). Loss of weight (and area) by individual leaves measured through time yields estimates of litter breakdown rates. When biological activity increases in late spring and summer, rapid rates of loss are found. It is not clear whether these rapid losses are due to the separation of large fragments from the leaf, or the unbagged rates allow for larger fauna to attack the decomposing leaf, or both. The simultaneous use of both techniques yields estimates of breakdown rates that doubtless bracket the true values. Blair et al. (1992) used "litter baskets," which confine an entire block of mineral soil along with the L, F, and H layers, within wire mesh to study litter decomposition and nutrient transport down the soil profile. A further recent modification of the technique is that of "litter sandwiches" (Binkley, 2002), in which fiberglass mesh is placed each year on the annual accumulation of litterfall on a defined area of forest floor. Each subsequent year of decomposition can then be measured over an extended period of time. Binkley (2002) found that 80% of litter organic matter decomposed over 10 years in a loblolly pine forest, yielding $k = 0.1655$; the data predicted the steady-state forest floor mass within 10% of the actual value.

5.6 PATTERNS OF MASS LOSS DURING DECOMPOSITION

A graph of mass retained in litterbags during decomposition of leaf litter in a temperate deciduous forest reveals a three-phase curve (Fig. 5.9). Initially, following autumnal leaf drop, there is a rapid decrease in weight, due to loss of rapidly metabolizable compounds or simply

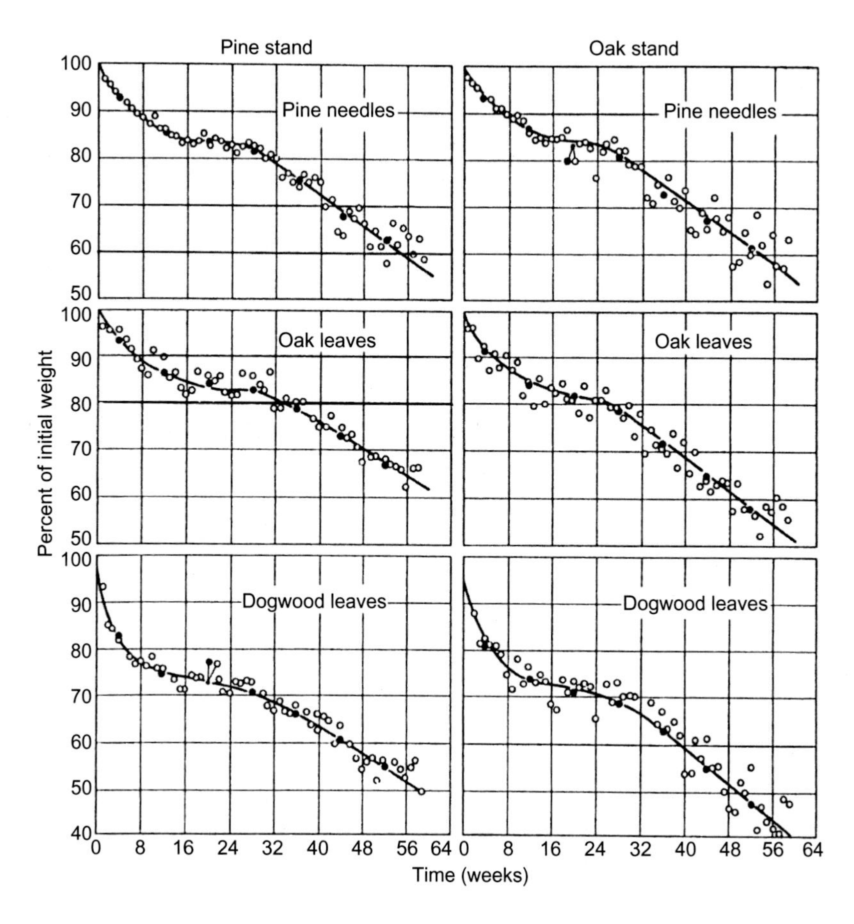

FIGURE 5.9 Three-phase mass loss curve. Weights of decaying leaves in litterbags, expressed as a percentage of initial weight of litter, through time. Three leaf species in two stands. Hollow circles are individual measurements; solid circles are averages for 8-week cycles. Lines fitted by eye. *Source: From Olson, J.S., 1963. Energy storage and the balance of producers and decomposers in ecological systems. Ecology, 44, 322–331.*

readily leachable substances. This initial phase is followed by a slow rate of loss during winter months. During late spring, rates again become accelerated as microclimates become more favorable for biological activity. During winter some microbial and faunal attack occurs, but the major abundance of fauna and microbes is found in litterbags during spring and summer.

Although rates do vary with season, the model using a single exponential constant (k) provides a good fit to these data (Fig. 5.10). The coefficient of determination (r^2) for these

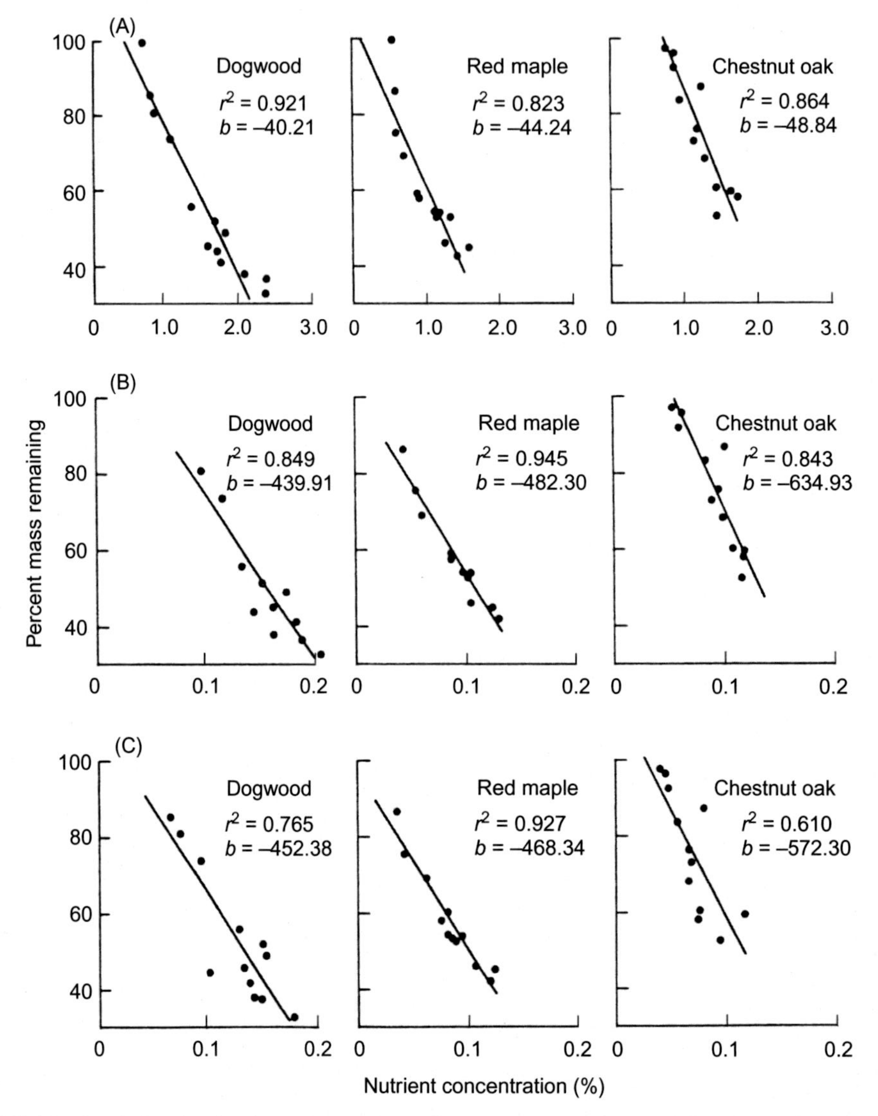

FIGURE 5.10 Single k value for decay (regressions of % mass remaining over (A) nitrogen, (B) sulfur, and (C) phosphorus concentrations in the residual litter for flowering dogwood, red maple, and chestnut oak). *Source: Reprinted from Blair, J.M., 1988a. Nitrogen, sulfur and phosphorus dynamics in decomposing deciduous leaf litter in the southern Appalachians. Soil Biol. Biochem. 20, 693–701, with permission.*

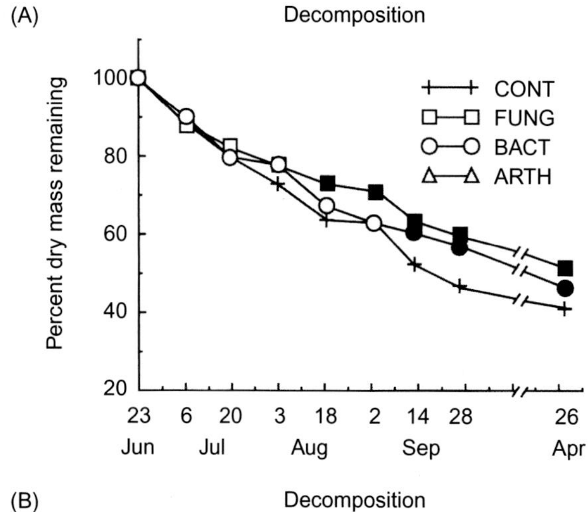

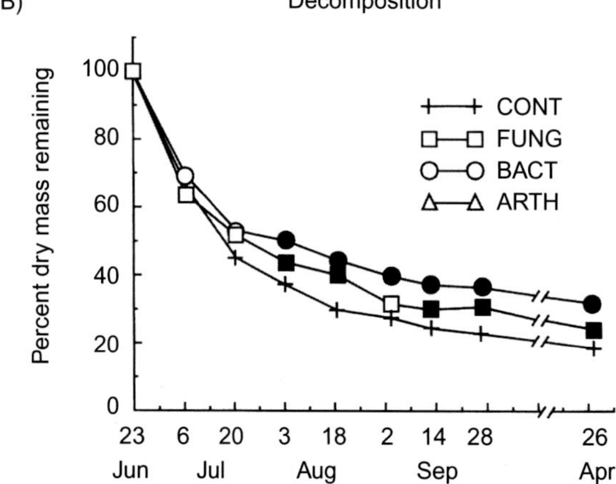

FIGURE 5.11 Mass loss rates for surface (A) and buried (B) rye litter over 320 days. *CONT*, Control situation; *FUNG*, fungicide, with about one-half fungal population biomass; *BACT*, bactericide (oxycarboxin); *ARTH*, arthropod repellant (naphthalene). Filled symbols are significantly different from controls ($P < 0.05$) on a given date. *Source: Modified from Beare, M.H., Parmelee, R.W., Hendrix, P.F., Cheng, W., Coleman, D.C., Crossley, D.A., Jr., 1992. Microbial and faunal interactions and effects on litter nitrogen and decomposition in agroecosystems. Ecol. Monograph., 62, 569–591, with permission.*

curves usually exceeds 0.85. The constant k conceals seasonal dynamics but is a useful means for comparing leaf types or habitats, or geographical regions. More precision can be gained by calculating k from the spring–summer values alone. Some typical breakdown rates for forest litter are given in Table 5.3.

Breakdown rates in agricultural systems are generally more rapid than in forested systems: crop residues, as a rule, tend to have fewer recalcitrant components. Fig. 5.11 shows mass loss rates for rye litter from litterbags either placed on the soil surface (no-tillage) or buried following plowing (conventional tillage) (Beare et al., 1992). Loss rates were faster under conventional tillage ($k = 0.03$ per day) than in no-tillage soils ($k = 0.02$ per day). Usually, buried residues decompose more rapidly than surface residues because of more intimate contact with mineral soil and microbes and because of more moderated microclimate beneath the soil surface. In any case, rates for rye litter were much faster than for forest tree leaf litter.

To establish the extent of impact of faunal-moderated decomposition on plant residues, Tian et al. (1995) used gnotobiotic microcosms with residues of five plants from tropical agroecosystems (*Dactyladenia barteri*, *Gliricidia sepium*, and *Leucaena leucocephala* prunings, maize stover, and rice straw), placed on the surface of large pots filled with an Oxic Paleustalf in Nigeria. The soil was defaunated by sun drying and the larger fauna removed by hand. Eighteen mature earthworms (*Eudrilus eugeniae*) and/or three millipedes (Spirostreptidae) were then added to a subset of pots. The experiment was run for 10 weeks, with an intermediate sampling of mass losses occurring after 4 weeks. Both earthworms and millipedes contributed more to the breakdown of the low-quality litter (*Dactyladenia*, maize stover and rice straw), compared to the decomposition of the higher-quality leguminous tree prunings (Tian et al., 1995). This effect has been observed in a number of forest ecosystems, as noted later in this chapter.

5.7 EFFECTS OF FAUNA ON LITTER BREAKDOWN RATES

The association of soil fauna with litter decomposition is an ancient one. Labandeira et al. (1997) reviewed the evidence concerning associations of soil fauna in the geologic record. The incidence of oribatid mite feeding in coal deposits from Illinois and Appalachian sedimentary basins occurred in all major plant taxa in Pennsylvanian coal swamps. Virtually every type of plant litter tissue was used by the mites. Evidence for termites and holometabolous wood-boring insects dates back to the early Mesozoic. The illustrations published by Labandeira et al. (1997) provide striking evidence of the importance of detritivores in these primitive forests.

In more modern times, the Russian soil scientist Galina Kurcheva (1960, 1964) found that naphthalene (an insecticide) applied to oak leaf litter would drastically reduce the rate of breakdown. In the succeeding decades various biocides and other techniques have been used to suppress various components of the soil biota, as a measure of their importance in leaf litter breakdown (e.g., Beare et al., 1992; Parker et al., 1984). The upshot of these experimental manipulations has been to demonstrate that bacterial, fungal, and faunal members of the soil biota all have significant effects on litter breakdown. Given that actual breakdown and decomposition rates are a function of the *interaction* among the various biota and with substrate quality and climate, the rate estimates derived from manipulations must be accepted with caution. The main effects, however, seem clear.

Seastedt (1984b) suggested that the equation describing litter breakdown might be partitioned into components, so that the constant k could be considered as the sum of several ks:

$$dX/dt = -kX = -(k_{\text{bacteria}} + k_{\text{fungi}} + k_{\text{fauna}})X$$

Seastedt reviewed studies in which microarthropods had been suppressed and found that a variable percentage of breakdown rates could be attributed to microarthropod activities. Table 5.4 provides the results of the Seastedt equation applied to forest tree litter in a floodplain forest in Athens, Georgia, United States. Litterbags with a 1 mm mesh size were used, so that macrofauna were excluded from the bags. Naphthalene applications were used to reduce microarthropod populations in some of the litterbags. The results show that the importance of microarthropods varied with litter quality. Microarthropod activities were least significant for the more rapidly decomposing litter species (dogwood,

TABLE 5.4 Percent of Leaf Litter Decomposition (Mass Loss) Attributable to Soil Fauna[a]

Leaf Species	k_T	k_{NA}	k_F	Percent Due to Fauna
Dogwood	−0.00248	−0.00089	−0.00159	64.1
Sweetgum	−0.00248	−0.00089	−0.00159	71.4
Tulip-poplar	−0.00229	−0.00113	−0.00116	50.7
Red maple	− 0.00125	−0.00069	−0.00056	44.8
Water oak	−0.00174	−0.00037	−0.00137	78.7
White oak	−0.00216	−0.00076	−0.00140	64.8

[a]*Loss rate due to faunal activities calculates as total rate (k_T) minus naphthalene rate (k_{NA}) equals rate due to fauna (k_F). Percent difference calculated as (k_F/k_T) × 100.*
From Crossley Jr. (unpublished).

tulip-poplar) and were most important for the slowest, most recalcitrant litter type (water oak). In a carefully controlled experiment, Couteaux et al. (1991) measured decomposition of litters of various qualities, namely with C:N of 75 (low quality) versus 40 (higher quality). The soil fauna contributed more to the decomposition of the low-quality substrate, and the effect was significantly greater at later stages of incubation in the 24-week experiment, with greater faunal complexity accounting for greater amount of dry mass loss and total CO_2 evolution per unit time.

Experimental approaches such as these must be interpreted with caution. Usually more than one component of the system is modified by manipulations, be they chemical or physical ones (Crossley et al., 1992). Other approaches, such as tracer methods and laboratory microcosms, need to be used in conjunction with manipulative experiments. Biocides such as naphthalene may alter other system components, sometimes to a large extent. Naphthalene, for example, may suppress microbes. However, González et al. (2001) reported higher microbial biomass when naphthalene was used to exclude soil fauna.

5.8 NUTRIENT MOVEMENT DURING DECOMPOSITION

Soils contain many of the same elements as found in their underlying rock parent material, but the proportions differ greatly. Elements such as Ca, Mg, K, and Na are lost as soluble cations during weathering, depending on climatic conditions (especially precipitation). Some other elements, such as Fe and Al, are resistant to leaching losses and their proportions may increase compared to rocks. Movements of cations are governed by the exchange properties of the soil, which are dependent upon the nature of the clays and amount and type of organic matter. Exchangeable cations in soils include Ca^{2+}, Mg^{2+}, K^+, NH_4^+, and Na^+, affinities for exchange sites (i.e., energy of adsorption) decreasing approximately in that order. Certain anions are not as tightly held in soils, again depending on the nature of clay colloids and on soil pH. Phosphate ions, multiply charged, are more tightly fixed by anion exchange properties than are singly charged ions such as nitrate (Bowen, 1979; Foth, 1990).

During the decomposition process elements are converted from organic to inorganic forms (i.e., mineralized) and may enter the exchangeable pools, from which they are available for plant uptake or microbial use. Cellulose and hemicellulose account for more than 50% of carbon in plant debris and help to fuel microbial processes, such as transformations of nitrogen (Fig. 5.12) and sulfur (Fig. 5.13), which gradually reduce the C:N and C:S ratios in decomposing materials.

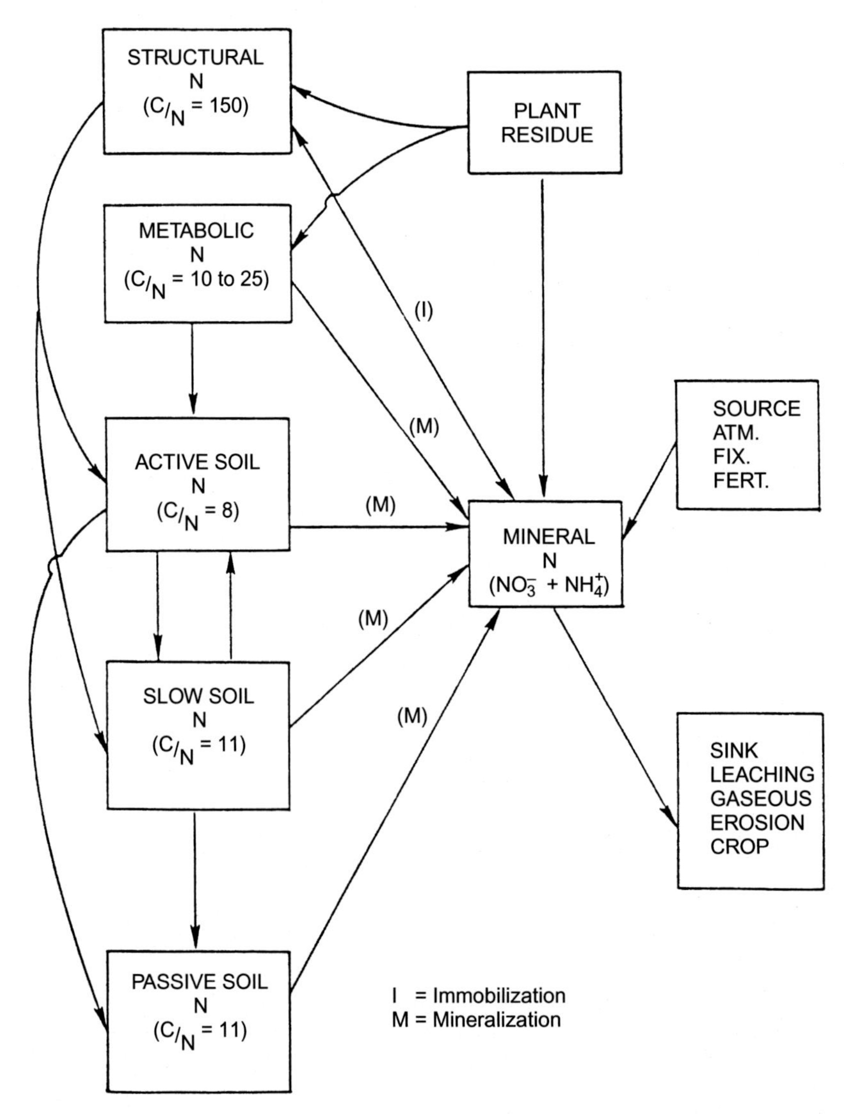

FIGURE 5.12 Soil nitrogen (N) cycle showing active (1–1.5 years), slow (10–100 years), and passive (100–1000 years) fractions. Flow diagram for the nitrogen submodel of the CENTURY model. *Source: From Parton et al. (1987), with permission.*

As plant litter decomposes, the elemental mix changes because of differential mobility and biological fixation (i.e., assimilation of constituent elements into tissues). Carbon is lost through microbial respiration, as cellulose and other labile organic compounds are hydrolyzed and utilized in growth and maintenance. Potassium is highly mobile until it encounters exchange sites, where it can become fixed. Sodium ions, which are more mobile in soils, are not accumulated in plants but are essential for animals. The "herbivore exclusion hypothesis" (McNaughton, 1976; McNaughton et al., 1998) proposes that plants discriminate against sodium and thereby limit herbivory. Sodium does accumulate in food chains, often increasing by a factor of 2–3 between trophic transfers. Calcium is lost from decomposing litter at about the same rate as mass is lost (Cromack, 1973).

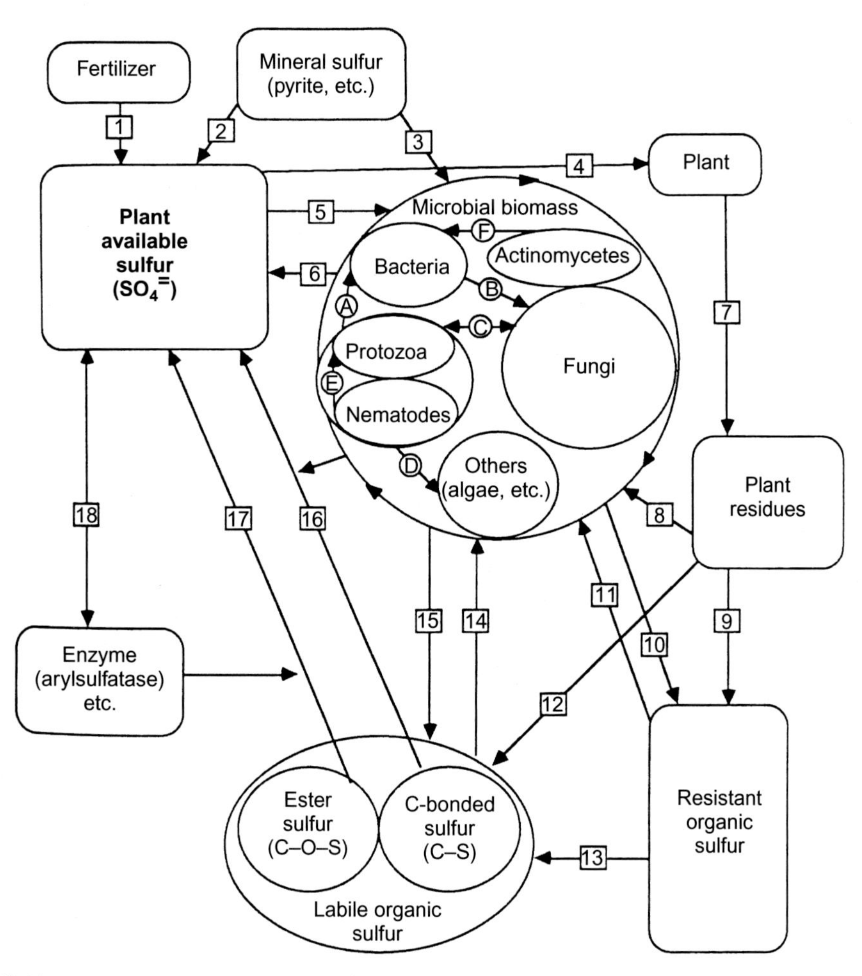

FIGURE 5.13 Conceptual model for microbial and faunal components of the sulfur cycle in soil. *Source: From Gupta, V.V.S.R., Germida, J.J., 1989. Influence of bacterial–amoebal interactions on sulfur transformations in soil. Soil Biol. Biochem. 21, 921–930.*

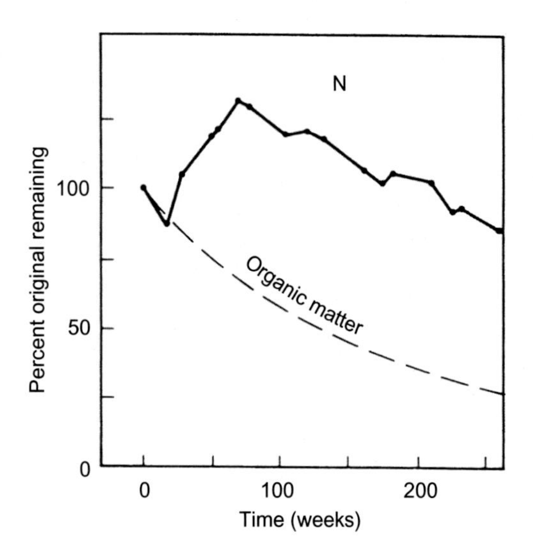

FIGURE 5.14 Rates of nutrient immobilization and mineralization from decomposing Scots pine (*Pinus sylvestris*) litter over a 5-year period. Note initial influx of nitrogen (N) into litter. *Source: From Staaf and Berg (1982).*

The nitrogen content of decomposing litter increases during the initial stages of decomposition, and then declines (Fig. 5.14) (Berg and Staaf, 1981; Berg, 2014). Nitrogen is mineralized during decomposition and is simultaneously immobilized by microbes, resulting in an increase in the concentration of nitrogen in the litter, and in the absolute amount of N if it is transported into the litter from soil via fungal hyphae (Frey, 2000) or by atmospheric N-fixation. As decomposition proceeds, the C:N ratio declines until the substrate becomes more suitable for microbial action. In some forests the period of nitrogen increase may extend for 2 years or more (Fig. 5.15) (Blair and Crossley, 1988). Phosphorus and sulfur also show increases in absolute amounts during decomposition of some species of tree leaf litter (Fig. 5.16) (Blair, 1988a), even though mass is being lost. Calcium and magnesium concentrations in decomposing litter change only slightly through time. There may be an initial decrease in concentration followed by a slight increase (Blair, 1988b). Thus, the absolute amounts of these elements during decomposition approximately track the loss of mass. Potassium is not a structural element, and is lost via solubilization more rapidly than mass is lost from decomposing leaf litter. Decomposing woody litter, in contrast, accumulates calcium and phosphorus, evidently as a result of fungal invasion and translocation from soil.

The nitrogen pool in decomposing litter is a dynamic one. Although nitrogen is accumulating, there is evidently a large amount of turnover taking place. When tracer amounts of ^{15}N (as $(NH_4)_2SO_4$) were added to leaf litter, significant losses of tracer took place even as total nitrogen accumulated (Fig. 5.17) (Blair et al., 1992). These authors suggested that nitrogen became incorporated from exogenous sources, in amounts greater than those lost through biotic factors. Inputs of nitrogen via rainfall or canopy throughfall are a potential source of added nitrogen. However, these could not account for the amount of nitrogen immobilized in litter. Fungal translocation from lower layers (F, H, or mineral soil) is another possibility (Frey et al., 2000). Finally, lateral transport to and from "hot spots" in the forest floor may contribute to the dilution of tracers.

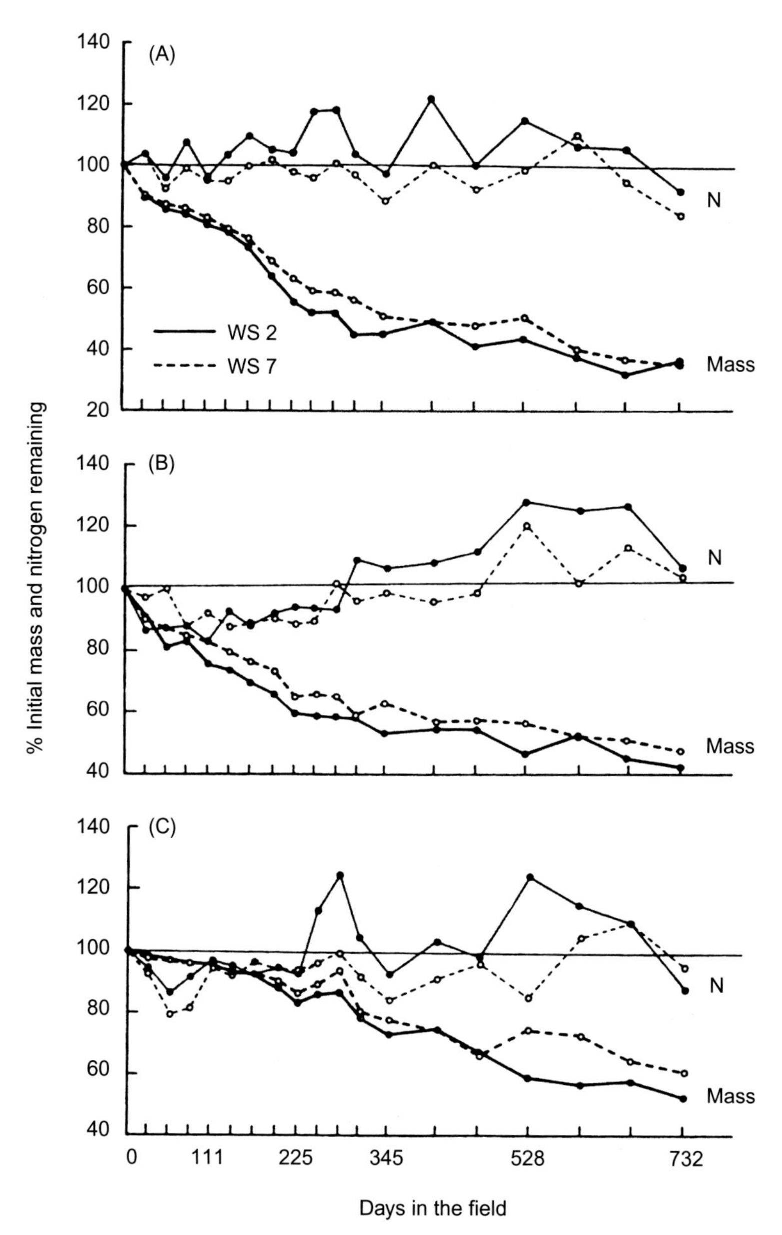

FIGURE 5.15 Mean percentage of initial mass and nitrogen remaining over time in (A) *C. florida*, (B) *Acer rubrum*, and (C) *Quercus prinus* litter on uncut WS 2 (solid line) and clear-cut WS 7 (dashed line) at Coweeta Hydrologic Laboratory from January 1975 to January 1977. *Source: From Blair, J.M., Crossley, D.A., Jr., 1988. Litter decomposition, nitrogen dynamics and litter microarthropods in a southern Appalachian hardwood forest 8 years following clearcutting. J. Appl. Ecol. 25, 683–698.*

Nitrogen addition effects on microbial biomass were assessed in a meta-analysis of 82 published field studies. Treseder (2008) found that microbial biomass declined an average of 15% under nitrogen fertilization, but fungal and bacterial diversities were not significantly altered. Interestingly, soil CO_2 emissions responded in concert with the biomass. In contrast, in a more extensive study in space and time, Geisseler and Scow (2014) made a meta-analysis of 107 datasets from 64 long-term trials worldwide. They found that there

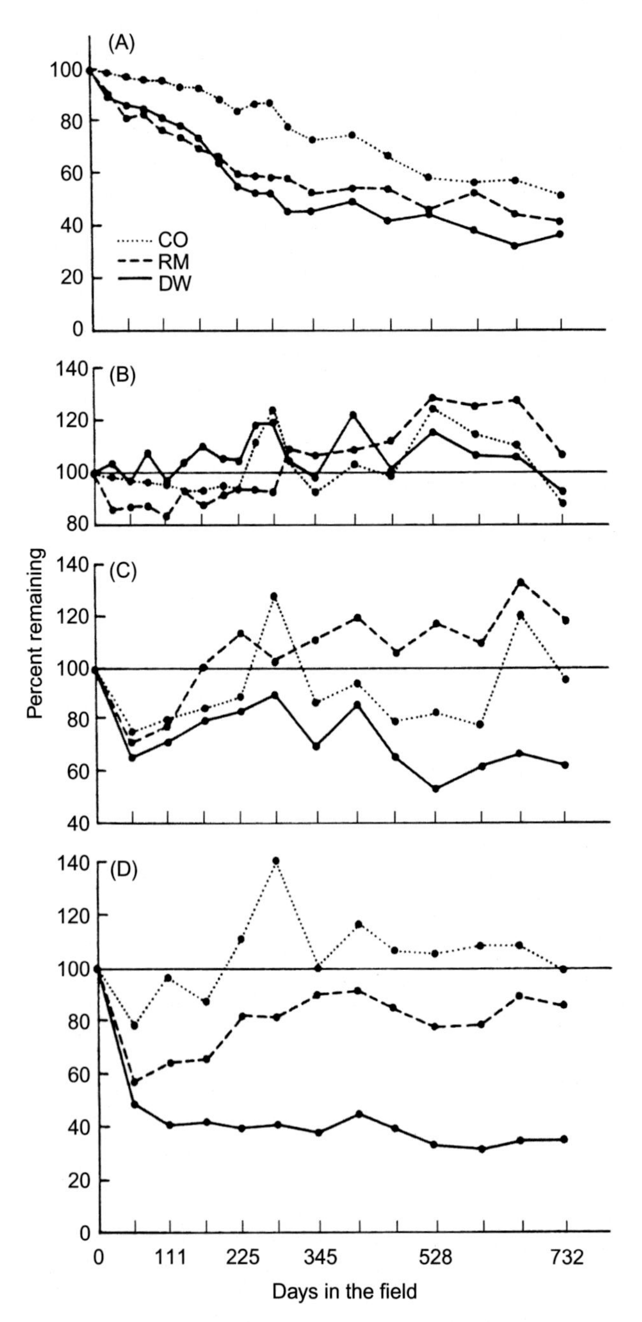

FIGURE 5.16 Mass, nitrogen, phosphorus, and sulfur changes in (A) mass and absolute amounts of (B) nitrogen, (C) phosphorus, and (D) sulfur in flowering dogwood (DW), red maple (RM), and chestnut oak (CO) litter decomposing over a 2-year period. *Source: Reprinted from Blair, J.M., 1988a. Nitrogen, sulfur and phosphorus dynamics in decomposing deciduous leaf litter in the southern Appalachians. Soil Biol. Biochem. 20, 693–701, with permission.*

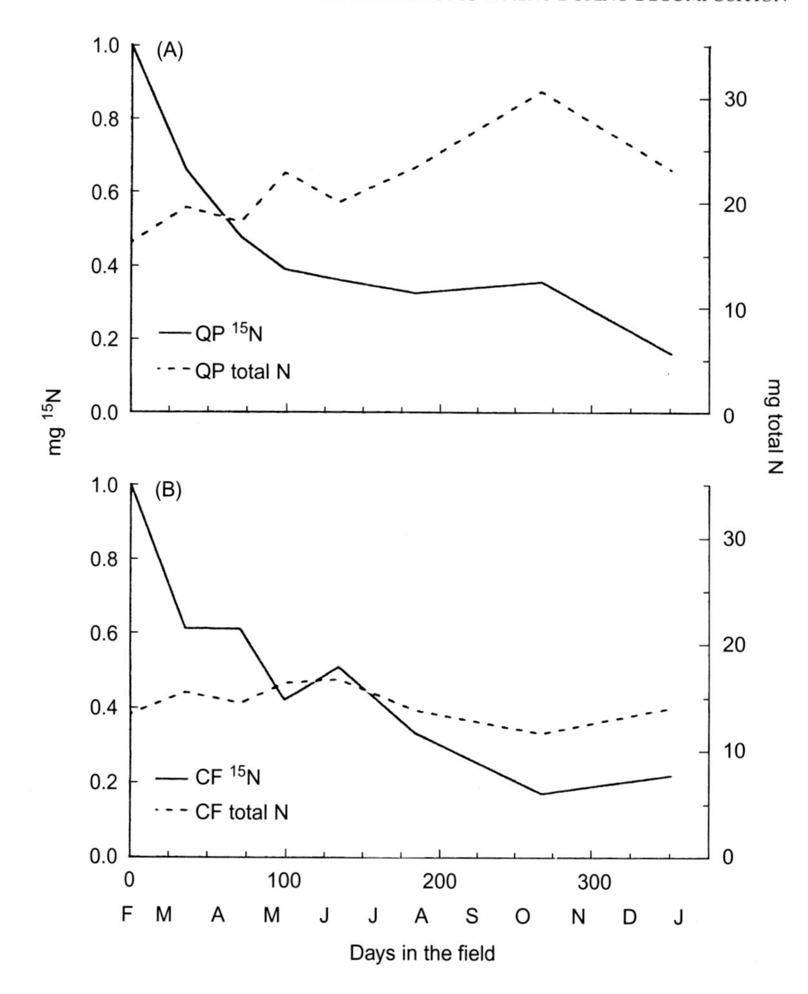

FIGURE 5.17 ^{15}N and total N. Changes in the amount of added ^{15}N recovered in the litter (solid line) versus changes in the total amount of nitrogen (N) (dashed line) over time in (A) *Quercus prinus* (QP) and (b) *C. florida* (CF) litter from control litter baskets. *Source: Reprinted, with permission, from Blair, J.M., Crossley, D. A. Jr., Callaham, L.C., 1992. Effects of litter quality and microarthropods on N dynamics and retention of exogenous* ^{15}N *in decomposing litter. Biol. Fertil. Soils 12, 241–252.*

was a 15.1% increase in microbial carbon (C_{mic}) above that in unfertilized controls. The main finding was that soil organic C also increased proportionally and was a major factor in increase of C_{mic}. The greatest impact occurred in trials running more than 20 years in duration. Significantly, there were no increases in C_{mic} at pH values less than 5.

As noted in Chapter 2, Primary Production Processes in Soils: Roots and Rhizosphere Associates, one of the main sources of particulate organic matter to soils is that from decomposing roots. Researchers have often used root litterbags, and followed dry matter and nutrient loss over several months to a few years. Unfortunately, preparation of the root tissues for decomposition studies represents a significant departure from in situ conditions. Dornbush et al. (2002) developed an intact-core technique that retains natural rhizosphere associations, maintaining in situ decay conditions. Cores (15 cm × 5.3 cm diameter) were taken under monospecific stands of silver maple, maize and winter wheat, and covered at the top and bottom with 160 μm mesh polyethylene caps. The same mesh was used to make litterbags, to hold an amount of roots similar to that in the soil cores.

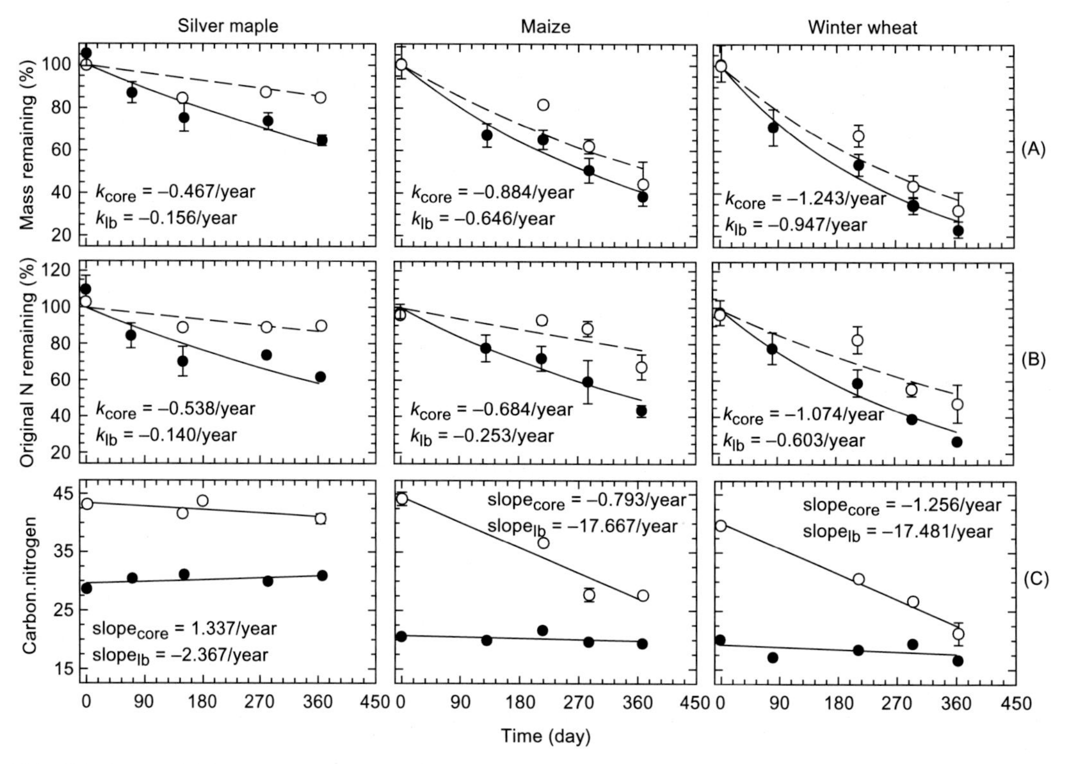

FIGURE 5.18 (A) Mass loss, (B) nitrogen loss, and (C) changes in carbon–nitrogen ratio for fine roots of silver maple, maize, and winter wheat decomposed in a riparian meadow in central Iowa in the United States. Solid circles represent intact cores, and open circles represent litterbags (lb); k values are based on exponential decay models. In the upper six panels, the intact-core data are normalized such that 100% equals the $t = 0$ intercept of the exponential decay regression; error bars are ±1 SE. *Source: From Dornbush, M.E., Isenhart, T.M., Raich, J.W., 2002. Quantifying fine-root decomposition: an alternative to buried litterbags. Ecology, 83, 2985–2990.*

After reinstallation in the field sites, cores and bags were retrieved seasonally at time intervals up to 1 year and the decay rates compared. After 1 year, mass loss was 10%–23% greater, and N release was from 21% to 29% higher within intact cores, than in litterbags (Fig. 5.18). Dornbush et al. (2002) attributed a majority of the differences to alterations in litterbag-induced dynamics of decomposer organisms and unavoidable changes to fine-root size-class composition (≤1 mm diameter) in the bags.

Nutrient movement through the L, F, and H layers and through mineral soil has been studied intensively by both biologists and organic geochemists. Tracking nutrient movements in soil is difficult because of nonlinearities of flow due to the existence of preferential flow paths compared to the bulk soil. For example, in a carefully instrumented study in a mixed beech and spruce forest in Switzerland, Bundt et al. (2001) measured organic C concentrations from 10% to 70% higher in the preferential flow paths than in the soil matrix. In addition, organic N concentrations, effective cation exchange capacity, and the base saturation were all increased in the preferential flow paths. DNA concentrations and direct cell counts showed similar patterns, but there were no changes in domain-specific organisms such as Eukarya or Archaea. However, *Pseudomonas* showed increased abundances in the preferential flow paths, indicating that it responds to increased organic matter status in this "hot spot" similarly to that in other ones, such as the rhizosphere (Bundt et al., 2001).

It is essential to keep in mind the marked spatial separation that exists in distribution of decomposer microbes in forest floors. In an analysis of bacteria and fungi in litter and top-soil of a *Picea abies* forest, Baldrian et al. (2012) measured RNA (indication of active fraction) versus DNA (total amounts present), and found that fungi dominate in the litter horizon, and there were similar amounts of fungal and bacterial biomass in the organic horizons. Cellulose decomposition was mediated by highly diverse fungal populations, largely distinct between soil horizons (Baldrian et al., 2012).

In studies of SOM breakdown and the activities of primary decomposers (prokaryotes and fungi), it is important to recognize that all conditions in similar sites are not always equivalent. In a pioneering comparison of cross-site decomposition rates in forest, mountain meadow, and grassland sites in eastern Wyoming and northern Colorado, Hunt et al. (1988) noted a "home-field advantage" (HFA) (greater decomposition rate) occurred for leaf litters decomposing in their own site versus being translocated to other ones to be decomposed a few hundred km away, but in similar climatic conditions. In an extensive comparison of forest leaf litter mass losses in a wide range of sites in North America, South America, and Europe, Ayres et al. (2009) corroborated this, and found that the HFA averaged an increased 8% at home versus in foreign sites. They noted that although climate and initial litter quality explain up to 70% of the total variability, HFA accounts for several percent more. Although the HFA effect is observed in many situations, it is not a general rule, as Freschet et al. (2012) pointed out, and these authors posit that HFA generally only occurs when foreign litter and the typical native litter at a particular site are quite different. Their analysis suggests that other factors such as functional efficiency of decomposer communities across a spectrum of substrate types and qualities determine the outcomes of reciprocal transplant experiments such as those used to demonstrate HFA effects.

Nutrients and organic matter also move through soils in soluble forms, e.g., as dissolved organic matter (DOM). In general, sorptive interactions between DOM and mineral phases contribute to the preservation of SOM. However, the situation is considerably more complex than this, due to the existence of several structural features in the soil profile. In a comparison of movement in the profiles of seven forest soils, Guggenberger and Kaiser (2003) considered the location of OM within the soil profile. They contrasted the amount and concentrations of OM in soil water with that on "fresh" or exposed mineral surfaces, versus that on "natural" soil surfaces, which have a considerable amount of microbially derived biofilms. Based on their study, they produced a conceptual model of entities in the soil profile that differed in their biological activity. In the soil solution, soil microorganism density is low, OM concentration is low, and there is little biodegradation of DOM. On the "fresh" mineral surfaces, there is OM sorption to minerals, including complexation of functional groups, changed conformation, incalation in small pores, and sorptive stabilization. On the "natural" soil surfaces with high microorganism density, there is OM sorption into biofilms; the sorption concentrates OM, which appears to be a prerequisite for decomposition (Fig. 5.19) (Guggenberger and Kaiser, 2003). Research is currently under

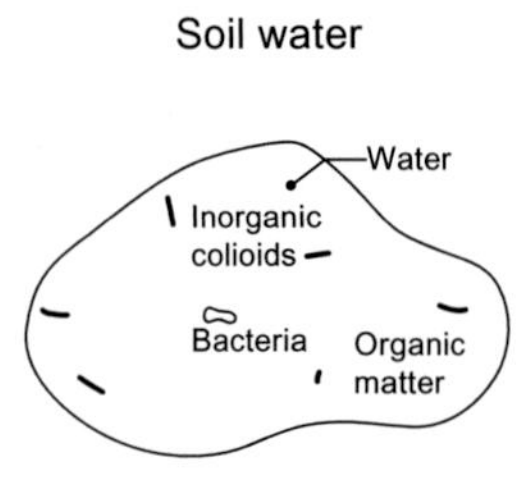

FIGURE 5.19 Conceptual model of the fate of organic matter in the soil solution and sorbed to soil surfaces differing in their biological activity. *Source: From Guggenberger, G., Kaiser, K., 2003. Dissolved organic matter in soil: challenging the paradigm of sorptive preservation. Geoderma, 113, 293–310.*

way to determine the dynamics of exchange phenomena between DOM and biofilms. The OM input enhances the heterotrophic activity in the biofilm, converting the DOM into either organic compounds by resynthesis or inorganic mineralization products. Iron hydrous oxides embedded within the biofilms may serve as both a sorbent and a shuttle for dissolved organic compounds from the surrounding aqueous media.

McDowell (2003) notes that our knowledge of the ecological significance of dissolved organic nitrogen relative to that of dissolved organic carbon is yet in its infancy. We also need more information on how fluxes of these dissolved substances are altered in human-dominated environments. New analytical techniques are being developed to better quantify dissolved organic compounds in soils and their effects on microorganisms, particularly saprophytic fungi and mycorrhizae (Blagodatskaya and Kuzyakov, 2013). These new approaches are giving us better insights into soil ecosystem function, as noted in Chapter 3, Secondary Production: Activities of Heterotrophic Organisms—Microbes, and Chapter 8, Future Developments in Soil Ecology.

5.9 NUTRIENT CYCLING LINKS IN SOIL SYSTEMS

In addition to the translocation abilities of saprophytic fungi mentioned previously, there is a rapidly growing literature on the roles of ectomycorrhizal and other symbiotic fungi in mobilizing N and P from organic pools (Bending and Read, 1996; Chalot and Brun, 1998; Northup et al., 1995; van der Heijden et al., 2008). This work has extended to the clearly demonstrated roles of ectomycorrhizal fungi in mineral weathering, i.e., mobilizing inorganic N as ammonium from the interstices of feldspar minerals, and solubilizing P from volcanic rocks (Landeweert et al., 2001). As noted about the nutrition of oribatid mites in Chapter 4, Secondary Production: Activities of Heterotrophic Organisms—The

Soil Fauna, the fungal mat-like structures formed by ectomycorrhizal fungi (e.g., *Hysterangium*, *Hydnellum*, and *Gautieria* spp.) at the interface of the surface humus layer and upper A horizon may cover several m^2 of forest floor. The mineral soil within this concentrated mass of mycorrhizal hyphae is more strongly weathered than the surrounding soil, as a result of the excretion of oxalic acid by the fungi. Within the mat, calcium oxalate crystals are abundant and decomposition rates and nutrient availability are increased relative to the nearby soil (Entry et al., 1992). The calcium oxalate crystals are a readily available source of Ca ions for the mites, as was demonstrated elegantly by Cromack et al. (1988). The fact that this inorganic "hot spot" serves as a possible source of both inorganic and organic nutrients for the microbivorous fauna is further proof of the impressive nutrient feedback loops operating in soils. For more extensive coverage of nutrient dynamics in soil profiles over centuries and millennia, see the extensive synthesis of Richter and Markewitz (2001).

Much progress has been made on the role of Biochar-C in soils. This category of substances, produced by pyrolysis of feedstocks, such as wood chips, has been studied under experimental conditions worldwide in both temperate and tropical ecosystems. Using feedstocks produced at 400°C and 550°C, Singh et al.(2012) found that, when incubated in a vertisol with a predominantly C4 vegetation source, there were mean residence times (MRTs) from 90 to 1600 years. Biochar-C MRT is likely longer in lower moisture, temperature, and greater nutrient limitation conditions. Similar findings were reported by Santos et al. (2012) in temperate forest soils. With the prevalence of forest and wildland fires worldwide, the explicit category of Biochar-C should be included for more realistic SOM models.

5.10 ROLE OF SOIL FAUNA IN ORGANIC MATTER DYNAMICS AND NUTRIENT TURNOVER

For the last several decades there has been interest in the role of soil fauna in litter and organic matter turnover in ecosystems. The pioneering studies of Darwin (1881) and Müller (1878) emphasized the prominent signs left in many temperate forest and grassland communities by earthworm, mesofauna, and biotic activities in general. A very prescient account of the "biotic" structure of soils was given by Jacot (1936). Signs of faunal activity include coating of mineral grains, which has a significant effect on promoting the formation of aggregates (Kubiëna, 1938). Termites in semitropical and tropical regions have similar functions, as well (e.g., Lee and Wood, 1971; Wood et al., 1983). Only a few ecologists are aware, however, that often the soil meso- and microfauna are vastly more numerous; and usually more active in terms of respiratory activity, than the large soil fauna (Coleman, 1994b; Coleman and Wall, 2015; Wolters, 1991).

Our concerns as ecosystem researchers should include both an understanding of which organisms are present and the major processes that they carry out in a wide range of terrestrial ecosystems. Following the flow of energy and nutrients in the system (as noted by Volobuev, 1964) will enable us to concentrate on key processes that occur, avoiding the pitfall of singling out for study only what is obvious to the naked eye. We must get to the appropriate level of resolution to ascertain the roles of participants in soil processes

(Coleman, 1985; Macfadyen, 1969). This requires exploring the myriad of surfaces and volumes that occur in a few mm^3 of soil and organic matter (Elliott, 1986; Elliott and Coleman, 1988; Beare et al., 1995).

Fauna are members of the "organism" category in the Jenny (1941) factors of soil formation (recall from Fig. 1.5: s.f. = f (cl, o, r, p, t), where cl = climate, o = organisms, r = relief, p = parent material, and t = time). As noted by Crocker (1952), only a few of these factors are independent variables, so we are dealing with a multiple-causation, interdependent subset of a terrestrial ecosystem. To simplify matters, let us consider organisms alone, i.e., vegetation, organic matter inputs therefrom, and the array of heterotrophic organisms feeding and decomposing organic detritus. The factors plus ecosystem processes acting over time lead to ecosystem properties (Coleman et al., 1983; Elliott et al., 1994).

The immediate result of faunal-feeding activity is the production of fecal pellets, some of which can be identified as species or group-specific (FitzPatrick, 1984; Jongerius, 1964; Kühnelt, 1958; Pawluk, 1987; Rusek, 1975; Zachariae, 1965). For example, the fecal pellets of collembola and oribatid mites are surrounded by a chitin-rich layer, called the "peritrophic membrane" (Krantz, 1978) which acts to retard the rate at which the fecal pellet disappears (Fig. 5.20). A comprehensive review (Bal, 1982) of soil fauna activities in soil refers to "zoological ripening" as faunal movement of organic matter and mineral materials in previously uncolonized soil. This soil maturation and development process has been of great significance in Dutch polder regions, and has been demonstrated in Canadian (Nielson and Hole, 1964) and New Zealand (Stockdill, 1966) soils as well. These processes are reviewed extensively in Brussaard and Kooistra (1993).

In addition to physical signs, there are chemical indicators of faunal presence and activity. For example, in certain cool, moist New Zealand tussock grassland soils, nearly 10% of the organic P is comprised of phosphonates (C–P bonded) (Newman and Tate, 1980; Tate and Newman, 1982), as contrasted with the more prevalent phosphate esters. Phosphonates are produced by ciliates, and their subsequent rates of input and flow

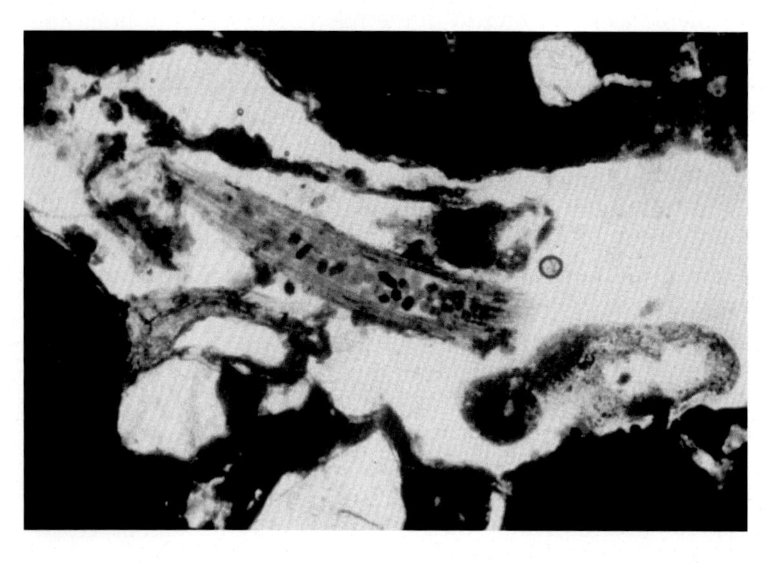

FIGURE 5.20 Fecal pellets in the soil profile from the Horseshoe Bend agroecosystem site, Athens, GA, United States, at a depth of 5–10 cm. *Source: From Larry, T. West, personal communication.*

through the soil P cycle remain unknown (Stewart and McKercher, 1982). More recent studies of temperate pasture soils have measured 1%–3% of total P as phosphonates, and trace amounts of polyphosphates (Turner et al., 2003).

Several authors have reviewed work on experimental pedogenesis (soil formation), examining roles of primary colonizing plants, including dissolution of rock minerals by lichens and fungi, as well as faunal impacts on mineral or soil movement, and organic matter transformation (Bal, 1982; Hallsworth and Crawford, 1965; Landeweert et al., 2001). Webb (1977) studied the effects of particle size, and decomposability of macrofaunal and microfaunal fecal pellets. There are differing effects of comminution (breaking up) of leaf litter by large and small fauna, and they play different roles in facilitating further leaf litter decomposition. Webb (1977) noted that fecal pellets of *Narceus annularis* (Diplopoda: Spirobolidae) had a lower surface-to-mass ratio, whereas those of microarthropods, such as oribatids, had a greater surface-to-volume ratio than the original leaf litter. This should lead to greater decomposition per unit time (Fig. 5.21) (see above comments about the peritrophic membrane).

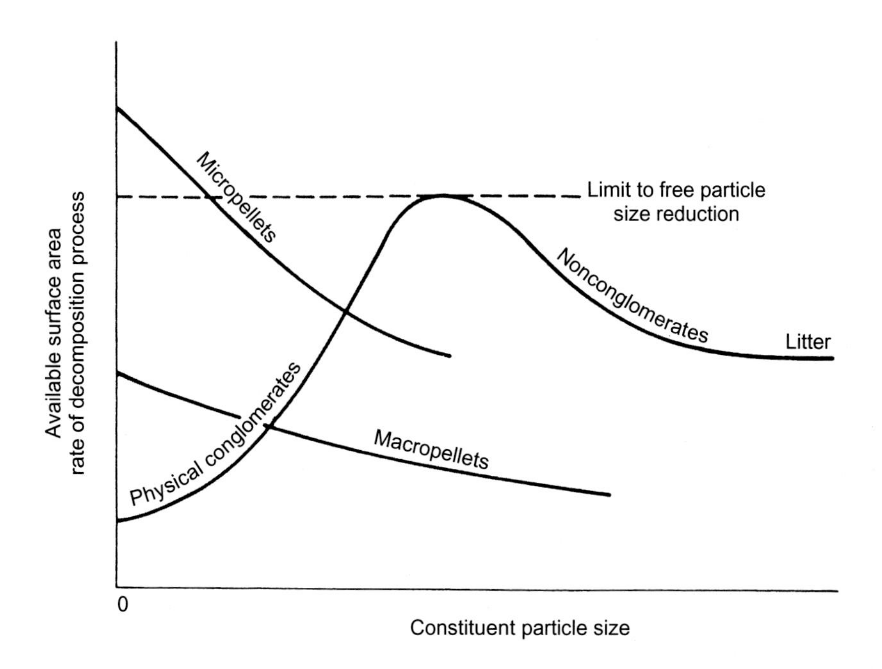

FIGURE 5.21 Graphical representation of physical conglomerate feces differentiation theory. As particle size of litter (right to left) is reduced, surface area and decomposition increase until constituent particles are small enough to aggregate into more stable conglomerates (limit to free particle size reduction). Physical conglomerates increase in size as constituent particle size decreases, but arthropod pellets decrease in size because of the direct relationship of body size to degree of pulverization and pellet (conglomerate) size. Micropellets are therefore able to maintain a much smaller conglomerate size and break the limit to free particle size reduction. *Source: From Webb, D.P., 1977. Regulation of deciduous forest litter decomposition by soil arthropod feces. In: Mattson, W.J. (Ed.), The Role of Arthropods in Forest Ecosystems. Springer, New York, NY, pp. 57–69.*

Physical interpretation of organic matter decomposition should be tempered with careful observation of life history details, such as likelihood of localized aggregation of mite or collembolan fecal pellets which may decompose locally at a much slower rate than hypothesized from in vitro laboratory studies. Substrate quality plays an important role here. Dunger (1983) noted that macroarthropods ingest mineral soil along with litter material. Kilbertus and Vannier (1981) and Touchot et al. (1983) demonstrated ingestion of argillic (clay) material by *Tomocerus* and *Folsomia* sp. (Collembola), a trait that was particularly evident when they ingested polyphenol-rich *Quercus* leaves. This detoxification process presumably led to greater decomposition of the leaf material, with enhanced bacterial growth in the pellets with clay particles versus those without the clay adsorbent material. The impact of collembola is greatest in mor soils, which may have entire layers in the F or H horizon filled with collembolan fecal pellets (Pawluk, 1987).

Research over the last 8–10 years has shown a significant impact of root-associated organisms on nutrient dynamics of phosphorus and nitrogen in experimental microcosms. These studies are reviewed in Coleman et al. (1983) and Anderson et al. (1981a,b). The use of both microcosm and mesocosm (i.e., $>m^2$-sized field enclosures) in soil ecological studies has proliferated in recent years and greatly increased our understanding of biological effects on nutrient cycling in soils (Beare et al., 1992; Ingham et al., 1985; Ingham et al., 1986a,b; Ingham et al., 1989; Moore et al., 1996; Parmelee et al., 1990).

In the laboratory, groups of rhizosphere bacteria, fungi, and microbivorous nematodes were grown singly or in combination, all with growing seedlings of the shortgrass prairie grass, *Bouteloua gracilis* (blue grama). In all treatments that had the root, microbe and microbial grazer (*Pelodera* sp. as bacterial feeder) and *Aphelenchus avenae* as fungal feeder, there was an enhanced shoot growth and dry-matter yield, when compared to the plant-alone control (Ingham et al., 1985).

Other work using mesofauna (nematodes, Ingham et al., 1986a,b) and macrofauna (isopods, Anderson et al., 1985) has shown significant enhancement of nutrient cycling (nitrogenous compounds) in field experimental situations. Thus an enhanced (20%–50%) nutrient return (mineralization) occurs in the presence of the fauna, compared with experiments in which they are present in very low numbers, or completely absent (Anderson et al., 1983). This work was further amplified by simulation models of detrital food webs, which showed a significant (c. 35%) contribution to mineralization of nitrogen by microfauna (amoebae and flagellates) and bacterial-feeding nematodes (De Ruiter et al., 1993; Hunt et al., 1987; Moore et al., 1996; Moore and de Ruiter, 2012). More detailed studies using ^{13}C and ^{15}N tracers in microcosms with varying degrees of organic matter accumulation ("hot-spots") and microbes alone, microbes and protozoa, or nematodes and microbes with both faunal groups in combination (Bonkowski et al., 2000) have revealed similar patterns to earlier microcosm studies (e.g., Ingham et al., 1985). Rye grass seedlings significantly increased in dry-matter and N content, with protozoa and nematodes and protozoa present (Bonkowski et al., 2000). Interestingly, the pattern of decomposition of labeled litter closely followed the nitrogen dynamics, with protozoa and nematodes and protozoa showing significantly more $^{13}CO_2$–C respiration between weeks 2 and 4 of the 6-week long experiment. This study was conducted concomitantly with a detailed analysis of microbial community composition. Griffiths et al. (1999) found significant selectivity of protozoa for species of soil bacteria, with a definite preference shown for several gram-positive species.

Macrofauna (e.g., earthworms) can also have profound influences on the disappearance and cycling of surface residues in ecosystems, and these were discussed in detail in Chapter 4, Secondary Production: Activities of Heterotrophic Organisms—The Soil Fauna, and Chapter 8, Future Developments in Soil Ecology.

5.11 GLOBAL CHANGE EXPERIMENTS, SOIL FAUNA EFFECTS ON DECOMPOSITION

Global experiments and syntheses have continued to address the quantification of the role of soil fauna in ecosystem processes and, in particular, have led to increased evidence for their contribution to C cycling. Global multisite experiments show that soil fauna are key regulators of decomposition rates at biome and global scales (Makkonen et al., 2012; Powers et al., 2009; Wall et al., 2008). Garcia-Palacios et al. (2013) conducted a meta-analysis on 440 litterbag case studies across 129 sites to assess how climate, litter quality, and soil invertebrates affect decomposition. This analysis showed fauna were responsible for 27% average enhancement of litter decomposition across global and biome scales.

5.12 PRINCIPAL CHARACTERISTICS OF SOIL SYSTEMS GOVERNING SOM DYNAMICS

There has been a virtual revolution in the perception of how SOM exists in soil profiles, and also how it turns over (Paul, 2016). In addition to the proven changes in SOM turnover with depth in the soil profile (Rumpel and Koegel-Knabner, 2011), the manner and extent of accessibility to it makes a very marked difference indeed. Thus it is imperative to consider accessibility and not recalcitrance of SOM moieties. This has been incorporated in new modeling approaches to the dynamics of SOM with considerable success (Dungait et al., 2012). This approach was foreshadowed in a major review paper, which noted a key fact in SOM dynamics: persistence of SOM is primarily not a molecular property (existence of heterocyclic rings, etc.) but is rather an ecosystem property. Thus the physicochemical and biological conditions of the soil matrix, and macroaggregates and their substituent microaggregates are the principal controlling variables governing the persistence of SOM moieties in the soil profile (Schmidt et al., 2011).

Needed improvements in approaches to research and modeling, over decadal time spans, if possible are presented in Table 5.5.

5.13 MAJOR SYNTHESES OF INTEGRATIVE PLANT AND SOIL FUNCTIONS

There has been a long-term need for an integration across major biomes of soil carbon inputs and outputs by plants and associated soil heterotrophs. Net soil C sequestration depends on quantity and quality of plant and heterotroph C pools. This is encapsulated in DeDeyn et al. (2008) (Fig. 5.22). An additional integrating feature is a consideration of soil

TABLE 5.5 Representation of Soil Carbon in Ecosystem Models and Recommendations for Potential Improvements

Insight	Properties of Most Published Models	Recommendations
1. Molecular structure	Decay rate of all pools keyed to substrate (or texture in CENTURY-type models[a]) and modified by moisture and temperature as constant Q_{10} above 0°C	Model decay rate as function of substrate properties and positions in microenvironment, microbial activity, and soil conditions including pH, temperature, and moisture. See 4, 5, 6, 8
2. Humic substances	Have a cascade of increasing intrinsic recalcitrance due to decomposition and synthesis	Replace the cascade with cycling of organic matter into and out of microbial biomass. See 1, 8
3. Fire-derived carton	Do not include fire residues as inputs or SOM. Do not represent decay of analogous substrates	Add input pathway for fire-derived carbon. Add aromatic compounds to SOM types
4. Roots	Parameterize litter quality with leaf/needle chemistry. Have simplified root and dissolved organic carbon inputs	Use separate characterizations for belowground and aboveground inputs. See 6
5. Physical heterogeneity	Lack physical processes, such as aggregation (some have tillage factor), spatial heterogeneity, or processes that would produce priming effect[b]	Nonnormal probability distributions, density-dependent terms for organic matter and microbial biomass. Parameters from 3D, fine-resolution models
6. Soil depth	No change in processes or rate constants with depth of soil or carbon input. Site-level tuning required to reproduce long turnover times	Representations of mineral associations, root and dissolved organic inputs, and physical disconnections. Explicit depth resolution for decomposition and transport
7. Permafrost	Lack processes governing permafrost soil carbon cycling. Lack fully coupled methane biogeochemistry	Add O_2 limitation and freezing effects on CO_2 and CH_4 production. Develop soil columns to represent inundation, permafrost thaw and thermokarst
8. Soil microorganisms	Treat microbial biomass as pool of active carbon. Lack effects of microbial community or enzymes on rates and decomposition products	Create and model microbial functional types, analogous to plant functional types. Introduce full soil nitrogen cycle coupled to carbon cycle

[a] *Some models use texture (clay content) to determine the amount of carbon in the slowest-cycling pool. However, soils with the same texture differ twofold in carbon stock and turnover time owing to differences in mineralogy, for example. One improvement would be to replace texture with reactive iron and aluminum or mineral surface charge density, estimated globally from a pseudotransfer function.*
[b] *Priming effect means that carbon input rate has positive effect on decomposition rate.*
Shown are properties of the soil carbon component of published ecosystem models used for global change and carbon cycle analysis, and recommendations for potential improvements. Globally implemented land models, such as Orchidee, LPJ, IBIS, CASA, and CLM, are based on CENTURY, CN, or RothC soil models.
From Schmidt, M.W.I., Torn, M.S., Abiven, S., Dittmar, T., Guggenberger, G., Janssens, I.A., et al., 2011. Persistence of soil organic matter as an ecosystem property. Nature, 478 (7367), 49–56.

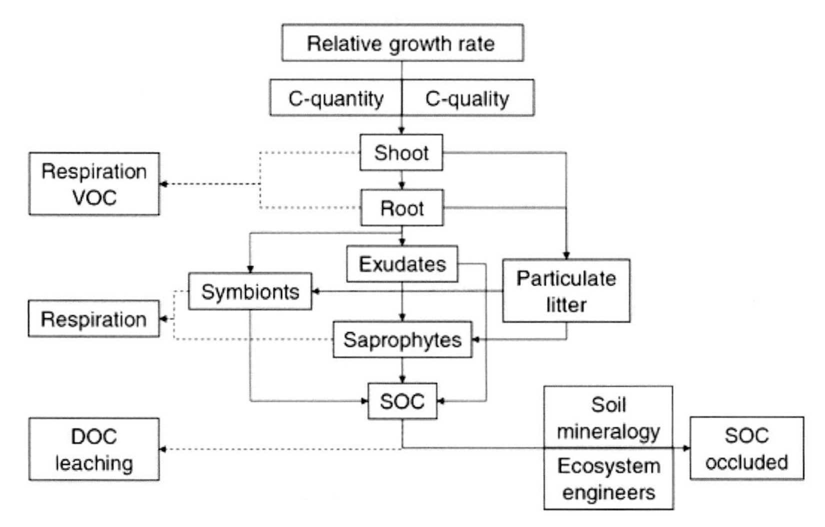

FIGURE 5.22 Soil carbon (C) in- and output by plants and associated soil heterotrophs. Net soil carbon sequestration depends on the quantity and quality of the plant and heterotroph carbon pools, which determine carbon use efficiency and soil carbon residence time. High growth rate generally corresponds with large, fast fluxes and relatively small soil carbon pools, slow growth rate with slower fluxes, more recalcitrant, and more persistent carbon pools. Occlusion of carbon in soil minerals, often enhanced by rhizodeposits and ecosystem engineers, enables long residence time. Solid lines indicate carbon incorporation and dotted lines soil carbon loss. *SOC*, Soil organic carbon; *VOC*, volatile organic carbon; *DOC*, dissolved organic carbon. *Source: From deDeyn, G.B., Cornelissen, J.H.C., Bardgett, R.D., 2008. Plant functional traits and soil carbon sequestration in contrasting biomes. Ecol. Lett. 11, 516–531.*

organic C pools and plant C sequestration traits across biomes, with characteristic mean annual temperatures (MAT) and mean annual precipitation (MAP) regimes (Fig. 5.23) (De Deyn et al., 2008). This pair of figures enables us to take a truly holistic view across all the major terrestrial biomes and their likelihood to be net sources or sinks for C across the coming decades.

5.14 FAUNAL IMPACTS IN APPLIED ECOLOGY—AGROECOSYSTEMS

There are several areas in the interface between theoretical and applied ecology where our knowledge of soil physics, chemistry, and biology can, and should, be put to good use. One of these is in the area of agroecosystem studies. The essentials of decomposition and nutrient dynamics in temperate agroecosystems were reviewed by Andre'n et al. (1990), Hendrix et al. (1992), Coleman et al. (1993, 2006), and in Africa by Vanlauwe et al. (2002).

It is generally acknowledged that zero, or reduced, tillage has several effects on abiotic and biotic regimes in agroecosystems. Retention of litter keeps the surface of the soil cooler and moister than in a conventionally tilled plot (Fenster and Peterson, 1979; Phillips and Phillips, 1984), and also leaves more substrate available in the 0–7.5 cm depths for

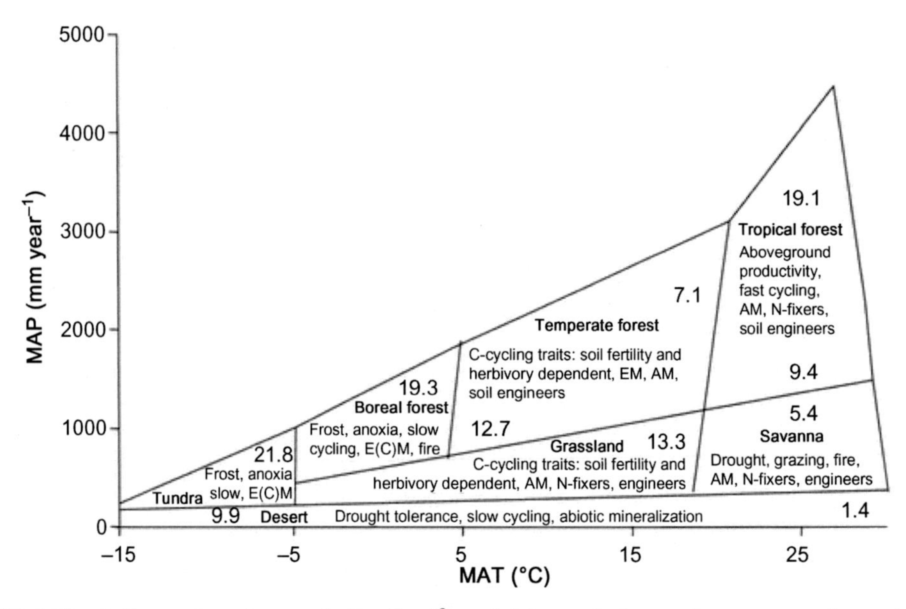

FIGURE 5.23 Soil organic carbon pools (kg C m^{-2}) and drivers of plant carbon sequestration traits across biomes with characteristic MAT and MAP. Lower and higher values within biomes represent warm- versus cool-temperate forest, respectively, and drier versus wet (peaty) tropical forests. *EM*, ecto-; *ECM*, ericoid-; and *AM*, arbuscular mycorrhizal fungi. *Source: From DeDeyn et al. (2009).*

nitrifiers and denitrifiers (Doran 1980a,b). This abiotic buffering seems to promote a slower N cycle, one that continues over a longer time span, but at a lower rate per unit time (Elliott et al., 1984; House et al., 1984). Soil invertebrate populations, particularly microarthropods (House et al., 1984; Stinner and Crossley, 1980), and earthworms (Parmelee et al., 1990) are enhanced as well (Table 5.6) (Hendrix et al., 1986). The microarthropods are undoubtedly responding to increased populations of litter-decomposing fungi, which tend to concentrate nitrogen by hyphal translocation (Holland and Coleman, 1987). In fact, dominant families of fungivorous mites responded by markedly decreasing in numbers in field mesocosm plots treated with captan, which brought fungal populations down to c. 40% of normal levels (Mueller et al., 1990).

A total of 22 agroecosystem components and processes were compared in no-till and conventional tillage in Georgia. In many instances, there was greater resilience in the no-till system, as shown by greater invertebrate species richness, greater SOM, and ecosystem N turnover time (Table 5.7) (House et al., 1984).

These findings were confirmed and extended by Elliott et al. (1984), who examined dynamics in long-term stubble-mulch and no-till plots on a silty-loam soil in eastern Colorado that underwent alternate crop and fallow regimes. These plots had been under cultivation for more than 75 years, and no-till had been an experimental treatment for nearly 20 years. Nitrate accumulated to a greater extent in the fallow than in the cropped rotation (Table 5.8). Ammonium-N was usually at very low levels ($\sim$1.0 μg NH_4–N g^{-1} soil), but on one date the concentration reached 4.6 μg NH_4–N g^{-1} soil in the top 2.5 cm of the no-till plots just prior to the highest rate of NO_3–N accumulation in the no-till than in

TABLE 5.6 Numbers and Estimated Biomass of Soil Fauna in Conventional Tillage (CT) and No-Tillage (NT) Agroecosystems at Horseshoe Bend

	Numbers m^{-2}		mg dry wt m^{-2}	
	CT	**NT**	**CT**	**NT**
***NEMATODES*[a]**				
Bacterivores	1836[e]	909	237	117
Fungivores	227[e]	500	14	31
Herbivores	945[e]	1064	93	104
	3008	2473	344	252
***MICROARTHROPODS*[b]**				
Mites	41,081[e]	78,256	118	303
Collembola	6244[e]	14,684	17	40
Insects	2105	2548	–	–
	49,430	95,489	135	343
***MACROARTHROPODS*[c]**				
Ground beetles	7[e]	33	6	30
Spiders	1[e]	17	1	14
Others	6[e]	28	–	–
	14	78	7	44
***ANNELIDS*[d]**				
Earthworms	149[e]	967	3129	20,307
Enchytraeids	1867	520	592	17
	2016	1487	3721	20,324
Total	**54,468**	**99,526**	**4207**	**20,980**

[a]*Means of samples from June–October 1983; numbers are* $\times 10^{-3}$.
[b]*Means of samples from May–December 1983.*
[c]*Means of samples from April–June 1983.*
[d]*Means of samples from April 1983.*
[e]*For numbers of organisms, tillage treatments differ significantly at* $P = 0.05$.
From Hendrix, P.F., Parmelee, R.W., Crossley Jr., D.A., Coleman, D.C., Odum, E.P., Groffman, P., 1986. Detritus food webs in conventional and no-tillage agroecosystems. Bioscience, 36, 374–380.

the stubble-mulch treatments. However, it is possible that there was more mineralization in the stubble-mulch plots earlier in the year before the first sample date, and this mineralized N was moved below the sampling depth (20 cm) as NO_3–N during a rainfall event. Interactions between modification of system structure and major nutrient processes need more study. Certainly soil fauna are sensitive to increased nutrient inputs from fertilizers and manures, and this needs to be considered in experimental work (Coleman et al., 2012; Hendrix et al., 1992; Marshall, 1977).

TABLE 5.7 Comparison of Agroecosystem Components and Associated Agroecosystem Processes From Conventional Tillage (CT) and No-Tillage (NT) Systems

Component or Process	CT Versus NT
Crop yields	NT = CT (except during drought)
Crop biomass	Decreasing in both CT and NT
Weed biomass	NT > CT
Plant nitrogen dynamics	CT > NT (nitrogen flux)
Shoot to root ratios	CT > NT
Nitrogen fixation	NT > CT (?)
Surface crop and weed residues	NT ≫ CT
Litter decomposition rates	CT > NT
Surface litter (%N)	NT > CT
Soil total N	NT > CT in upper soil layer
Nitrification activity	NT > CT in upper soil layer
	CT > NT in middle soil layer (?)
Soil organic matter	NT > CT
Soil moisture	NT > CT
Ground water leaching (nitrate-N)	CT > NT (?)
Foliage arthropods	CT = NT
Crop herbivory by insects	CT > NT
Nitrogen content of crop foliage	CT > NT
Arthropods species diversity	NT > CT
Soil arthropods (no. of individuals)	BT ≫ CT
Nitrogen contained in arthropods:	
Soil	NT > CT
Foliage	CT > NT
Ecosystem N turnover time	NT > CT
Ecosystem N efficiency	NT > CT (?)

From House, G.J., Stinner, B.R., Crossley Jr., D.A., 1984. Nitrogen cycling in conventional and no-tillage agroecosystems: analysis of pathways and processes. J. Appl. Ecol., 21, 991–1012.

TABLE 5.8 Faunal Carbon, Microbial Biomass, Mineralized Carbon and Nitrogen, Nitrogen and Phosphorus Under Stubble Mulch, and No-Till Treatments of the Fallow Phase of Dryland Wheat Plots

Variable	Treatment	Date				
		8 June	6 July	2 August	23 August	13 September
FAUNAL CARBON[a]						
Collembola	Stubble mulch	6.69	1.03	2.18	0.61	1.49
× 100	No till	5.58	8.29	4.01	1.60	1.49
Acari	Stubble mulch	3.77	0.47	0.91	0.40	0.92
× 10	No till	3.60	3.31	1.24	1.32	0.54
Holophagous nematodes	Stubble mulch	0.88	0.46	1.12	0.51	0.56
× 10	No till	0.48	1.69	0.96	0.57	0.32
Protozoa	Stubble mulch	1.76	0.67	0.74	1.96	1.92
× 1	No till	2.20	0.96	0.93	2.29	1.92
MICROBIAL BIOMASS[b]						
Carbon	Stubble mulch	245	204	271	186	255
	No till	329	273	299	194	256
Nitrogen	Stubble mulch	81	57	53	60	56
	No till	92	72	49	62	48
Phosphorus	Stubble mulch	5.7	7.5	10.1	4.6	7.2
	No till	5.5	6.8	10.1	4.9	7.1
MINERALIZED C AND N[c]						
Respired C	Stubble mulch	52	77	57	42	68
(0–10 days)	No till	97	110	98	41	67
Respired C	Stubble mulch	49	42	56	53	24
(10–20 days)	No till	80	69	84	57	24
Mineralizable N	Stubble mulch	10.63	6.39	5.79	2.62	1.95
(0–20 days)	No till	12.68	5.79	8.90	−3.3	−1.38
N AND P[d]						
NH^{+}_{4}–N	Stubble mulch	3.1	1.0	2.0	1.8	0.8
	No till	4.0	1.8	4.6	1.2	1.4
NH^{-}_{3}–N	Stubble mulch	7.2	10.6	16.3	17.8	16.1
	No till	7.8	23.8	21.6	45.1	46.2

[a]*Soil fauna biomass C (kg C ha^{-1} to 10 cm) for four categories. (Note differences in the multiplier for each category).*
[b]*Microbial biomass C, N, and P (kg element ha^{-1} to 10 cm).*
[c]*Mineralizable C (CO_2–C) and NO_3–N as kg ha^{-1} to 10 cm in unchloroformed 0–10 or 10–20 day incubations of soils sampled from the field.*
[d]*NH_4–N, NO_3–N, and extractable inorganic and organic P amounts (kg ha^{-1}) in the top 10 cm of soil.*
From Elliott, E.T., Horton, K., Moore, J.C., Coleman, D.C., Cole, C.V., 1984. Mineralization dynamics in fallow dryland wheat plots, Colorado. Plant Soil, 76, 149–155.

5.15 APPLIED ECOLOGY IN FORESTED ECOSYSTEMS

There are some interesting comparisons and analogies to be drawn between no-till agriculture and forested ecosystems of the "mor" type, which have a distinct stratification of L, F, and H layers (O_i, O_e, and O_a in the US terminology). It is generally recognized that abundance of fungi and fungivorous arthropods is greater in these soils than in soils with a less-pronounced litter layer (Blair et al., 1992; Kühnelt, 1976; Pawluk, 1987; Wallwork I976). However, it is important to determine the amount of activity occurring in these surface layers, as well. Ingham et al. (1989) and Coleman et al. (1990) have shown significantly greater fungal biomass and microarthropod biomass in L, F, and H layers under *Pinus contorta* (lodgepole pine), compared with mountain meadow. There was also a greater amount of fungal activity, as demonstrated by FDA-positive fungal hyphae (Ingham and Klein, 1982; Söderström, 1977). This forest experience is corroborated by Verhoef and De Goede (1985), who noted greater activity of Collembola in pine forests in Holland, contrasted with habitats which had a thin or nonexistent litter layer.

Because energy flow is fundamental to the function of decomposer organisms and ecosystems, energetics could provide some fundamental constraints on soil C dynamics (Currie, 2003). Often, C is considered as a surrogate for energy in studies of detrital decay and C turnover in soils. By testing relationships between C and energy across samples of forest detritus above and below ground, across decay stages, and between a deciduous and coniferous forest at the Harvard Forest, United States, Currie (2003) found that energy and C concentrations were closely related (within 10%), as were ratios of heterotrophic energy dissipation to C mineralization across types of detritus (within 16%). These relationships should be borne in mind when we explore the energetics of detrital food webs, in Chapter 6, Soil Food Webs: Detritivory and Microbivory in Soils.

Other areas of interest in applied soil ecology include revegetation of mine spoils. Extensive studies in the United Kingdom, Germany, and elsewhere have been made of decomposition ecology, and microbial parameters in strip-mined coal lands (Bentham et al., 1992). Intentional manipulations (especially introductions) of earthworm populations have been used to enhance productivity of crop and pasture lands (Lee, 1995) and to speed organic waste decomposition via vermicomposting (Edwards, 1998), as discussed in Chapter 4, Secondary Production: Activities of Heterotrophic Organisms—The Soil Fauna. An excellent synthesis and review of the role of soil biota in reclamation of mined lands is given by Frouz (2015).

Several researchers (Jastrow and Miller, 1991; Jastrow et al., 1998; Rothwell, 1984; Six et al., 2004; Tisdall and Oades, 1982) have investigated the roles of saprophytic and arbuscular mycorrhizal (AM) fungi in stabilizing micro- and macroaggregates. Rothwell (1984) suggests that there is a biochemical coupling reaction between glucosamines in the hyphal walls of the fungus with phenolic compounds released during lignin degradation from leaf and root tissues. An additional possibility, little investigated yet, is the apparently widespread occurrence of interspecific physical linkages, enabling transfers of nutrients via mycorrhizae of various annual and perennial plants (Chiariello et al., 1982; Kaiser et al., 2015; Read et al., 1985; Read, 1991). Physical, chemical, and biological contacts may be operating simultaneously in mycorrhizal-mediated interactions.

Some of the latter examples may seem a bit removed from the general theme: role of soil fauna in soil processes. However, it is apparent from studies by Warnock et al. (1982), Moore et al. (1985), and Curl and Truelove (1986) that soil mesofauna, e.g., Collembola, show considerable preference for, and have an impact on AM fungal growth, just as they do for saprophytic fungi (Newell, 1984a,b) and plant pathogenic fungi (Lartey et al., 1994). This impact undoubtedly extends to nematodes (Ingham et al., 1985) and soil amoebae, as well (Chakraborty and Warcup, 1983; Chakraborty et al., 1983; Gupta and Germida, 1988; Gupta and Yeates, 1997; ref. from Griffiths, c. 2010).

5.16 SUMMARY

The major lesson to be learned, as soil ecologists, is one of paying attention to details, yet considering them in a holistic perspective. Certainly we are past the time when measurements of the "soil biomass" (referring to the microbial biomass) alone, by whatever method, is considered adequate (Blagodatskaya and Kuzyakov, 2013; Coleman, 1994a). Small groups of organisms, perhaps highly aggregated within the ecosystem, may be facilitating (or retarding) turnover of other organisms, or major nutrients, such as nitrogen, phosphorus, and sulfur. In fact, as Darwin observed over a century ago, these seemingly small biological processes operating over long time periods and large spatial scales can make profound changes in the world around us, including the formation of soil.

Decomposition rates, along with nutrient dynamics, soil respiration, and formation of soil structure, are integrating variables. They are generalized measurements of the functional properties of ecosystems, and they summarize the combined actions of soil microflora, fauna, abiotic variables and resource quality factors. Litter breakdown rates can be compared using simple first-order models, so that rate variations between ecosystems or between different substrates may be compared. Litter breakdown rates are easily measured using bagged leaf litter ("litterbags"). Decomposition per se is due to microbial activities, but experiments show that fauna have a strong influence on litter breakdown rates, especially for more resistant substrates. The interaction between microflora and fauna is especially important for nutrient cycling mechanisms. Organic matter dynamics are strongly influenced by soil fauna. Termites and earthworms are well known for their influences on nutrient dynamics, SOM, and soil structure. But the entire soil fauna is involved in these processes, and through their interactions with soil microbes, must be considered in a holistic perspective.

CHAPTER

6

Soil Food Webs: Detritivory and Microbivory in Soils

6.1 INTRODUCTION

The traditional studies of food webs and food chains began with pioneering efforts of Summerhayes and Elton (1923), in Spitsbergen. This early study explicitly linked detrital biotic interactions with other parts of the terrestrial and aquatic food web (Fig. 6.1). Work on detrital food webs progressed slowly for the next 20 years, although Bornebusch (1930) carried out some pioneering studies of detrital food webs and their energetics. Further insights were gained from the studies of Lindeman (1942), who developed the concept of trophic levels.

Building on the soil ecology studies funded by the US Atomic Energy Commission in the late 1950s (Auerbach, 1958) and into the early 1960s, a clear need was recognized for a more holistic study of energetics and interactions of organisms in ecosystems. This led to the ambitious effort known as the International Biological Program (IBP). The overall intent was to bring working groups together, addressing how carbon and energy flow in a wide range of terrestrial and aquatic ecosystems, with the ultimate goal being a better understanding of how ecosystems work and could be manipulated for the benefit of mankind (Blair, 1977). The main findings of the IBP were for a wide range of grassland, desert, and forested ecosystems, the net flow into the aboveground grazing (consumer) component is only 5% or less, with the remainder entering the detrital–decomposer food web (Coleman et al., 1976; Coleman, 2010). This research led to several post-IBP studies in North America and Europe, to follow up on the initial results.

In the late 1970s and 1980s, a series of investigations of detrital food webs were carried out in the semiarid and arid grasslands and desert lands of Colorado and New Mexico (Coleman et al., 1977, 1983; Hunt et al., 1987; Moore et al., 1988; Parker et al., 1984; Whitford et al., 1983) (Fig. 6.2). These studies and several in the Netherlands (Brussaard et al., 1990; De Ruiter et al., 1993; Moore and De Ruiter, 2000), Germany (Koller et al., 2013), Sweden (Andrén et al., 1990; Bååth et al., 1981; Persson, 1980), and in the United Kingdom (Anderson et al., 1985) found that microbial/faunal interactions have significant impacts on nutrient cycles of the major nutrients, namely nitrogen, phosphorus, and sulfur (Gupta and Germida, 1989). Some of these studies used assemblages of a few species in

Fundamentals of Soil Ecology
DOI: http://dx.doi.org/10.1016/B978-0-12-805251-8.00006-5

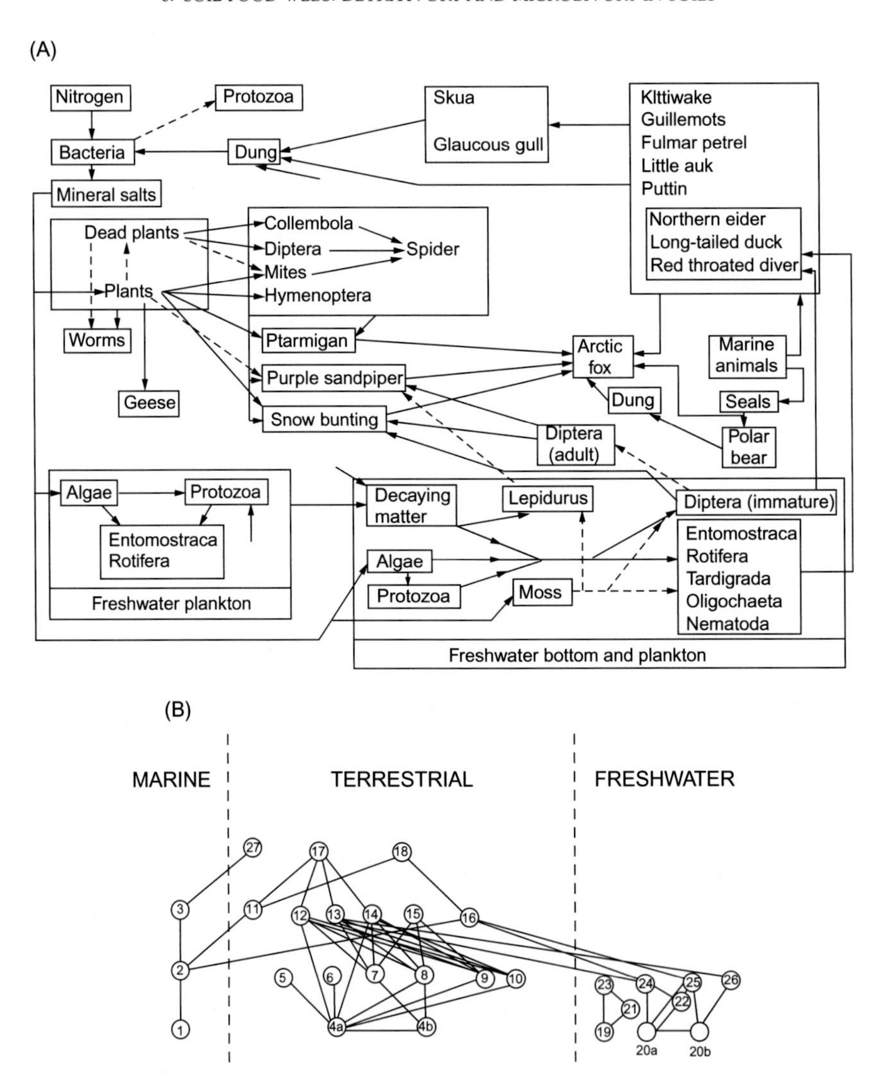

FIGURE 6.1 Arctic food web (A) as described by Summerhayes and Elton (1923) and as diagrammed (B) by Pimm (1982). (B): (1) plankton, (2) marine animals, (3) seals, (4a) plants, (4b) dead plants, (5) worms, (6) geese, (7) Collembola, (8) Diptera, terrestrial, (9) mites, (10) Hymenoptera, (11) seabirds, (12) snow bunting, (13) purple sandpiper, (14) ptarmigan, (15) spiders, (16) ducks and divers, (17) arctic fox, (18) skua and Glaucous gull, (19) planktonic algae, (20a) benthic algae, (20b) decaying matter, (21) protozoa, (22) invertebrates, (23) Diptera, freshwater, (24) other invertebrates, (25) Lepidurus, and (27) polar bear. Source: *From Pimm (1982) and Pimm and Lawton (1980).*

microcosms, but were beginning to delineate the mechanisms that are important in soil systems in general. Among the fauna, the protozoa were often overlooked, despite the findings by Cutler et al. (1923) that there are important predator–prey interactions between protozoa and bacteria in soils. Clarholm (1985) noted that soil protozoa are avid microbivores, and turn over an average of 10–12 times in a growing season, in contrast to

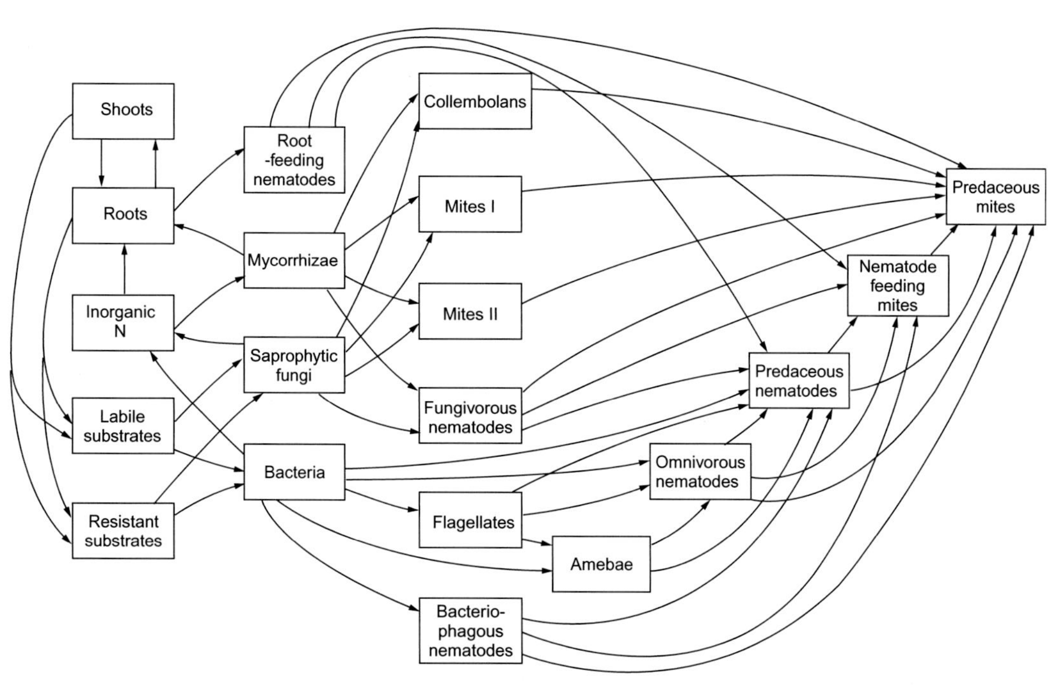

FIGURE 6.2 Representation of detrital food web in shortgrass prairie. Fungal-feeding mites are separated into two groups (I and II) to distinguish the slow-growing Oribatids from faster-growing taxa. Flows omitted from the figure for the sake of clarity include transfers from every organism to the substrate pools (death) and transfers from every animal to the substrate pools (defecation) and to inorganic nitrogen (N) (ammonification). *Source: From Hunt, H.W., Coleman, D.C., Ingham, E.R., Ingham, R.E., Elliott, E.T., Moore, J.C., et al., 1987. The detrital food web in a shortgrass prairie. Biol. Fertil. Soils, 3, 57–68, reprinted with permission.*

many other members of the soil biota, which may turn over only once or twice in a c. 120- to 140-day growing, or activity season. These findings were further extended (Kuikman et al., 1990), with the observation that N uptake by plants may increase from 9% to 17% when large inocula of protozoa are present. The demographics and microbial/faunal interactions provide much of the driving force in the models of nitrogen turnover in semiarid grasslands (Hunt et al., 1987) and arable lands (Moore and de Ruiter, 1991, 2000). The interactions between arbuscular mycorrhizal fungi (AMF) and protozoa in grasslands have been noted in experimental studies of Koller et al. (2013), with the interdependence of AMF and protists being enhanced during situations of N scarcity. Studies of the varying contributions to plant growth by the conjoint activities of the assemblage of all organisms, from prokaryotes to mycorrhiza and soil fauna noted that the overall impact of the multiple trophic interactions was generally positive, but only slightly so, in grassland microcosms (Ladygina et al., 2010).

Recent studies have noted the more complex nature of food webs when detrital components are included (Hall and Raffaelli, 1993; Polis, 1991; Scheu and Setälä, 2002). DeAngelis (1992) in his treatise on nutrient cycling devoted an entire chapter to nutrient interactions of detritus and decomposers. His ideas have provided insights into decomposition/nutrient cycling processes. This chapter addresses several aspects of soil biota and nutrient cycling in soils, namely demography and "hot spots of activity," sensu Kuzyakov and Blagodatsky (2015) which are often overlooked in energetics studies of soil systems. These factors are crucial to understanding how organisms and soils interact, and contribute to ecosystem function.

On a more general level, one can envision plant–soil interactions in general across a global perspective. As a rule, most direct, intraspecific and intragroup interactions seem to be primarily negative (van der Putten et al., 2013). However, as these authors note, there may be more positive plant–soil feedbacks interactions in nature, which might be revealed when testing the concept in the field and over a longer period, e.g., as resulting from plant species-specific effects on decomposition, the so-called home field advantage effect that has been shown for some plant species. Other examples of positive feedback often involve changes in density of symbiotic mutualists, such as nitrogen-fixing rhizobacteria and mycorrhizal fungi.

6.2 PHYSIOLOGICAL ECOLOGY OF SOIL ORGANISMS

Given the physiological ecology of the microbes and fauna involved, are long food chains energetically possible? There are several theoretical reasons why long food chains could be expected. Let us take, as an example, the energetically most dominant interactions between microbes and fauna, which occur in many terrestrial ecosystems, summarized by Hunt et al. (1987) (Fig. 6.2). The flow of organic carbon or nitrogen moves from initial organic substrates (labile or resistant) to the primary decomposer, either bacteria or fungi, and then on into microbivorous microfauna (flagellates and amoebae), or microbivorous mesofauna (feeding on fungi) and in turn to omnivorous or predaceous nematodes, and on to nematode-feeding mites, and predaceous mites. Further predation upon the mites by ants (E.O. Wilson, personal communication), or lithobiomorph Chilopods (Centipedes) is possible, although not explicitly represented by Hunt et al. (1987). There are at least eight links in the bacterial-based detrital food chain, with considerable evidence of omnivory. For example, many fungivorous mites require a nematode "supplement" to complete their life cycles (Walter et al., 1991). Note that this is a rather ecosystem-specific diagram. One could draw another for decomposition in a coniferous or oak/beech forest, with a significant proportion of the total decomposition being mediated by ectotrophic mycorrhizae, operating perhaps in competition with the saprophytic fungi (Gadgil and Gadgil, 1975).

For desert and estuarine food webs, reviewed by Hall and Raffaelli (1993), the detrital food chain length noted earlier is comparable to the average length of 5–7 (Polis, 1991), with maxima recorded of 8. In contrast, Hairston and Hairston (1993) assert that the usual food chain length in detrital systems seldom exceeds three. As noted later, these long chain lengths of 5–7 links are not only feasible, but also thermodynamically possible at several times and in several locations in the soil matrix, particularly the rhizosphere, and other "hot spots" of activity. What levels of taxonomic resolution are both most useful and appropriate for detrital food web studies? Our inability to sort out the details of microbial taxonomy in situ (see Furlong et al., 2002, and other references in Chapter 3: Secondary

Production: Activities of Heterotrophic Organisms—Microbes for insight into molecular probing techniques in agroecosystems), and limited knowledge of many of the soil invertebrates, particularly the immature stages (Behan-Pelletier and Bissett, 1993), requires use of rather coarse functional groups for taxonomy of the soil biota. Interestingly, this sort of separation enabled Wardle and Yeates (1993) to identify competition and predation forces operative in an assemblage of detritus/microbial/ nematode trophic groups in an agricultural field. Using a correlation analysis, they noted that predatory nematodes reflected most closely the changes in primary production, and the microbivorous nematodes seemed to be more dependent on substrate quality in the microbial (bacterial and fungal) community.

6.3 ENERGY AVAILABLE FOR DETRITAL FOOD CHAINS AND WEBS

The energy available for detrital food chains is considered next. If one considers the variance, or range around the mean values of the assimilation and production efficiencies of the biota (Table 6.1), the amount of energy can be calculated which will move from primary decomposers all the way up the food chain. There is indeed energy to spare for such elaborate food chains. Using the maximal values for production efficiency, such as that for bacteria of 70% (Payne, 1970), and 80% for soil amoebae and flagellates (Humphreys, 1979), moving on to protozoan-consuming nematodes (Fig. 6.3), which doubtless occurs in certain hot spots, e.g., at the zone of elongation of a growing root, then there is an adequate amount of carbon available for passage through the four- and five-membered detrital food chains of interest. Considerable omnivory is prevalent in these soil systems (DeAngelis, 1992). The protozoa and nematode-feeding pathway highlighted in Fig. 6.3 (Hunt et al., 1987) accounted for 37% of the total nitrogen mineralization, and some 82% of the total mineralization due to soil fauna. Similar percentages were obtained for a wide range of agroecosystems in the United States and Europe (de Ruiter et al., 1993; Moore and de Ruiter, 2000).

The relative contributions of the soil fauna to microbial turnover and nutrient mineralization are directly related to the demographics of the soil biota (Coleman et al., 1983, 1993; Coleman and Wall, 2015), as noted for average standing crops and energetic parameters and turnover times per year for microorganisms, micro-, meso-, and macrofauna in a grassland and a no-tillage agroecosystem (Coleman et al., 1993) (Table 6.2). Thus the protozoa, and naked amoebae in particular, turn over 10 or more times per season, and consume several times their mass of living microbial tissues. The microbes and several other faunal groups have much lower turnover rates, on average. Although the amoebae are considered to be primarily bacterial feeders, there are important instances when other amoebal species will feed on protoplasm in fungal hyphae, or even on the fungal spores themselves (Chakraborty et al., 1983; Chakraborty and Warcup, 1983). When considered in combination with the information in Table 6.1, on the range of assimilation and production efficiencies, the impacts of these small organisms are very marked. It should be noted that extensive studies in Sweden on arable lands (Andrén et al., 1990) have reached similar conclusions. The increasing miniaturization of sensors, so that one can carry out

TABLE 6.1 Physiological Data on Major Biotic Groups in Soil

	Fraction of Food Assimilated			Production–Assimilation Ratio		
Trophic Group	**max**	$\overline{X}$	**min**	**max**	$\overline{X}$	**min**
Bacteria	?	1.0	<0.01?	0.7	0.4	<0.01?
Saprophytic fungi		1.0		0.7	0.4	
AM		?		0.8?	0.4	
Amoebae		0.95		0.8	0.4	
Flagellates		0.95		0.8	0.4	
Nematodes phytophagous		0.25		0.5?	0.37	
Nematodes fungivorous		0.38		0.5?	0.37	
Nematodes bacterivorous		0.6		0.5?	0.37	
Nematodes omnivorous/predaceous		0.55		0.5?	0.37	
Mites fungivorous (*r*)		0.5		0.5?	0.35	
Mites fungivorous (*k*)		0.5		0.5?	0.35	
Mites nematophagous		0.9		0.5?	0.35	
Mites predaceous		0.6		0.5?	0.35	
Collembola		0.5		0.5?	0.35	
Enchytraeids		0.28			0.4	
Earthworms		0.2			0.45	
Termites	↓	0.4?	↓		0.15?	↓

k, Slow growth strategy; *r*, rapid growth strategy.
Modified from Humphreys, W.F., 1979. Production and respiration in animal populations. J. Anim. Ecol. 48, 427–453. Hunt, H.W., Coleman, D.C., Ingham, E.R., Ingham, R.E., Elliott, E.T., Moore, J.C., et al., 1987. The detrital food web in a shortgrass prairie. Biol. Fertil. Soils, 3, 57–68. Payne, W.J., 1970. Energy yields and growth of heterotrophs. Annu. Rev. Microbiol. 24, 17–52. De Ruiter, P.C., Moore, J.C., Zwart, K.B., Bouwman, L.A., Hassink, J., Bloem, J., et al., 1993. Simulation of nitrogen mineralization in the below-ground food webs of two winter wheat fields. J. Appl. Ecol. 30, 95–106.

microcalorimetry (Battley, 1987) at localized microsites will enable us to measure direct energetic transformations more readily in situ.

The practical implications of soil food webs in agroecosystems have been of interest to researchers in several countries, notably the Netherlands, Sweden, and the United States. In a major synthesis of several research papers, including some of which are cited earlier, Bloem et al. (1997) calculated the impact of microbivorous invertebrate fauna in agroecosystems. Using a combination of experimental results and simulation modeling runs, they calculated that in fields that had greater additions of organic matter (OM), including manure, average N mineralization was 30% higher than in fields that did not have such OM additions. This reflected the activities of protozoa and nematodes, which were 64% and 22% higher numbers, respectively, in the fields with organic additions. N mineralization was performed mainly by the bacteria, which dominated in these fields, but the N

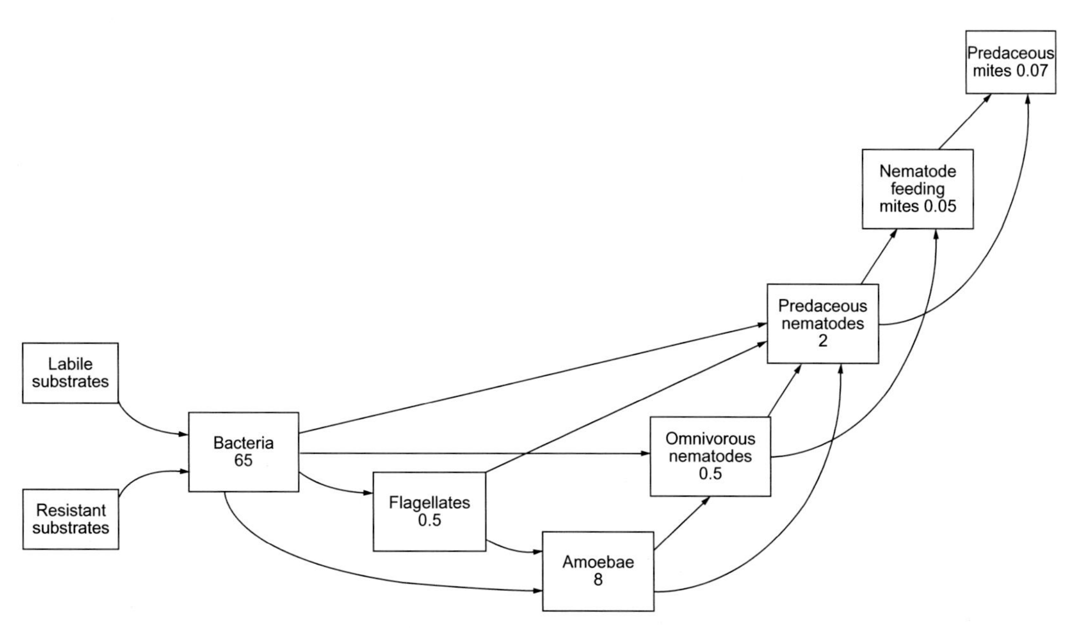

FIGURE 6.3 Calculations of annual carbon flows along a bacteriophagic food chain assemblage, exhibiting considerable omnivory. Average standing crops are indicated. Flows via protozoan feeding are estimated as probably two to three times greater than via nematodes in that ecosystem. Proportions may be reversed in lower-pH forested systems, and more flows via fungi (not shown). Source: *From Hunt, H.W., Coleman, D.C., Ingham, E.R., Ingham, R.E., Elliott, E.T., Moore, J.C., et al., 1987. The detrital food web in a shortgrass prairie. Biol. Fertil. Soils, 3, 57–68, reprinted with permission.*

mineralization was increased by protozoa by 30%, on a growing season average. Interestingly, the protozoa did not enhance C mineralization, as their impacts, as were those of the nematodes, were by direct grazing upon and lysing the bacterial cells (Bloem et al., 1997).

6.3.1 Arenas of Interest

Soils can be considered most profitably as the extremely heterogeneous entities they are. This requires that we "let the soil work for us" (Elliott and Coleman, 1988), and stratify, in a statistical sense, the regions of the soil which are "hot spots" of activity. These zones include the rhizosphere, aggregates, litter and organic detritus, and the "drilosphere," which is that portion of the soil volume influenced by secretions of earthworms (Bouché, 1975) (Fig. 6.4). Each region is a relatively small subset of the total soil volume, but may contain a preponderance of numbers, and more importantly, activity of the soil biota (Beare et al., 1995). Examples include the 5%–7% of the total soil which was root-influenced or rhizosphere, in extensive pot trials of Ingham et al. (1985), contained a majority (greater than 70%) of the bacterial and fungal-feeding nematodes. Ingham et al. (1985) also measured higher biomasses of rhizosphere bacteria in microcosms with large numbers of microbivorous nematodes (greater than 4000 per g rhizosphere soil) than in

TABLE 6.2 Average Standing Crop and Energetic Parameters for Microorganisms, Mesofauna, and Earthworms in a Lucerne Ley and Georgia No-Tillage Agroecosystem[a]

	Naked Amoebae	Flagellates	Ciliates	Bacteria	Fungi	Microbivorous Nematodes	Collembola	Mites	Enchytraeids	Earthworms
Typical size in soil	30 μm	10 μm	80 μm	0.5–1 × 1–2 μm	Ø 2.5 μm 1.0–5.5 μm	Ø ~40 μm	Ø 5000 μm	Ø 1000 μm	Ø 1000 μm	Ø 5000 μm
Mode of living	In water films on surfaces	Free swimming in water films		On surfaces	Free and on surfaces	In water films, free, and on surfaces	Free	Free	Free	Free in soil
Biomass (kg dw ha^{-1})	95%	5%	<1%	500–750[c]	700–2700[d]	1.5–4[e]	0.2–0.5[e]	2–8[e]	1–8[e]	25–50[e]
		50[b]								
% active	0–100			15–30	2–10	0–100	80–100	80–100	?	0–100
Estimated turnover times, $season^{-1}$		10		2–3	0.75	2–4	2–3	2–3	?	3
No. of bacteria $division^{-1} \times 10^{-3}$	3–8	0.6–1	20–2000							
		2–4								
Minimum generation time in soil (hours)				0.5	4–8	120	720	720	170	720

[a]*Modified from Clarholm, M., 1985. Possible roles for roots, bacteria, protozoa and fungi in supplying nitrogen to plants. In: Fitter, A.H., Atkinson, D., Read, D.J., Usher, M.B. (Eds.), Ecological Interactions in Soil: Plants, Microbes and Animals. Blackwell, Oxford, pp. 355–365, Hendrix, P.F., Crossley Jr., D.A., Coleman, D.C., Parmelee, R.W., Beare, M.H., 1987. Carbon dynamics in soil microbes and fauna in conventional and no-tillage agroecosystems. INTECOL Bull., 15, 59–63. Beare, M.H., Parmelee, R.W., Hendrix, P.F., Cheng, W., Coleman, D.C., Crossley, D.A. Jr., 1992. Microbial and faunal interactions and effects on litter nitrogen and decomposition in agroecosystems. Ecol. Monograph. 62, 569–591. Reprinted with permission from Coleman, D.C., Reid, C.P.P., Cole, C.V., 1983. Biological strategies of nutrient cycling in soil systems. Adv. Ecol. Res. 13, 1–55. Copyright Lewis Publishers, an imprint of CRC Press, Boca Raton, FL.*
[b]*MPN technique.*
[c]*Direct counts plus size class estimations.*
[d]*Direct estimation of total hyphal length and diameter.*
[e]*Extractions and sorting.*

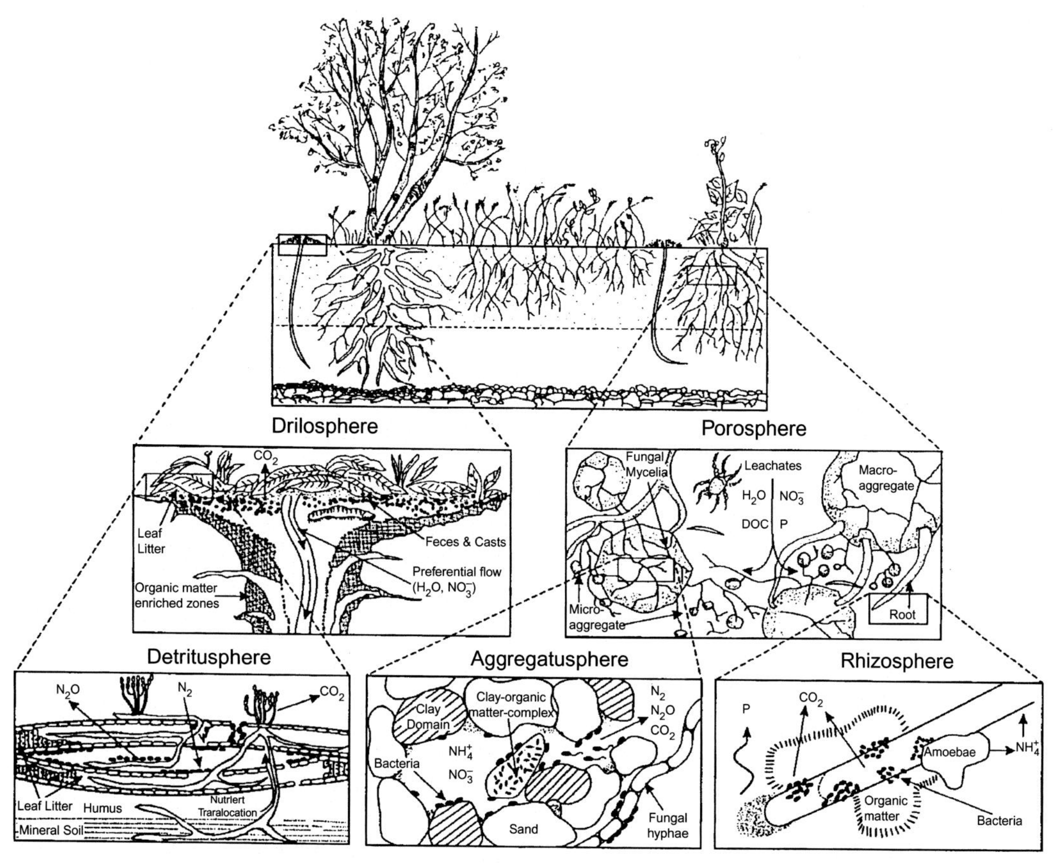

FIGURE 6.4 Arenas of activity in soil systems. These "hot spots" of activity may be less than 10% of the total soil volume, but represent more than 90% of the total biological activity in most soils worldwide. Source: *From Beare, M.H., Coleman, D.C., Crossley Jr., D.A., Hendrix, P.F., Odum, E.P., 1995. A hierarchical approach to evaluating the significance of soil biodiversity to biogeochemical cycling. Plant Soil, 170, 5–22, reprinted with permission.*

microcosms without these nematodes. Yet the extent of mineralization of nitrogen in the microcosms with nematodes reflected that they were ingesting large quantities of microbes, as well. Thus there was a net enhancement of microbial production, in a fashion similar to that measured by Porter (1975), who found a net stimulation of phytoplankton growth after the cells had undergone transit through the guts of *Daphnia* sp. in freshwater incubations. As an example of the dynamic nature of shifting hot spots, Griffiths and Caul (1993) found that more nematodes were active in the rhizosphere, and they moved readily to new concentrations of fresh OM (leaf litter) in short-term trials. Other examples of hot spots that have shown enhanced microbial activity include the drilosphere and worm castings, which show enhanced carbon and nitrogen (Daniel and Anderson, 1992; Syers et al., 1979) and phosphorus mineralization (Lavelle et al., 1992; Syers et al., 1979). *Lumbricus terrestris* "middens" (small patches of plant litter and casts gathered around the burrow entrance) in experimental field sites in Ohio were found to be functionally different, with enhanced acetate incorporation and microbial cell synthesis, compared with surrounding nonearthworm-influenced soil (Bohlen et al., 2002). Another center of activity is the aggregatusphere (Fig. 6.4), or region of micro- and macroaggregates (Beare et al., 1995; Elliott, 1986; Elliott and Coleman, 1988; Six et al., 1999). This zone of influence is less well studied energetically, and is a major source of some of the dynamical, yet highly patchy behavior found in soils. Foster (1985) and Foster and Dormaar (1991) have demonstrated, using electron microscopy (Fig. 3.3), amoebal pseudopodia extending into very small pore-necks and pores (only a few tenths of micrometers diameter) in well-structured soil, attacking bacterial colonies which seemed to be inaccessible to the smallest nematodes and amoebae or other protozoa. A study using a combined approach to rhizosphere and soil cracks for locations of "hot spots" of labile OM was used by van Noordwijk et al. (1993) to good effect.

Some recent studies of the roles of soil fauna in detrital food webs appear later. Some further insights into soil faunal impacts come from Verbruggen et al. in Johnson et al. (2017); see later.

Fungal biomass C storage, turnover (Hobbie and Agerer, 2010), and necromass decomposition (Fernandez et al., 2016) vary with fungal mycelial exploration type (Agerer, 2001). Thus any induced shift in mycorrhizal or saprotrophic fungal community composition may alter the rates of these processes. For example, cord-forming hydrophobic hyphal structures, specific to medium-distance fringe and long-distance exploration types, tend to persist in soils longer compared with other types (Fernandez et al., 2016). In a field experiment Kanters et al. (2015) reported a massive decline of *Amphinema byssoides* and *Cortinarius* sp. through the actions of collembola, and these ectomycorrhizal (ECM) species are characterized by the above-mentioned mycelial type. In microcosm experiments, they found that *Protaphorura armata*, a Collembolan prevalent in their forests, consumed significant amounts of fungal mycelium and had significant feedback effects on total litter decomposition rates in their studies. Using birch or Scots pine seedlings, the authors measured changes in $^{14}CO_2$ evolution in the presence of collembolan grazers, or in their absence, measuring the metabolic activity of several ECM fungal species. There was a significant plant/ECM species × collembola interaction ($F7,20 = 3.17$; $P = 0.020$), with collembola significantly reducing $^{14}CO_2$ release from three of the eight plant/ECM species combinations (*P. involutus*/Scots pine, *P. fallax*/Scots pine and *P. fallax*/birch). In one combination (*Suillus variegatus*/Scots pine), collembola significantly increased $^{14}CO_2$ release. However, in five out of eight instances, there was no significant difference between experimental versus control replicates. The presence of collembola led to considerable variation and a reduction in the C:N ratio of the litter, which was 264:1 ±105 (mean ± SE) in the bags with collembola compared to 83:1 ± 18 in the controls ($P = 0.033$). There was no change in total C concentration, but a minor decrease in the N concentration of litter from 0.73% ± 0.11% to 0.54% ± 0.10% in the presence of collembola, but this difference was not significant. What does this mean, in the larger scheme of things? For one thing, a combined approach, using a suite of mesofauna, would likely lead to more marked differences between decomposition patterns with and without fauna.

Another microcosm experiment examined the effects of protozoa grazing on bacteria on the foraging efficiency of AMF for mineral N from OM in soil. As has been demonstrated by others (Bonkowski and Clarholm, 2012), the effect of metabolized N from protozoan ingested AMF markedly enhanced the N uptake of host plants (*Plantago lanceolata*) in microcosms with roots present, or with OM patches. Koller et al. (2013) found that protozoa mobilized N by consuming bacteria, and the mobilized N was translocated via AMF to the host plant. The presence of protozoa in both the OM and root compartment stimulated photosynthesis and the translocation of C from the host plant via AMF into the OM patch. This stimulated microbial activity in the OM patch, plant N uptake from OM, and doubled plant growth. Again, this was demonstrated in a microcosm experiment, which is useful to demonstrate main effects in controlled circumstances.

This may lead to lower decomposition rates of fungal necromass in collembolan-shaped mycorrhizal communities. Such effects could be further amplified by the reduction of plant contributions to soil C via mycorrhizal hyphal networks in the presence of soil mesofauna due to hyphal feeding or disruption (Johnson et al., 2005). The outcome of interactions between mycorrhizal fungi and soil saprotrophs may vary with animal density (Steinaker and Wilson, 2008) and environmental conditions (Yang et al., 2015). In addition, competitive interactions of mycorrhizal and saprotrophic fungi might be modulated by animal decomposers via preferential feeding of the two guilds (Gange, 2000; Tiunov and Scheu, 2005). Although new research has significantly advanced our understanding of the effects of soil fauna and mycorrhizal fungi on soil C stocks, the complex interactive effects between mycorrhizal fungi and soil fauna warrant further investigation (Van der Wal et al., 2013).

The take-home message is, for soil fauna, we now have the ability to make gut content analyses which, in combination with sequencing, can reveal feeding preferences, which enables establishing real trophic links based on observations, which can then be visualized as networks of feeding interactions. The fate of carbon flow through the soil food web can be traced by using stable isotopes combined with sequence-based techniques. This provides insight into trophic connections and interaction strength (Morrien, 2016).

6.4 A HIERARCHICAL APPROACH TO ORGANISMS IN SOILS

Because of the need to deal with soil heterogeneity in space and time, arenas of interest, noted earlier, are represented in Fig. 6.4 (Beare et al., 1995) showing the volumes and biotic groups of concern. The aggregatusphere shows bacteria, amoebae, and some nematodes, having varying degrees of success in gaining access to the prey biota of interest (Vargas and Hattori, 1986). Moving up to a coarser level of resolution, to the rhizosphere, a few millimeters or less in scale one sees the microbes and fauna associated with them, and the considerable feeding and activity which has been documented numerous times. The activities are strongly influenced by abiotic, i.e., wetting and drying events, and the intrusion of new organic substances from growing root tips (Cheng et al., 1993; Kuzyakov, 2002; Kuzyakov and Blagodatskaya, 2015), or deposited feces from microarthropods, enchytraeids, or other mesofauna. The next level of resolution expands from many centimeters to several meters across the landscape, when any of the macrofauna, such as earthworms or burrowing beetles, come into play. There is then a qualitative shift, brought about by the ingestion of soil, which includes considerable amounts of micro- and mesobiota, e.g., protozoa and nematodes (Piearce and Phillips, 1980; Yeates, 1981) as food.

Interestingly, even with earthworms, the drilosphere sensu stricto is only 2–3 mm in thickness (Bouché, 1975) but the burrow extends laterally for many centimeters or meters through the soil. As a consequence of this activity, there can be major short-term decreases in viability of the existing biota, but possibly longer-term stimulation by enhanced microbial activity, as noted earlier, and also from the considerable input of mucopolysaccharide-containing mucus (Marinissen and Dexter, 1990).

An additional aspect of altered species makeup of bacteria in earthworm-influenced soils has been explored using molecular probing techniques. Using 16S rRNA probes and "libraries" of soil bacteria at the Horseshoe Bend site in Athens, GA, Furlong et al. (2002) and Singleton et al. (2003) found enhanced percentage occurrences of Actinobacteria, Firmicutes, and gamma-Proteobacteria in castings of *Lumbricus rubellus*, an epigeic earthworm.

In addition, considerable amounts of ammonia and urea, as nitrogenous end products of metabolism, may be voided either externally, through nephridiopores, or internally into the gut cavities of earthworm genera which have that mode of nitrogen excretion (Lavelle et al., 1992). In tropical regions, certain endogeic earthworms will process and assimilate end products of the breakdown (from 2% to 9%) of soil OM in a wide range of ecosystems (Lavelle and Martin, 1992).

Similar sorts of activities may be catalyzed by certain termites, particularly those in the advanced family Termitidae, which are truly geophagous. These geophages utilize soil OM, deriving significant amounts of nutrition from this low-quality substrate by processing the OM in a high-pH chemical milieu in the region between the midgut and the first proctodaeal segment (see Section 4.4.1.7.4) (Bignell, 1984; Fouquet et al., 2011; see Chapter 8: Future Developments in Soil Ecology references). The additional influence of microbial enzymes on insect digestive processes and indeed enhancement of nitrogen fixation in downed branches and logs (Martin, 1984) are well known. Finally, the impacts of ant and termite nests are significant and certainly have an influence at the landscape scale. The impacts of the macrofauna, sometimes termed "ecosystem engineers" (Jones et al., 1994; Lavelle et al., 2016), can extend for many meters beyond the immediate zones that they occupy. It has been contrasted with the impacts of smaller fauna, with smaller fauna more influential in energy flow and immediate nutrient recycling, noted earlier, versus the longer-term effect of the "engineering" by the macrofauna (Scheu and Setälä, 2002; Fig. 6.5). Scheu and Setälä (2002) and Wardle (2002) note that "trophic cascades," the term denoting the effects of predation on the biomass of organisms at least two trophic levels removed, occur in soil systems. Although developed principally for systems with living net primary production as the energy base, there are numerous examples in soil systems, particularly ones dominated by fungi. Scheu and Setälä (2002) comment on the limited number of studies of trophic cascades in soil systems to date, and that the fungal-based energy channel may be much more prone to trophic cascades than the bacterial-based channel. This assertion is certainly a candidate for further experiments in the future.

In an extensive synthesis of soil food web studies, Bradford (2016) noted that there is a growing consensus that bacterial versus fungal-based food webs may need to be reconsidered in the light of the many crossover routes or avenues that food sources follow in soils. Thus the influence of root-derived carbon-containing compounds, whether from root exudates or exfoliates, and their influences on mycorrhizal and saprophytic fungi and the

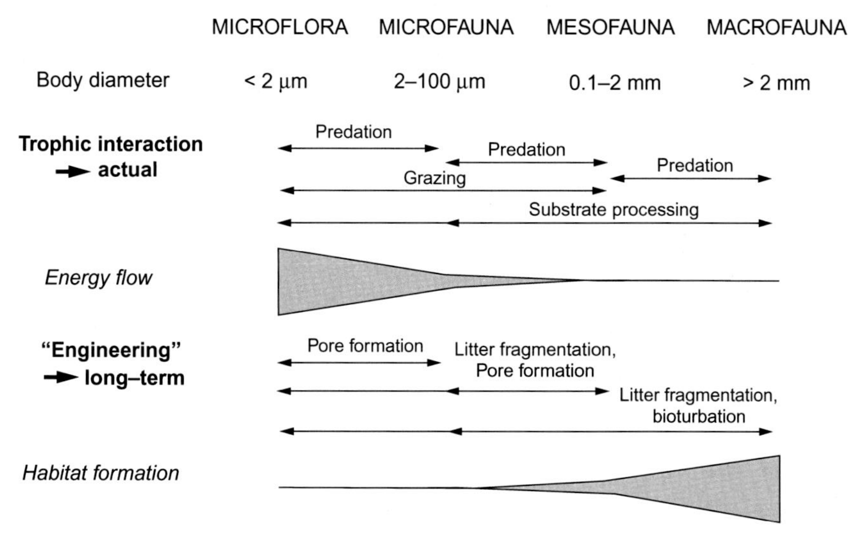

FIGURE 6.5 Size-dependent interactions among soil organisms. Trophic interactions and interactions caused by "engineering" are separated; both are indicated by arrows. Note that trophic interactions and interactions caused by engineering are strongly size dependent but complement each other (tapering and widening triangles). Both function at different scales: trophic interactions drive the current energy flow, engineering sets the conditions for the existence of the soil biota community in the long term. Source: *From Scheu, S., Setälä, H., 2002. Multitrophic interactions in decomposer food-webs. In: Tscharntke, B., Hawkins, B.A. (Eds.), Multitrophic Level Interactions. Cambridge University Press, Cambridge, MA, pp. 223–264.*

associated bacteria have a far-greater impact on how the system functions than following bacterial or fungal food webs, *per se*. This helps to unify the focus of the studies, while complicating matters in a fashion that more closely follows what is indeed occurring in nature.

We will pursue some of the implications for this revised approach to soil food webs when we address the effects of trophic interactions on longer-term phenomena in soils, leading ultimately to what we are considering an "evolutionary pedology."

Conceptualizations of detrital food webs are undergoing a considerable shift early in the third millennium. Following up on earlier ideas of Wardle and Yeates (1999) and Lavelle et al. (1999), Pokarzhevskii et al. (2003) note that a definite nested element exists such that different compartments feed into others. For example, the bacteria–algae–protozoa compartment is nested inside a fungi–microarthropod compartment, and this in turn is contained with an earthworm–rhizosphere compartment. Animals at higher levels consume communities of the lower levels as a whole (Pokarzhevskii et al., 2003; Fig. 6.6). This arises from the dependence of all animals on microorganisms for their supply of proteins and scarce minerals. The concept of "ecological stoichiometry," the roles of interactions between several major nutrients, e.g., N, P, and/or S, has been discussed at length by Sterner and Elser (2002). Much of Pokarzhevskii et al.'s (2003) paper discusses the need to consider the effects of limiting nutrients, which may be in shorter supply than the carbon or energy that characterize the outlook of many of the previously developed detrital food webs.

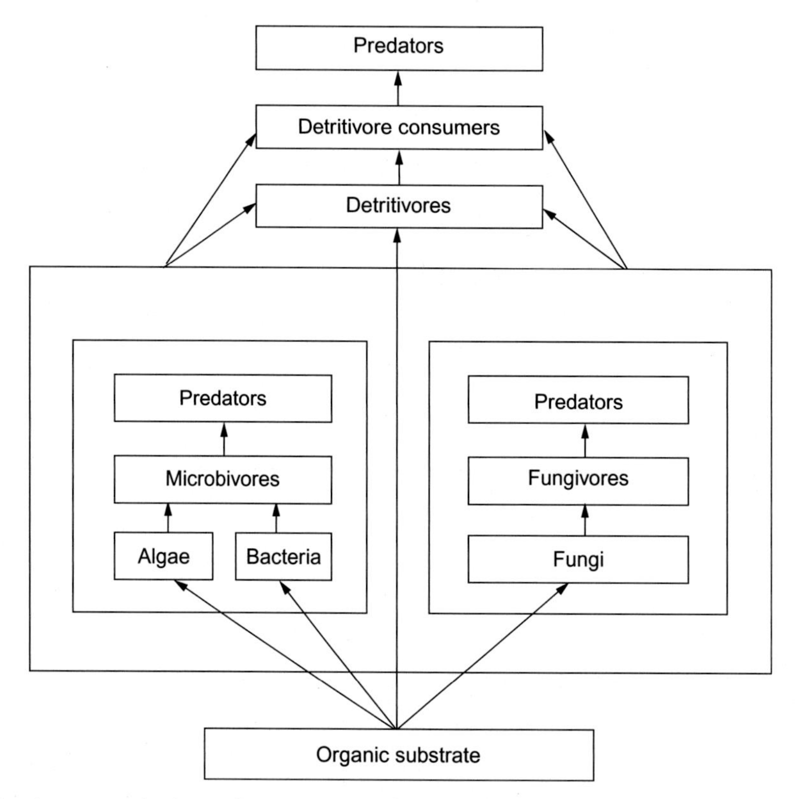

FIGURE 6.6 A conceptual scheme illustrating the nested structure of detrital food webs. A distinction is made between bacteria–algae–protozoa communities (left), fungi–microarthropod communities (right), and earthworm–plant communities (top). Communities of the higher levels consume communities of the lower levels as a whole (indicated by arrows). Source: *From Pokarzhevskii, A.D., van Straalen, N.M., Zaboev, D.P., Zaitsev, A.S., 2003. Microbial links and element flows in nested detrital food-webs. Pedobiologia, 47, 213–224.*

6.5 FUTURE RESEARCH PROSPECTS

It is becoming more and more imperative to bring small working groups, or teams of investigators, together to make further progress in food web studies. The real breakthroughs are certain to come from efforts which include the more transitional fauna between above- and belowground, such as ants, dipteran larvae and ground beetles, or cryptozoans, such as the isopods, centipedes, and millipedes, linking them to the truly belowground fauna and microbes.

Various techniques noted in several chapters in this volume should be extended, as well. Stable isotopes, introduced in an initially enriched substrate, such as labeled glucose or acetate, will be useful in delineating food webs. The effective use of carbon-13 and nitrogen-15 was reviewed extensively by Scheu (2002). An innovative use of ^{15}N tagging in a microcosm study detected significant predation on springtails by an ECM fungus, *Laccaria bicolor* (Klironomos and Hart, 2001). The ECM fungus immobilized the animals before infecting them. Springtails (*Folsomia candida*) alive or already dead labeled with ^{15}N were added to the microcosms containing mycorrhizal or nonmycorrhizal *Pinus strobus* plants. Only the fungus and not the roots made contact with the animals. Amounts of nitrogen were determined in plant tissues and extraradical fungal hyphae over a 2-month period. Up to 25% of plant nitrogen was derived from springtails when they were in the presence of *L. bicolor*. At the end of the experiment, less than 10% of the number of animals were present compared to at the start. Using the same system, growing *P. strobus* seedlings with a different ECM fungus, Klironomos and Hart (2001) measured less than 5% of plant N acquired from the springtails. This experiment demonstrates a much greater range of possible interactions between mycorrhiza and fungal grazing animals, yet demonstrates another example of the tight linkages existing in forest nutrient cycling.

Opportunities for use of radiotracer ^{14}C must also be kept in mind. For example, Kisselle et al. (2001), Garrett et al. (2001), Fu et al. (2001), and Coleman et al. (2002) described the detrital food web and its dynamics in an agroecosystem, as a function of the impacts of aboveground experimentally induced herbivory. They measured increased microbial biomass production in no-tillage treatments that experienced moderate levels of aboveground herbivory (grasshoppers grazing on corn leaves). This was transmitted up the food chain to bacterial-feeding nematodes, with significantly more ^{14}C activity being taken up in the low grazing-intensity treatments, similar to the findings of Holland et al. (1996) (Fig. 6.7). Another notable finding was the higher ^{14}C activity in microarthropods extracted from rhizospheres of weed plants, compared to that of corn (Fig. 6.8). Garrett et al. (2001) suggest that weed rhizospheres may be more important than crop rhizospheres in supporting soil food webs. This might be expected, because crop plants are selected to maximize their aboveground NPP unlike weeds. If this pattern is general, weeds may be a significant factor for the protection of soil biodiversity, especially in conventionally tilled agroecosystems. The linkage between above- and belowground food webs is an exciting new topic for the first decade of the 21st century (Hooper et al., 2000; Wolters et al., 2000).

Hall and Raffaelli (1993) suggest two major areas of food web research, which could be most profitably followed: (1) focusing on community assembly and (2) documenting the strength of trophic interactions between elements in webs. Examining the latter objective, Neutel et al. (2002) studied interaction strengths organized in trophic loops (defined as the product of interaction strengths in a food web. Using seven documented soil food webs,

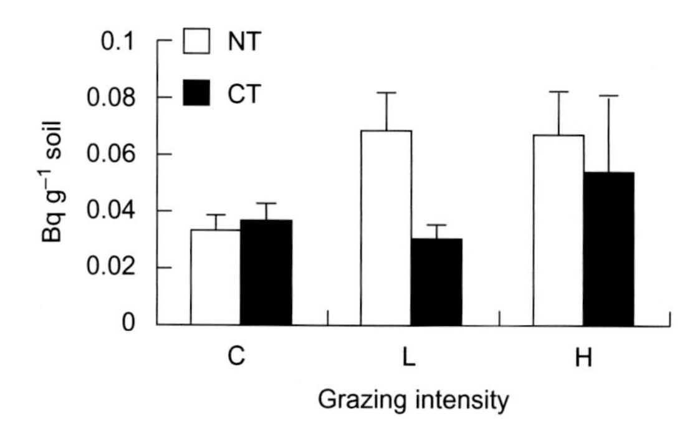

FIGURE 6.7 The ^{14}C-specific activity (Bq g^{-1} soil) of soil nematodes under different levels of aboveground herbivory. *C*, Control, no grazing; *L*, low-grazing level (four grasshoppers per plant); *H*, high grazing level (eight grasshoppers per plant); *NT*, no-tillage; *CT*, conventional tillage. Source: *From Fu et al. (2001).*

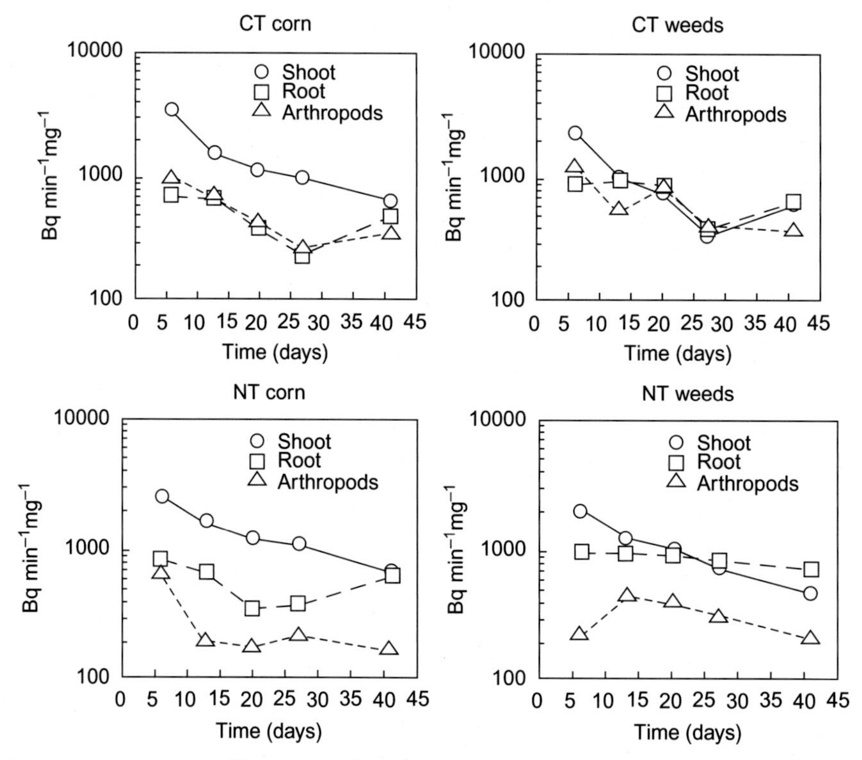

FIGURE 6.8 Concentrations of ^{14}C (Bq min^{-1}mg^{-1}) in shoots, roots, and microarthropods from rhizospheres of corn and weed plants grown under no-tillage (NT) and conventional tillage (CT) regimes. Source: *From Garrett, C.J., Crossley Jr., D.A., Coleman, D.C., Hendrix, P.F., Kisselle, K.W., Potter, R.L., 2001. Impact of the rhizosphere on soil microarthropods in agroecosystems on the Georgia piedmont. Appl. Soil Ecol. 16, 141–148.*

Neutel et al. (2002) introduced the term "loop weight," which is the geometric mean of the absolute values of the interaction strengths in the loop (Fig. 6.9). This enables one to compare loops of different lengths and to use the maximum of all loop weights as an indicator for matrix stability. They used the conservative figure of 0.1 for trophic transfer efficiencies, from one trophic level to another. They innovatively compared the community matrix, including the patterned interaction strengths ("real matrices") with several randomizations of this matrix ("randomized matrices"). This was done by randomly exchanging predator–prey pairs of interaction strengths, keeping these pairs intact and preserving the sign structure of the matrix. Stability was measured as the minimum degree of relative intraspecific interaction needed for matrix stability (*s*). Matrices with a smaller *s* were considered "more stable."

Loop weights of the longer loops were low in the real matrix and tended to be heavier in the randomized matrices than the shorter loops (Fig. 6.9A and B). Interestingly, although absolute values of effects of predators on their prey are generally two orders of magnitude larger than effects of prey on their predators, and was so shown in the randomized matrices, in the real matrices, the long loops with many top-down effects had a

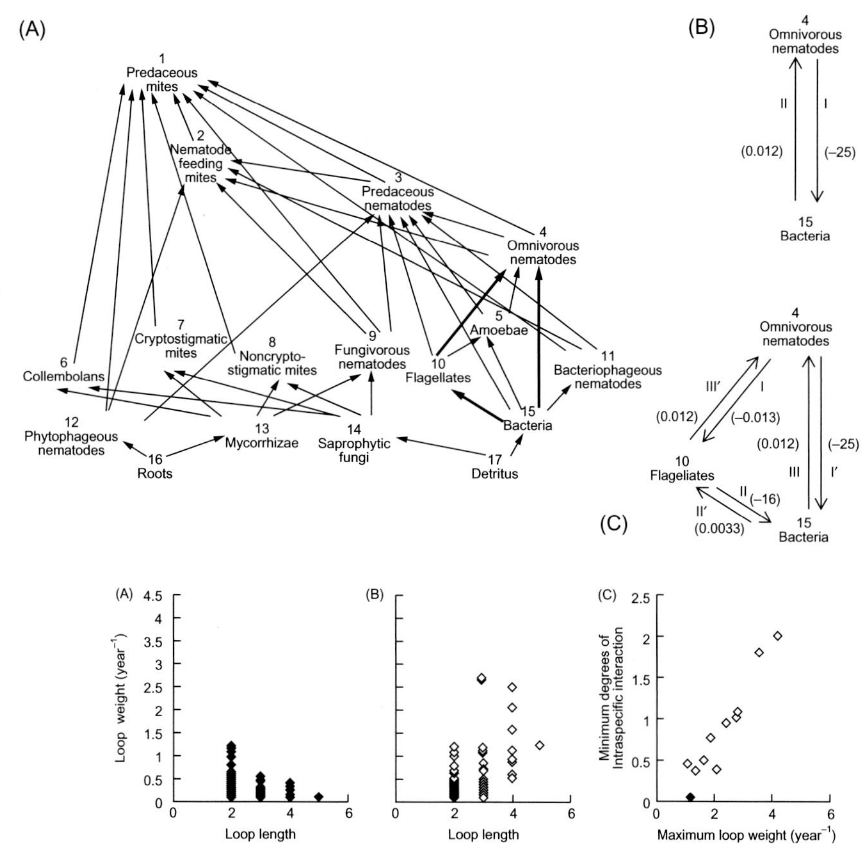

FIGURE 6.9 Loop length, loop weight, and stability in the Central Plains Experimental Range (CPER) food web and randomizations of this matrix. (A) Loop weight versus loop length in the real matrix. (B) Loop weight versus loop length in a randomized matrix (a typical example). Long loops with a relatively small weight—those with many bottom-up effects—are not shown because they are not relevant for maximum loop weight. (C) Maximum loop weight and stability of the real matrix (solid diamond) and of 10 randomized matrices (open diamonds). Stability was measured as the value *s* that leads to a minimum level of intraspecific interaction strength needed for matrix stability. In a sensitivity analysis, variation in the parameter values within intervals between half and twice the observed value led to only a small variation in stability. Source: *From Neutel, A-M., Heesterbeek, J.A.P., de Ruiter, P.C., 2002. Stability in real food webs: weak links in long loops. Science, 296, 1120–1123.*

relatively low weight. This revealed that not all top-down effects were equal. With the maximum loop weight in the real matrix being markedly lower, the real matrix was much more stable (Fig. 6.9C). Neutel et al. (2002) explored the ramifications of omnivory. For a three-species omnivorous interaction (Fig. 6.10), the omnivore feeds on two prey types, which are at different trophic levels. Assuming that it feeds according to prey abundance, and that the biomass of the prey on the lower trophic level is significantly larger than that of the prey on the higher trophic level, then the omnivore feeds largely on the lowest trophic level. Consequently, it exerts a relatively large top-down effect on its lowest prey and relatively small top-down effect on its higher prey, because the top-down effect is the

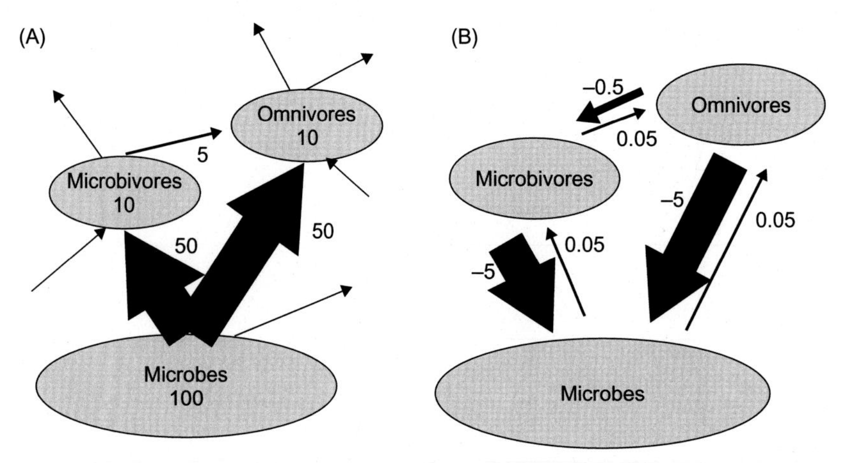

FIGURE 6.10 Interaction strengths and loop weights in an omnivorous food web. (A) Equilibrium feeding rates and population sizes. Feeding rates were assumed to be proportional to the population sizes of the prey. (B) Interaction strengths. In the example, efficiencies were assumed to be 0.1 for all species. The loop weights of the two loops of length 3 are $([-0.5] \times [-5] \times 0.05)^{1/3} = 0.5$ (anticlockwise loop, starting with the omnivores) and $([-5] \times 0.05 \times 0.05)^{1/3} = 0.23$ (clockwise loop). The relatively small top-down effect (−0.5) keeps the weight of the loop with two top-down effects relatively low. Source: *From Neutel, A-M., Heesterbeek, J.A.P., de Ruiter, P.C., 2002. Stability in real food webs: weak links in long loops. Science, 296, 1120–1123.*

feeding rate per unit of predator biomass. This approach was extended further to a wide range of published food webs, and their findings held true even for aboveground orientated food webs.

In very seminal reviews, Moore et al. (2003) and Moore and de Ruiter (2012) noted that predators within the rhizosphere alter the interactions between microbes and plants in two contrasting but probably equally important ways. Predators regulate their prey in a traditional "top-down" fashion but in doing so, they alter the release of nutrients that may limit plant productivity and thereby affect plant growth in a "bottom-up" fashion as well. They note that the interdependence between the aboveground and belowground realms can be explained in terms of the patterning of trophic interactions within the rhizosphere and the influence of these interactions on the supply of nutrients and rates of nutrient uptake by plants.

We suggest that a useful approach will include a melding of the earlier two objectives, in terms of documenting the extent of soil food webs, and the relative impacts of the trophic interactions at the various hierarchical levels of organization, and location in the landscape (Coleman and Schoute, 1993; Hooper et al., 2000). For example, does the soil system in the absence of earthworm or termite activity operate at more or less of a background or maintenance level? When the macrofauna move through the soil matrix, literally consuming, and chaotically reassembling it, does this represent a more intensive level of activity? Certainly it is at a different level of resolution, but one dependent upon the myriads of interactions of the microbes, micro- and mesofauna.

Evidence on how the role of soil fauna in food webs will respond to environmental change and influence aboveground processes has advanced considerably (see Bragazza et al., 2013; Cheeke et al., 2012; de Vries et al., 2012; Wall et al., 2012). Loss of species due to varying management practices, erosion, pollution, urbanization, and resulting effects on ecosystem function and services are becoming widely recognized and related to larger issues of biodiversity loss, desertification, and elevated greenhouse gas concentrations (Koch et al., 2013). This has resulted in international attention to the importance of soils and, in particular, to soil biodiversity. An analysis (see the European Atlas of Soil Biodiversity, Jeffery et al., 2010, and the Global Soil Biodiversity Atlas, Orgiazzi et al., 2016 cited in Chapter 8: Future Developments in Soil Ecology) includes maps showing the potential vulnerability of soil biodiversity and ecosystem services to various environmental changes, including land use change. We are now able to answer basic questions about soil fauna including topics such as whether there are cosmopolitan versus endemic fauna species; what is the local and global biogeographic distribution and range of faunal species; what are the factors influencing the distribution of key species; and whether the loss of species affects ecosystem function (Cock et al., 2012; Wall et al., 2012). Food web ecology, with its emphasis on community assembly and disassembly, has the potential to act as an integrating concept across conservation biology and community and ecosystem ecology, as well as for provision of ecosystem services (de Vries et al., 2012; Moore and de Ruiter, 2012; Thompson et al., 2012; Wall et al., 2012).

6.6 SUMMARY

There is an interesting convergence occurring in aboveground and belowground portions of detrital food webs. In both locations, particularly in arid habitats (i.e., in deserts) the food webs are long (seven to eight membered) and show extensive amounts of omnivory. By including members of the microfauna (protozoa) and mesofauna (microbivorous nematodes) that have been overlooked often in the past, there are ample amounts of food, as secondary production passes up the food chains. Production efficiencies may reach or exceed 70%, and trophic transfer efficiencies may exceed 20% in various "hot spots" such as rhizospheres, drilospheres, or in any other concentrations of reduced, labile OM. The diversity of nutrient retention and recycling strategies in soil systems continues to increase as more innovative experiments are carried out using a variety of isotopic tracer techniques.

CHAPTER

7

Soil Biodiversity and Linkages to Soil Processes

7.1 INTRODUCTION

In the last two decades, there has been a rising current of interest in soils as reservoirs of biodiversity. It is important to define biodiversity, which is an inclusive concept. Biodiversity encompasses a wide range of functional attributes in ecosystems in addition to being concerned with the numbers of species present in a given ecosystem. Within terrestrial ecosystems, soils may contain some of the last great "unknowns" of many of the biota (Wall et al., 2012). This includes such relatively well-studied fauna as ants (Hölldobler and Wilson, 1990), as well as the more numerous and less studied mesofauna, such as microarthropods (Behan-Pelletier and Newton, 1999) and nematodes (Coleman and Wall, 2015; Ettema and Yeates, 2003), that interact with elements of the microbiota such as mycorrhiza in several ways, including mutualistic ones (Wall and Moore, 1999). Much has been learned about prokaryotic genetic diversity in soils; see reviews by Hugenholtz et al. (1998) and Fierer and Jackson (2006).

7.2 BIODIVERSITY IN SOILS AND ITS IMPACTS ON TERRESTRIAL ECOSYSTEM FUNCTION

There is increasing concern among biologists in the fates of the very diverse array of organisms in all ecosystems of the world. What do we know of the full species richness, particularly in soils, to make even any educated guesses about the total extent of the organisms, or how many of them may be in an endangered status (Coleman, 2001; Coleman et al., 1994b; Hawksworth, 1991a, 2001; Orgiazzi et al., 2016)? Soil biodiversity is best considered by focusing on the groups of soil organisms that play major roles in ecosystem functioning. Spheres of influence of soil biota are recognized, such as the root biota, the shredders of organic matter, and the soil bioturbators. These organisms influence or control ecosystem processes and have further influence via their interactions with key soil biota (e.g., plants) (Coleman, 2001; Lavelle et al., 2016; Wardle, 2002). Some organisms, such as the fungus and litter-consuming microarthropods, are very speciose. For example,

Fundamentals of Soil Ecology
DOI: http://dx.doi.org/10.1016/B978-0-12-805251-8.00007-7

there are up to 170 species in one order of mites, the oribatida, in the forest floor of one watershed in western North Carolina. Hansen (2000) measured increased species richness of oribatids as she experimentally increased litter species richness in experimental enclosures from one to two, four, and finally seven species of deciduous tree litter. This was attributed to the greater physical and chemical diversity of available microhabitats, which is in accord with the mechanisms suggested earlier by Anderson (1975a).

Only 30%–35% of the oribatids in North America have been adequately described (Behan-Pelletier and Bissett, 1993), despite many studies carried out in recent decades. They suggest that there may be more than 100,000 undescribed species of oribatid mites yet to be discovered. Particularly in many tropical regions, oribatids and other small arthropods are very little known in both soil and tree canopy environments (Behan-Pelletier and Newton, 1999; Nadkarni et al., 2002). This difficulty is compounded by our very poor knowledge of identities of the immature stages of soil fauna, particularly the Acari and Diptera. Solution of this problem may require considerable application of molecular techniques to more effectively work with all life stages of the soil fauna (Behan-Pelletier and Newton, 1999; Coleman, 1994a; Freckman, 1994; Morrien, 2016). We concur with Behan-Pelletier and Bissett (1993): "Advances in systematics and ecology must progress in tandem: systematics providing both the basis and predictions for ecological studies, and ecology providing information on community structure and explanations for recent evolution and adaptation." Chapin et al. (2000) note that 12% of birds and nearly 20% of mammals are considered threatened with extinction, and that from 5% to 10% of fish and plants are similarly threatened. With many of the soil invertebrates yet undescribed, it is impossible to affix a numerical value to losses of these members of the biota.

There are currently over 98,000 species of saprotrophic fungi described (van der Wal et al., 2013) (Table 7.1). By assuming that a constant ratio of species of fungi exists to those plant species already known, Hawksworth (1991b, 2001) calculated that there may be a total of 1.5 million species of fungi described when this mammoth classification task is completed. However, the assumption of a constant ratio of fungal to plant species has

TABLE 7.1 Comparison of the Numbers of Known and Estimated Total Species Globally of Selected Groups or Organisms

Group	Known Species	Estimated Total Species	Percentage Known
Vascular plants	220,000	270,000	81
Bryophytes	17,000	25,000	68
Algae	40,000	60,000	67
Fungi	98,000[a]	1,500,000	5

[a]*From Van der Wal, A., Geydan, T.D., Kuyper, T.W., De Boer, W., 2013. A thready affair: linking fungal diversity and community dynamics to terrestrial decomposition processes. FEMS Microbiol. Rev. 37, 477–494.*

Modified from Hawksworth, D.L., 1991b. The fungal dimension of biodiversity: magnitude, significance and conservation. Mycologic. Res. 95, 641–655.

recently been challenged with molecular data, and this has complicated the estimation of the actual number of fungal species extant on Earth (Tedersoo et al., 2015).

Indeed, it may be possible to gain insights into biotic functions belowground by considering a "universal" set of functions for soil and sediment biota that include degradation of organic matter, cycling of nutrients, sequestration of carbon, production, and consumption of trace gases, and degradation of water, air, and soil pollutants (Groffman and Bohlen, 1999).

What are the consequences of biodiversity? Does the massive array of hundreds of thousands of fungi and probably millions of bacterial species make sense in any ecological or evolutionary context? As was noted in Chapter 3, Secondary Production: Activities of Heterotrophic Organisms—Microbes, on microbes, the numbers of bacterial species are greatly underestimated, because most investigations have relied on culturing isolates and examining them microscopically. There have been two key developments in studies of microbial diversity. Firstly, the use of signature DNA sequences has greatly increased the numbers of identified taxa, with hundreds of novel DNA sequences being identified yearly (Kuzyakov and Blagodatskaya, 2015). On the other hand, DNA-based estimates may have overstated the actual diversity due to relic DNA persisting in soil and causing the appearance of greater diversity (Carini et al., 2016). Thus, the definitive estimate of the "real" number of bacterial taxa may remain elusive for quite some time into the future. Two bacterial divisions, which appear to be abundant and ubiquitous in soils but have very few cultured representatives, are Acidobacterium and Verrucomicrobium (Hugenholtz et al., 1998). Secondly, we have only recently come to an appreciation of the incredibly wide distribution of prokaryotes (both archaea (methanogens, extreme halophiles living in hypersaline environments, and hyperthermophiles living in volcanic hot springs, and in midsea oceanic hot-water vents) and bacteria) worldwide. Prokaryotes constitute two of the three principal domains, or collections of all organisms, with eucarya consisting of protists, fungi, plants, and animals (Fig. 7.1) (Coleman, 2001; Pace, 1999). The total numbers of bacteria on earth in all habitats is truly mind-boggling: $4-6 \times 10^{30}$ cells, or 350–550 Pg (10^{15} g) of carbon (Whitman et al., 1998). The amount of the total bacteria calculated to exist in soils is approximately 2.6×10^{29} cells, or about 5% of the total on earth. A majority of bacteria exist in oceanic and terrestrial subsurfaces, especially in the deep mantle regions, extending several kilometers below the earth's surface. Some of these organisms, which are the most substrate-starved on earth, may have turnover times of centuries to millennia (Whitman et al., 1998).

What is the implication of the apparent "excess" of species diversity of soil microflora, where many species exist at a very low frequency, and in an inactive state? If considerable species richness and accompanying large genetic pools are maintained in soils, what are the impacts on the evolution of new taxa? What are the implications for ecosystem function if this degree of redundancy exists; does it imply that some of the organisms are somehow vestigial remnants or relics of bygone conditions (Coleman et al., 1994b)? What are the functional roles of such hidden or apparently cryptic organisms? Are they performing some essential, but unknown functions, perhaps at microsites that we don't observe or work with? One approach that may show promise is the use of reporter genes linked to gene promoters, in order to measure in situ the activity of specific enzymes related to defined processes (Wilson et al., 1994). We need to link specific methods such as those noted here with soil thin-section studies, such as those of Tippkötter et al. (1986), Postma

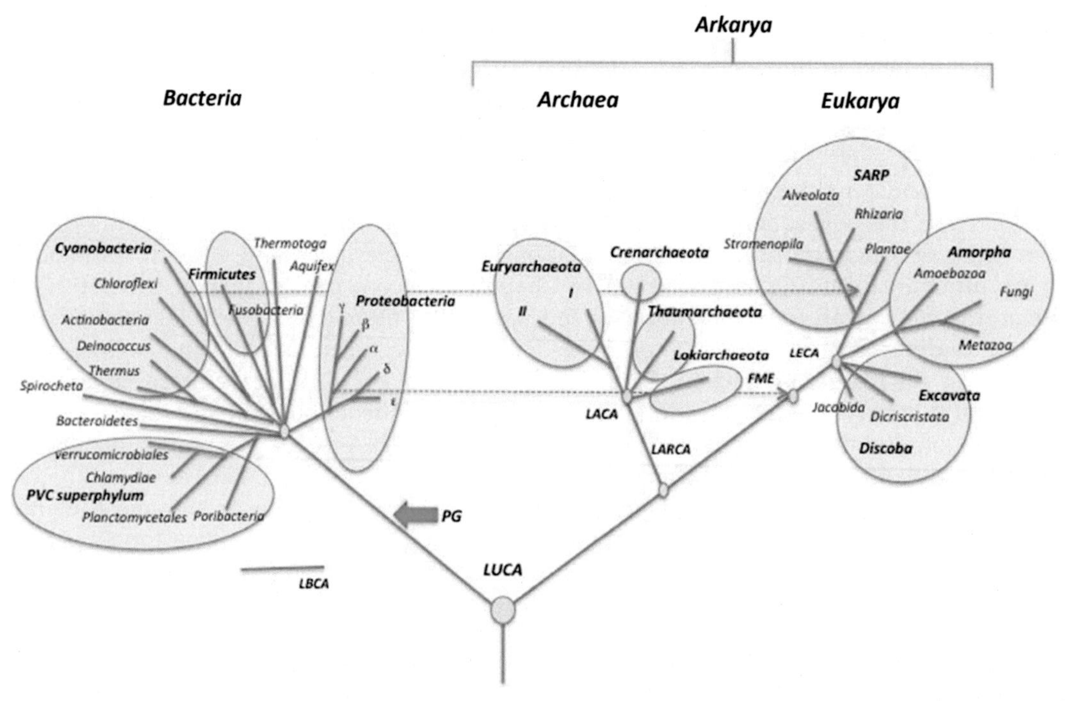

FIGURE 7.1 Schematic simplified universal tree of life updated from Woese et al. (1990). *DNA*, (Blue arrows) introduction of DNA; *FME*, first mitochondriate eukarya; *LACA*, last archaeal common ancestor; *LARCA*, last Arkarya common ancestor; *LBCA*, last bacterial common ancestor; LUCA, last universal common ancestor; *LECA*, last eukaryotic common ancestor; *PG*, peptidoglycan; *SARP*, stramenopila, alveolata, rhizobia, plantae; *blue circles*, mesophilic ancestors. Source: *From Forterre, P., 2015. The universal tree of life: an update. Front. Microbiol. 6, 717.*

and Altemüller (1990), and Foster (1994). Such means will enable the inclusion of spatial dimensions to soil ecological studies; addition of a temporal one provides the much-needed aspect of time as well. Soils are rife with historical signs and legacies, as has been made evident by studies using radiotracers (Trumbore, 2009), and stable isotope studies (Gaudinski et al., 2000; Maraun et al., 2011; Nadelhoffer et al., 1985; Stout et al., 1981).

What are the linkages between biodiversity and ecosystem function (Bardgett and van der Putten, 2014)? It should be possible to look for natural "experiments," such as regions with low species richness, e.g., on an island, versus sites at similar latitudes which are on continents, where one can measure key ecosystem processes, such as rates of decomposition or nutrient cycling. Under such conditions, all of the major abiotic factors are held reasonably similar, allowing study of the impacts of species richness of key indicator microflora or fauna on ecosystem processes of interest. Such experiments are certainly performable and might yield some surprising results. Studies of the interactions of climate change and biological diversity have been reviewed by Vitousek (1994) and in a modeling context, by Smith et al. (1998). By using species/area curves from island biogeography it is possible to estimate the fraction of species whose loss is entrained by loss of habitat (land use change), even without knowing how many species exist (Wilson, 1992, cited in Vitousek, 1994). Unfortunately, there is little information available on soil organisms to conduct such a comparative study.

More than 20 studies of the empirical evidence of relationships between ecosystem processes and different components of plant diversity (species richness, functional richness, and functional composition) were followed in natural and synthetically assembled groups of grassland species worldwide (Diaz and Cabido, 2001). The linkage was found to be neither simple nor universal, but some significant trends were noted. The range and more particularly the values of functional traits carried by plants, (e.g., whether they are nitrogen-fixing, warm-season grasses, or rosette forbs) are generally strong drivers of ecosystem processes. These studies combined simplified microcosms and natural field sites, so extrapolation from them is limited (Table 7.2). However, it is noteworthy that most of the studies showed that species richness and functional composition had positive effects on aboveground biomass.

Numbers of species aboveground and belowground may be correlated when taxa in both habitats respond similarly to the same or correlated environmental driving variables, in particular across large gradients of disturbance, climate, soil conditions, or geographical area. Differentiating between simple correlation and causation may be problematic, however. High diversity in plant species can result in high diversity of litter quality or types of litter entering the belowground system. This resource heterogeneity can lead to a greater diversity of decomposers and detritivores (Hooper et al., 2000). In contrast, a high diversity of resources and species in soil could feed back to a high diversity aboveground, where certain species or functional groups are closely linked to groups belowground. A useful example of this was noted by van der Heijden et al. (1998), who found a positive correlation between the diversity of endomycorrhizal species and plant diversity, perhaps because different species of fungi infect different species of plants to different degrees, although alternative explanations have been offered for these patterns (Wardle et al., 1999). Interestingly, Hartnett and Wilson (1999) and Smith et al. (1999) working in a Kansas tallgrass prairie showed that mycorrhiza promoted obligately mycorrhizal C_4 grasses, resulting in competitive exclusion of facultatively mycorrhizal C_3 species, reducing overall plant species diversity. A similar mechanism seems to operate in tropical rainforests, in which ectomycorrhizal (ECM) tree species competitively exclude arbuscular mycorrhizal (AM) species (Connell and Lowman, 1989, cited in Wardle, 2002). It should be noted that at the level of functional types of mycorrhiza, this pattern does not hold: low diversity AM can be associated with high diversity of plants, and high diversity ECM communities can be associated with low diversity of plants (Allen et al., 1995).

In an extensive experiment carried out under field conditions, Porazinska et al. (2003) tested aboveground–belowground diversity relationships in a naturally developed tallgrass prairie ecosystem by comparing soil biota and soil processes occurring in homogeneous and heterogeneous plant combinations of C_3 and C_4 photosynthetic pathways. Some bacterial and nematode groups were affected by plant characteristics specific to a given plant species, but no uniform patterns emerged. Interestingly, invasive and native plants were quite similar with respect to the measured soil variables (e.g., phospholipid fatty acids, protozoa, and nematode functional groups). Contrast these results with those of Evans et al. (2003) given toward the end of this chapter.

TABLE 7.2 Empirical Evidence of Relations Between Ecosystem Processes and Different Components of Plant Diversity[a]

Ecosystem	Experimental Setup	Ecosystem Processes[b]	Positive Effects Reported[c]			Functional Types (Sensu Lato)
			Species Richness	Functional Richness	Functional Composition	
SYNTHETIC ASSEMBLAGES						
Serpentine grassland, United States	Plant mixtures planted in the field	N retention in ecosystem	NA	No	Yes	Bunchgrasses N-fixers, early- and late-season annual forbs
		Aboveground biomass	NA	No	Yes	
		Inorganic N pools in soil	NA	Yes	Yes	
Savannah grassland, United States	Plant mixtures planted in the field	Aboveground biomass, light penetration, and plant % and total N	No	Yes	Yes	C_3 grasses, C_4 grasses, legumes, forbs, and woody plants
Mesic grassland, United Kingdom	Plant mixtures planted in the field	No. of invading species and total biomass of invasives	No	NA	Yes	Perennial grasses and forbs
Grasslands, Germany, Portugal, Switzerland, Greece, Ireland, Sweden, United Kingdom	Plant mixtures planted in the field	Total aboveground biomass	Yes	Yes	Yes	Grasses, legumes, herbs
Annual grassland, France	Plant mixtures planted in the field	No. of invasives from soil seed bank and survival of seedlings of the exotic and annual forbs *Conyza bonariensis* and *C. canadensis*	No	No	Yes	Annual grasses, annual legumes, and annual Asteraceae
Acid grassland, United Kingdom	Plant mixtures planted in the field	Decomposition of standard material	Yes	No	No	Grasses, legumes, and herbs
		Decomposition of litter mixtures	No	No	Yes	
Grassland on old fields, Switzerland and Sweden	Plant mixtures planted in the field	No. of leafhoppers (Cicadellidae)	No	No	Yes	Grasses, legumes, land forbs
			No	Yes	Yes	

(*Continued*)

TABLE 7.2 (Continued)

Ecosystem	Experimental Setup	Ecosystem Processes[b]	Positive Effects Reported[c]			Functional Types (Sensu Lato)
			Species Richness	Functional Richness	Functional Composition	
		No. of wingless aphids (Aphididae)				
		No. of hymenopteran parasitoids	No	No	No	
		No. of grasshoppers (Acrididae) and slugs (Gastropoda)	No	No	No	
		No. of carabid beetles (Carabidae) and spiders (Araneae)	No	No	Yes	
Calcareous grassland on old field, Switzerland	Plant mixtures planted in the field	Preference by voles[d]	Yes	No	Yes	Grasses, legumes, and forbs
		Earthworm biomass[d]	Yes	Yes	No	
		Plant aboveground biomass, soil microbial biomass, LAI, plant light absorbance per unit ground area	Yes	Yes	Yes	
		Mesofauna feeding activity	No	No	No	
		Decomposition of standard material	No	No	Yes	
		Soil moisture	No	No	Yes	
Grassland, Greece	Plant mixtures planted in the field	Total aboveground biomass	Yes	NA	Yes[e]	Annuals and perennial grasses, geophytes, and legumes
Serpentine grassland, United States	Plant mixtures planted in the field	Aboveground biomass of invasive forb *Centaurea solstitialis*	No	Yes	Yes	Annual grasses, perennial grasses

(*Continued*)

TABLE 7.2 (Continued)

Ecosystem	Experimental Setup	Ecosystem Processes[b]	Positive Effects Reported[c] Species Richness	Functional Richness	Functional Composition	Functional Types (Sensu Lato)
		Impact of invader on aboveground biomass of resident species and whole-system evapotranspiration	Yes	Yes	Yes	Bunchgrasses, early-season and late-season annual forbs
Grasslands on old fields, Czech Republic, the Netherlands, United Kingdom, Sweden, and Spain	Plant mixtures planted in the field	Total aboveground biomass	Yes	NA	Yes	Grasses, forbs, and legumes
		Suppression of natural colonizers	Yes	NA	Yes	
Grassland, United States	Plant mixtures planted in greenhouse microcosms	Aboveground biomass N retention	Yes No	NA NA	Yes Yes	C_3 grasses, C_4 grasses, legumes, and forbs
Annual grassland, France	Plant mixtures in greenhouse microcosms	Invasibility (establishment of the forb *Echium plantagineum)*	No	No	Yes	Grasses, legumes, and rosette dicots
Prairie grassland, United States	Plant mixtures in greenhouse microcosms	Above- and belowground biomass, light transmission, and water retention in soil	Yes	Yes	Yes	Grasses, legumes, and forbs
		Decomposition of standard material	No	No	Yes	
Prairie grassland, United States	Plant mixtures planted in the field and in greenhouse microcosms	Resistance to invasion (total biomass of invasive)	Yes	NA	Yes	C_3 grasses, C_4 grasses, legumes, and forbs
Grassland-crop site, New Zealand	Litterbags placed in the field	Decomposition rate of, rate of N release from, and active microbial biomass on litter	No	NA	Yes	Grasses, weedy forbs, forbs from grasslands, and trees

(*Continued*)

TABLE 7.2 (Continued)

Ecosystem	Experimental Setup	Ecosystem Processes[b]	Positive Effects Reported[c] Species Richness	Functional Richness	Functional Composition	Functional Types (Sensu Lato)
Grasslands, United Kingdom	Litterbags placed in indoor soil microcosms	Soil microbial biomass	No[f]	NA	Yes	Dominant species in intensively managed fertile grasslands, or traditionally managed unfertilized grasslands
MANIPULATION OF NATURAL COMMUNITIES						
Grassland, Argentina	Mostly perennial grassland in neighboring paddocks under different grazing regimes	Aboveground net primary production	No	No	Yes	Cool-season graminoids, warm-season grasses, cool-season and warm-season forbs
Boreal forest, Sweden	Vegetation on islands of different area, subjected to different frequencies of wildfires	Aboveground biomass, litter decomposition, N mineralization, and humus accumulation	No	NA	Yes	Early versus late successional species
Savannah grasslands, India	Vegetation along a productivity, diversity, and disturbance gradient, with different burning and grazing experimental treatments	Resistance to compositional change across communities	No[g]	NA[g]	Yes[g]	Not explicit, communities dominated by the grasses *Cymbopogon flexuosus* or *Aristida setacea*
		Resistance to species turnover across communities	Yes[g]	NA	No	
		Resistance to compositional change and to species turnover within communities	No[g]	NA	No	
Calcareous grasslands, United Kingdom	Contrasting grassland subjected to temperature and precipitation manipulations in the field	Resistance of total aboveground biomass and species compositions	No	NA	Yes	Communities dominated by fast-growing early successional species or by slow-growing, stress-tolerant perennial grasses and sedges

(*Continued*)

TABLE 7.2 (Continued)

Ecosystem	Experimental Setup	Ecosystem Processes[b]	Positive Effects Reported[c] Species Richness	Functional Richness	Functional Composition	Functional Types (Sensu Lato)
Mediterranean shrublands, Greece	Sites naturally differing in species diversity and growth-form composition	Aboveground biomass	Yes	NA	Yes	*Cistus* sp., other shrubs, and herbs
Sand-prairie grassland, United States	Experimental removal from natural communities on old fields	No. of individuals and cover of invaders	NA	Yes	Yes	C_3 graminoids, C_4 graminoids, and forbs
		Light transmittance through canopy	NA	Yes	Yes	
		Soil moisture, soil extractable N, and aboveground biomass	NA	No	Yes	
Dairy grasslands, New Zealand	Grasslands differing in climate and seasonal vegetation, subjected to experimental extreme temperature and rainfall events	Stability of biomass production after extreme events	No	NA	Yes	C_3 or C_4 species
Sand-prairie grassland, United States	Old-field communities subjected to removal of different functional types	Total aboveground biomass	NA	No	Yes	C_3 graminoids, C_4 graminoids, and forbs
		Community drought resistance	NA	No	Yes	

[a]*Only studies assessing the impact of at least two components of plant diversity on ecosystem processes, and published in 1995 or later, were considered. Comparisons are qualitative and should be taken with caution, because unless a study explicitly has a test for species richness, functional richness, and functional composition in its design, it might lead to underestimation or misrepresentation of different components of diversity. Field studies differ markedly among themselves and with synthetic assemblages studies in approach, design, and intervening factors and thus strict comparison is not possible.*

[b]*LAI, leaf area index; N, nitrogen.*

[c]*In the case of species and functional richness, only positive effects were considered: No, either no effect or a negative effect. In the case of functional composition: Yes, any significant (positive or negative) effect; NA, not assessed.*

[d]*Species: vole,* Arvicola terrestris; *earthworms,* Octolasion cyaneum, Nicodrilus longus, Allolobophora rosea, Allolobophora chlorotica, Lumbricus terrestris, *and* Lumbricus castaneus.

[e]*Species richness effect obvious only when annuals were included in analysis.*

[f]*Effect of increasing litter diversity on soil microbial biomass was not unidirectional: two- and four-species litter treatments decreased it, whereas five- and six-species treatments increased it.*

[g]*Shannon Diversity Index.*

From Diaz, S., Cabido, M., 2001. Vive la différence: plant functional diversity matters to ecosystem processes. Trends Ecol. Evol. 16, 646–655.

Viewed more generally across terrestrial ecosystems worldwide, van der Putten et al. (2013) have considered the array of plant–soil interactions as being in two categories: (1) direct and (2) indirect. Much of the microbial–mycorrhizal–faunal interactions fall in the latter category. Plant–soil community feedback or weighted community effects can also be compared against direct feedback effects by growing species in mixtures versus monocultures. An example of positive plant–soil community feedback effects is the over-yielding in species-rich plant communities compared to monocultures, which has been proposed to explain plant diversity–productivity relationships (van der Putten et al., 2013).

7.3 HETEROGENEITY OF CARBON SUBSTRATES AND EFFECTS ON SOIL BIODIVERSITY

A stepwise process for the ways in which increased heterogeneity of carbon (C) substrates from aboveground will positively influence belowground diversity is as follows (Fig. 7.2) (Hooper et al., 2000): (1) diversity of primary producers leads to diversity of C inputs belowground, (2) carbon resource heterogeneity leads to diversity of herbivores and detritivores, and (3) diversity of detritivores or belowground herbivores leads to diversity of organisms at higher trophic levels in belowground food webs. The critical point is the nature and extent of trophic interactions (Bradford, 2016; Hooper et al., 2000). There are three general categories of interactions by which organisms in one compartment can affect biodiversity in another one: (1) obligate, selective interactions (one-to-one linkage), through mutualism, for example; (2) one-to-many species linkages, via keystones and dominants, and (3) causal richness, or many-to-many linkages. The nature and extent of these interactions varies a great deal depending on the systems studied and the spatial scales at which the mechanisms are being considered.

There is a strong interaction between ecosystem function, organismal abundance and diversity, and the nature of humus forms in soil. Ponge (2003, 2013) compared more than 20 ecosystem attributes, and the nature of the processes and organisms occurring in mull, moder, and mor soils (Table 7.3) (Ponge, 2003). The table is a useful means of comparing many soil attributes, across a broad range of physical, chemical, and biological traits. It shows a marked gradient from high (mull) to low (mor) biodiversity and rapid to slow and very slow rates of humification. Not surprisingly, a key determinant of litter decomposability, phenolic content, varied inversely across the same sequence of three humus types. Of course, we have yet to see how well these generalizations hold up when including a detailed analysis of the microbial communities in all three humus types.

Studies of biodiversity should include assessments of the nature and extent of anthropogenic disturbance. In a multistate and province-wide study of snail distributions and diversity in 443 sites, anthropogenic disturbance was found to be a major factor in decreases in species richness in forested ("duff") versus grassland ("turf") sites. This indicates that the conservation of faunas in the former will require protection of the soil surface architecture (Nekola, 2003).

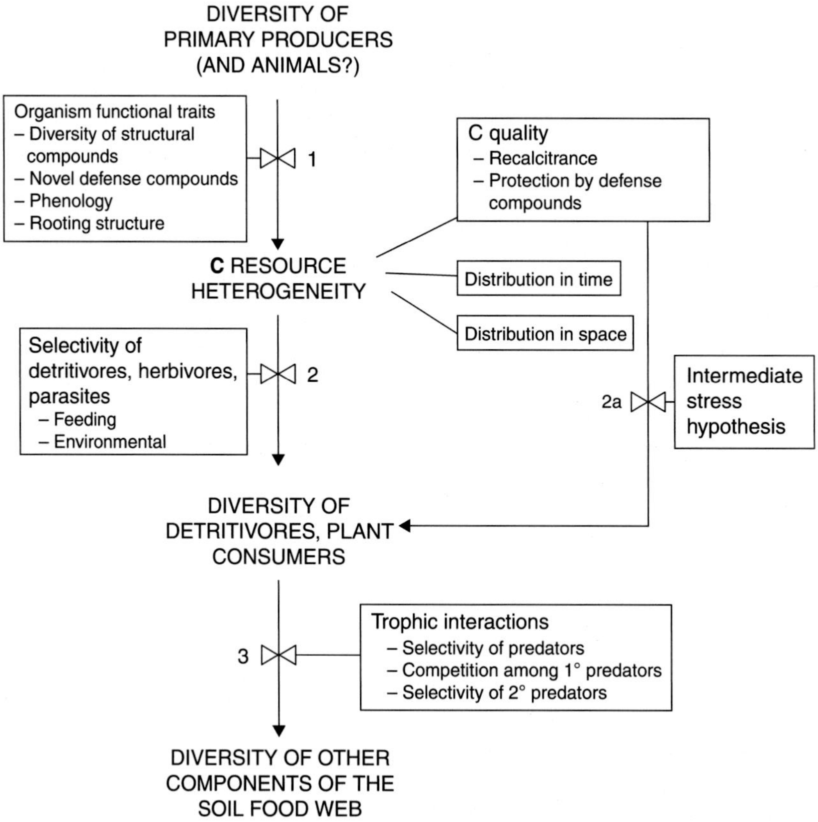

DIVERSITY OF PRIMARY PRODUCERS (AND ANIMALS?)

Organism functional traits
– Diversity of structural compounds
– Novel defense compounds
– Phenology
– Rooting structure

1

C quality
– Recalcitrance
– Protection by defense compounds

C RESOURCE HETEROGENEITY

Distribution in time

Distribution in space

Selectivity of detritivores, herbivores, parasites
– Feeding
– Environmental

2

2a

Intermediate stress hypothesis

DIVERSITY OF DETRITIVORES, PLANT CONSUMERS

3

Trophic interactions
– Selectivity of predators
– Competition among 1° predators
– Selectivity of 2° predators

DIVERSITY OF OTHER COMPONENTS OF THE SOIL FOOD WEB

FIGURE 7.2 Steps in the hypothesis that increased heterogeneity of carbon (C) substrates from aboveground organisms will positively influence belowground diversity. This mechanism postulates strong bottom-up control of diversity in belowground communities; it should be tested in the context of other potential (e.g., top-down) controls (Hunter and Price, 1992). Step 1. Diversity of primary producers leads to diversity of C inputs belowground. Step 2. Carbon resource heterogeneity leads to diversity of herbivores and detritivores. (Alternative Step 2. Carbon resource quality, rather than heterogeneity, leads to diversity of detritivores.) Step 3. Diversity of detritivores or belowground herbivores leads to diversity of organisms at higher trophic levels in belowground food webs. Source: *From Hooper, D.U., Bignell, D.E., Brown, V.K., Brussaard, L., Dangerfield, J.M., Wall, D.H., et al., 2000. Interactions between aboveground and belowground biodiversity in terrestrial ecosystems: patterns, mechanisms, and feedbacks. BioScience 50, 1049–1061; see paper for more details of the complex interactions involved in aboveground and belowground diversity.*

7.4 IMPACTS OF SPECIES RICHNESS ON ECOSYSTEM FUNCTION

Studies of Wall and colleagues in the McMurdo Dry Valleys of Antarctica may offer some insights into the impacts of species richness. The dry valley ecosystems contain only three species of nematode: one bacterial feeder, one microbial feeder, and one omnivore-predator that are present in very low numbers (2–5 kg^{-1} soil) (Wall and Virginia, 1999). These systems have very low precipitation (~10 cm rainfall equivalent $year^{-1}$), and make

TABLE 7.3 Main Biological Features of the Three Main Humus Forms

	Mull	**Moder**	**Mor**
Ecosystems	**Grasslands, Deciduous Woodlands with Rich Herb Layer, Mediterranean Scrublands**	**Deciduous and Coniferous Woodlands with Poor Herb Layer**	**Heathlands, Coniferous Woodlands, Sphagnum Bogs, Alpine Meadows**
Biodiversity	High	Medium	Low
Productivity	High	Medium	Low
Litter horizons	OL, OF	OL, OF, OH	OL, OM
Soil type	Brown soils	Gray-brown podzolic soils medium	Podzols
Phenolic content of litter	Poor		High
Humification	Rapid	Slow	Very slow
Humified organic matter	Organo-mineral aggregates with clay–humus complexes	Holorganic fecal pellets	Slow oxidation of plant debris
Exchange sites	Mineral	Organic (rich)	Organic (poor)
Mineral weathering	High	Medium	Poor
Mineral buffer type	Carbonate range	Silicate range	Iron/aluminum range
Impact of fire	Low (except in Mediterranean ecosystems)	Medium	High
Regeneration of trees	Easy (permanent)	Poor (cyclic processes)	None (fire needed)
Dominant mycorrhizal types	Arbuscular mycorrhizae	Ectomycorrhizae	Ericoid and arbutoid mycorrhizae
Mycorrhizal partners	Zygomycetes	Basidiomycetes	Ascomycetes
Nitrogen forms	Protein, ammonium, nitrate	Protein, ammonium	Protein
Nutrient availability to plants	Direct (through absorbing hairs)	Indirect (through extramatrical mycelium)	Poor
Nutrient use efficiency	Low	Medium	High
Fauna	Megafauna, macrofauna, mesofauna, microfauna	Macrofauna (poor), mesofauna (rich), microfauna	Mesofauna (poor), microfauna (poor)
Faunal group dominant in biomass	Earthworms	Enchytraeids	None
Microbial group dominant in biomass	Bacteria	Fungi	None
Affinities with polluted condition	Low	Medium	High

OF, Fermentation layer; *OH*, humification layer; *OL*, litter layer, *OM*, matted organic matter just above the mineral soil.
Modified from Ponge, J-F., 2003. Humus forms in terrestrial ecosystems: a framework to diversity. Soil Biol. Biochem. 35, 935–945.

the usually harsh climate of the Chihuahuan desert of New Mexico seem like an oasis, with 7 plant parasites, 10 genera of microbivores, 2 omnivore genera, and 3 genera of predators (Fig. 7.3) (Wall and Virginia, 1999). The latter system contains numerous vascular plants, with considerable organic inputs both above- and belowground. In the McMurdo Dry Valleys, the sources of organic matter are restricted to allochthonous inputs from algae in nearby lakes or streams, or small amounts of indigenous soil algae and cyanobacteria. Although depauperate in species, their distributions spatially are markedly different, and highly correlated with differences in tolerances to desiccation and salinity, with the omnivore-predator and bacterivore being more water-requiring, concentrating in streambeds, and the microbivorous (bacteria and yeast spp.) endemic species *Scottnema lindsayae* restricted to the drier uplands (Treonis et al., 1999). Although complicated in terms of life-history details, the fact that the number of species is so small, it seems likely that a fuller understanding of microbial and faunal interactions related to diversities is possible.

The role of redundant species and the functional roles played by them is crucial to understanding the interplays between biodiversity and ecosystem function. Without detailed knowledge of the biology of species involved, it can be difficult to decide how

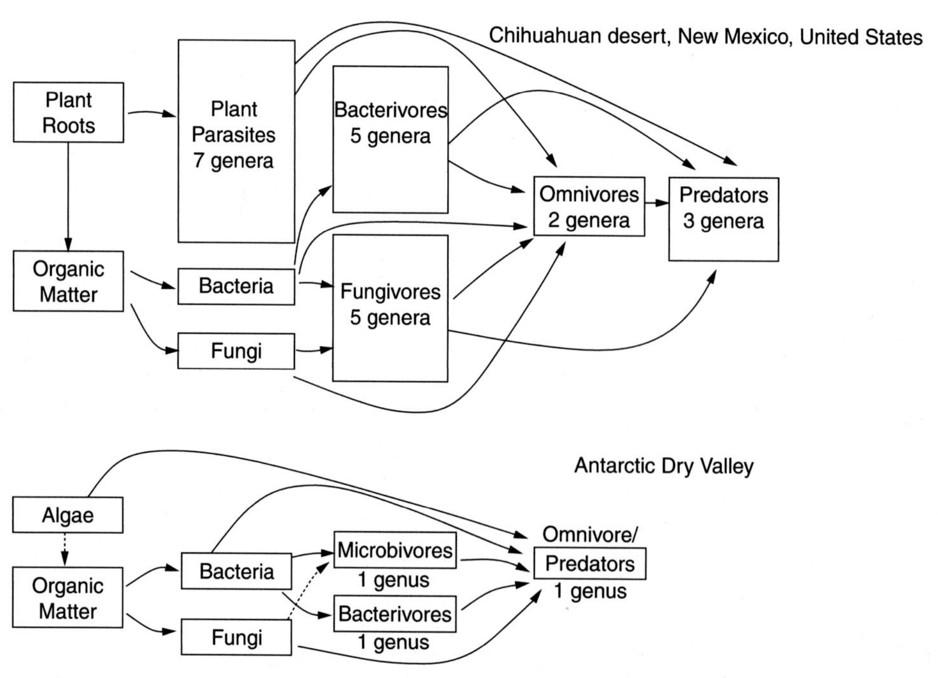

FIGURE 7.3 Complexity of soil nematode food webs in a hot desert (Chihuahuan, Jornada Long-Term Ecological Research (LTER), New Mexico) with 22 nematode genera, and a cold desert (Taylor Valley, McMurdo LTER in Antarctica) with three genera. For the nematodes, the height of the boxes illustrates the number of genera. The Antarctic Dry Valley has one species of a microbivore, *S. lindsayae*, that feeds on bacteria and yeast; one bacterivore, *Plectus antarcticus*, that feeds on bacteria; and an omnivore-predator, *Eudorylaimus antarcticus*, that probably feeds on algal cells, bacteria, yeast, fungi, nematodes, and other small fauna. Source: *From Wall, D.H., Virginia, R.A., 1999. Controls on soil biodiversity: insights from extreme environments. Appl. Soil Ecol. 13, 137–150.*

many functional types are present in a system or determine the functional roles of individual species (Bolger, 2001). Pathogen protection benefits of AM may be as significant as the nutritional benefits to many plants growing in temperate ecosystems (Bolger, 2001).

7.5 MODELS, MICROCOSMS, AND SOIL BIODIVERSITY

Hunt and Wall (2002) modeled the effects of loss of soil biodiversity, viewed from a functional group perspective, on ecosystem function. They constructed a model for carbon and nitrogen transfers among plants, and functional groups of microbes and fauna. They used 15 functional groups of microbes and soil fauna, comprised of bacteria, saprophytic and mycorrhizal fungi, root-feeding, bacteria-feeding, fungal-feeding, omnivorous and predaceous nematodes, flagellates and amoebae, collembola, r- and k-selected fungal-feeding mites, nematophagous, and predaceous mites (Fig. 6.2) (Hunt et al., 1987). The 15 functional groups were deleted one at a time and the model was run to steady state. Only six of the 15 deletions led to as much as a 15% change in abundance of a remaining group, and only deletions of bacteria and saprophytic fungi led to extinctions of other groups. By this analysis, no single faunal group had a significant effect on subsequent ecosystem behavior. However, the authors caution that, despite numerous compensatory mechanisms that occurred, it is premature to assume that the system is inherently stable even with the loss of several faunal groups. In fact, analyses of similar food webs by Moore et al. (1993) and Moore and De Ruiter (2000, 2012) showed that loss of top predators had much greater impacts on lower trophic levels than their low biomasses might indicate.

An interesting follow-up to studies of agroecosystem food webs in the Netherlands was a study of the carbon and nitrogen mineralization patterns occurring in an old field, which underwent succession for 22 years after abandonment of the cultivation process. Holtkamp et al. (2011) noted that calculated contributions to mineralization of organisms at trophic level 1 increased during secondary succession following land abandonment. The fungal decomposition channel contributed more to N mineralization than the bacterial decomposition channel, whereas both channels contributed equally to C mineralization rates. Direct contributions by higher trophic levels to mineralization decreased during secondary succession. However, higher trophic levels had a direct influence on N mineralization and indirect for both C and N mineralization due to their effect on biomass turnover rates of groups at lower trophic levels.

Another approach to biodiversity and its linkages to soil processes is by use of experimental microcosms. Building on results of earlier studies of Setälä et al. (1997), Liiri et al. (2002) established microcosms with litter, humus and mineral layers, and controlled access from the outside soil allowed by using either 45 μm or 1 mm mesh screens on the side of the microcosms. The microcosms were then half-buried to the top of the mesh in the side of the funnel, and then the upper portion left open to provide light for the pine seedling in the microcosm (Fig. 7.4) (Liiri et al., 2002). Microcosms were watered at regular intervals, or alternatively run through drought cycles, and leachates drawn off from the collecting bottle underneath, to analyze for inorganic N and organic C in them. The experiment was run for 152 weeks, or nearly 3 years. The authors followed microbial community composition using phospholipid-derived fatty acids (PLFA) and Biolog to differentiate

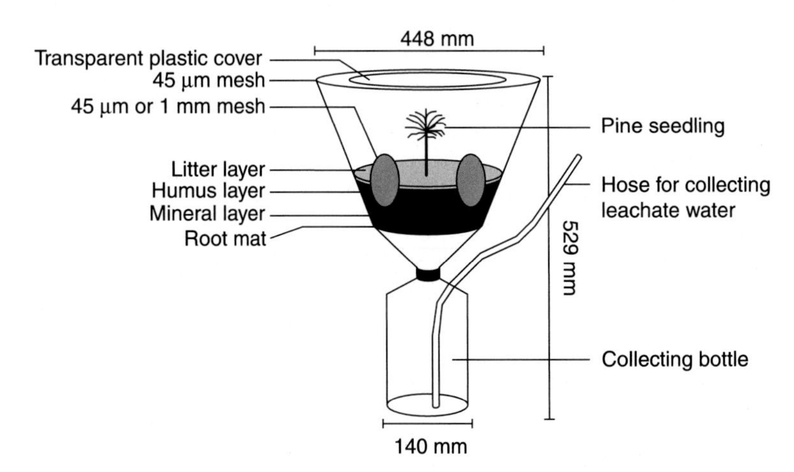

FIGURE 7.4 A scheme of the lysimeter used with forest soils. Source: *From Liiri, M., Setälä, H., Haimi, J., Pennanen, T., Fritze, H., 2002. Soil processes are not influenced by the functional complexity of soil decomposer food webs under disturbance. Soil Biol. Biochem. 34, 1009–1020.*

between bacteria and fungi, and sampled periodically for nematodes, enchytraeids, and microarthropods. They varied pH regimes by applying wood ash to some microcosms, and not others. They observed significant decreases in microarthropod numbers in the first year, followed by gradual increases in numbers of organisms with small body sizes. Enchytraeid numbers followed similar patterns. Nematodes had ready access to all microcosms, and were quite numerous, ranging from 67 to 191 g $soil^{-1}$ in the controls, and from 98 to 545 g^{-1} soil in the ash-treated microcosms. The ash had significant effects on the microbial community makeup inside, but not in the soil outside the microcosms. The main effects on pine seedling growth and nutrient dynamics were governed by the abiotic factors of pH and availability of water. These seemed to govern the overall dynamics, in spite of the functional complexity of the soil biota. It would be most instructive to see this experimental design repeated in other habitats and biomes, to ascertain the generality of the findings. In addition, it would be useful to compare the fauna at least to the family or genus level, to see if finer-grained responses to the experimental manipulations occurred during this nearly 3-year long experiment.

In an extensive comparison of seven food webs of native and agricultural soils, de Ruiter et al. (1998) modeled energetics and stability, evaluating the roles of various groups of organisms and their interactions in energy flow and community stability. They measured feeding rates, interaction strengths, and impacts of the interactions on food web stability (percent) arranged according to trophic position in the seven belowground food webs from Central Plain, CO, two tillage manipulations at Lovinkhoeve in the Netherlands, two tillage manipulations at Horseshoe Bend, Athens, GA, and no fertilizer and fertilizer additions at Kjettslinge in southern Sweden. De Ruiter et al. (1998) found that only a fraction of the species manipulations had a strong effect on food web structure. Also there was an absence of correlation between the impacts on stability and feeding rates, meaning that interactions representing a relatively low rate of flow of materials can have a large impact on stability, and interactions having a high rate of material flow can have a small impact. Thus the higher-level predatory mites and nematodes had an impact far out of proportion to their biomass, and the contrary was true of the high biomass organisms, namely bacteria and fungi. De Ruiter et al. (1998) urge that future research be focused on the energetic properties of the organisms forming the basis of the patterning of interaction strengths. This is a big order, and one that will require innovative experiments under both laboratory and field conditions. The stakes are high however, as these studies should help to provide further insights into the nature of biodiversity and ecosystem function.

7.6 EXPERIMENTAL ADDITIONS AND DELETIONS IN SOIL BIODIVERSITY STUDIES

Additional studies of biotic roles of soil fauna and bacteria and fungi have been approached in two ways. One approach is by gamma irradiating sieved soils and inoculating them with suspensions of full strength, 10^2, 10^4, and 10^6 dilutions of soil organisms (Griffiths et al., 2001). The other approach subjects unsterile soils to chloroform fumigation and incubation (Griffiths et al., 2000), and following the subsequent changes in functional variables, such as ammonium, nitrate, soil respiration, etc., in relation to microbial biomass and diversity, as measured by DNA patterns on denaturing gel electrophoresis. Results were divergent in the two studies, with the chloroform fumigation simplification of community biomass and species diversity having a direct impact on the functional stability, as measured by the physiological response variables. In contrast, although there were progressive declines in biodiversity of the soil microbial and protozoan populations, there were no consistent changes in functional parameters. Some functions showed no trend (thymidine and leucine incorporation, nitrate accumulation, respiratory growth response), some a gradual increase with increasing dilution (substrate induced respiration), some declined only at the highest dilution treatment (short-term respiration from added grass, potential nitrification rate, and community level physiological profile), while others varied even more idiosyncratically. At no stage were any of the physiological functions eliminated completely. The final commentary on this by Griffiths et al. (2001) is that within any realistic sort of range of changes in biodiversity to be experienced by soils, there will be no direct effect on any soil functional parameters measured. Other authors, e.g., Wardle et al. (1999), suggested that it is possible to overcome selective species effects by (1) measuring the effects of all species in monoculture and (2) by species removal experiments. Neither of these approaches is feasible with current technology, so this problem awaits the attention of a future generation of soil ecologists.

Another approach to microcosm studies was taken by the large group working in the Ecotron-controlled environment facility at Silwood Park, UK. Constructing analogs of a temperate, acid, sheep-grazed grassland in the northern Britain, Bradford et al. (2002) established terrestrial microcosms of graded complexity, with soil, plant, and microorganisms, and then assemblages of microfauna, micro- and mesofauna, and then micro-, meso-, and macrofauna. This functional group approach provided a range of metabolic rates, generation times, population densities, and food size. The microcosms were maintained in the Ecotron for a period of 8.5 months. Bradford et al. (2002) found significant increases in decomposition rate in the most complex faunal treatment, but both mycorrhizal colonization and root biomass were less abundant in the macrofauna treatments. Interestingly plant growth was not enhanced in these treatments, despite higher nutrient (N and P) availability. Contrary to initial hypotheses, neither aboveground net primary production (NPP) (plant biomass) nor net ecosystem production (net CO_2 uptake) were enhanced in the most complex microcosms. Bradford et al. suggested that respiration was most likely buffered by the combined stimulatory effect of both mesofauna and macrofauna on microbes (see Chapter 4: Secondary Production: Activities of Heterotrophic Organisms—The Soil Fauna and Chapter 5: Decomposition and Nutrient Cycling), which served to maintain microbial activity at a level equivalent to that in the micro- and mesofauna communities. This study has served as a benchmark in large-scale microcosm studies, but as Bradford et al. (2002) note, it is not a substitute of longer-term in situ field studies, as difficult to conduct and interpret as they may be.

7.7 PROBLEMS OF CONCERN IN SOIL BIODIVERSITY STUDIES

An alternative to the above functional approaches is taken by André et al. (2002), who note that most investigators use inadequate sampling designs or sample too shallowly in the soil profile to get a complete sample of microarthropods to provide the information used in the models noted earlier. In an extensive survey of the worldwide literature on microarthropods, they claim that on average, at most 10% of the soil microarthropod populations have been explored and 10% of the species described, due to the use of inefficient extraction procedures. This is supported by Walter and Proctor (2000) who suggest that perhaps only 5% of the species of mites worldwide are described so far. André et al. (2002) make the very valid point that ecologists need to be aware of the numerous pitfalls and possible flaws inherent in many extraction procedures, i.e., none of them is 100% efficient. In Chapter 9, Laboratory and Field Exercises in Soil Ecology, we explore some of these concerns more extensively.

There is an understandable concern that some quantifiable relationship be given to the relationship between ecosystem function and diversity. This is portrayed in Fig. 7.5 (Bengtsson, 1998), which contrasts two curves of ecosystem function as a function of increasing numbers of species. Type 1, a continually ascending curve, represents the hypothesis that all species are important for ecosystem function. Type 2, initially convex and then flat, represents the species redundancy hypothesis. Bengtsson (1998) argues that it is more informative to consider specific functions in ecosystems, namely decomposition, nutrient mineralization, or primary production, thus focusing on phenomena that are more amenable to scientific inquiry. Bengtsson (1998) argues strongly that diversity does *not* play a role in ecosystem function. He goes so far as to assert that: "correlations between diversity and ecosystem functions—which may very well exist—will be mainly noncausal correlations only." As we are trying to show in this chapter, the truth may well lie in some midpoint between these extremes. The fact that certain functions may be linked to just a few genera or species, such as autotrophic and heterotrophic nitrifiers, for example, means that this might well be a "pressure point," for concern about long-term ecosystem function. The "natural insurance capital" concept of Folke et al. (1996), also discussed in detail by Bolger (2001), suggests that it is essential to retain as much species richness as

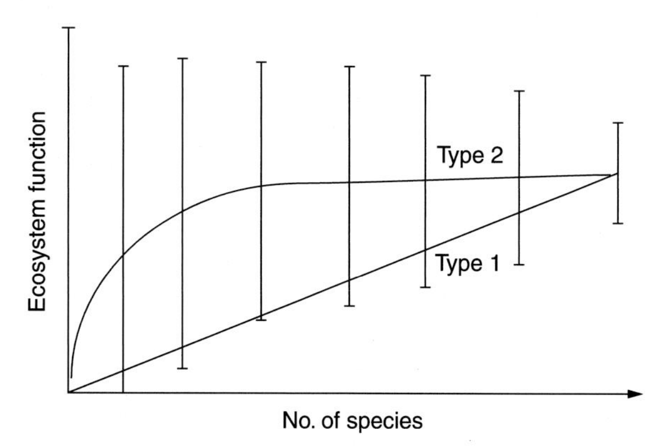

FIGURE 7.5 A hypothetical example of an attempt to quantify the form of the relationship between ecosystem function and diversity. Type 1 curve represents the hypothesis that all species are important for ecosystem function, while Type 2 is the species redundancy hypothesis. The bars indicate the range of responses as different numbers of species are randomly drawn from a source pool of species, given that species' effects on ecosystem function are mainly species-specific (idiosyncratic) and not related to diversity. Note that although an average response may be observed, it neither allows a distinction between the two different hypotheses (Type 1 and Type 2), nor does this average response allow any useful prediction of what will happen in individual cases of species deletions. Source: *From Bengtsson, J. 1998. Which species? What kind of diversity? Which ecosystem function? Some problems in studies of relations between biodiversity and ecosystem function. Appl. Soil Ecol. 10, 191–199.*

possible to ensure that complete ecosystem services exist as human needs or environmental changes occur.

It is essential to examine specific details of biodiversity and ecosystem services, particularly in the context of agroecosystems, where biological control has long been of concern. The following example, noted in a review article by Cavigelli et al. (2012) cites at least six steps needed to demonstrate if organic management in these systems supports biodiversity that is functionally significant with regard to pest control. Thus organic farming often results in greater soil biodiversity (step 1); that biodiversity must reflect an increase in the number of beneficial organisms (step 2); which must cause higher mortality among pests (step 3); found in similar abundance in organic and conventional systems (step 4); and reduce pest damage (step 5); leading to yield increases compared to conventional methods, such as pesticide application (step 6).

A study by Crowder et al. (2010) deserves close scrutiny because it addresses at least five of the six steps. Thus evenness of predators and soil pathogens, though the latter was not statistically significant, of the potato beetle, *Leptinotarsa decemlineata*, was greater on organic than conventional potato farms (steps 1 and 2). *L. decemlineata* abundance was similar in the two systems (step 3). Potato plant growth rate and *L. decemlineata* mortality increased with increasing evenness of both predators and soil pathogens (steps 4 and 6; plant size is related to yield in potato). These impacts translated into an 18% decrease in pest densities and 35% increase in plant size for plants grown in similar environments (Crowder et al., 2010). Keeping in mind the semiartificial environment, Cavigelli et al. (2012) note that this study is an exemplary one in examining a key ecosystem service, biocontrol.

As in all areas of ecology, there is a spatial dimension to the biodiversity of soil organisms. It is essential to know not only which species are present, but also where the counted species occur in relation to one another. Do species occur together at every microsite, or do they occur mostly individually in separate sites? This has an important bearing on competition and other interactions, with functional consequences for the ecosystem. Ettema and Yeates (2003) measured patterns of small (cm) and intermediate (m) scales in nematode communities in a forest compared to a pasture system on a similar soil type in New Zealand. Using geostatistical techniques and mathematical calculations of species turnover, they compared nematode genera in forestland, which was assumed to have greater variation in vegetation, and hence belowground inputs, on small and intermediate scales than in the pasture. Thus they hypothesized that nematode genera are more strongly aggregated (occurring in "hot spots") in the mixed forest than in the ryegrass/white clover pasture. Applying an optimization method for sampling in geostatistical studies called spatial simulated annealing of Van Groenigen and Stein (1998), Ettema and Yeates (2003) sampled along 40 m long transects for the m scale, with distance classes of 3 m, reflecting the scale of tree spacing. The cm scale transect was one-tenth of the large scale, or 4 m. The total number of nematodes per soil core volume was more than five times higher in the pasture (2800 ± 1234) than in the forest (430 ± 252), but the average number of genera in the forest (23.7 ± 3.3) was higher in the forest than in the pasture (19.1 ± 2.5). Also, many more nematode genera occurred in the forest (53) than in the pasture (37). Dissimilarity analysis showed that generic turnover was significantly greater in the forest than in the pasture, both at the small and intermediate scales (Fig. 7.6) (Ettema and Yeates, 2003). Because increasing distance in the forest led to increasing dissimilarity between communities, and no plateau was reached, it is possible that there is additional species turnover on scales larger than those explicitly sampled. The amount of work required for larger-scale studies would be much greater, and should be kept in mind when considering work on spatial scales even with soil fauna of relatively small size.

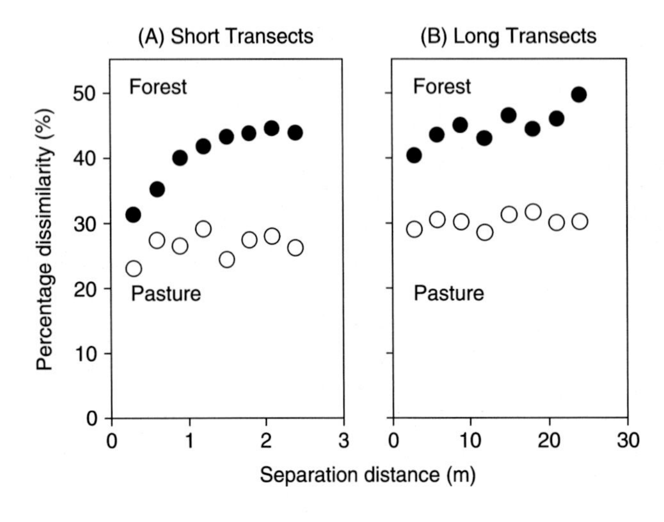

FIGURE 7.6 Mean dissimilarity (%) as a function of separation distance along the short (A) and long (B) transects. Each point is an average of $P = 35–42$ observation pairs. Open symbols represent pasture, closed symbols forest. Source: *From Ettema, C.H., Yeates, G.W., 2003. Nested spatial biodiversity patterns of nematode genera in a New Zealand forest and pasture soil. Soil Biol. Biochem. 35, 339–342.*

7.8 WHY IS SOIL DIVERSITY SO HIGH?

We arrive at the end of this chapter with the phenomena of high species and functional diversity well appreciated (e.g., Anderson, 1975b, 2000; Orgiazzi et al., 2016), but their root causes are yet unknown. As noted by Wardle (2002), the belowground environment provides numerous niche axes in the Hutchinsonian (1957) hyperspace, concerning numerous microhabitats, microclimatic properties, soil chemical properties, and phenologies of the organisms themselves. When one adds in the fact that many of the organisms may exist in quiescent or dormant stages (Coleman, 2001, 2008, and noted in Chapter 4: Secondary Production: Activities of Heterotrophic Organisms—The Soil Fauna), there is considerable niche space for the impressive belowground species diversity.

7.9 BIOGEOGRAPHICAL TRENDS IN DIVERSITY OF SOIL ORGANISMS

Interestingly, with the exception of termites, whose diversity declines significantly over a large geographical gradient, numerous taxa of soil organisms, ranging from ciliate protozoa (Foissner, 1987a,b) to earthworms (Hendrix, 1995) to oribatid mites (Maraun et al., 2007) do not decrease from 0 to 60 degrees latitudinal range. Wardle (2002) suggests two principal reasons: (1) with increasing latitude there is a general trend for greater amounts of organic matter accumulation, and higher amounts of carbon and nutrients are stored in the soil relative to the amount of plant biomass present. Greater humus depth may provide greater habitat heterogeneity and greater amounts of nutrients present in the soil. Secondly, diversity of soil organisms may be governed by local factors rather than by regional pool size. If one adds in the fact that numerous smaller soil organisms (soil microfauna and microflora (both fungi and bacteria)) can be transported by wind currents and macrobiota over intercontinental distances, one would expect to see pandemic distributions, and this is what is observed (Wardle, 2002), particularly for microorganisms $<20\ \mu m$ in size (Wilkinson et al., 2011). Maraun et al. (2007) point out that at least for oribatids (but likely for other soil taxa as well), the lack of increased diversity moving from temperate forests toward the tropics might be explained by their generalist feeding behavior which would limit the strength of coevolutionary relationships between mites and plants (i.e., increased plant diversity in the tropics would not necessarily lead to increased mite diversity). Furthermore, these authors note that although plant diversity is greater in the tropics the chemical composition of the resulting litter in forest floors is not much different from that in temperate forests, and thus there may not be much in the way of niche diversity distinguishing these habitats, and therefore no difference in the mite diversity evolved to inhabit them.

CHAPTER

8

Future Developments in Soil Ecology

8.1 INTRODUCTION

There are several areas of rapid change that are of interest to soil ecologists in the 21st century. The effects of soil processes and soil biota on global change, particularly with relation to global greenhouse gases, are of concern to resource managers and globally oriented ecologists (Coleman et al., 1992; Orgiazzi et al., 2016). More recently, as noted in Chapter 7, Soil Biodiversity and Linkages to Soil Processes, there has been a rising current of interest in soils and biodiversity (Coleman, 2001; Wall et al., 2012). Within terrestrial ecosystems, soils may contain some of the last great "unknowns" of many of the biota. This includes relatively well-studied fauna such as ants (Hölldobler and Wilson, 1990) as well as the more numerous and less studied mesofauna, such as microarthropods and nematodes. The role of soils in the ecology of invasive species is an area of rapidly increasing findings. We also consider the roles of soils and the "Gaia" mechanism, and finally, ways to evaluate soil "quality."

8.2 ROLES OF SOILS IN CARBON SEQUESTRATION

Soils are probably the last great frontier in the quest for knowledge about the major sources and sinks of carbon in the biosphere. The direct effects of deforestation on global patterns of carbon cycles are relatively minor; the effects of changed sink strengths, with deforestation decreasing rates of CO_2 uptake, may be much larger. Another source of carbon input to the atmosphere has come from the oxidation of soil organic matter (SOM) during cultivation of native lands, such as the Great Plains region of North America, and also the Eurasian steppes of eastern Russia (Houghton et al., 1983; Wilson, 1978). The standing stocks of soil carbon are twice as large as all of the standing crop biomass of all of the terrestrial biomes combined (Fig. 8.1) (Post et al., 1990; Anderson, 1992; Wall et al., 2012). However, the plant/soil systems are strongly coupled, and the rates of inflows and outflows are significantly controlled by the rates of above- and belowground herbivory in forests (Pastor and Post, 1988) and in grasslands (Schimel, 1993). The feedback effects of the principal greenhouse gases, namely carbon dioxide, methane, and nitrous oxide are very large (Mosier et al., 1991; Rogers and Whitman, 1991), with the effects of carbon dioxide being some 56% of the total impact (Anderson, 1992). However, the rate of increase of methane is almost twice that of CO_2 (Anthony et al., 2016; Conrad, 2009) and is being closely observed by

Fundamentals of Soil Ecology
DOI: http://dx.doi.org/10.1016/B978-0-12-805251-8.00008-9

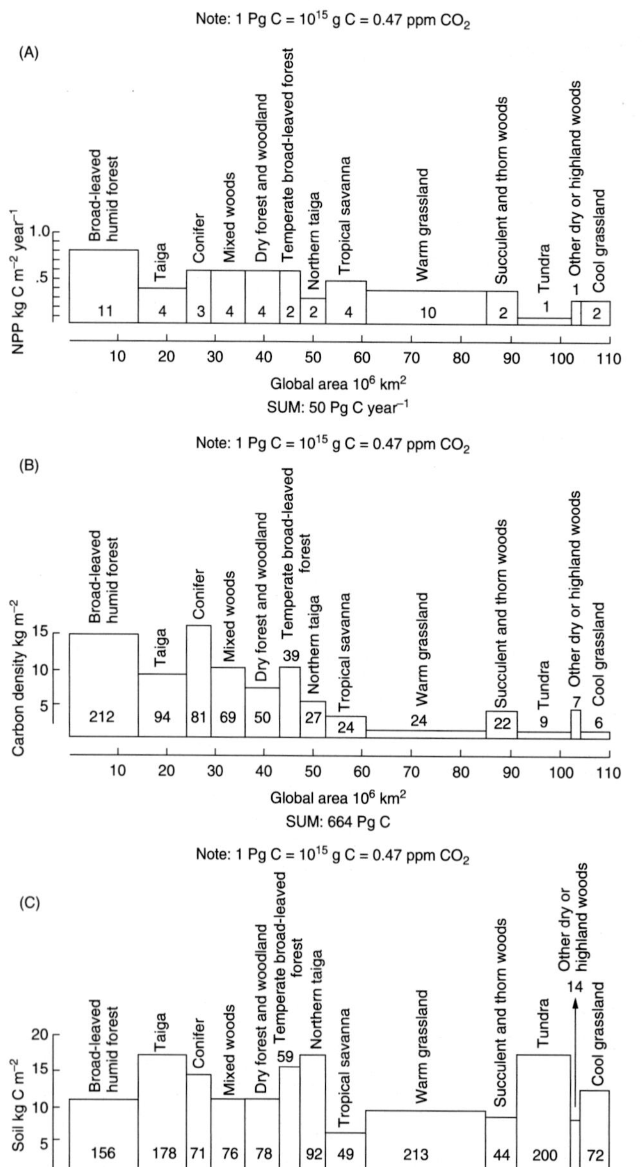

FIGURE 8.1 Pools and fluxes of carbon in major terrestrial ecosystem types: (A) distribution of NPP, (B) biomass, and (C) soil carbon pools. The total area occupied by each ecosystem type is represented by the horizontal axis with flux or density of the vertical axis; the area is therefore proportional to the global production or storage in each ecosystem type. Note: The numbers inside the boxed areas are measured in petagrams C (Pg C). *Source: From Anderson, J.M., 1992. Responses of soils to climate change. Adv. Ecol. Res. 22, 163–210.*

atmospheric scientists. One of the major concerns of scientists interested in global change is the extent of involvement by soils and soil processes in the evolution of greenhouse gases, and the largely unknown roles of soil biota and organic matter in the global carbon cycle (Lubbers et al., 2013). This concern is justified by a meta-analysis covering studies from nine biomes across the globe which showed that soil respiration (i.e., CO_2 efflux) increased with increased temperature up to 25°C, and that arctic and subarctic soils, with their large stocks of C and lower annual temperatures would be more responsive and serve as greater sources of CO_2 for a longer period of time into the future (Carey et al., 2016).

We examine next the ways in which soils operate over ecological and geological time spans, and how they may be influenced by, or have an effect on global change processes. Soil development and change may be viewed as the result of the basic processes of additions, removals, transformations, and translocations (Anderson, 1988). A given landscape will experience runon, runoff, transformations, and transfers up and down in the profile, and additions and losses either aerially or pedologically (Fig. 8.2). These processes may be

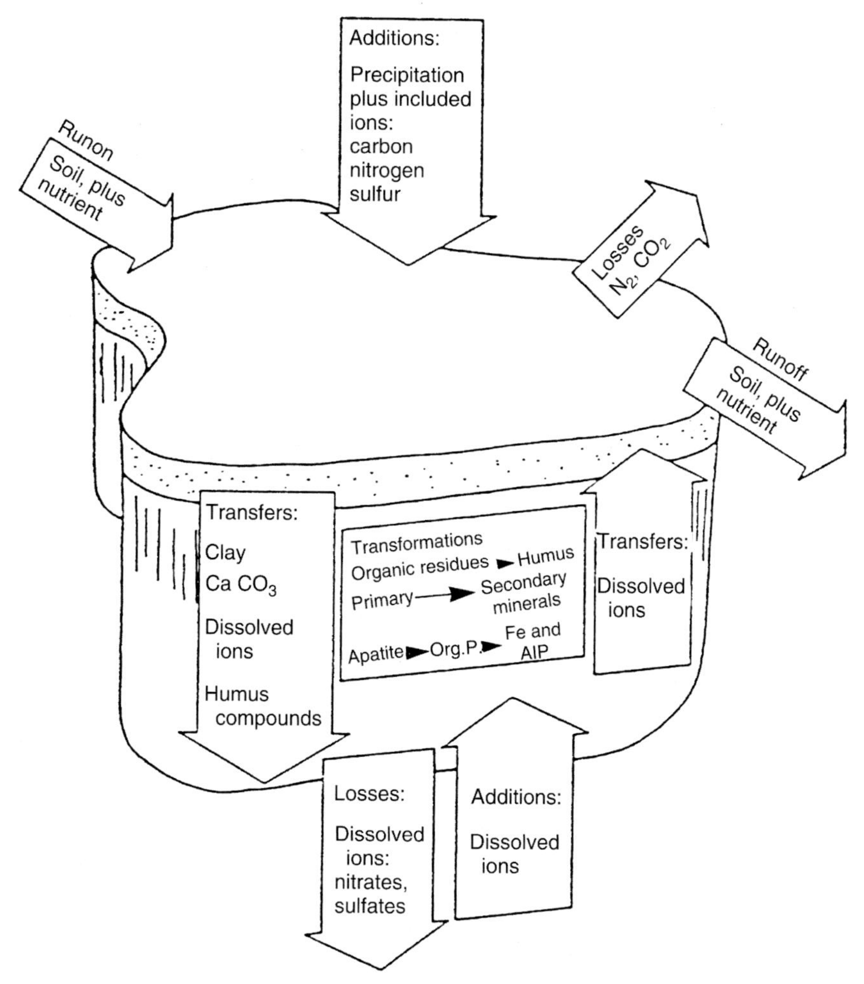

FIGURE 8.2 Soil-forming factors based on the concepts of Simonson (1959) as described by Anderson (1988). *Source: From Stewart, J.W.B., Anderson, D.W., Elliott, E.T., Cole, C.V., 1990. The use of models of soil pedogenic processes in understanding changing land use and climatic change. In: Scharpenseel, H.W., Schomaker, M., Ayoub, A. (Eds.), Soils on a Warmer Earth. Elsevier, Amsterdam, pp. 121–131.*

TABLE 8.1 Grouping of Soil-Related Processes and Components Based on Time

Highly Dynamic	Dynamic	More Static, Slow
Soluble nutrients	Adsorbed nutrients	Nutrient reserves in minerals
Active or soluble organic matter	Labile organic matter adsorbed to clay	Chemically stabilized organic matter
Solution and movement of soluble components	Weathering of carbonate minerals	Weathering of silicates and clay minerals
Microbial growth	Microfauna and mesofauna plant growth	Vegetation, i.e., forest

From Stewart, J.W.B., Anderson, D.W., Elliott, E.T., Cole, C.V., 1990. The use of models of soil pedogenic processes in understanding changing land use and climatic change. In: Scharpenseel, H.W., Schomaker, M., Ayoub, A. (Eds.), Soils on a Warmer Earth. Elsevier, Amsterdam, pp. 121–131.

very dynamic for processes such as movement of soluble salts, which vary within seasons, or be measured in thousands of years, e.g., for clay weathering processes. The microbial portion of the organic matter cycle will have mean net turnover times of 1–1.5 years, whereas organic matter stabilization processes, such as the interactions of clay/organic compounds, may be considered intermediate (centuries) in timescale (Stewart et al., 1990) (Table 8.1). These processes can be envisioned readily via the carbon, nitrogen, and phosphorus submodels of the Century model (Fig. 8.3). This model was developed to simulate the additions and losses in agricultural lands and grasslands worldwide (Parton et al., 1987, 1989a) and includes Century with a daily time interval (DayCent et al., 1994), but has now been extended to a wide range of ecosystems, including tundra and taiga (Smith et al., 1992) and tropical ones as well (Del Grosso et al., 2002, 2005; Parton et al., 1989b, 2016; Schimel et al., 1994; Smith et al., 1998).

8.3 ROLES OF SOILS IN THE GLOBAL CARBON CYCLE

What patterns and processes of global change are most likely to affect the global C cycle in soils? What are the effects of climate change on vegetation; are there possible changes in sink strengths (pools of organic matter, active roots, etc.) in various parts of the globe? Do we know enough about the dynamics of carbon in the 13 or more major biomes that comprise the terrestrial biosphere? For example, consider the size of the live biomass in broad-leaved humid forest, which amounts to 212 petagrams (Pg $= 10^{15}$ g), versus warm grasslands that have only 24 Pg live biomass. When comparing the amounts of soil carbon stored, versus carbon in live biomass, there is relatively less storage in the humid broad-leaved forest (156 Pg), giving a ratio of 212/156, or 1.36 more in the live biomass (Fig. 8.1) (Anderson, 1992). Warm grasslands, with 213 Pg in SOM, have a ratio of 24/213, or 0.11 in biomass versus that in the SOM. Tundra, with only 9 Pg in live biomass versus 200 in the SOM, has a ratio of only 0.05 in living biomass versus SOM (Anderson, 1992). What are the climatological versus plant physiological and microbiological implications of such

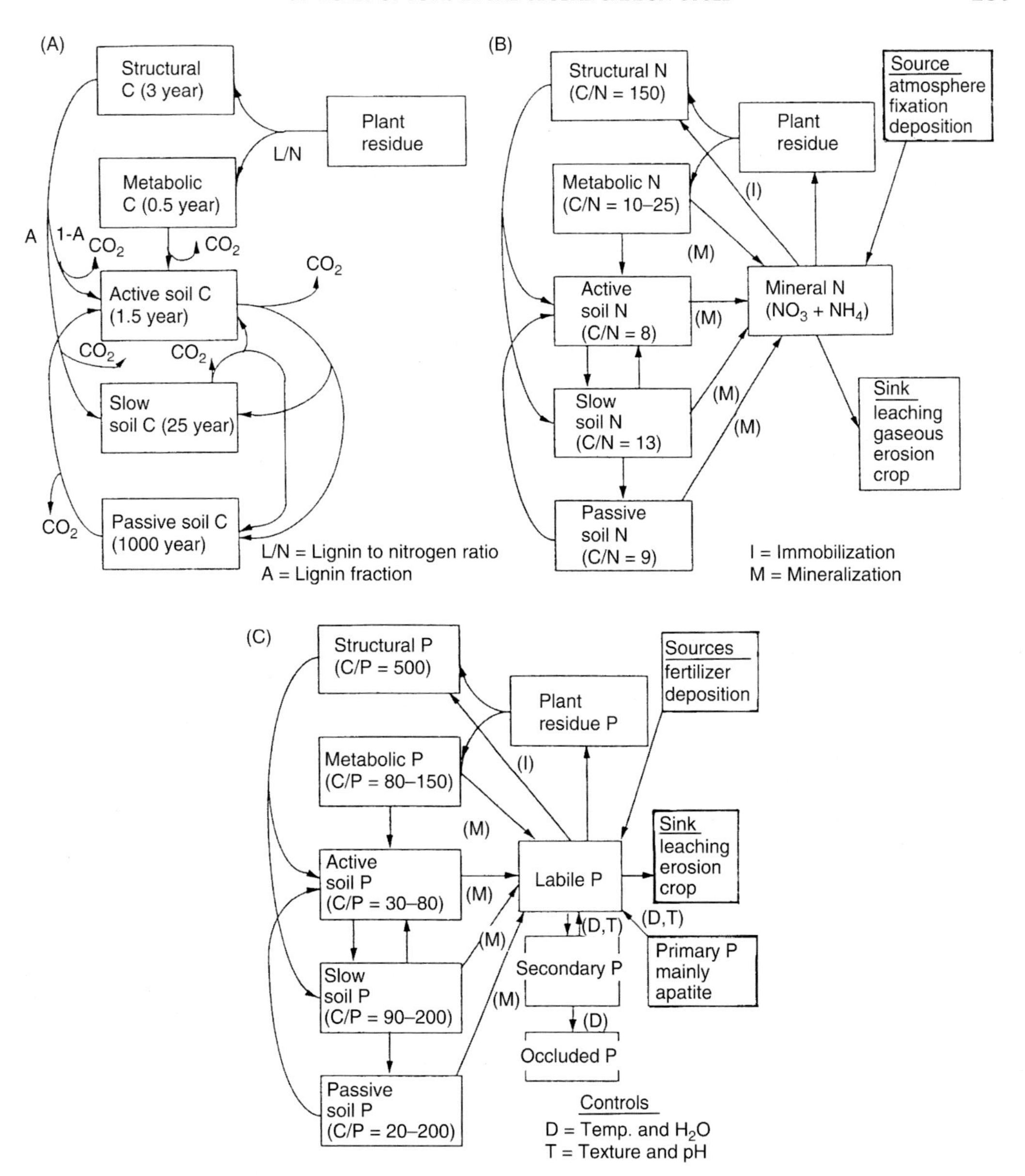

FIGURE 8.3 Flow diagram for the (A) carbon submodel, (B) nitrogen submodel, and (C) phosphorus submodel of Century. *Source: Adapted from Parton et al. (1987). Parton, W.J., Sanford, R.L., Sanchez, P.A., Stewart, J.W.B., 1989b. Modeling soil organic matter dynamics in tropical soils. In: Coleman, D.C., Oades, J.M., Uehara, G. (Eds.), Dynamics of Soil Organic Matter in Tropical Ecosystems. University of Hawaii, Honolulu, HI, pp. 153–171. From Stewart, J.W.B., Anderson, D.W., Elliott, E.T., Cole, C.V., 1990. The use of models of soil pedogenic processes in understanding changing land use and climatic change. In: Scharpenseel, H.W., Schomaker, M., Ayoub, A. (Eds.), Soils on a Warmer Earth. Elsevier, Amsterdam, pp. 121–131.*

differences in these widely different biomes? Research in this area requires considerable effort in soil science and also microbial ecology, because we are faced with problems of measuring substrate quality, covered earlier, and its feedback effects on future primary production and nutrient dynamics. Of course the modes of growth of grasses versus trees are also influential, because more of the total growth effort is invested belowground in both grassland and tundra soils.

Several reviews have addressed key aspects of the terrestrial carbon cycle: carbon fixation by primary production, and then mechanisms for either sequestering the carbon during organic matter decomposition and transformation processes, or mechanisms for mineralization via human-induced or natural processes (e.g., Houghton, 2003; Lal, 2002; Scharlemann et al., 2014; Paul, 2016). A central concern for ecologists and soil scientists is that soil organic carbon is the second largest pool in the terrestrial organic carbon cycle, with about 1415–1550 Pg involved.

Concerns about imbalances in the global carbon cycle are not new; rapidly increasing amounts of carbon dioxide entering the atmosphere from human activities, including burning of fossil fuels, were noted first one century ago (Arrhenius, 1896). Since then, interest in the rates of flow of carbon, and amounts sequestered in various pools in the biosphere, has waxed and waned. For example, Plass (1956) expressed concern about the amounts of carbon dioxide being released by the burning of fossil fuels worldwide. An additional contribution to increased global carbon dioxide is contributed by the relatively large amounts of SOM being "mined" by extensive cultivation throughout the major "breadbaskets" of the world. In several regions, e.g., the North American Great Plains, the former Soviet Union, and Canada, the loss is quite large, perhaps up to 40% of the surface layers (Coleman et al., 1984; Haas et al., 1957; Wilson, 1978). Mann (1986) concluded in a survey of 625 soils studied pairwise, cultivated versus noncultivated on the same soil type, that 20% or more of carbon was lost over decadal time spans from soils with high amounts of carbon (ranging from 6 to 16 kg/m^2). Interestingly, she noted that modest gains occur in soils that are initially very low in soil organic carbon, such as very sandy-textured ones, if they are put into cultivation. A significant amount of carbon fixation, and subsequent movement into the SOM, in the surface-to-30 cm depth, will occur over several years' time span. If extensive application of fertilizers is required to achieve these gains, then overall the global carbon balance is still toward the positive side, in terms of carbon costs for fossil-fuel-derived nitrogen, for example (Vitousek et al., 2002).

In an extensive review of SOM models and global estimates of changes in soil organic carbon under a $2 \times CO_2$ climate, Post et al. (1996) ran simulations with the Rothamsted model over a 100-year time period. They found that global SOM change was only about one-third of the way toward an eventual equilibrium under the enriched CO_2 regime. The change predicted with the UKMO (UK Meteorological Office), shows a small net sequestration of carbon in soil early in the climate transition period resulting from increases in tropical ecosystem soil carbon. However, this is followed by large carbon releases from arctic and boreal soils later in the century-long climate transition period. The largest net release of C from soil occurred at the end of the 100-year climate transition, after which the net releases decreased gradually as the soil carbon pools approached equilibrium under the double CO_2 regime (Post et al., 1996). For an extensive synthesis of modeling of SOM, see Campbell and Paustian (2015).

Houghton (2003) noted that the carbon balance of the world's terrestrial ecosystems is uncertain. Several top-down (atmospheric) and bottom-up (forest inventory and land-use change) approaches are in use and difficult to compare, as they contain incomplete accounting inherent in their methods. After a brief discussion of the methods and their inherent limitations, we consider the possible resolution of the uncertainties arising from use of these methods. Of the top-down estimates, the first uses concentrations of O_2 and CO_2 to partition atmospheric sinks of carbon between land and ocean. Using this assessment, terrestrial ecosystems were globally a net sink for carbon, averaging 0.2 ($\pm$ 0.7) PgC $year^{-1}$ and 1.4 ($\pm$0.7) PgC $year^{-1}$ in the 1980s and 1990s, respectively. The reason for the large increase between decades is unknown. A second top-down method is inverse modeling, which uses atmospheric transport models, together with spatial and temporal variations in atmospheric concentrations of CO_2 obtained through a network of flask air samples, to infer surface sources and sinks of carbon. The budget will not reflect accurately any changes in the amount of carbon on land or in the sea if some of the carbon fixed by terrestrial plants or used in weathering minerals is transported by rivers to the ocean and respired or released to the atmosphere there. The two top-down methods based on atmospheric measurements yield similar global estimates of a net terrestrial sink of approximately 0.7 ($\pm$0.8) PgC $year^{-1}$ for the 1990s (Houghton, 2003). Two bottom-up estimates have been used to estimate terrestrial sources and sinks over large regions: analyses of forest inventories and analyses of land-use change. One recent synthesis of forest inventories, which included converting wood volumes to total biomass and accounting for the fate of harvested products and changes in pools of woody debris, forest floor and soils, found a net northern midlatitude terrestrial sink of from 0.6 to 0.7 PgC $year^{-1}$ for the years around 1990 (Goodale et al., 2002; cited in Houghton, 2003). The estimate is only one-third of that calculated from atmospheric data corrected for river transport. Houghton (2003) noted that accumulation of C belowground, not directly measured in forest inventories was underestimated and might account for the difference in estimates. Because the few studies that have measured the accumulation of C in forest soils have consistently found soils account for only a small proportion (5%–15%) of measured ecosystem sinks, Houghton (2003) concluded that, despite the fact that soils worldwide hold from two to three times more carbon than biomass, there is no evidence as yet that they account for a significant terrestrial carbon sink. The second sort of bottom-up estimate, analyses of land-use change, calculated that globally, all factors of land-use change averaged 2.0 and 2.2 PgC $year^{-1}$ respectively in the 1980s and 1990s (Houghton, 2003). In contrast to the unknown biases of atmospheric methods, analyses based on land-use change have deliberate biases built into them. These latter analyses consider only the changes in terrestrial carbon resulting directly from human activity. In other words, there may be other sources and sinks of carbon not related to land-use change such as those caused by CO_2 fertilization or changes in climate, that are considered by other methods but ignored in analyses of land-use change. The terrestrial sources and sinks of carbon in PgC $year^{-1}$ as estimated by different methods are given in Table 8.2 (Houghton, 2003).

A major concern noted by Houghton (2003) was the unknown rate of turnover of carbon belowground. One of the major sources of carbon inputs in all terrestrial ecosystems has been attributed to fine roots. Uncertainties in estimates of root longevity have markedly hampered proper quantification of net primary production (NPP) and belowground C

TABLE 8.2 Terrestrial Sources (+) and Sinks (−) of Carbon (Pg C $year^{-1}$) Estimated by Different Methods

Region	Inversions Based on Atmospheric Data and Models	Analysis of Land-Use Change	Forest Inventories
Globe	−1.4 (±0.8)	2.2 (±0.8)	
North	−2.4 (±0.8)	−0.03 (±0.5)	−0.65 (±0.05)
Tropics	1.2 (±1.2)	2.2 (±0.8)	
South	−0.2 (±0.6)	0.02 (±0.2)	

From Houghton, R.A., 2003. Why are estimates of the terrestrial carbon balance so different? Global Change Biol. 9, 500–509.

allocation, particularly in forests. In a comparison of fine root carbon inputs in two field sites, a hardwood forest (sweetgum *Liquidambar styraciflua* L.) in Tennessee and a loblolly pine (*Pinus taeda*) forest in North Carolina, Matamala et al. (2003) measured the ^{13}C isotopic signatures of live roots before and after carbon enrichment was applied in free air carbon enrichment (FACE) experiments. There was a marked difference in tree species, with mean residence time (MRT) in roots between 1 and 2 mm being 5.7 years for pine and only 3 years for sweetgum roots of the same diameter. Matamala et al. (2003) note that any estimates of belowground carbon inputs need to pay careful attention to the species of vegetation involved, and also to study root turnover rates by at least two different methods, as we noted in Chapter 2, Two Primary Production Processes in Soils: Roots and Rhizosphere Associates, on studies of root turnover. In another attempt to describe the turnover of belowground C in forests of North America, McFarlane et al. (2013) conducted a study to evaluate the importance of temperature and moisture (at the continental scale) on these processes. These authors found that temperature and moisture were not straightforward predictors of soil C turnover, and that the influences of macrofaunal activities (e.g., earthworm feeding and burrowing) and soil physical characteristics (such as texture) were at least as important in determining turnover rates. The implications of much longer MRTs of forest tree fine roots and the context dependency of soil physical and biological characteristics are quite profound, and deserving of further study by terrestrial ecologists.

8.4 PROBLEMS IN MODELING SOIL CARBON DYNAMICS

A more general problem yet faces soil ecologists. One of our current needs is to "model the measurable," rather than "measure the modelable" (Elliott et al., 1994). There are pools in models such as Century, mentioned earlier, which are more easily conceptualized than actually measured. A more readily measurable entity is the labile pool, consisting primarily of the microbial biomass. The intermediate and long-term pools, existing from decades to millennia, are very difficult to measure directly, and much work is under way to more effectively isolate and characterize these pools by a variety of methods (Six et al., 2002a,b). This problem requires integration across several levels of resolution, dealing with numerous human activities in sociology and economics that have a direct impact on soil management. These include the concept of the effectiveness of management of C resources, which

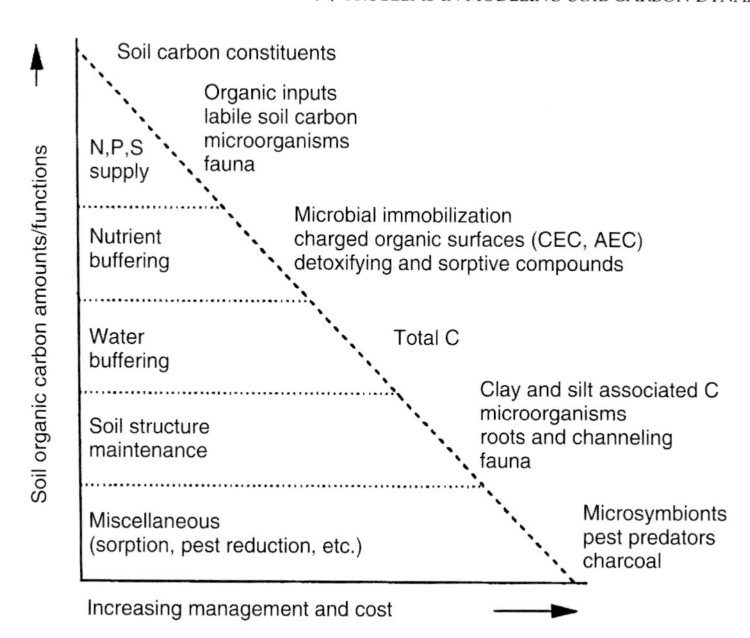

FIGURE 8.4 The functional role of SOM within an ecosystem depends on the intensity with which that system is managed. *Source: From E.T. Elliott, personal communication; modified from Woomer, P.L., Swift, M.J., 1994. Report of the Tropical Soil Biology and Fertility Programme. Tropical Soil Biology and Fertility Programme, Nairobi, Kenya.*

is inversely related to the cost of subsidizing the lost functions of organic matter (Fig. 8.4) (Woomer and Swift, 1994). The effectiveness of C resource management decreases with sequential loss of constituents and subsequent loss of function as land use intensifies without subsidizing lost organic matter. Elliott (1994) and colleagues urged soil ecologists to isolate functional SOM fractions and determine their roles in soil processes in order to understand the mechanisms controlling soil processes. This includes the mechanisms and processes involved in the formation and turnover of macro- and microaggregates in a wide range of soil types worldwide (see, for example, Beare et al., 1994a,b; Paul, 2016; Six et al., 1999, 2002a). Indeed, as noted in Chapter 3, Secondary Production: Activities of Heterotrophic Organisms—Microbes, chemical, microbiological, and macrobiological characterization of physically isolated fractions may provide the best opportunity for identifying functional pools of SOM. For example, each major category of soil biota has a significant effect on one or more aspects of soil structure, including production of organic compounds that bind aggregates, and hyphal entanglement of soil particles (microflora), producing fecal pellets and creating biopores (meso- and macrofauna) (Hendrix et al., 1990; Linden et al., 1994). A complete list of influences of soil biota is given in Table 4.14 (Hendrix et al., 1990).

Major developments have been made in conceptualizing SOM dynamics, which should have a considerable impact on the ways in which soils are viewed and managed for carbon sequestration. There are three principal stabilization mechanisms of SOM that are strongly related to the ways in which SOM pools are protected: They are (1) physically stabilized, or protected from decomposition through microaggregation; (2) intimate association of SOM with silt and clay particles; and (3) biochemical stabilization through the formation of recalcitrant SOM compounds (Six et al., 2002a). The protective capacity of soil has been represented graphically, in an ascending series (Fig. 8.5) (Six, 2002a), showing silt

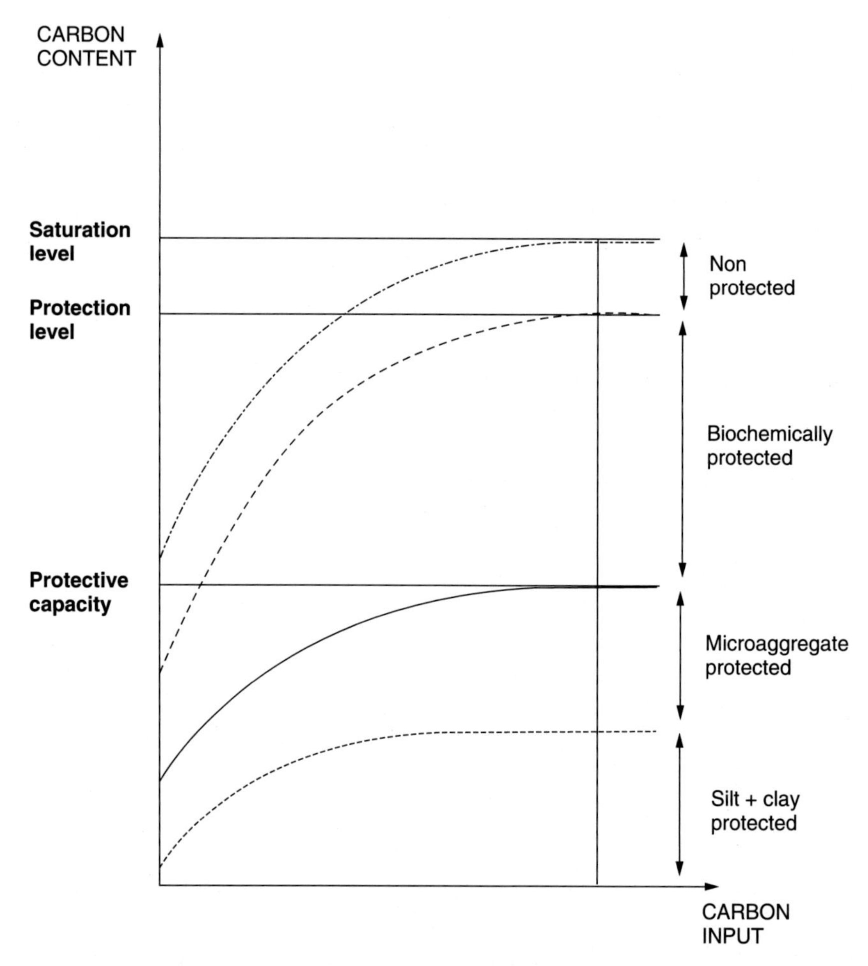

FIGURE 8.5 The protective capacity of soil (which governs the silt- and clay-protected carbon and microaggregate-protected carbon pools), the biochemically stabilized carbon pool, and the unprotected carbon pool define a maximum carbon content for soils. The pool size of each fraction is determined by its unique stabilizing mechanism. *Source: From Six, J., Conant, R.T., Paul, E.A., Paustian, K., 2002a. Stabilization mechanisms of soil organic matter: implications for C-saturation of soils. Plant Soil 241, 155–176.*

and clay at an asymptotic maximum that is the maximum protection possible, as anything above that is considered nonprotected. A conceptual model of SOM dynamics with the aforementioned measurable pools follows a sequence beginning with above- and belowground inputs into unprotected soil C, moving either into microaggregate-associated soil C by aggregate turnover, or via adsorption/desorption into soil and clay-associated soil C. It then moves subsequently into nonhydrolyzable soil C via condensation and complexation reactions, producing the biochemically protected soil C (Fig. 8.6) (Six et al., 2002a). In contrast with earlier conceptual and simulation models, this scheme is indeed measurable, hence meeting the criterion of Elliott et al. (1994) of "modeling the measurable." It also has the virtue of reflecting realities in lightly versus heavily weathered soils. The former with

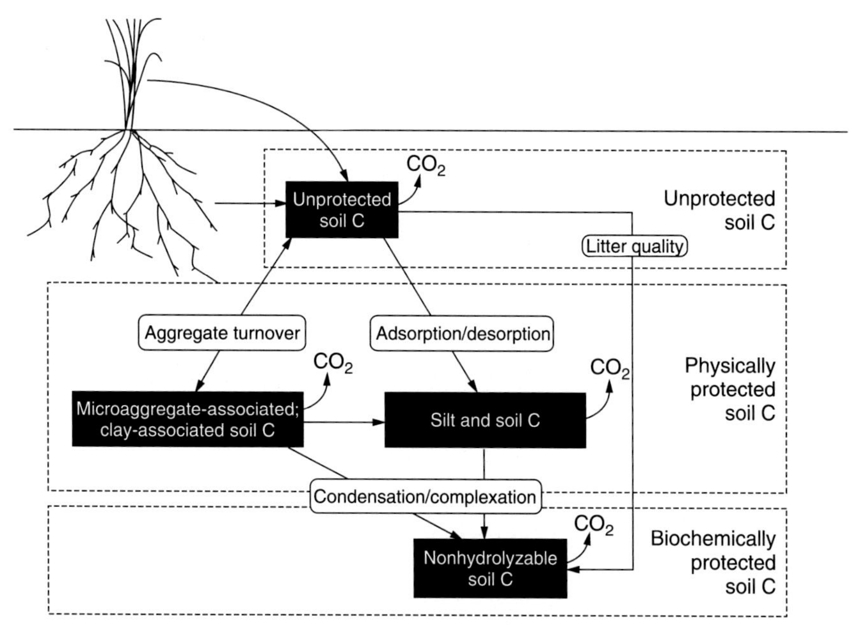

FIGURE 8.6 Conceptual model of SOM dynamics with measurable pools. The soil processes of aggregate formation/degradation, SOM adsorption/desorption and SOM condensation/complexation, and the litter quality of the SOM determine the SOM pool dynamics. *Source: From Six, J., Conant, R.T., Paul, E.A., Paustian, K., 2002a. Stabilization mechanisms of soil organic matter: implications for C-saturation of soils. Plant Soil 241, 155–176.*

a greater proportion of 2:1 clay mineral dominated soils have a greater silt- and clay-protected C pool than the 1:1 clay mineral dominated soils. The latter minerals, e.g., kaolinite and gibbsite, tend to dominate in the typically heavily weathered tropical soils (Theng et al., 1989).

Six et al. (2002b) proceeded to test some of the assumptions in the conceptual model given earlier by performing a major synthesis of SOM dynamics in a wide range of temperate and tropical soils worldwide, comprising over 32 sites. Six et al. (2002b) found a 1.8 times longer average MRT of C in the soil surface of temperate versus tropical soils (63 ± 7 vs 35 ± 6 years). This indicates that there is generally a faster C turnover in tropical than temperate soils. Interestingly, the range of MRT values was similar for both temperate and tropical soils, being 14–141 vs 13–108 years, respectively. The higher turnover rate for tropical soils is due primarily to faster turnover rates of the slow C pool in tropical soils (Feller and Beare, 1997, cited in Six et al., 2002b).

Returning to a theme we are emphasizing throughout this book, we cite the following prescient comment by a pioneer in pedology, Prof. T.W. Walker (1965): "...the soil cannot be properly studied apart from its biota; the ecosystem, the totality of its plant and animal life, is the best theoretical unit of study." We hasten to add that the involvement of prokaryotic organisms, bacteria and archaea, must be considered integral with the biota listed earlier. Schmidt et al. (2011) have noted a key new understanding: recent analytical and experimental advances have demonstrated that molecular structure alone does not control SOM stability; in fact, environmental and biological controls predominate. Indeed, our concepts of what the actual nature of SOM might be is a currently a matter of some debate. Some authors have effectively argued that the existence of "humic substances" in soil (i.e., large molecular weight compounds) is questionable, and that models which incorporate such compounds as "slow" cycling pools of organic matter may be obsolete (Lehmann and Kleber, 2015). One thing that is not in question, however, is the critical role of soil fauna and microbes in the process of organic matter decomposition. Clarification of the interactions between fauna, microbes, and the role of the fauna in physical processes that regulate both protection of organic material from and access of organic material to microbes is a major area of future research in the field of SOM dynamics.

8.5 BIOLOGICAL INTERACTIONS IN SOILS AND GLOBAL CHANGE

Perhaps the principal element of the global change scenario is the steadily increasing annual temperature, which rises about 0.1°C, annually. As this increase occurs, there should be a perceptible increase in loss of soil carbon (Jenkinson et al., 1991; Scharpenseel et al., 1990; Schimel et al., 1994). A countervailing tendency will exist with effects of CO_2 fertilization, enhancing plant primary production. However, several authors have noted the further constraints of other limiting resources to growth, such as mineral nutrients (Pastor and Post, 1988; Schimel, 1993). More holistic modeling efforts of changes in SOM, particularly ones which include soils, plants, herbivores, and detritivores together, are more realistic in their outcomes than those which model the plants, heterotrophs, and soil carbon pools separately (Schimel, 1993; Tinker et al., 1996).

Global change effects on relationships to soil biota (aboveground to belowground) can be modeled as a nested set of control variables. Morphological features of dominant life forms determine engineering activities at the ecosystem level, physicochemical properties of plant functional groups, modifying the provision of nutritional resources at the community level, and biological properties of individual species controlling direct interactions at the population level (Fig. 8.7) (Wolters et al., 2000). Changes in ecosystem functions created by plant-induced alterations in the disturbance regime, to and by resource consumption rates of soil organisms, should be confined to situations where essential traits of the vegetation are drastically changed. Such a change is most likely when the strength of environmental change overrides all other factors controlling plant assemblage structure, when plants with key attributes or functions invade or become extinct, and when species-poor environments are affected.

Much of the experimental research on ecosystem responses has focused on individual species-level responses, and seldom concerned with multifactor system-level responses. An ongoing study of ecosystem-scale manipulations in a California annual grassland has now progressed past 3 years' duration. Shaw et al. (2002) measured system-level responses in control and elevated carbon dioxide plots, using the Jasper Ridge Global Change Experiment (JRGCE). The JRGCE imposed four global change factors at two levels: (1) CO_2 (ambient and 680 parts per million), (2) temperature (ambient and ambient plus $80\ W\ m^{-2}$ of thermal radiation), (3) precipitation (ambient and 50% above ambient plus 3-week-long growing season prolongation), and (4) N deposition (ambient and ambient plus 7 g of nitrate $N\ m^{-2}\ year^{-1}$), in a complete factorial design. In the third year of

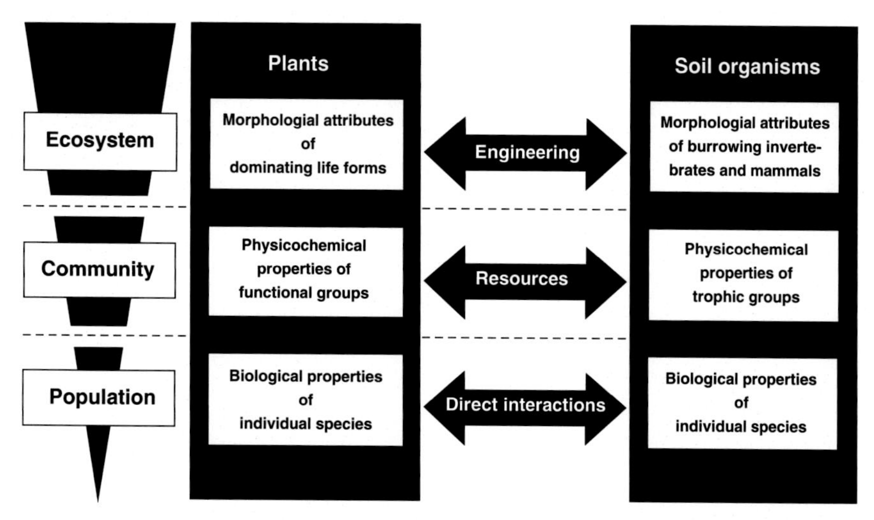

FIGURE 8.7 Hierarchy of plant effects on soil biota and vice versa. *Source: From Wolters, V., Silver, W.H., Bignell, D.E., Coleman, D.C., Lavelle, P., van der Putten et al., 2000. Effects of global changes on above- and below-ground biodiversity in terrestrial ecosystems: implications for ecosystem functioning. BioScience 50, 1089–1098.*

manipulations of the JRGCE, elevated CO_2 stimulated the production of aboveground biomass in the treatments with all of the other factors at ambient levels. Aboveground biomass increased >32%, which is comparable to that in other single-factor CO_2 enrichment experiments (North Carolina pine plantations = 25%, and Arizona free air CO_2 enrichment experiment (FACE) of 20%–43% greater). Interestingly, although each of the treatments involving increased temperature, N, or precipitation increased aboveground biomass and NPP, but elevated CO_2 consistently shrank these increases. In fact, with the three other factors and ambient CO_2 increased NPP by 84%, but increased CO_2 more than halved this, down to 40%. Belowground NPP was even more suppressed, with an average effect across all treatments of minus 22%. This study by Shaw et al. (2002) is an object lesson in the need to pursue multifactorial studies over many years to ascertain the full effects of the manifold variables involved in global change phenomena.

8.5.1 Changes in Soil Biodiversity in Relation to Global Climate Change

Belowground feedbacks and responses as driven by global change include: Climate change impacts soil biodiversity directly (1) through changes in temperature and moisture, and indirectly (2) through shifts in resource supply from plants. Combined, these cause changes in the physiology and growth of individual soil organisms, leading to changes in the diversity and composition of soil communities through altered functional responses and biotic interactions (3). As a result, selection for new traits and life histories within soil communities will take place, which in turn drives eco-evolutionary dynamics of aboveground communities (4) and ecological feedbacks to ecosystem processes, including

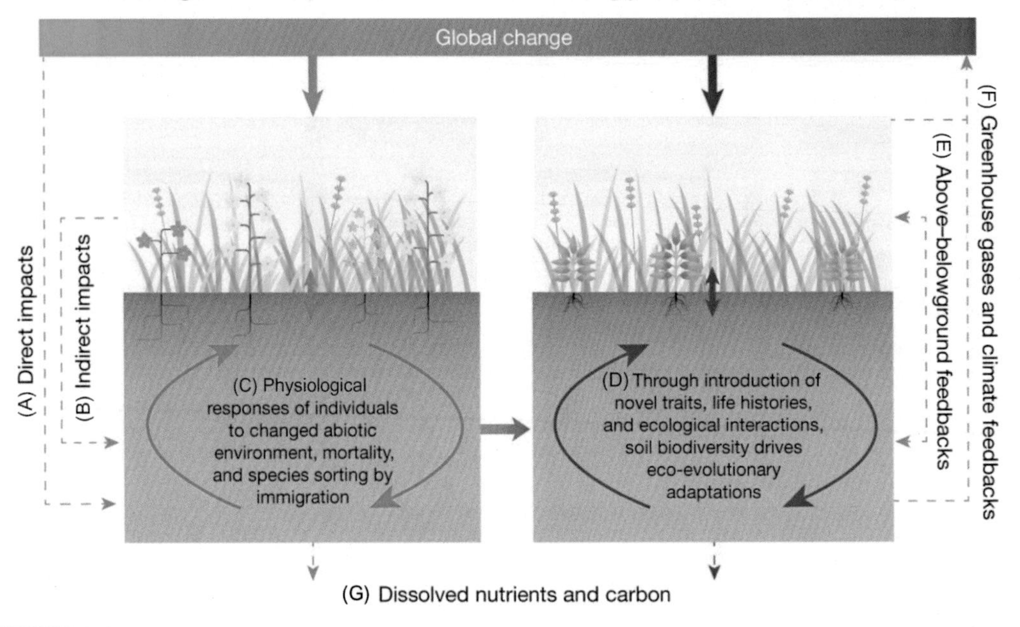

FIGURE 8.8 Climate change impacts soil biodiversity (A) directly, through changes in temperature and moisture, and (B) indirectly, through shifts in resource supply from plants. Combined, these cause changes in the physiology and growth of individual soil organisms, leading to changes in the diversity and composition of soil communities through (C) altered functional responses and biotic interactions. As a result, selection for (D) new traits and life histories within soil communities will take place, which in turn drives eco-evolutionary dynamics of (E) aboveground communities and ecological feedbacks to ecosystem processes, including (F) greenhouse gas emissions and (G) leaching of dissolved carbon and nutrients from soil. *Source: Modified from Bardgett, R.D., van der Putten, W., 2014. Belowground biodiversity and ecosystem functioning. Nature 515, 505–511.*

greenhouse gas emissions and leaching of dissolved carbon and nutrients from soil (5) (Bardgett and van der Putten, 2014) (Fig. 8.8).

A similar control hierarchy arises in the soil system as follows: attributes of burrowing macroinvertebrates and mammals establish different dynamic equilibria for soil at the ecosystem level; physicochemical properties of trophic groups affect the vegetation at the community level; and biological properties of individual species regulate direct interactions at the population level (Fig. 8.7) (Wolters et al., 2000).

Land-use change is probably one of the greatest agents of change in soil biology and ecology, and given the fact that it is so pervasive on all continents now, one can readily agree with Wolters et al. (2000) that land-use change rapidly and persistently alters all levels of above- and belowground interactions and acts on a large scale.

Climate change, of the magnitude envisioned over the next century (an average global temperature increase of 2.5°C), will lead to a major shift in the boundaries of ecological systems. There will be climate-induced alteration in the makeup of many plant communities and changes in litter quality due to changes in species composition. For example, as mixed spruce-hardwood forests in the southern boreal region are replaced by hardwood due to global warming, the anticipated higher-quality litter will provide increased availability of resources to the soil organisms. Of course, in more northerly climes, as major climate changes occur, there will be significant alterations in ecosystem functions when it affects organisms that carry out functions performed by few other organisms. Schimel and Gulledge (1998) predicted that in areas where episodic drying and rewetting of soil associated with climate change becomes more severe, populations of cellulolytic and ligninolytic fungi may be reduced, resulting in a decrease in litter decomposition greater than would be predicted by considering only the changes in soil and litter moisture.

Some recent studies of experimental soil warming have noted some aspects that require further consideration. Numerous soil warming experiments have focused on the surface 50–100 cm depth only. In a deep soil warming experiment, to a depth of 4 m, Pries et al. (2017) found that carbon dioxide production from all depths increased with 4°C warming. In fact annual soil respiration increased by 34%–37% over that of control soils. This greater soil respiration response bears remembering when considering effects of long-term global warming, which is expected to lead to global soil warming by an average of 4°C by 2100 (IPCC, 2013).

In sum, as climate change and other forms of global change force plants and animals to occupy new habitats, we should expect to see plant–soil–animal–climate combinations that have not previously existed. The area of inquiry which examines this eventuality falls under the umbrella term of "no-analog" ecosystem ecology (Williams and Jackson, 2007; Veloz et al., 2012).

8.6 ECOLOGY OF INVASIVE SPECIES IN SOIL SYSTEMS: AN INCREASING PROBLEM IN SOIL ECOLOGY

One of the primary concerns of ecology in the 21st century has been the increasing numbers of invasive species in ecosystems around the world. The publicity concerning invasive species in aquatic systems has been extensive. Case studies of lampreys invading the Great Lakes of North America via the Welland Canal and later via the St. Lawrence Seaway, and the rapid spread of the zebra mussel in lakes and streams over much of the Western Hemisphere are noteworthy examples. There is a less obvious but growing literature documenting the effects of introduced plants and animals displacing or outcompeting native species in soils in numerous ecosystems of the world. Some examples follow.

Eastern deciduous forests in North America have been invaded by two species of plants that are often dominant in the understory vegetation. *Berberis thunbergii* is a woody shrub that often forms dense thickets. *Microstegium vimineum*, a C_4 grass, forms dense carpets. The two invasives co-occur often. In a series of laboratory and greenhouse experiments in New Jersey, Ehrenfeld et al. (2001) found that the soil under these plants was increased in available nitrate and had elevated pH as well. The two invasive plants have different mechanisms to achieve a similar end result. *Berberis* combines large biomasses of N-rich roots with N-rich leaf litter, whereas *Microstegium* clumps combine small biomasses of N-rich roots with small biomasses of N-poor litter that leave much of the surface soil with few roots. Changing key chemical characteristics of soil, e.g., changed nitrate and pH, are undoubtedly only two of numerous ways in which invasive plant species alter the arena or playing field in contesting for dominance of patches of soil.

In the same research sites that were used by Ehrenfeld et al. (2001), Kourtev et al. (2002) measured alteration of microbial community structure and function by exotic plant species (Japanese barberry = *B. thunbergii* and Japanese stilt grass = *M. vimineum*), compared to a co-occurring native species (blueberry = *Vaccinium* spp.). They found in both bulk and rhizosphere soils that phospholipid fatty acid (PLFA) profiles, enzyme activities, and substrate-induced respiration (SIR) profiles of microbial communities were significantly altered under the two exotic species. The PLFA profiles provided only an index of community structure rather than specific information about what species were active. A correlation of structure (PLFA) and function, namely enzymes, showed that a particular set of species is associated with a particular pattern of enzyme activities but does not provide information about which of the species were responsible. Kourtev et al. (2002) found that profiles of enzymatic and catabolic capacity in the soil definitely differed with different microbial communities.

One of the more noted plant invasions of the past century was that of the annual grass, *Bromus tectorum* L., which has a current range of 40,000,000 ha, notably in wide regions of Washington, Oregon, Idaho, and Utah. Evans et al. (2001) measured litter biomass and C:N and lignin:N ratios to determine the effects on litter dynamics in a site that had been invaded in 1994. Long-term soil incubations (415 days) were used to measure potential soil microbial respiration and net N mineralization. Plant-available N was measured for 2 years with ion-exchange bags, and potential changes in rates of gaseous N losses were measured using denitrification enzyme activity. *Bromus* invasion significantly increased litter biomass, and its litter had significantly greater C:N and lignin:N ratios than did native species. The changes in litter quality and chemistry decreased potential rates of N mineralization in sites with *Bromus* by decreasing N available for microbial activity. Evans et al. (2001) suggest that *Bromus* may cause a short-term decrease in N loss by decreasing substrate availability and denitrification activity, but over the long term, N losses are likely to be greater in invaded sites because of increased fire frequency and greater N volatilization during fire. This mechanism, in conjunction with land-use change, will set into play a set of positive feedbacks that will decrease N availability and alter species composition.

In a companion study to that of Evans et al. (2001), Belnap and Phillips (2001) studied the effects of invasion by *B. tectorum* in three study sites in the Canyonlands of southwestern Utah. They measured litter and soil changes in sites that had been dominated previously by *Hilaria jamesii*, a fall-active C_4 grass, and *Stipa comata* and *Stipa hymenoides*, predominantly spring-active C_3 species. Belnap and Phillips (2001) measured the abundances of a wide range of microbes, microarthropods, and macroarthropods under *Hilaria* and *Stipa* communities, as well as in those that had been invaded by *Bromus* in 1994 (Fig. 8.9). There were significant changes in numbers and diversity, due in part to changes in amounts and qualities of litter. In the *Bromus* invaded plots, litter quantity was 2.2 times higher in *Bromus* + *Hilaria* than in *Hilaria* alone, contrasted with *Stipa* and *Bromus* which was 2.8 times greater than in the *Stipa* alone. Soil biota responded generally in opposite manners in the two perennial + annual grass plots. Active bacteria decreased in *Hilaria* versus *Hilaria* + *Bromus*, and increased in *Stipa* versus *Stipa* + *Bromus*. Most higher trophic-level organisms increased in Hilaria + Bromus relative to Hilaria alone, while decreasing in *Stipa* + *Bromus* relative to *Stipa* alone. The soil and soil food web

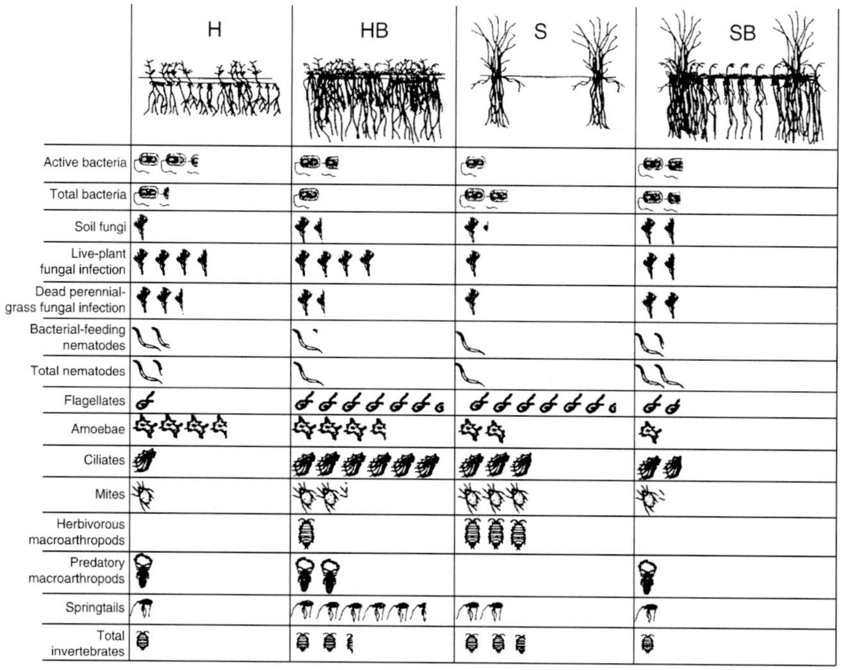

FIGURE 8.9 Diagram of soil food web structure in the four different grassland communities, showing the relative abundance within a given functional group. The numbers of icons on a given line are relative to each other; thus twice as many icons indicate that organisms are twice as abundant. *B*, Bromus; *H*, Hilaria; *S*, Stipa. *Source: From Belnap, J., Phillips, S.L., 2001. Soil biota in an ungrazed grassland: response to annual grass (*Bromus tectorum*) invasion. Ecol. Appl. 11, 1261–1275.*

characteristics of the newly invaded sites included the following: (1) lower species richness and numbers of fungi and invertebrates; (2) greater numbers of active bacteria; (3) similar species of bacteria and fungi as those invaded over 50 years previously; (4) higher levels of silt (hence greater water-holding capacity and soil fertility); and (5) a more continuous cover of living and dead plant material. The authors note that food web architecture can vary widely from what had existed previously within the same vegetation type, depending on the reactions to the invasive species relative to the previous uninvaded condition. Addition of a common resource can shift conditions significantly, and careful attention to the effects of species by season by site is definitely warranted.

Another form of biological invasion playing out in soils around the globe is that of earthworms being introduced either accidentally or intentionally (Hendrix et al., 2008). These large soil organisms are particularly problematic by virtue of their status as ecosystem engineers (*sensu* Jones et al., 1994), and the sometimes profound physical impacts that their feeding and burrowing activity can produce (see Chapter 4: Secondary Production: Activities of Heterotrophic Organisms—The Soil Fauna). There are several dozen "peregrine" earthworm species that have been transported around the globe, with approximately 45 species established in North American soils (Hendrix et al., 2008). Importantly, discoveries of new species introductions are being made virtually whenever systematic sampling efforts are conducted (Carrera-Martínez and Snyder, 2016; Callaham et al., 2016). When nonnative earthworms become established, they have been shown to significantly alter ecosystem properties ranging from plant population dynamics (e.g., Gundale 2002) to biogeochemical cycling (e.g., Cameron et al., 2015). The establishment of exotic earthworms in North American soils has been hypothesized to be related to general patterns of human-related disturbances, and the earliest records of European and Asian earthworms are from agricultural and roadside soils (Reynolds, 1995). These patterns suggest that movement of materials associated with agriculture, horticulture, and with recreational uses of earthworms as fish-bait are responsible for the extensive transport and introduction of European and Asian earthworms (Callaham et al., 2006). In soils where no native earthworms exist (i.e., those that were glaciated in the Wisconsinan Glaciation), introduced earthworms meet no resistance, and past disturbance seems not to be so important in terms of where they occur (Hale et al., 2005; Frelich et al., 2006), but in soils where native species persist, a disturbance displacing native species may provide an advantage to the introduced species (Hendrix et al., 2008). Still, human activities are strongly related to nonnative earthworm distributions in less disturbed habitats, for example, in the northern boreal forests of Canada, where Cameron and Bayne (2009) found that the time since road construction was a strong predictor of earthworm invasion. In a similar vein, Costello et al. (2011) found that exotic earthworms were passively transported along stream corridors from harvested portions of watersheds into undisturbed habitats downstream. Indeed, human disturbance can be very diffuse and still result in earthworm establishment, as Chaffin et al. (Lee Frelich, personal communication) found that recreational canoe trails in the roadless areas of the Boundary Waters Canoe Area Wildernesss were strongly associated with populations of nonnative earthworms.

A much different example of a soil invasion is the movement of the predatory New Zealand flatworm, *Arthurdendyus triangulatus* into Western Europe, where it preys upon earthworms. It was established in Great Britain by unknown means, but probably on soil associated with nursery stock. Boag and Yeates (2001) suggest that the avenue of introduction was rather circuitous: initially the flatworms spread from botanic gardens to horticultural wholesalers, then to domestic gardens, and finally invading agricultural lands. They note that it is one of twelve alien terrestrial planarians in Britain considered to be a pest, and hence seems to obey the "rule of tens" (sensu Williamson, 1996), in which only one in ten invasives assumes an outbreak or pest status. The indigenous flatworms, including *Arthurdendyus*, are not a problem in New Zealand because New Zealand has a drier and warmer climate than in the west of Scotland, where the outbreaks are the most severe. Interestingly, under minimum tillage practice in New Zealand, where crop residues provide refuges (Yeates et al., 1999), flatworms have the potential to reduce lumbricid earthworm populations.

In general, numerous theoretical studies have usually supported Elton's (1958) biotic resistance hypothesis, in which more diverse communities better resist invading species (see Byers and Noonburg, 2003, and references cited therein). In a mathematical overview of more general aspects of biotic resistance to invasive species, Byers and Noonburg (2003) demonstrate that invasibility is influenced not only by the number of native species present, but also by the number of resources present in a given ecosystem. Building on a

Lotka–Volterra competition model, Byers and Noonburg's model predicts that increasing invasibility with native diversity across large scales is the result of decreasing mean interaction strength as resources increase. The strength of the positive relationship between native and exotic species diversity and relative contribution of factors extrinsic to the community depend on whether niche breadth increases with the number of available resources. Interestingly, the same mechanism—the sum of interspecific competitive effects ($\Sigma\alpha_{ij}n_j$)—drives the opposite pattern of decreasing invasibility with native richness at small scales because resource numbers are held constant. As a consequence, Byers and Noonburg (2003) conclude that Elton's biotic resistance hypothesis, interpreted as a small-scale phenomenon, is consistent with large-scale patterns in exotic species diversity.

8.7 SOILS AND "GAIA": POSSIBLE MECHANISMS FOR EVOLUTION OF "THE FITNESS OF THE SOIL ENVIRONMENT?"

As was mentioned in the beginning of our book (Chapter 1: Introduction to Soil: Historical Overview, Soil Science Basics, and the Fitness of the Soil Environment), there are many positive feedback mechanisms in soils, in which organisms have arisen and/or evolved together. These include roots and arbuscular mycorrhiza (AM), and many of the genera and families of soil fauna. The following discussion is based on the very insightful and stimulating article by van Breemen (1993), entitled: "Soils as biotic constructs favouring net primary productivity." Van Breemen asks the central question: have soils merely been influenced by biota, or have biota created soils as natural bodies with properties favorable for terrestrial life? He presents hypotheses or postulates related to the overarching theme: (1) there are soil properties "favorable" for terrestrial life in general; (2) biota, including plants and the soil dwelling organisms, are able to affect those soil properties; (3) on a scale of ecosystems and a global ("Gaian") scale, biotic action makes the outermost (1–100 cm) layer of the earth's crust more favorable for terrestrial life in general than it would have been in their absence; (4) at an ecosystem-level scale, biota tend to offset the effects of unfavorable properties of the soil or soil parent material by modifying those soil properties; and (5) modification of soil properties may play a role in species competition.

Following from the ideas of Odum and Biever (1984), there should be some positive or donor–recipient controls on interactions between primary producers and other biota, with the AM being a prominent example. As we have noted earlier, feeding on detrital organic matter in the soil is generally the principal energy flow in terrestrial ecosystems. Therefore feedback loops arising in the soil community (such as detrital food webs, Chapter 6: Soil Food Webs: Detritivory and Microbivory in Soils) should have a major effect on net primary productivity. Thus, soil–biota interactions may be a most fruitful area to investigate and test hypotheses about positive effects of biota on the environment.

For favorable soil properties, changes and general improvements in soil porosities, and SOM status are prime examples of general improvements in soil characteristics, by cumulative interactions of the soil biota. This is not a simple linear progression however; there are examples of surface-feeding and surface-casting earthworms, which remove enough of the surface leaf litter material, or selectively feed on smaller soil particles and cast these to the surface, to cause a greater amount of soil erosion in their presence than in their absence (Johnson, 1990; Duarte et al., 2014; Jouquet et al., 2013).

In the areas of soil texture and structure, as well as soil chemical properties, there are numerous examples of soil biotic interactions having a generally beneficial effect in the top meter of soil material. One example of this is provided by Gill and Abrol (1986) who described how planting *Eucalyptus tereticornis* and *Acacia nilotica* on an alkali soil (pH 10.5) markedly decreased pH and salinity within 3–6 years. These changes were probably caused by a suite of factors, including increased water permeability, following the development of root channels and the accumulation of organic matter in the upper 20–50 cm of the soil profile. Other biota, notably termites, can promote higher salt content in soils, as detected by measurements in inhabited and abandoned termite hills compared to the surrounding soil (de Wit, 1978). Many of these processes tend to increase the amounts of heterogeneity within soil profiles, which has been well reviewed by Sileshi et al. (2010).

At both ecosystem and global scales, there are significant effects of biota on rock and soil weathering. The early pioneering researches of Vernadsky (1944, 1998) and Volobuev (1964) in particular originated and made popular the concept of "organic weathering." The able partnership of roots and microbes in mineral translocation is noteworthy; e.g., removing the interlayer K from phlogopite (vermiculitization) within the first 2 mm of the rhizosphere. For other references on biological impacts on mineral weathering, see Schlesinger (1996). As noted in Chapter 6, Soil Food Webs: Detritivory and Microbivory in Soils, the soil physical effects of earthworms on soil structure, formation of heterogeneous pores, and high structural stability is one of the hallmarks of soil-by-biota interactions over long time intervals.

There are several examples of transformations that counteract unfavorable soil properties. These arise principally from the influence of the biota on translocation and concentration of nutrients in the upper 1 m of the earth's mantle, the living soil. In general, biota tends to invest more in increasing nutrient supply under nutrient-poor than in nutrient-rich conditions. Root production and activity, as a fraction of total NPP, tends to be higher in the nutrient-poor conditions (Odum, 1971). This should be considered against a background of the generally slow growth rates and nutrient fluxes which occur in many wild plants on low-nutrient soils (Chapin, 1980). There is also an intriguing nutrient conservation process that occurs in many low-nutrient ecosystems. Development of mor humus types, characterized by thick organic horizons, is typical for "poor" (low-productivity) sites, and may represent nutrient conservation brought about as a result of the poorly decomposable litter formed in the surface layers (Vos and Stortelder, 1988). This in turn may lead to further inhibition of decomposition and NPP, so is an example of a positive feedback effect, which may require occasional fires or other disturbances to act as a suitable "reset" over millennial time spans. Soil phosphorus, in its various inorganic and organic forms, is perhaps the most limiting element in terrestrial ecosystems (van Breemen, 1993). Storage of P by secondary iron and aluminum phases is partly under biotic control and may be regarded as part of tight biotic cycling of P for three reasons: (1) secondary iron and aluminum oxides result from biologically mediated weathering of primary minerals; (2) they are often precipitated under the influence of iron oxidizing bacteria; and (3) can be kept in a mostly amorphous form by interaction with humic substances, from which P can be extracted by plants more efficiently than from crystalline oxides.

A further development in assessment of soil genesis and ecosystem condition is a quantitative assessment of forest humus forms, on a scale ranging from 1 (Eumull) to 7 (Dysmoder), which is called the *humus index* (Ponge et al., 2002; Ponge, 2013).

In the 72 sites studied, they were arranged as follows:

1. Eumull (crumby A horizon, Oi horizon absent, Oe horizon absent, Oa horizon absent)
2. Mesomull (crumby A horizon, Oi horizon present, Oe and Oa horizons absent)
3. Oligomull (crumby A horizon, Oi horizon present, Oe horizon 0.5 cm thick, Oa horizon absent)
4. Dysmull (crumby A horizon, Oi horizon present, Oe horizon 1 cm or more thick, Oa horizon absent)
   - **5a.** Amphimull (crumby A horizon, Oi, Oe, and Oa horizons present)
   - **5b.** Hemimoder (compact A horizon, Oi and Oe horizons present, Oa horizon absent)
6. Eumoder (compact A horizon, Oi and Oe horizons present, Oa horizon 0.5 or 1 cm thick)
7. Dysmoder (compact A horizon, Oi and Oe horizons present, Oa horizon more than 1 cm thick)

This index is well correlated with several morphological and chemical variables describing forest floors and topsoil profiles: thickness of the Oe horizon, depth of the crumby mineral horizon, Munsell hue, pH_{KCl} and pH_{H2O}, H and Al exchangeable acidity, percentage base saturation, cation-exchange capacity, exchangeable bases, C and N content, and available P of the A horizon (Fig. 8.10, from Ponge et al., 2002). Used in concert with the Ponge (2003) concept of humus forms as a framework of soil biodiversity, this approach should provide a more general comparative tool for assessing the chemical and biological conditions of a wide range of soil systems worldwide.

All of the foregoing perhaps raises more questions than answers. However, the general trend is for the number of species and individuals with positive effects to increase, both in successional sequences and over evolutionary time. In essence, the property of an individual that improves the environment for that individual, or increases its reproductive success, will benefit both it, and its competitors, as well. The selective advantage for such a trait(s) is probably small, viewed in a classical Darwinian context. If viewed in more general contexts, such as enhancement of site qualities, then this can be considered more general application of community and ecosystem development. Van Breemen (1993) notes that development of a trait in an earthworm, allowing it to better control the moisture content and CO_2/O_2 balance of its immediate surroundings would redound to the benefit of other organisms and site properties. If requirements of plants or a plant species happen to match those of the earthworm, then coevolution of the plant and worm might be possible, too. Wilson (1980) suggested that one might envision further development and evolution of a community of microbes, which could coevolve with the earthworm, to better enhance nutrient cycling processes. This is an evolutionary example of significant processes at "hot spots," as noted in Chapter 6, Soil Food Webs: Detritivory and Microbivory in Soils. The scenario is speculative, but serves as an example of where

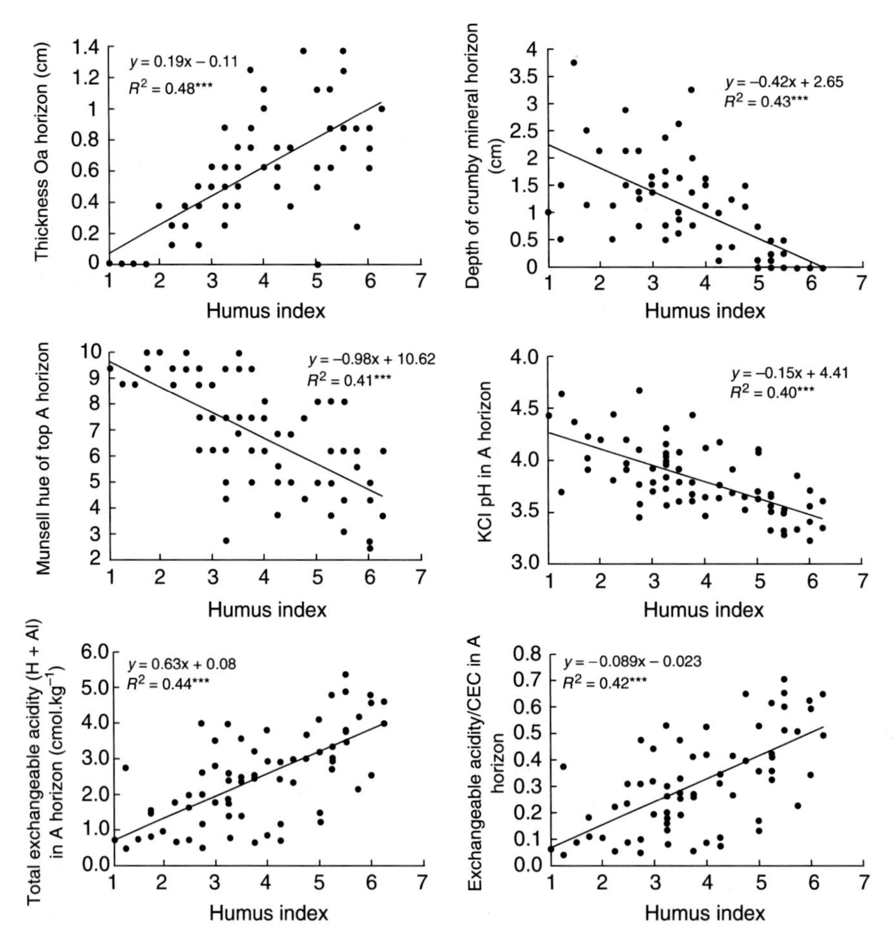

FIGURE 8.10 Correlation of the humus index with some variables measured in the topsoil profiles (means of four replicates). *Source: From Ponge, J-F., Chevalier, R., Loussot, P., 2002. Humus index: an integrated tool for the assessment of forest floor and topsoil properties. Soil Sci. Soc. Am. J. 66, 1996–2001.*

we may be expecting to see additional breakthroughs occurring in the cryptic and fascinating world of soil ecology.

8.8 IMPORTANCE OF NATURAL HISTORY COLLECTIONS TO INFORMING GLOBAL CHANGE STUDIES IN THE FUTURE

One of our coauthors (Crossley) is currently an emeritus curator at our University's museum of natural history, and he is working to identify and catalog soil mite specimens collected more than 50 years ago in the northern reaches of the Alaskan Brooks Range.

The existence of this collection (among other natural history collections) is of immeasurable value, as it represents the composition of the mite community at the site at a particular point in time. Given the likelihood of changes to soil ecosystems, and disruptions to biological linkages and processes in northern latitudes with global climate change, having this record of community as a baseline to make future comparisons will be crucially important. There can be no replacement for such collections, and we encourage the use of any soil-related archival material as a means of gaining insight into long-term processes. These collections can even include archives of soil samples, and such have been used to great effect (e.g., Richter et al., 2007). With an ever-expanding suite of analytical tools being developed, the importance of having archived samples available for examination cannot overstated.

8.9 SOIL ECOLOGY IN THE THIRD MILLENNIUM

We now come full circle to some issues that were raised in Chapter 1, Introduction to Soil: Historical Overview, Soil Science Basics, and the Fitness of the Soil Environment, rapidly increasing human population growth, providing ever-increasing pressure on a finite base of natural resources. As noted by Daily (1997), the direct substitution cost of a hydroponic plant production system for 1 ha of soil is the equivalent of US$850,000, and still rising. When one adds to that the cost of cleansing and recycling, this is a sizable fraction of the >$30 trillion dollar cost of annual goods and services provided by ecosystems globally (Costanza et al., 1997).

In dealing with environmental remediation and environmental assessment in general, what is a "healthy" soil? Is there a simple one-sentence definition of "soil quality," sensu Doran (2002)? Is there a clean soil similar to either clean air or water? The short answer is no. The longer answer is quite complex, but informative, if one takes an ecosystem-level approach (Coleman et al., 1998; Coleman, 2008). A general definition of soil quality is "the capacity of a soil to function within ecosystem boundaries to sustain biological productivity, maintain environmental quality, and promote plant and animal health" (Doran and Parkin, 1994). This is a beginning, but the healthy activity of all organisms, including microorganisms, should be considered explicitly (Coleman et al., 1998). As noted in Chapter 7, Soil Biodiversity and Linkages to Soil Processes, the state of our knowledge of microbial diversity, indeed that of a majority of the organisms active in soil, is still at a rudimentary stage. As a heuristic concept, soil quality has been useful for both education and assessment. These education and assessment tools encourage land managers to examine biological, chemical, and physical properties and processes occurring within their soil resources and to use that information as a framework for helping to make adaptive soil management decisions (Karlen et al., 2001). Soil quality has not been embraced universally, because some soil scientists have been concerned that value-based decisions could supplant value-neutral science and thus lead to premature interpretations and assertions of soil quality before the concept has been thoroughly and analytically challenged (Sojka and Upchurch, 1999).

It is possible to examine the health of the litter-soil subsystem of terrestrial ecosystems by utilizing indicator indices. One example is the use of the ratio of microbial biomass C to soil organic C (C_{mic}/C_{org}). This index is related to soil C availability and the tendency for a soil to accumulate or lose organic matter. It has been used successfully in evaluating the status of restored ecosystem, e.g., restored coalmine lands (Insam and Domsch, 1988).

A wide range of soil quality indices (SQI) has been calculated, related to specific groups of microbes and fauna (Coleman et al., 1998). These include nitrogen mineralization, soil respiration, respiration to microbial biomass ratios, faunal populations, and rates of litter decomposition (Knoepp et al., 2000). Considerably less attention has been paid to ecosystem-level analyses. The following is an overview of several studies, undertaken in two agroecosystems in the Georgia Piedmont, an aggrading forested ecosystem in western North Carolina, and an agroecosystem in Nebraska.

In the agroecosystem study, a wide range of biological, chemical, and physical factors were measured in two field sites, in which alternative poultry-litter management practices were compared. Multivariate statistical techniques were used to determine the smallest set of chemical, physical, and biological indicators that accounted for at least 85% of the variability in the total data set at each site. This set was defined as the minimum data set (MDS) for evaluating soil quality (Andrews and Carroll, 2001). The efficacy of the chosen MDS was evaluated by performing multiple regressions of each MDS against numerical estimates of environmental and agricultural management sustainability goals (e.g., net revenues, P runoff potential, metal contamination, and amount of litter disposed of). Coefficients of determination ranged from 0.35 to 0.91, with an average R^2 of 0.71. Each MDS was then transformed and combined into an additive SQI. SQI varied between the two sites, but Andrews and Carroll (2001) noted that this "designed SQI" enabled the indices to be tailored to local conditions.

In the forest ecosystem study, a combination of chemical and biological indices was used to measure soil quality in five watersheds arranged along an elevational gradient at the Coweeta Hydrologic Laboratory, in the southern Appalachian mountains of western North Carolina. The selected characteristics of the elevation gradient stands are presented in Table 8.3 (Knoepp et al., 2000). The sites represented a gradient in vegetation and elevation and included xeric oak-pine (OP), cove hardwood (CH), mesic mixed-oak at low and high elevations (MO-L; MO-H), and mesic northern hardwood (NH) vegetation. The sites were then ranked on a range of soil chemical characteristics, N availability, litter decomposition rates, and forest floor mass and coarse woody debris standing crop, soil oribatid mite populations as numbers and total species and Shannon–Wiener biodiversity index (Table 8.4) (Knoepp et al., 2000), and then several measures of soil C availability: CO_2 flux, microbial carbon, qCO_2 (μg C g $soil^{-1}$), and qC_{mic} ($\mu g C_{mic}$ gC $total^{-1}$). Note that all sites had approximately equal diversity of overstory tree species, with H values ranging between 1.93 for the NH and 2.25 for OP.

The five sites were compared for overall soil/site quality, ranked using biological and chemical or physical quality, and the aboveground indices wood production, net primary productivity, and biodiversity. Overall, soil biological quality was highest for OP and MO-L, with the highest scores in N and C availability, and fauna population indicators. Based on soil chemical and physical properties, NH ranked highest with the greatest cation, C and N concentration, and lowest bulk density. In sum, the highest quality site is dependent on the goal desired for that site. In terms of wood production, MO-H was the highest quality site. Both mixed oak sites had the highest productivity using the total litterfall index. If one

TABLE 8.3 Selected Characteristics of the Elevation Gradient Stands[a]

Site	*OP*	*CH*	*MO-L*	*MO-H*	*NH*
Elevation (m)	782	795	865	1001	1347
Aspect (degree)	180	340	15	75	20
Slope (degree)	34	21	34	33	33
Vegetation type	Oak-pine	Cove hardwoods	Mixed-oak	Mixed-oak	Northern hardwoods
Dominant species	*Kalmia latifolia, Quercus prinus, Q. rubra, Carya* spp.	*Liriodendron tulipifera, Quercus rubra, Tsuga candensis, Carya* spp.	*Rhododendron maximum, Quercus coccinea, Q. prinus*	*Rhododendron maximum, Quercus rubra, Q. prinus*	*Betula allegheniensis, Liriodendron tulipifera, Quercus rubra*
Moisture regime	Xeric	Mesic	Mesic	Mesic	Mesic
Soil series and subgroup(s)	Evard/Cowee, Chandler, Edneyville/ Chestnut, Typic Hapludults, Typic Dystrochrepts	Saunook, Tuckaseegee, Humic Hapludults, Typic Haplumbrepts	Trimont, Humic Hapludults	Chandler, Typic Dystrochrepts	Plott, Typic Haplumbrepts

[a]*Data compiled from Coweeta Long-Term Ecological Research Program records.*
From Knoepp, J.D., Coleman, D.C., Crossley Jr., D.A., Clark, J., 2000. Biological indices of soil quality: an ecosystem case study of their use. Forest Ecol. Manage. 138, 357–368.

TABLE 8.4 Soil Oribatid Mite Populations in Five Representative Sites in the Coweeta Hydrologic Laboratory Basin

Site/Rank[a]	N[b]	#Spp.[c]	J[d]	H[e]
OP/3	2237	78	0.752	3.28
MO-L/4	2234	96	0.746	3.41
MO-H/4	2192	92	0.761	3.44
CH/1	570	64	0.876	3.64
NH/2	1454	81	0.793	3.48

[a]*Site rank for oribatid mite populations.*
[b]*Abundance of individuals collected.*
[c]*Total number of oribatid mite species identified.*
[d]*Pielou's evenness index.*
[e]*Shannon–Wiener biodiversity index.*
After Lamoncha, K.L., Crossley Jr., D.A., 1998. Oribatid mite diversity along an elevation gradient in a southeastern Appalachian forest. Pedobiologia 42, 43–55; from Knoepp, J.D., Coleman, D.C., Crossley Jr., D.A., Clark, J., 2000. Biological indices of soil quality: an ecosystem case study of their use. Forest Ecol. Manage. 138, 357–368.

desired to maximize biodiversity, both aboveground and in the soil, all sites ranked highly (Knoepp et al., 2000). The overall take-home message is important for land-use managers and ecologists in general: the site quality really depends on the objectives of the users, and the context in which the sites (in this case, situated in two different watersheds) exist. There is ample room for further investigation in this important area in which scientists and managers from a variety of disciplines will collaborate. This is especially true when comparisons are made across wide continental gradients, e.g., across ecoregions.

The foregoing examples involved very extensive sampling and analytical regimes that might preclude their wide adoption in soil quality studies. An innovative study in Nebraska employed the fact that differences in electromagnetic (EM) soil conductivity and available N levels over a growing season can be linked to feedlot manure/compost application and use of a green winter cover crop (Eigenberg et al., 2002). A series of soil conductivity maps of a research cornfield were generated using global positioning system and EM induction methods. The study was conducted over a 7-year period. Image processing techniques were used to establish EC treatment means for each of the growing season surveys. Sequential measurement of profile weighted soil electrical conductivity (EC_a) was effective in identifying the dynamic changes in available soil N as affected by animal manure and N fertilizer treatments, during the corn-growing season. This real-time monitoring approach shows considerable promise in enabling farmers to more efficiently use N sources in cropping management systems, and in minimizing N losses to the environment.

It is imperative to have a robust, quantitative, and universally applicable metric for soil quality. Considering the 4.5×10^9 ha that are tropical soils, use of an updated fertility capability soil classification system (Sanchez et al., 2003) should be helpful for soil ecologists. It employs quantitative topsoil attributes including percent total organic C saturation (van Noordwijk et al., 1998) compared with undisturbed or productive site and soil taxonomy. The top three soil constraints in the tropics include moisture limitations, low nutrient capital reserves, and high erosion risk. Because many smallholders in tropical regions depend on organic sources for nutrient inputs to their crops, this becomes an ideal situation to practice sound organic agriculture. This approach has been promoted ably by the Tropical Soil Biology and Fertility Programme, which has a network of research sites throughout eastern and southern Africa, India, and southeast Asia (see van Noordwijk et al., 1998; Swift, 1999; cited in Sanchez et al., 2003; Palm et al., 2001).

Promoting the ecological complexity and robustness of soil biodiversity through improved management practices is a truly underutilized resource with the ability to improve human health. Wall et al. (2015) offer some basic guidelines for management of soil biodiversity. A new approach for land use and management is required that acknowledges that soil biota act in concert to provide multiple benefits, even if these benefits are not easily observed. Moreover, increased soil food web complexity promotes resistance and resilience to perturbation and may buffer the impacts of extreme events.

Agroecological practices that enhance SOM content and soil biodiversity can promote nutrient supply, water infiltration, and well-structured soil. Effective management options for cropping systems include reduced tillage with residue retention and rotation, cover crop inclusion, integrated pest management, and integrated soil fertility management (such as the combination of chemical and organic fertilizer). Expanding plant species diversity in crop and/or land rotations and adding organic amendments to pastures can increase soil biodiversity and mimic better the natural soil food web. Additionally, maintenance of soil biodiversity at the landscape level can be enhanced through buffer strips and riparian zones and land rotations. Drainage water management can reduce the movement of pollutants, agrochemicals, and other contaminants to nearby landscapes. Likewise, several forestry practices exist that promote soil biodiversity: reestablished mixed deciduous forest stands in Europe were shown to have higher soil biodiversity than pure coniferous stands (Wall et al., 2015).

A recent synthesis by Lavelle et al. (2016) explores the interfaces between soil food webs, ecosystem engineers, and the wider provision of ecosystem services at scales greater than 100–1000 m on the landscape. This provides the entry point into consideration of how biodiversity and ecosystem function evolve, across centuries and millennia. For example, across a range of soil ages in a chronosequence ranging from young to old, namely incipient, young, mature, and old weathered soils, earthworms and ants followed somewhat similar population densities, while termites became most numerous, on average, in the old weathered soils, which were presumably tens of thousands of years old (Fig. 8.11) (Lavelle et al., 2016). We suggest that this sort of broad thinking, past the immediate activities in field plots, may well lead us to new insights into how the biota and soils evolve together over time. It is worth remembering that pioneers in our field, including Darwin (1881) and White (1789), launched our field on its current trajectory, but it has taken explicit consideration of the biota and pedology links that are giving us some useful insights into how our soils work, particularly following up on the self-organization theme as a leit motif, as it were, for future research.

By working with geomorphologists, soil ecologists can greatly expand their purview of how soils evolve over geologic time. Thus rates of bioturbation can be as rapid as sustained maximum rates of tectonic uplift. In concert with surface geomorphic processes, bioturbation alters fundamental properties of soil, including particle-size distribution, porosity, the content of carbon and other nutrients, and creep flux rate (Wilkinson et al., 2009). As should be evident to our readers by now, the key element in the soil formation factors equation often overlooked, time, plays an essential role in providing the fourth dimension for the sweeping but gradual changes that occur in soil pedogenesis.

We suggest that a holistic consideration of soil biology and pedology, as shown in Fig. 8.11 (Lavelle et al., 2016), sets the framework for future studies in soil ecology sensu lato. The figure depicts how some of the macrofauna, namely earthworms, ants, and termites, are distributed over a wide range of soils of different ages, ranging from incipient, to young, mature and then old soils.

We agree that not all soil types are amenable to significant alterations by soil biological activity. However, favorable conditions for amelioration of soil properties exist in 12 out of 39 diagnostic soil horizons, as defined by the World Resource Base of the Food and Agricultural Organization (Targulian and Krasilnikov, 2007). In contrast, in 27 out of the 39 diagnostic soil horizons, the biota adapts to the environment rather than improving it (Targulian and Krasilnikov, 2007).

For those who are interested in pursuing practical, hands-on studies in soil ecology, Chapter 9, Laboratory and Field Exercises in Soil Ecology, contains some selected field and laboratory exercises that should be of use in both research and teaching activities.

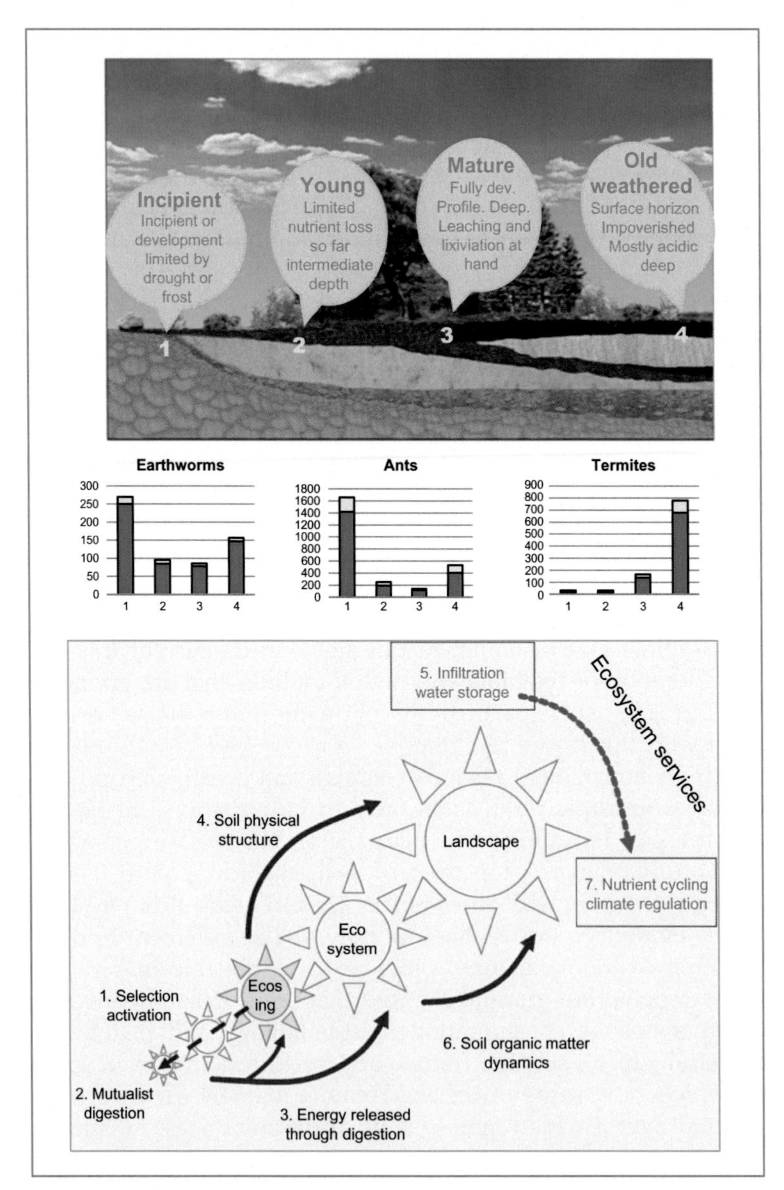

FIGURE 8.11 Top panel: Changes in mean population densities (individuals m^{-2}) (SE in yellow) of the main invertebrate soil ecosystem engineers (earthworms, ants, and termites) across a chronosequence of soil maturation sites from (1) incipient, to (2) young, (3) mature, and (4) old weathered (Orgiazzi et al., 2016).
Bottom panel: Interactions among self-organized units across scales and the delivery of ecosystem services. Note that additional scales such as biome and biosphere may be added when ecosystem services are considered at the global scale. Ecosystem engineers select and activate microbial communities in their (1) functional domains. (2) A mutualist digestion system releases (3) energy used by ecosystem engineers to build biogenic structures and organize (4) soil physical structure with resulting effects on (5) water services (infiltration and storage). Combination of mineralization processes during the digestion process and sequestration of organic compounds in biogenic structures affects (6) SOM dynamics, and hence, (7) nutrient cycling and climate regulation. *Source: From Lavelle, P., Spain, A., Blouin, M., Brown, G., Decaëns, T., Grimaldi, M., et al., 2016. Ecosystem engineers in a self-organized soil: a review of concepts and future research questions. Soil Sci. 181, 91–109.*

CHAPTER

9

Laboratory and Field Exercises in Soil Ecology

9.1 INTRODUCTION

The following are some exercises that we have found to be useful to better acquaint our students with techniques used to conduct field research in soil ecology. They provide a combination of process and taxonomic identification work, both of which are necessary to make some headway in our field.

9.2 ROOT-RELATED EXERCISES

Principle: Observation of plant root systems in the laboratory and in the field. Measurements include root growth, turnover, and production. Methods can be destructive or nondestructive.

9.2.1 Construction of a Simple Root Observatory (i.e., a Rhizotron)

Procedure: One simple way to facilitate the observation of roots (or other soil biological activity) is to construct a small "rhizotron" for laboratory use. These root observation frames consist of three pieces of wood, two sheets of transparent glass or plastic, hardware to hold everything together, some soil and seeds (Fig. 9.1). In a relatively short time, it is possible to construct one of these rhizotrons, plant some seeds (maize or other grass crops are well suited for the small space), and begin root observations. Make sure the plants have all the resources they'll need including reasonably fertile soil, water, and plenty of light. Over the course of a semester, students can use a digital camera to collect images at regular time intervals (e.g., weekly) for analysis (or trace roots on clear plastic sheets affixed to the face of the rhizotron) to track the production of roots in the laboratory. It is important to keep the sides of the rhizotrons covered with some opaque material (card stock or black plastic) when the frames are not being measured to prevent light from impacting root development.

Fundamentals of Soil Ecology
DOI: http://dx.doi.org/10.1016/B978-0-12-805251-8.00009-0

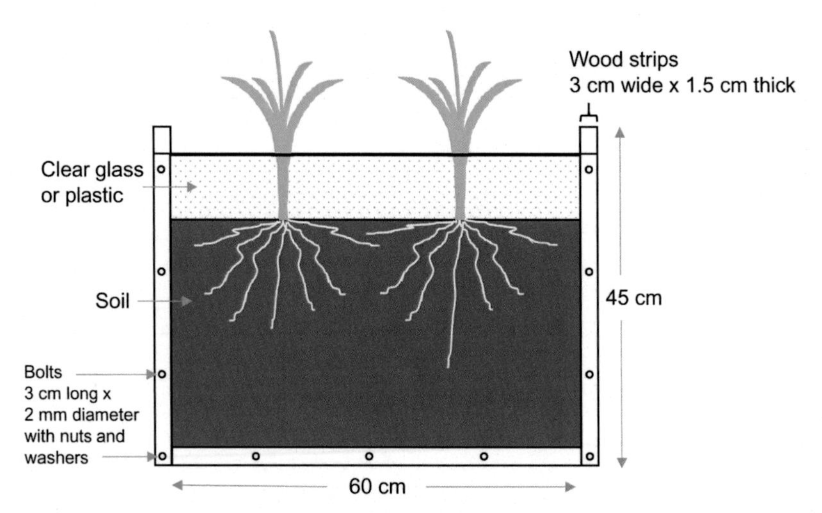

FIGURE 9.1 Simple schematic for constructing a rhizotron for laboratory use. The drawing is not to scale, and dimensions are only recommendations. These soil observation frames can be used for root studies, and are also useful for observing the influence of earthworms (or other burrowing invertebrates) on the movement of soil and litter. The contents of the frame can be modified to suit the particular interest of the observer.

9.2.2 Minirhizotron Studies

A more advanced option for root observation involves the use of a minirhizotron root imaging system. Obviously, this requires the instructor to have access to one of these systems. The advantage of the minirhizotron is that multiple viewing tubes can be installed with comparatively little disturbance into remote field sites. Measurements are made in situ by taking the camera to where the roots are growing.

Description of a minirhizotron: A clear polycarbonate tube 5 cm in diameter and 1.8 m in length is used as a minirhizotron. Each tube is marked with a groove and depths are stamped with length increments externally at 10 cm intervals. A single groove etched along the length of the tube is used as a reference for orienting the camera. The section of tube exposed above the soil surface is wrapped with black plastic tape and the top is plugged with a rubber stopper to keep light from entering the minirhizotron. The bottom end of each tube is sealed with a watertight polycarbonate stopper.

Installation of the minirhizotrons: Holes are drilled by an auger mounted on a tractor in the field plot at an angle of 20–45 degrees from the vertical along the direction of the crop row. Holes are cleaned with a sharp-ended stainless steel tube of the same diameter as the outside of the polycarbonate tube. A minirhizotron is then driven into the hole forming a close tube/soil interface.

Enough minirhizotrons should be installed to enable one to obtain standard error estimates that are <10% of the mean. This usually requires between six and eight tubes to be installed per treatment (Cheng et al., 1990; Fahey et al., 1999; Pregitzer et al., 2002).

Observation and recording: Observation of roots intersecting minirhizotrons are accomplished using a digital camera, a small video monitor, and a datalogger. The image produced by the camera is observed in the monitor and logged simultaneously. During image acquisition, the camera (or a reflector) is moved from the bottom of the tube upward, recording a 2 cm wide image at each increment. This allows observation and counting of roots and minimizes field operation time.

Data acquisition from images: There are several types of data that can be collected from minirhizotron images. The most basic of these is the simple counting of roots. The number of roots in the 2 cm wide strip observed is recorded for each 10 cm length of the tube. Counts are independent of the length or diameter of the root at the interface. If a root branches it receives one count for the main root and one for each branch. If a root crosses the depth indication groove it should receive one count for each depth interval.

Root counts can be converted to root length densities (RLD), as $cm\ cm^{-3}$, using the equation: $RLD = (N \times d)/(A \times d)$, where N is the number of intersecting roots, d the outside diameter of the tube, and A the area of tube observed.

Tracing technique: More detailed data can be extracted from minirhizotron images using a technique wherein the length and diameter of each root in each image is traced using a digital interface.

Root length can be converted to RLD using the equation:

$RLD = Lr/C \times A \times D$, where Lr is the root length, C the magnification factor, A the area of tube observed, and D the distance from the outside surface of the tube which can be observed in soil (usually assumed to be 0.2 or 0.3 cm in the literature) (Fahey et al., 1999).

Automated root length measures: Although minirhizotron technology has been in use for at least 30 years, the processing of images and extraction of data from these images has been a monumental challenge. Procedures to automate the acquisition of data from images have been developed, but these are still typically quite laborious, and require a great deal of human input in terms of tracing roots in each image. However, improvements in image analysis will continue to improve the ability of interested researchers to study root dynamics. Indeed, Vincent et al. (2017) have developed a technique whereby only a subset of images need be traced manually in order to "train" an image analysis program to analyze further images with good accuracy. Advances such as these bring real promise to future detailed studies of root dynamics using minirhizotron tubes.

9.2.3 Sequential Sampling of Root Cores

This technique requires that soil cores be collected at a set interval across a semester (or growing season). Materials required include a soil coring tool (preferably at least 5 cm in diameter), appropriate sample storage and labeling equipment, facilities for washing roots from the samples, sieves (0.5–2.0 mm), and facilities for drying and weighing the roots after washing.

Procedure:

1. Collect soil cores at a field site of interest (crop fields can be useful examples), in large enough replicate numbers to minimize the variability between samples. Depending on the spatial variability of root systems in the sampled area, sampling may need to be adapted to take spatial factors into account (i.e., sampling exclusively within the row of

crop plants rather than both within and between rows). Label all samples with appropriate information.
2. Transport samples to the laboratory or root washing facility (this may be outdoors). If transport times are long, or the samples will not be washed immediately, then samples should be stored in a cold room (at 2–4°C).
3. Soil cores are placed into a sieve (e.g., 0.5 mm), broken apart gently, and as many large roots as possible picked out by hand. The remainder of the sample should be washed with a stream of water until all mineral soil particles have been washed through the sieve. All roots remaining on the sieve are picked out and placed with larger roots into a paper bag or other oven-safe container.
4. Place the samples into a 50–75°C drying oven for 48 hours, or until samples cease to lose weight. Once a steady weight has been achieved, weigh each sample on a top-loading balance and record the weight for each sample. Data can be converted to a mass per unit area (e.g., g root m^{-2}), and this represents the standing stock of root mass on the particular sampling date.
5. Repeat steps 1–4 on a biweekly (or more frequent) basis for the semester, or growing season. Root mass data from each sample date can be plotted across time in order to determine patterns of root production or loss over the course of sampling.

9.2.4 Root In-Growth Cores

This technique is a variation on the previous sequential coring technique wherein soil that has had all roots removed is placed into a mesh cylinder and placed into the field to allow roots to grow into the soil inside the cylinder. Materials needed will be the same as those listed in Section 9.2.3, as well as some wire mesh or heavy plastic mesh for in-growth cylinder construction. Although somewhat hindered by the disturbance to soil during packing and installation of the mesh cores, the destructive sampling of these in-growth cores over a set time interval allows an estimate of root production.
Procedure:

1. Collect soil from the field site of interest, pass through a 2 mm sieve, and remove all roots. A volume of soil sufficient to fill the total number of desired root in-growth cylinders will be needed. Removal of roots can be laborious and time consuming, and may involve a number of techniques depending on the amount of roots present, and the desire of the investigator to have completely root-free soil for in-growth cores. For instructional purposes, it may be sufficient to remove roots during the sieving process with forceps and an equal-effort timed search and removal of roots for each volume of soil.
2. Construct cylinders, 5 cm diameter and 15 cm depth, from wire mesh or heavy plastic mesh (Fig. 9.2). Fill these cylinders with a known mass of root-free soil (ideally an amount that will approximate the bulk density of the soil at the field site).
3. Return to the field site and remove soil cores of the same dimensions of wire in-growth cylinders, and replace these with the root-free in-growth cylinders.
4. On harvest dates, return to the field site, use a shovel to dig up the root in-growth cores with a block of soil surrounding the core. Carefully remove the soil around the core, being careful not to pull out roots. Cut roots emerging around the margins of the

FIGURE 9.2 Example of a root in-growth core. The core is filled with root-free soil and placed into the field where new roots are allowed to grow through the mesh and into the soil inside. These are harvested at set time intervals to estimate root production rates. *Source: Photo by Mac Callaham.*

in-growth cores with scissors or a knife. Label cores with the appropriate information (e.g., date, location, treatment, depth sampled), place in a plastic bag, and put into cold storage until further processing.

5. Follow the root washing and drying procedures outlined in Section 9.2.3. Repeat sampling on a set interval and express data as in Section 9.2.3.

9.3 PROCESS-RELATED EXERCISES

Process level studies allow investigators to evaluate rates and fluxes of important gasses and matter. These studies can be particularly interesting because the measured process integrates several levels of organization and function in soil ecosystems. For example, soil respiration is a function of both heterotrophic and autotrophic respiration, and it responds to changes in temperature, moisture, and to the activity of fauna. As a result, it is easy to design experiments which manipulate any combination of these variables, and using respiration as the response variable, to demonstrate the effects in the field or laboratory.

9.3.1 Soil Respiration Studies

Principle: A dilute solution of alkali (typically 1 M NaOH) is placed in an open glass jar above the soil surface. A metal cylinder (usually 10 cm diameter, irrigation pipe will do) is installed well into the A horizon, carbon dioxide evolution is measured by absorption in the alkali solution for 24 hours, and then the CO_2 amount absorbed is measured by back-titration of the excess alkali remaining. When expressed on a per unit area basis, the soil respiration data are comparable with literally hundreds of values from the literature about many different ecosystems.

Soil respiration is one of the most commonly used methods of determining metabolic activities of organisms in soil. As noted in Chapter 3, Secondary Production: Activities of Heterotrophic Organisms—Microbes, there is interest in determining the relative contributions of carbon dioxide evolved by the secondary consumers (microbes and fauna) as differentiated from that originating from respiring roots. For the purposes of this laboratory exercise, we will rely on comparisons of respiration from different ecosystems, such as an arable field, forest, or grassland. Soil respiration, reflecting all of the biotic activity, is often measured to compare and contrast the side effects of chemicals such as pesticides and heavy metals. Soil respiration can be determined directly in the field. By measuring soil temperature and percent water in the soil, the relative contributions of these key abiotic variables can be calculated, which is useful in comparative ecosystem studies (e.g., Coleman, 1973). For more background on this method, also consult Alef and Nannipieri (1995) and Luo and Zhou (2006).

Materials and supplies needed:

1. Metal or plastic cylinders with one end beveled for ease in insertion into soil
2. Plastic lids to seal the cylinders, available from Sinclair & Rush, Inc., St. Louis, MO
3. Aluminum foil, to cover cylinders and lids, to minimize any "greenhouse effect"
4. Screw-capped glass jars (4–6 cm diameter 7–8 cm high)
5. Tripods made of metal or plastic
6. NaOH solution (0.5 or 1.0 M)
7. Barium chloride solution (1 M)
8. HCl solution, or definitive (determined) molarity, ideally 1.0 M
9. Thymolphthalein indicator: 1 g thymolphthalein is dissolved in 100 mL 95% ethanol. Other references (e.g., Alef and Nannipieri, 1995) describe the more often used phenolphthalein, but it has a less-definitive end point, changing from dark to light pink at the end point. In contrast, the thymolphthalein changes from blue to colorless at the end point very rapidly, which gives the definitive end point so desired in this titration.

Procedure: About a week after the respiration chambers are installed, the CO_2 flush from the soil disturbance caused by the installation should have diminished. It is now possible to measure the amount of CO_2 respired from a known surface area (the diameter of the chamber is 10 cm), by trapping it into jars of 1 M NaOH during 24 hours.

Pipette 20 mL of NaOH solution into the glass jar, place it on the tripod in the center of the selected cylinder. Immediately cover the cylinder with the plastic snug-fitting lid, and cover the entire setup with aluminum foil. Note time zero.

After 24-hour incubation, retrieve the jar, pipetting 20 mL 1 M $BaCl_2$ into it before covering with screw-on lid. Take the jars to the laboratory for titration, using the color indicator noted above, and the standard 1 M HCl.

Note that control treatments are performed by incubating sealed jars of NaOH solution in the field.

Calculations:

CO_2 evolution rates are calculated as follows:

$$CO_2 - \text{Carbon (mg)} = (B - X)\,\text{M E}$$

where B is the HCl (mL) needed to titrate the NaOH solution from the controls, X is the HCl (mL) needed to titrate the NaOH solution in the experimental jars, exposed to the soil atmosphere, M = 1.0 (HCl molarity), and E is the equivalent weight (22 for CO_2, and 6 for C). The data are thus expressed as milligrams of CO_2 or CO_2−carbon per square meter per day.

9.3.2 Litter Decomposition Studies

Principle: Weight loss by bagged leaf litter has been a useful method for measuring leaf litter breakdown for more than 50 years (Bocock and Gilbert, 1957). Known masses of litter enclosed in mesh bags or envelopes may be exposed in field sites and then retrieved at later times, for remeasurement of mass. Disappearance or breakdown of bagged litter is a valuable means of comparing substrates such as leaf litter, twigs, or roots. Different habitats or geographic regions may also be compared. Litterbags have been successfully used in hardwood and conifer forests, deserts, agroecosystems, and arctic soil situations. In prairies, in-growth of grass may present a problem. Litterbags have been used to sample a subset of soil microarthropods, in order to discover which species are active in a given stage of decomposition or to follow the development of microarthropod communities. Litterbags retrieved from field experiments are extracted with Tullgren "Berlese" funnels before estimation of mass loss. Fractions of litterbag material may be removed also for nematode extraction. In this experimental context, different mesh sizes may be used to exclude macroarthropods, microarthropods, or microfauna. Breakdown rates are more rapid in bags with larger mesh and slowest in fine mesh bags that admit only microfauna. Litterbags may be treated with insecticides or fungicides to manipulate specific groups of soil biota for studies of their effects on decomposition.

Litterbags consistently underestimate decomposition rates; thus, they are properly used in a comparative context. Such underestimation is most extreme for rapidly decomposing substrates—leaves of *Cornus florida* or (in agricultural systems) leguminous foliage. For more recalcitrant litter types—*Rhododendron* or some *Quercus* species—mass loss from bagged litter more closely approximates that of unconfined leaves.

As an alternative method, a group of leaves secured by a nylon string attached to their petioles, may be used in conjunction with litterbags. Loss of leaf area as well as mass loss may be measured (Hargrove and Crossley, 1988). Since entire fragments may be broken off and lost, this string method overestimates decomposition rates and is viewed as a comparative method. Decomposition of rapidly decaying substrates is characteristically overestimated. Some leaf litter species (such as sweetgum) tend to become detached from their petioles early in litter breakdown.

Another method of measuring leaf area loss was employed by Edwards and Heath (1963), who buried leaf disks of known area in litterbags. When they were retrieved the leaf area loss was estimated using a grid, and soil animals were enumerated.

Litterbag construction: Various types of netting have been used for litterbags. Mosquito netting, nylon drapery mesh, and plastic window screen are suitable materials. Cloth bags must be sewed together. Plastic screen bags may be constructed by using a soldering iron to melt and seal the edges of the bag. Cloth bags are more flexible, conform to the shape of the forest floor and admit more fauna because the mesh openings are more flexible. Window screen bags are more rigid and openings are fixed, but they are much easier to construct. Size of bags has ranged from 0.25 m^2 down to 10 mm; 10 cm × 10 cm (1 dm) is a frequently used dimension. Actual dimensions of 12 cm × 12 cm yield an effective area of about 100 cm^2. A flap on the open end of the bag allows it to be closed with a safety pin.

Leaf litter substrates should be air dried if possible before insertion into the litterbags. We use contrasting litters with different palatabilities, C/N ratios, etc., such as dogwood, chestnut oak, or rhododendron to demonstrate the differences in their decay rates. Decimeter litterbags will contain about 1.5 g of dry deciduous leaf material before breakage becomes a problem. After mass determination, an identifying label may be placed inside the litterbag. (Aluminum tags attached outside the bags tend to attract large mammals.)

Place the bags out in field sites, such as old fields, forest floor, or agricultural fields. On each sample date, randomly select four replicate litterbags from each of several treatments. Upon retrieval, litterbags should be placed in plastic bags and returned to the laboratory as soon as possible. The bag may be opened and a small increment removed for nematode extraction (with suitable estimation of litter mass). The intact litterbag may then be extracted on a Tullgren funnel (see microarthropod extraction section below). Finally, the litter substrate is removed from the bag for mass determination.

- Clean the outside of the litterbag (brush off any soil and litter particles).
- Remove the metal label, and record its number, the collection date, treatment codes, and any other pertinent information on a new label.
- Weigh beaker (with tape label on it) and record on weighing sheet. Above a piece of waxed paper, carefully take all the litter material out of the litter bag and put it into the beaker. If there was any soil transported into the litterbag, make sure you remove the soil before you put the litter in the beaker. Put the beaker in the 60°C Oven and dry the litter for a minimum of 2 days.
- After >2 days, record the dry weight (beaker + litter dry weight) and transfer the litter to labeled paper bag.

9.4 SOIL MICROBE AND PROTOZOAN EXERCISES

9.4.1 Quantifying Mycorrhizal Fungi Colonization of Roots

Principle: Fine roots are observed and the total root length that is colonized by mycorrhizal fungi is calculated. These methods are for the preparation and analysis of roots with the two most common types of mycorrhizal fungi associations: AM (arbuscular mycorrhizae) and ECM (ectomycorrhizae). Note: This exercise could easily have been included in the section on root studies, but we include it here as fungi are appropriately viewed as microbes.

Materials and supplies needed (all roots):

large, shallow dish;
small paint brush;
ethanol;
forceps;
automated counter;
scissors.

Additional materials required for assessing AM roots:

Beakers (100 mL or more) sized to accommodate root quantity
5% KOH solution
Hot plate
2% HCl solution
0.05% trypan blue solution in 2:1:1 lactic acid:water:glycerol
Microscope slides
Coverslips
Compound microscope with gridded eyepiece.

Additional materials required for assessing ECM roots:

Petri dish, marked with grid by marker or tape
Dissecting microscope.

Procedure: Root preparation—Fine roots (≤2 mm) should be gently removed from soil to minimize damage to the root tips. Very gently clean fresh roots to remove soil. To clean roots, place a section of the roots into a large, shallow dish half-filled with water and remove soil with a small paint brush, changing the water frequently. Cleaned roots may be stored in 95% ethanol at 4°C until analysis (Collins et al., 2016).

Procedure (AM roots): Arbuscular mycorrhizal roots need to be cleared of tannins and dyed before analysis. The exact time needed for clearing differs dramatically between roots of different species, ages, and diameters. Therefore, it is necessary to perfect this procedure by experimentation on "extra" roots before clearing the roots used for actual data collection.

To clear roots, start by rinsing them with deionized water. Cut roots into 2 cm fragments. Place roots in a beaker and half-fill it with a 5% potassium hydroxide (KOH) solution. All roots should be submerged. Cover the beaker with foil and heat it to 60°C. Roots are cleared when they appear white. If roots take longer than 1 hour to clear, replace the KOH solution in the beaker and again every hour until cleared. Rinse cleared roots in deionized water and then soak in 2% HCl for 3 minutes. To stain roots, cover them with 0.05% trypan blue solution (trypan blue in 2:1:1 lactic acid:water:glycerol) at 60°C for 15 minutes. Warning: trypan blue is toxic. Use necessary precautions to protect yourself and others. Rinse stained roots with deionized water and place in a destain solution (2:1:1 lactic acid:water:glycerol) overnight. Replace destain solution the next day and roots are ready for analysis (Collins et al., 2016).

To quantify mycorrhizal colonization, use the magnified intersections method described by McGonigle et al. (1990). Keep in mind that this method is adaptable and researchers have used several modifications, only one of which is described here. Methods should be optimized for specific questions. Mount sections of roots onto microscope slides and place coverslip on top. Use a dissecting microscope with a crosshair eyepiece at 40 × magnification to view mounted roots. Fungal tissues will appear blue. Move the microscope field of view to a predetermined number of locations on the slide. At each location, record whether the vertical crosshair line intersects with fungal hyphae, vesicles, and arbuscules (Fig. 9.3). Depending on the question, only one type of AM structure, or all three may be tallied. Keep a tally of desired data and use it to determine the proportion of colonized root length.

Procedure (ECM roots): Ectomycorrhizal fungal colonization is generally visible on root tips, so a lengthy clearing and staining procedure is not necessary. Ectomycorrhizal tissues are usually colored differently than nonmycorrhizal root tips and often in the shape of a "Y" or have a clustered appearance. To quantify ECM colonization, use the gridline intersect method (Brundrett et al., 1996). Mark a petri dish with a vertical and horizontal grid, typically 1 cm × 1 cm. Place sections of root (trimmed to fit on petri dish) in a single layer. Put a little water in the petri dish so that root tips are not stuck to the side or bottom of the dish. Do not overfill the dish with water as the roots will move during counting if they are floating. Use a dissecting microscope to look at each intersection of vertical and horizontal lines on the grid (Fig. 9.4). Record the number of intersections with roots and those with mycorrhizal roots. Dividing the number of intersections with mycorrhizal roots by the number with roots will give the proportion of colonization.

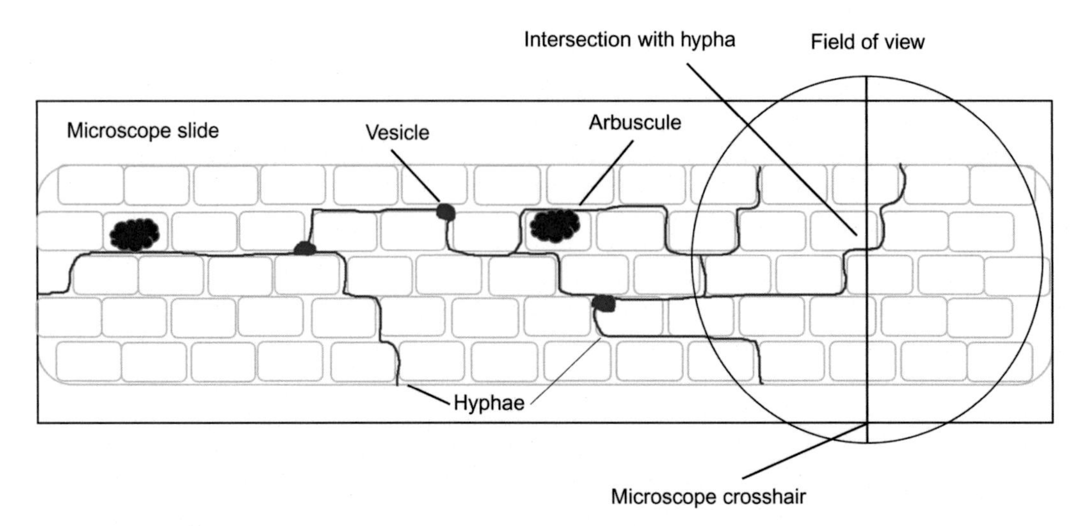

FIGURE 9.3 Schematic drawing of a mounted root fragment, stained for visualizing AM fungal structures inside the root. Source: *Drawing by Melanie Taylor.*

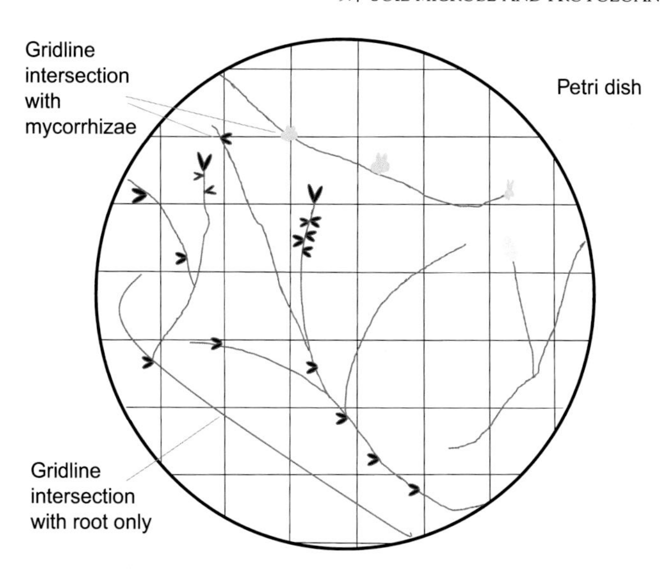

FIGURE 9.4 Schematic drawing of a gridded petri dish with root sections for visualizing and enumerating ECM fungal structures. *Source: Drawing by Melanie Taylor.*

9.4.2 Analyses for Soil Microbial Biomass

9.4.2.1 The Chloroform-Fumigation–K_2SO_4-Extraction Method

Principle: This procedure compares the amount of total organic carbon (TOC) in a chloroform-fumigated soil sample to a nonfumigated soil sample. In the chloroform-fumigated sample, TOC will be higher because the sample contains the cell contents of lysed microbial cells. Hence, the difference in extracted TOC between fumigated and non-fumigated samples will provide a measure of microbial biomass (Vance et al., 1987). Note that you can only assume that this TOC is of microbial origin if the soil samples have been picked free of roots, litter, earthworms, etc. (the microfaunal contribution to TOC is $<5\%$).

Fumigations are carried out for a period of 2 days in vacuum desiccators with distilled (alcohol-free) chloroform. Soil samples are extracted with 0.5 M K_2SO_4 and the filtrate is analyzed by atomic absorption for TOC. Analysis results need to be adjusted to a TOC/g dry soil value. It is important to refrigerate soil samples until the fumigation and K_2SO_4 extractions are performed.

Preparation and handling of potassium sulfate: Make 0.5 M K_2SO_4 by adding 87.13 g K_2SO_4 to a 1000 mL volumetric flask and bring to about two-thirds volume with deionized water. Place on a stir plate until K_2SO_4 is in solution and bring up to volume with deionized water. K_2SO_4 powder has a health rating of 1 (may cause irritation). Goggles and gloves should be used when handling this chemical.

Sample preparation:

- Remove all visible roots and leaf litter from the soil sample with a forceps. Roots may be dried and weighed to get a measure of root density of the soil sample.
- Mix the soil sample thoroughly to make it "homogeneous." Label two 125 mL Erlenmeyer flasks (one NF (nonfumigated) and one F (fumigated); include, of course, the sample site and number on the label) Note: Labels must be in pencil on special white tape, because marker ink and often glue dissolve in chloroform fumes.

- Weigh ≈25 g soil (fresh weight; or 20 g dry weight) into each flask (water content should be in the 20%–30% range for best fumigation results). Record weights for later correction to dry weight.
- Include at least four blanks (two F, two NF) (empty Erlenmeyer flasks).

Potassium sulfate extractions:

- Use the 50 mL dispenser to add 50 mL of 0.5 M K_2SO_4 to each NF flask. Cover flasks with parafilm and place them in a rotary shaker for 30 minutes (≈200 rpm).
- While samples are shaking, prepare your funnel and filter paper setup. Wearing powder-free gloves, fold Whatman 42 filter paper (15 cm diameter), so that it forms a cone, and place it into a plastic funnel. Fill each funnel with deionized water to rinse out any traces of carbon lingering in the paper.
- After 30 minutes, take flasks out of the shaker. For each flask: transfer white tape label to a 60 mL plastic bottle; place vial under funnel; gently shake extract to suspend the soil and pour it immediately onto filter paper. Do not try to get all the soil onto the filter. Store the extracts in a freezer until analysis on the TOC analyzer.

Chloroform fumigation:

- Clean the vacuum desiccators and cover the bottom with water (≈1 cm). Place the "F"-flasks inside the desiccator. As many flasks as will fit without spilling the soil or blocking the opening of another flask may be placed in each desiccator (about 13 flasks). Include one blank (empty Erlenmeyer flask) in each desiccator.
- Place a 100 mL beaker in each desiccator. Add a few glass beads and 40 mL of reagent grade purified chloroform. (Note: If your chloroform has ethanol added as a stabilizer, this ethanol must be removed in a distillation step.) Use gloves and work in the fume hood at all times when working with chloroform. Close the desiccator with the lid (use some Lubriseal or Vaseline for good sealing).
- Start the water aspirator, check the vacuum suction, then attach the hose to the desiccator. This will create a vacuum in the desiccator, causing the chloroform to boil and evaporate, and saturate the desiccator atmosphere. Allow the chloroform to boil vigorously for 5 minutes. Then FIRST close the vacuum route by sealing the desiccator (by turning the top), and then stop the water aspirator. The vacuum will suck water into the desiccator if you turn off the water before sealing the desiccator. Reapply the vacuum three times over the next hour. The chloroform may not boil after the first time. Allow the fumigation to proceed for 2 days.

Chloroform removal:

- After 2 days, release the pressure in the vacuum desiccators. Remove the beaker of chloroform, and store it in the back of the hood to let the remaining chloroform evaporate.
- As it is important that all residual chloroform is removed from the soil samples before proceeding with K_2SO_4 extractions, reapply the vacuum to the desiccator. However, do not seal it: simply detach the vacuum hose (while the tap is running) and allow the air to get back in and circulate. Repeat this procedure a few times to clear the chloroform from the flasks. Then, place the flasks near the opening of the fume hood, lower the

window to increase the velocity of air flowing over the flasks—the removal of chloroform will take about an hour.

- Do the potassium sulfate extractions (see earlier) for the F-flasks.

Microbial biomass carbon is calculated as: (TOC [F] − TOC [NF])/K_C, where $K_C = 0.38$ (Vance et al., 1987).

Notes:

1. Sample blanks should be run with all preparations and the results should be used to correct values above.
2. ALL glassware (volumetric flasks, Erlenmeyer flasks) and plasticware (Nalgene sample bottles) must be acid-washed (10% HCl) and dried before and after use.
3. The K_2SO_4 soil extracts can be used for analyses other than TOC, e.g., inorganic nitrogen (NO_3 and NH_4) (nonfumigated soil) on any continuous-flow spectrophotometric autoanalyzer.
   - Microbial N (after persulfate oxidation of both F and NF extracts); extractable N is then measured as NO_3−N on an autoanalyzer; microbial N is then calculated as (N [F] − N [NF])/K_N, where $K_N = 0.45$ (see Cabrera and Beare, 1993; Jenkinson, 1988).
   - For an overview of methods to calculate soil microbial biomass C × N × P and S, see Voroney et al. (2007).

9.4.3 Field Exercise for Soil Protozoan Activity and Biodiversity

Objective: To measure major protozoan groups in soils under field conditions.

Rationale: Soil protists are among the most numerous and active organisms among the Eukaryotes. They are occasionally overlooked by soil zoologists as they are more difficult to measure and study than their multicellular brethren. For details of the biology and ecology of these organisms, see Chapter 4, Secondary Production: Activities of Heterotrophic Organisms—The Soil Fauna. The reasons for measuring the dynamics of these fascinating unicellular Eukaryotes are that they are among the primary predators of bacteria and fungi. This protocol draws heavily upon that presented in Adl and Coleman (2005).

Materials and methods: Dissecting and compound microscopes (preferably inverted type) are necessary, as are sterile Whirl-Pak or similar bags and 15 mL. Falcon tubes. A Neubauer grid in a hemocytometer is also recommended.

Procedures: Take three core samples, 2 cm in diameter to 15 cm depth. These can be repeated across habitat types, e.g., after crop harvest in winter, and again during growth of a winter cover crop, such as crimson or other species of clover. Seal soil samples in Falcon tubes and transport them fresh to the laboratory. Samples should be processed soon after arrival in the laboratory.

Take known weights of soils (0.15 g for flagellates, and 3 g for ciliates). Each core should be subsampled five times into separate 3 mL suspensions, using 10 mL sterile deionized or distilled water in each tube. For small and large flagellates, use a hemocytometer to view them with phase contrast if possible, using Plan Apochromat objectives, at 200× and 400 × magnifications.

For ciliates, drain a soil suspension three times, and view the resulting "filtrate" in a sterile Petri plate, and scan it for ciliates and large flagellates. It is helpful to mark out grid lines on the bottom of the petri plates, to more easily count known areas and volumes of material.

For amoebae, abundances can be calculated from the suspension dilution factor. Naked amoebae abundances can be determined from soil suspension droplets placed on 2–3 mm thick plain (1.5%) agar. From each subsample, place up to ten 50 μL soil suspension droplets in a row on the agar. Incubate these in the dark at the temperature of the soil when it was sampled. Amoebae can be measured after 24 hours as they migrate to the edge of the droplet after incubation. Abundances can be determined by counting under phase contrast at 400 × magnification.

For truly motivated researchers, it is possible to measure bacteria in the soil suspensions, using a Neubauer grid of a hemocytometer for phase contrast optics. For ease of viewing, stain the preparations with 4′,6-diamino-2-phenylindole (DAPI, Sigma-Aldrich, United States) and count them under fluorescence at 1000 × magnification using a neofluor objective.

By sampling at wet and dry periods, it is possible to estimate the rates at which the protists become active from the encysted forms, and rates of reproduction by the active cells. For more details on this procedure, see Adl et al. (2008).

9.5 SOIL FAUNA EXERCISES

9.5.1 Sampling and Enumeration of Nematodes

Principle: Using a procedure to determine the numbers of free-living nematodes in litter and soil samples, we will be able to estimate the abundances of these important mesofaunal organisms in terrestrial ecosystems. When designing a sampling plan for nematodes, there are practical and theoretical considerations to be made, which are balanced with the amount of time and money available for the study.

1. *Goal of the study and required accuracy*: The goal of the study will greatly determine the degree of accuracy required. For qualitative studies such as diagnosis of a plant disease or taxonomic work, the requirement is relatively low. In an ecological study of nematode communities more accuracy is needed, as rare species should be included in the sampling. Even more effort is needed in sampling for disease control, e.g., to certify that a certain field is disease-free. In general, a higher accuracy costs more time and money.
2. *Variation in space (horizontal and vertical distribution*: One of the biggest problems in sampling is that nematodes have a patchy distribution, i.e., they are not randomly distributed in the soil. This pattern may have a biological basis (concentration around roots, or in islands of organic debris), an agronomic basis (cropping history; planting diseased material creating "hotspots" of disease), or a physical basis (e.g., texture gradients). If any of these factors are known, variability can be decreased by sampling and analyzing such areas separately ("*stratified* sampling"). In terms of vertical variation, most nematodes (and other soil organisms) can be found in the topsoil

(0–10cm). In agricultural fields, sample depth is often as deep as the plow layer (typically 15–18 cm), or deeper, e.g., when sampling tree roots. Again, the objectives of the study will dictate how deep the sampling will need to go.

3. *Variation in time*: Seasonal cycles and life cycles will influence the results of the sampling. Seasonal fluctuations in moisture, temperature, and food availability will influence abundances of different species in different ways. For sampling plant parasitic nematodes, knowledge of the life cycle is most relevant. For population studies of an endoparasite, sampling soil when the crop is in the field does not make sense, as most nematodes will be inside the roots.
4. *Statistical considerations*: By sampling we want to *estimate*, as well as we can, nematode abundance in a site. The "sampling error" (expressed as variance or standard deviation) is the sum of *systematic* and *random* errors, introduced at any stage in the sampling process (taking cores, mixing soil, extracting soil, counting subsamples, etc.). Systematic errors can be reduced by improving methods and working as carefully as possible. Random errors can be reduced by taking more and bigger samples, as much as resources allow. A general guide is that variability *within* a sample (or plot) is usually smaller than *between* samples (or plots). Thus, it makes more sense to take many samples than to analyze many *sub*samples (likewise, it is more efficient to sample many different plots than to take many samples within the same plot). Random sampling is best but not always most practical. Systematic sampling (e.g., taking samples along a transect, every few meters from a randomly chosen starting point) is more often used (Ettema and Yeates, 2003).

Sampling tools and precautions: Soil (and roots) can be sampled with any simple container such as a food can, a shovel, or special corers. When sampling to advise farmers on the presence/absence of plant parasitic nematodes, make sure the sampling tools are cleaned and disinfected after every soil sample. Always store the soil samples in plastic bags (in paper bags the soil will dry out), and keep them away from the sun (heat kills nematodes!). Handle the samples carefully: throwing them around, or any other mechanical disturbance (soil sieving, mixing) will kill nematodes as well. The sooner the samples are extracted, the better. For reliable results, the maximum storage time, at 4°C, is 1 week only.

9.5.1.1 Nematode Extraction: Baermann Funnel Method

Principle: Nematodes move out of the substrate toward the water and sink to the bottom of the funnel stem due to gravity. The method is suitable for extraction of mobile nematode from substrates like soils and sediments, plant material, and litter (if plant or litter material is cut into small pieces prior to extraction). Advantages of the method include the fact that it is cheap and simple; extraction efficiency is quite good if your sample is small relative to the funnel diameter (with only a thin layer of substrate on the sieve). One disadvantage of this method is that efficiency decreases very rapidly when the sample gets bigger, and a greater depth of soil is placed into the funnel. For alternatives, see Southey (1986).

Equipment: Funnel with a piece of rubber tubing attached to the stem, closed with a pinch clamp; funnel rack, Kimwipes, clean tap water (not deionized water), centrifuge tube for sample collection; small screen to support soil in funnel (e.g., aluminum window screen (avoid copper)).

Procedure:

1. Put the funnel in the rack. Make sure it is level. Fill it with clean water. Remove air bubbles by squeezing the rubber tube, or by draining water by opening the clamp.
2. *Carefully* mix the soil sample (excessive mixing will kill nematodes!). Take a 10 g subsample, spread it out on a double layer of Kimwipe tissue. Record sample weight. Place the sample on the screen in the funnel. Do this carefully, because you don't want small soil particles to get through the tissues and obscure your nematode sample later on.
3. The soil should be moist enough, but not totally submerged. Adjust the water level in the funnel using a spray bottle (don't spray *on* the sample, for same reasons as discussed under #2), or drain by opening the clamp. Cover the funnel with a Petri dish or wax paper to avoid contamination via dust or changes in the water level due to evaporation.
4. The extraction time is 2–4 days. The longer, the more nematodes you harvest, but the question arises whether this increase is because of catching slow moving nematodes, or due to hatchlings from eggs in the soil or larvae from fast reproducing species—some have a life cycle of 2 days (especially at laboratory temperature!). For that reason 48 hours is an optimal time. Check after the first 24 hours if the soil is still moist enough.
5. Harvest: By opening the clamp, get approximately 10 mL in a centrifuge tube (e.g., plastic Corning, 15 mL with screw cap). Store immediately under refrigeration, for maximum of 1 week. To kill and fix (5% formaldehyde) the nematodes, see Identification of nematode feeding groups article (Yeates et al., 1993).

9.5.1.2 Killing and Fixing Nematodes With Hot and Cold Formalin (5%) (Precautionary Note: Conduct This Work in a Fume Hood)

Using hot formalin, the nematodes are instantly killed and fixed. Immediately cooling down with cold formalin prevents overheating that could cause deformations. With this hot–cold formalin method, nematodes will be fixed in body curvatures characteristic for taxonomic groups, which makes routine identification easier. Another advantage is that most species keep their body transparency. This is in contrast to the method of fixing nematodes with cold formalin only, where nematodes die slowly instead of instantly, which causes deformations and loss of transparency.

Supplies needed:

1. Water aspirator (vacuum pump) with fine pipette tip attached to the hose
2. Nematode suspension in plastic 15 mL centrifuge tube width
3. Formalin 5% (10.8 mL formaldehyde 37%, in 89.2 mL deionized water), in a squeeze bottle.

Procedure:

1. Let the nematodes in the centrifuge tube settle to the bottom for at least 24 hours. Do this close to the place where your water aspirator (vacuum pump) is, so you don't need to move the sample a lot. Disturb as little as possible.

2. *Carefully* suck away the "supernatant" water till 2 mL is left (=the amount in the conical tip of the tube).
3. Add (approximately) 4 mL of 90°C dilute formalin and *immediately* cool down with (approximately) 4 mL cold formalin from the squeeze bottle.
4. Make sure the cap is tightened so formalin can't evaporate.

For a more extensive review of methods to extract soil nematodes, see Forge and Kimpinski (2008).

9.5.2 Sampling and Enumeration of Microarthropods

Principle: Using heat and dryness, microarthropods are "induced" to move out of litter and soil samples, into a collecting fluid, for enumeration and identification. With flotation, differential buoyancy of the microarthropods is used to advantage with a flotation medium.

9.5.2.1 Methods for the Study of Microarthropods

Sampling: Mites and collembolans are collected by extracting them from a sample of their soil habitat. Extraction procedures may be either Berlese funnel extraction of soil cores or litterbags, or alternatively, flotation (Bater, 1996). Total microarthropod populations are then estimated by extrapolating from the size of sample (weight or area) to field dimensions. For sampling protocols and considerations of sampling design, consult Hall (1996), Schinner et al. (1996), Coleman et al. (1999), and Larink (1997).

Soils vary greatly in structure, composition, pore size, moisture regime, and so forth; sampling methods need to be suited to the ecosystem under investigation. Most quantitative samples of microarthropods are taken from soil cores, 5–10 cm diameter by 5–10 cm deep. The smaller cores yield satisfactory results; a 5 cm × 5 cm core will contain several hundred mites and collembolans. A split core tool with a sharp beveled tip, designed to hold a sleeve for the soil sample, is preferable for most soils (Fig. 9.5). For many soils the great majority of microarthropods are found within 5 cm of the surface. In grassland soils and disturbed soils they may be distributed more deeply, and additional 5 cm increments may need to be extracted, to a depth of 15 cm or more. Sample cores should be extracted in a high-efficiency

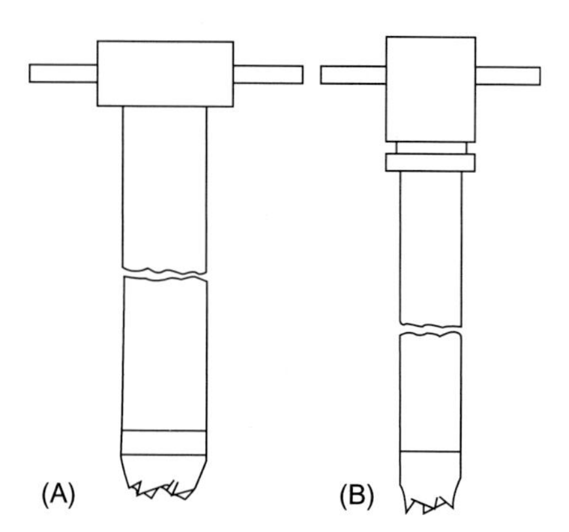

FIGURE 9.5 Soil coring device: (A) outer tube and (B) inner sleeve, typically made of aluminum. Effective core diameter: 5 cm. Source: *Modified from Gorny and Grüm (1993).*

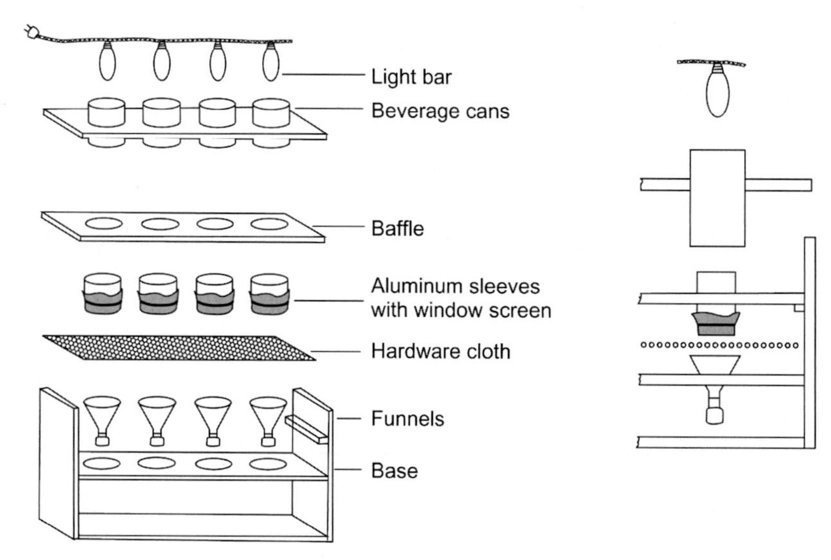

FIGURE 9.6 Design and assembly of the high-efficiency microarthropod extractor. Source: *From Crossley, D.A. Jr., Blair, J.M. 1991. A high efficiency, "low technology" Tullgren-type extractor for soil microarthropods. Agric. Ecosyst. Environ. 34, 182–187.*

extractor (Fig. 9.6) as soon as possible; storage for any significant period of time will result in lower numbers of microarthropods extracted.

Extraction of microarthropods from samples: There are many modifications of the basic Berlese funnel (Edwards, 1991) and most will yield satisfactory results. Heat is used to desiccate the sample, driving arthropods out and down into a collection fluid. Many designs are called Tullgren funnels, after the originator of the use of electric lights as heat sources. (the original Berlese funnels used steam as a heat source). Larger funnels (Fig. 9.7), used for extracting big samples of litter, can work effectively with small cores as well. Arrays of smaller funnels can handle more samples in a smaller space, and have become the most widely used piece of extraction equipment (Bater, 1996). Soil cores contained in their sleeves are extracted in an inverted position, surface layer down, so that arthropods can escape using natural channels in the soil. The upper (bottom) of the core should be moistened with water to improve extraction efficiency (T.R. Seastedt, personal communication). Seventy-percent ethyl alcohol is the usual collection fluid for the extracted arthropods. A 10% picric acid solution is preferred by some authors (e.g., Meyer, 1996). Care must be taken to keep mineral soil from falling into the sample with extracted microarthropods, as samples contaminated with soil are more difficult and time consuming to sort. To this end, a single layer of cheesecloth may be inserted between sample and funnel.

Litterbags (see later) are an alternative method to soil cores for sampling microarthropods. They offer several advantages: Microarthropods using different substrates, such as different litter species, may be detected (Hansen, 2000). Different stages of litter decomposition may be compared, and time sequences described. Litterbags are readily extracted, intact, on large Tullgren funnels.

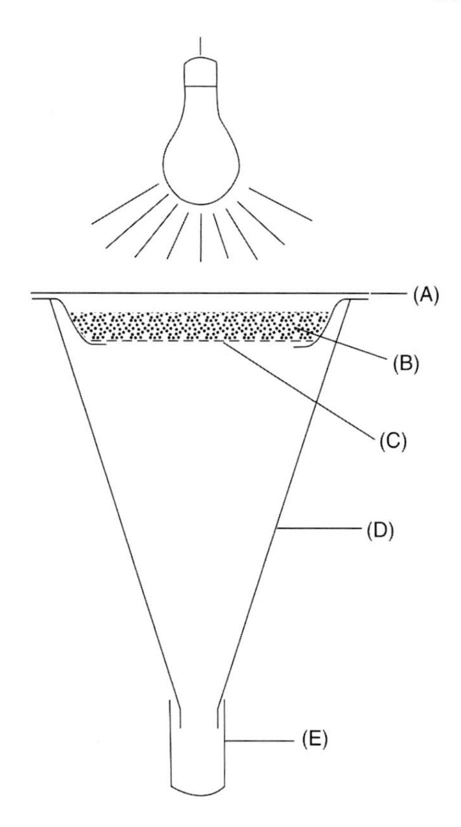

FIGURE 9.7 Schematic diagram of an extractor for soil macrofauna: (A) sample cover, (B) soil sample, (C) sample screen, (D) aluminum funnel, and (E) collection container with 70% alcohol or similar collection fluid. Source: *From Gorny and Grüm (1993).*

Flotation is another alternative method for extracting microarthropods from soil samples. Using organic solvents or saturated sugar solutions, arthropods may be separated from the soils, washed through a fine mesh screen and thus recovered (Fig. 4.10). In comparison with Tullgren extraction, flotation usually yields higher numbers of microarthropods. Some collembolans, such as members of the family Onychiuridae, respond poorly to Tullgren extraction; flotation is the method of choice for sampling these arthropods. The disadvantages of flotation are (1) the method is extremely laborious in comparison to Tullgren extraction and (2) it is not effective for samples with large amounts of organic matter. Finally, the use of organic solvents probably violates most laboratory health and safety considerations (Griffiths, 1996).

A more attractive procedure uses saturated sugar solutions instead of organic solvents (Snider and Snider, 1997). The method (Table 9.1) yielded much higher captures of microarthropods than did Tullgren extraction alone, in samples from two hardwood forest sites in upper Michigan. The method was extremely laborious; the authors report that 10–12 hours were required to sort a single sample. They note that financial constraints usually preclude labor-intensive procedures such as flotation, even though much more accurate population estimates were obtained.

Sample sorting and identification: A good dissecting microscope with magnification in the range of 10–40 × is essential. A preliminary sorting will separate collembola, mites, and

TABLE 9.1 Procedure for Flotation Extraction of Microarthropods From Soil Samples, Using Saturated Sugar Solutions

Step one	Place the soil core in a plastic bag; crumble it gently.
Step two	Transfer the crumbled core into a 1 L wide mouth jar. Wet it with distilled water.
Step three	Add saturated sugar solution, leaving approximately 3 cm of headspace.
Step four	Cover the jar with a lid, shake it gently, and let it stand for 2 h to allow organic matter to float to the surface.
Step five	Decant the solution through a number 200-mesh sleeve into a large bowl to trap the organic matter (do not include silt from the bottom of the jar).
Step six	Rinse the organic matter with distilled water, then wash it from the sieve into a sample jar containing 95% ethyl alcohol.
Step seven	Return the sugar solution to the bowl; return to step four.
Step eight	Repeat steps four to seven for three iterations, then combine all organic matter into one jar.

After Snider, R.M., Snider, R.J., 1997. Efficiency of arthropod extraction from soil cores. Entomol. News 108 (3), 203–208.

other microarthropods. The latter category includes the few tiny spiders, small beetles, and other insects (adults and larvae), which can usually be identified with the dissecting scope. Some may require slide mounts (see later). Collembolan and mite specimens can be transferred to sorting dishes with a fine-tipped pipette (such as a Pasteur pipette), a camel's hair brush trimmed to three to four lashes or a flattened, curved dissecting needle.

Collembolans: Identification of springtails almost always requires high magnification (400 × or greater) of cleared specimens, using a good phase contrast microscope. Christiansen (1990) recommends clearing specimens and observing them in temporary mounts; heavily pigmented forms may require more clearing than the mounting medium provides. He suggests the following reagents:

1. Potassium hydroxide, a 5% solution, for brief periods only.
2. Lactic acid, an excellent clearing agent; collembolans tolerate long exposure to it. May be mixed with an equal portion of glycerine.
3. André's fluid (40 cc chloral hydrate, 30 cc glacial acetic acid, 30 cc distilled water). Clears rapidly but may cause damage to specimens.
4. Bleaches. A 5.35% solution of sodium hypochlorite will clear heavily pigmented specimens, but is destructive to cuticles.

Christiansen (1990) further recommends the use of depression slides for study of cleared specimens, since weight of the cover glass may crush the arthropod. Slide-mounted specimens are convenient for reexamination and for reference specimens (see later for preparation of temporary and permanent mounts of microarthropods).

Mites: Preliminary sorting of mites into subgroups (Prostigmata, Mesostigmata, adult Oribatei, immature Oribatei, and Astigmatina) can be accomplished successfully with experience. Even so, slide mounts will be necessary for confirmation of these identifications.

In the sorting dish, the mites are separated into morphospecies and representatives are mounted on microscope slides, depending upon the group. Mesostigmata, Astigmatina, and Prostigmata are usually mounted in Hoyer's medium (see later), with preliminary clearing for large or heavily pigmented specimens. Oribatids require special consideration because their heavily pigmented and brittle exoskeleton is easily crushed by a coverslip.

Clearing agents for mites are similar to those used for collembolans. A popular one is lactophenol (Krantz and Walter, 2009):

Lactic acid	50 parts
Phenol crystals	25 parts
Distilled water	25 parts

Specimens may be left in lactophenol at room temperature for several days, or heated for more rapid action. Larger specimens may need to be punctured. After soaking, large mites such as trombidiids may be pressed with a flattened dissecting needle before mounting them. André's fluid (see earlier) is also recommended for mites as well as collembolans. Nesbitt's fluid (40 g chloral hydrate, 25 mL distilled water, and 2.5 mL concentrated acetic acid) is useful for specimens, which do not respond to milder clearing agents. Oribatids stored in lactic acid for a few weeks are usually satisfactorily cleared for study.

Most permanent or semipermanent mounting media used for mites (and collembolans as well) are aqueous, in that they contain, or are soluble in, water (Krantz and Walter, 2009). Gum arabic and chloral hydrate are the principal ingredients. Hoyer's medium is one of the most popular:

Distilled water	50 mL
Gum arabic	30 g
Chloral hydrate	200 g
Glycerine	20 mL

Clear crystals of gum arabic are preferred. Powdered gum arabic is difficult to wet but may be dissolved in alcohol, which is then allowed to evaporate (R.A. Norton, personal communication). If slide-mounted specimens are given gentle heat (40°C) for a few days, considerable clearing of the specimen will take place. Slide mounts using Hoyer's medium, if ringed, will last for some years but eventually deteriorate. Canada Balsam is not satisfactory for mites or collembolans because the refractive index of the medium is so similar to that of the cuticle (Christiansen, 1990). Permount is suitable but requires that specimens be dehydrated and mounted from xylene (Adl, 2003). Generally speaking, mite and collembolan specimens should be archived in 70% alcohol, although "constant vigilance" is necessary to guard against evaporation of the preservative (Christiansen, 1990).

As noted earlier, many specimens of oribatid mites cannot be mounted on slides in the usual manner, without crushing them and thus obscuring their features. Following clearing, specimens may be examined in depression slides, partially covered with a coverslip, in lactic acid or glycerin and manipulated with a fine needle. R.A. Norton (personal communication) recommends slide mounts using a procedure with a small cavity drilled into a microscope slide. A small drop of fluid is placed next to the tiny hole and allowed to flow into it. The mite is positioned and allowed to partially dry; then mounting medium and coverslip may be added. Norton further recommends a 50–50 mixture of Hoyer's medium and Nesbitt's fluid for preliminary clearing of oribatids.

9.5.3 Sampling and Enumeration of Macroarthropods

Principle: See comments about sampling in Chapter 4, Secondary Production: Activities of Heterotrophic Organisms—The Soil Fauna (Section 4.4, The Macrofauna).

Sampling: Macroarthropods are a diverse group, with representatives in several classes and orders of the Arthropoda—consequently, a variety of techniques must be employed to effectively sample these organisms which may occupy different parts of the soil. Most are visible to the unaided eye and hand collecting and sorting is a reasonable procedure for some of them. Many, such as the cryptozoa, are crepuscular or nocturnal, and even hand collecting may require special procedures such as flashlights, black lights, or baits.

Berlese or Tullgren extraction: Large Tullgren funnels will extract macroarthropods from samples of forest floor, but sample size is necessarily limited to 0.1–2.5 m^2 collections. Larger samples will be necessary for those taxa with densities less than 1–2 per m^2, and careful examination and hand sorting may be the preferable means of processing them. Forest floor materials can be removed from a measured area and then crumbled over a coarse screen atop a white enameled pan. Arthropods are captured as they are released from the sample. An alternate procedure involves crumbling the sample gently into a container of water, and collecting arthropods which float to the surface (Bater, 1996).

Flotation: In most terrestrial ecosystems macroarthropods are inhabitants of the mineral soil layers, and sampling them requires the use of coring tools. Soil cores of 5–25 cm diameter and 10–25 cm deep will recover the majority of macroarthropods in most systems. These monoliths may be crumbled and hand sorted, or washed through a set of sieves with running water (Edwards, 1991). The flotation procedure of Behre (1987) involves washing the soil through such a set of sieves of decreasing mesh sizes, with collecting bowls arranged in steps and connected to an overflow. The magnesium sulfate solution used for washing is thus collected and reused.

Emergence traps: Transient soil inhabitants such as cicadas and June beetles (Scarabaeidae) emerge from soil as adults and may be trapped to estimate their density and biomass. Emergence traps made from screen wire and covering a known area (Fig. 9.8) can be fitted against the soil, to collect adult macroarthropods when they appear (Callaham et al., 2003).

Pitfall trapping: Pitfall traps are a useful, inexpensive, and rapid method for assessing communities of macroarthropods. Pitfall traps have limited usefulness for assessing population sizes (Coleman et al., 1999), since catches reflect both density and mobility of arthropods. Still, pitfall traps are a valuable method for comparing habitats, assessing seasonal shifts in macroarthropod communities, and evaluating species richness. Species-area

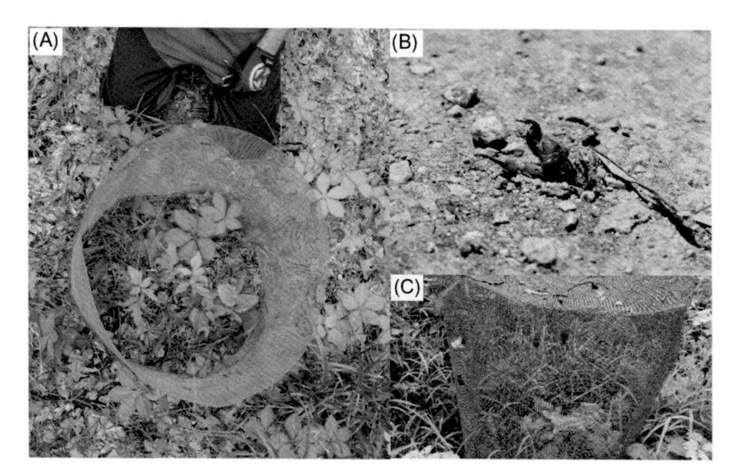

FIGURE 9.8 Emergence trap for collecting macroarthropods. (A) Trap cylinder under construction; (B) an emerging cicada; and (C) periodical cicadas inside an emergence trap. Source: *Photos (A) and (C) by Bruce Snyder; photo (B) by Roberto Carrera-Martínez.*

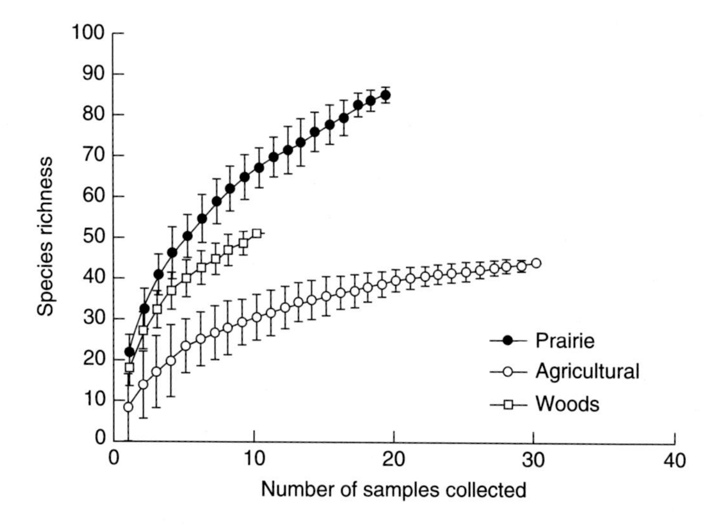

FIGURE 9.9 Species accumulation curves for ground beetles in tallgrass prairie, agricultural, and woodland habitats in northeastern Iowa based on annual pitfall trap samples collected over 5 years from each site and habitat. Source: *From Larsen, K.J., Work, T.T., Purrington, F.F., 2003. Habitat use patterns by ground beetles (Coleoptera: Carabidae) of northeastern Iowa. Pedobiologia 47 (3), 288–299.*

curves constructed from a series of trappings (Fig. 9.9) revealed the difference in species richness of a ground beetle community in Iowa (Larsen et al., 2003).

Pitfall traps consist of cans or jars set flush with the soil surface. These traps may be as simple as a plastic drinking cup with the lip flush to the soil surface, fitted with a funnel and vial of preservative. Other versions of the pitfall trap may be more elaborate (Fig. 9.10). Arthropods blundering into the traps are directed by a funnel into a vial with preservative. Alcohol and propylene glycol have been used as preservatives (Larsen et al., 2003). Propylene glycol is not subject to evaporative loss, but is poisonous to vertebrates and thus not recommended. If specimens are to be used for chemical analyses, a dry killing agent such as naphthalene or para-dichlorobenzene may be substituted. Pitfall traps should be emptied daily. Heavy rain may ruin the samples. Raised covers may be used to protect the pitfall traps and offer some protection from flooding by rainfall.

FIGURE 9.10 Example of a pitfall trap. Left: side view of the entire assembly; Top right: lip of the trap is set flush with the soil surface; Bottom right: a cover prevents flooding in the event of rain. Source: *Photos by Mac Callaham.*

FIGURE 9.11 Example of litter sampling in a forest setting. Left: a quadrat or frame of known size is placed on the soil surface and the sample is delineated by cutting along the sides with a blade; Right: the material is placed on a sheet of plastic for sorting. Source: *Photos by Evelyn Wenk.*

Hand sorting and pit digging: A combination of hand sorting known areas of surface litter and digging a pit of specific dimensions in mineral soil can be an effective and quantitative strategy for evaluating macroarthropod communities in many systems. A quadrat with the desired dimensions (typically 50 cm × 50 cm) is placed on the litter surface and a blade is used to cut around the edges, and material surrounding the sample area is moved aside (Fig. 9.11). The material remaining under the sample quadrat is then transferred to a sorting sheet (plastic or tarpaulin material), and carefully sorted for all macroarthropods encountered. As a practical matter, it is best to focus on collecting specimens that exceed 5 mm in length, as these are the only size class that can be expected to be efficiently sampled in a quantitative way using this technique.

9.5.4 Sampling and Enumeration of Enchytraeids

Principle: Because of their much smaller size than earthworms, enchytraeids are not effectively sampled by hand sorting from soil. Instead, behavioral methods, such as heat extraction are often employed. van Vliet (2000) reviews methods for sampling and extracting enchytraeids.

9.5.4.1 Collection of Enchytraeids

Quantitative sampling of enchytraeids is often done with cylindrical core samplers which keep the soil intact. Optimum sampler size is 5–7.5 cm in diameter (van Vliet, 2000). Because enchytraeids show clumped distributions, sufficient replicates need to be taken to estimate population density and composition. However, the size and number of the sampling units is mostly chosen as a compromise between accuracy of the abundance estimates and the amount of work involved (Didden, 1993).

The most commonly used method to extract enchytraeids from soil is the wet funnel method (similar to the Baerman funnel used for nematodes), using an extractor such as that shown in Fig. 9.12 (O'Connor, 1955). In this method, a thin soil sample is placed on a sieve in a funnel filled with water and exposed to light and heat. After about 3 hours, the

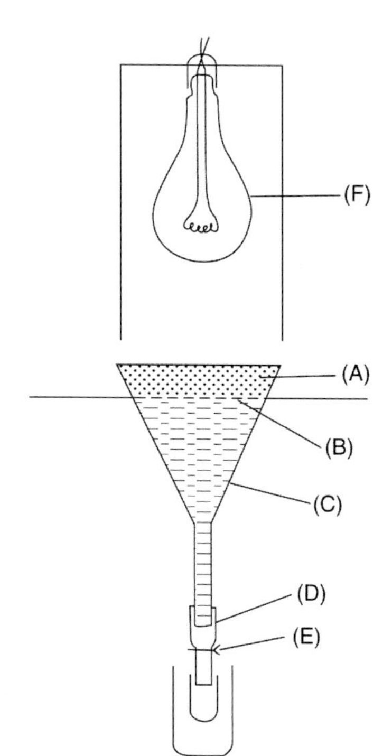

FIGURE 9.12 Modified Baermann funnel to extract enchytraeids: (A) soil sample, (B) wire-gauze sieve, (C) funnel, (D) rubber tube, (E) funnel outlet with spring clip, and (F) 25 W light bulb. Source: *From Górny and Grüm (1993).*

light intensity is increased gradually until the soil surface has reached a temperature of 45°C. Enchytraeids respond by moving downward away from the heat and pass through the sieve into the water below. A modified extraction method was described by Graefe (1973) and Schauermann (1983) in which enchytraeids are extracted from soil without heat. Extraction time is extended to several days for soils rich in organic matter and up to 2 weeks for mineral soils. The time of extraction is limited by the possibility of an oxygen deficit in the water, which may kill larger organisms.

Didden et al. (1995) compared the two methods and found more efficient extraction without heat. The length of the cold extraction period and the total extraction time had a significant positive influence on the extraction efficiency. The Graefe (1973) method is less expensive and easier to set up, but the long extraction time makes it more difficult to handle large numbers of samples. The O'Connor method is faster and can be modified with a longer initialization period before heat is applied.

9.5.4.2 Sampling and Enumeration of Earthworms

Principle: Earthworms may be sampled in a variety of ways, depending on behavioral traits and habitat preferences. Different groups of earthworms inhabit different portions of a soil volume and may be sampled accordingly. For example, some epigeic species (e.g., *Bimastos* spp., *Amynthas* spp.) may be collected by hand in the litter layer or in and under logs. Deep-burrowing, anecic species, such as *Lumbricus terrestris*, may be best sampled with chemical irritants that cause the worms to emerge to the soil surface where they may be hand collected. Many endogeic and epigeic species can be quantitatively collected by digging and hand sorting a known volume of soil. In general, the appropriate methods depend on the purpose of sampling (e.g., quantitative sampling vs qualitative biodiversity surveys). Collection and enumeration methods are reviewed in detail by Lee (1985), Edwards and Bohlen (1996), and Hendrix (2000).

Earthworm collection: Collection techniques can be classified as passive, behavioral, and indirect (Table 9.2).

Passive techniques: Hand digging and sorting, which is the most commonly used method for quantitative sampling of earthworms, involves digging pits of known volume (Fig. 9.13), breaking the soil apart by hand, and collecting all earthworms and cocoons found. Collected specimens are immediately preserved in 70% ethanol or 5% formalin for later counting and identification, or they may be kept alive in cool, moist media for use in experiments. Washing and sieving is an elaboration of hand sorting, in which the soil is dispersed in water, poured through a sieve, and the earthworms and cocoons hand-picked from the sieve contents. Bouché and Beugnot (1972) describe mechanical approaches to washing and sieving. Flotation of sieve contents in a high-density solution, such as 1.16–1.20 specific gravity $MgSO_4$, is another means of separating earthworms and other soil fauna.

Behavioral techniques: Several approaches have been taken to extracting earthworms from soil based on their behavioral response to certain stimuli. A number of chemical irritants have been used, including $HgCl_2$, $KMnO_4$, mustard, and formalin. Aqueous solutions of 0.165–0.550% formalin have commonly been used and are effective on *L. terrestris* when applied in three sequential doses of 18 L m^{-2}; but formalin may be less effective on other species (Satchell, 1969; Callaham and Hendrix, 1997), and can have long-lasting

TABLE 9.2 Descriptions of Methods for Collecting Earthworms

Method	Description	Advantages	Disadvantages
PASSIVE			
Hand sorting	Known volume of soil cut with spade or corer, broken apart, and worms removed by hand	Simple, reliable in the field; low cost	Laborious; may not collect deep-burrowing species, small earthworms, and cocoons
Washing and sieving	Known volume of soil cut with spade or corer, soaked in dispersant/preservative, and washed through sieve(s) by hand or mechanical device	Higher recovery of cocoons and small individuals	Laborious; may not collect deep-burrowing species
Flotation	Material from hand sorting or washing/sieving floated in high-density solution (e.g., $MgSO_4$)	Separates earthworms from soil and plant debris; cocoons and small individuals collected	Laborious; may not collect deep-burrowing species
BEHAVIORAL			
Chemical extraction	Soil saturated with chemical irritant (e.g., 0.2% formalin) causing earthworms to emerge onto soil surface	Simple; effective on deep-burrowing anecic species	Not effective on all species, in all soils or under all conditions
Heat extraction	Soil blocks or cores suspended under heat lamps in water into which earthworms migrate	Effective on dense root mats	Not effective on all species; inconvenient for field use
Electrical extraction	Metal rods inserted into soil and connected to AC electrical source	Useful for selective or comparative sampling	Highly variable; not convenient in the field; dangerous
Mechanical vibration	Stake or rod inserted into soil and vibrated with bow or flat iron	Simple; useful for selective or comparative sampling	Not effective on all species
Trapping	Pitfall or baited traps placed in soil and sampled at desired intervals	Simple; useful for selective or comparative sampling	Not effective on all species
Mark–recapture	Individuals tagged, released, and population sampled at intervals	Useful for estimating population density, dispersal, and mortality	Laborious
INDIRECT			
Cast counting	Surface castings enumerated and identified	Simple	Not a quantitative estimate of population density

Summarized from Lee, K.E., 1985. Earthworms: Their Ecology and Relationships With Soils and Land Use. Academic Press, Sydney, Australia; Edwards, C.A., Bohlen, P.J., 1996. Earthworm Biology and Ecology, third ed. Chapman and Hall, London; reproduced from Hendrix, P.F., 2000. Earthworms. In: Sumner, M.E. (Ed.), Handbook of Soil Science. CRC Press, Boca Raton, FL, pp. C-77–C-85.

FIGURE 9.13 Example of a macroinvertebrate (earthworms and macroarthropods) sampling pit. A pit is excavated in known dimensions, and the soil hand sorted on a plastic sheet and invertebrates collected. A typical pit might be 30 cm × 30 cm × 30 cm, as in this case. *Source: Photo by Evelyn Wenk.*

detrimental effects on other soil fauna in the volume of soil where the solution has been applied (Coja et al., 2008). Chemical extraction with aqueous mustard powder solution has been shown to be as effective as formalin in some cases, avoiding the use of toxic formaldehyde. Effectiveness varies with earthworm species and activity, soil water content, porosity, and temperature. Comparisons with hand sorting should be done before adopting extraction techniques for quantitative sampling.

Heat extraction is a modification of that used for enchytraeids (discussed earlier). Soil cores or blocks are placed in pans of water, exposed to heat from overhead light bulbs, and earthworms are collected from the water after several hours. This technique was more effective than hand sorting or formalin extraction on small earthworms in dense root mats (Satchell, 1969). As with hand sorting, it is not effective on deep-burrowing, anecic species such as *L. terrestris*.

Mechanical vibration employs a rod or stake driven into the soil, vibration for a few minutes with a bow or flat piece of metal, such as an automobile leaf spring, and collection of earthworms that emerge onto the soil surface. Some acanthodrilid species have been sampled with this technique (Hendrix et al., 1994; Mitra et al., 2009), but it is not effective on lumbricids and probably only useful for selective or comparative sampling of certain populations.

Electrical extraction of earthworms involves inserting metal rods into the soil, connecting them to a source of alternating current and collecting earthworms that come to the soil surface. Different voltages and amperages have been used with varying degrees of success; effectiveness of the technique is highly dependent on soil water content, electrolyte concentration, and temperature. As with mechanical vibration, the soil volume sampled is not known and therefore this method is best suited for qualitative or comparative sampling. However, a commercially available electrical sampler ("octet" device developed by Thielemann 1986) was evaluated by Schmidt (2001) and found to be highly effective for quantitative sampling of lumbricid species in pastures. Electrical extraction methods are potentially very dangerous and should only be used with extreme caution.

Two earthworm-trapping techniques have been described. Pitfall traps (open-top containers buried level with the soil surface and containing a fixative solution, such as picric acid) may be useful for sampling surface-active species in diurnal or seasonal studies. Arrays of traps are installed and sampled at 12 hours, 24 hours, or longer intervals. Baited traps, such as perforated clay pots containing manure or other attractants and inserted into the soil, may also be useful for collecting certain species. As with other behavioral methods, trapping is probably highly selective and best suited for qualitative or comparative sampling.

Mark, release, and recapture techniques have been widely used to study population dynamics of animals, including earthworms. Large numbers of individuals of desired species are collected, marked (e.g., with brands or nontoxic dyes), and released into the population of interest. Sampling over time and distance from the target site, and enumeration of tagged relative to untagged individuals, yields information on dispersal, mortality, and population density. Radioisotope and, more recently, immunofluorescent antibody techniques have been employed in earthworm mark–recapture studies.

For earthworm species that cast on the soil surface, such as *Aporrectodea longa*, numbers and identity of castings may be a useful index of population activity. Because casting is dependent on soil temperature and moisture, this technique is highly variable and not a quantitative estimate of population density.

In summary, digging and hand sorting or washing are probably the most reliable means of sampling earthworms. However, no single method will be adequate to sample earthworm populations in all situations. Combinations of methods will probably achieve reasonable results. For example, formalin or mustard solution can be applied to the bottom of pits previously excavated for hand sorting, to extract deep burrowing anecic forms not sampled by digging (Edwards and Bohlen, 1996). Combinations of various methods may be useful in other situations.

9.5.4.3 Sampling Exercise for Macroinvertebrates (Earthworms and Macroarthropods)

Principle: Known areas and volumes of litter and soil, respectively, are sampled and hand sorted for macroinvertebrate specimens. Once identified, the data resulting from such sampling can be used to calculate various metrics of community makeup.

Procedure:

1. Select a sampling site, or preferably a set of sampling sites that differ in terms of vegetative cover, land use, or other factor.
2. At each site, place a 50 cm × 50 cm quadrat (a piece of plastic, or a wooden frame works well) onto the soil surface, and use a blade to cut the leaf litter around the frame. Remove leaf litter from inside the quadrat to a sheet of plastic for sorting. It is usually most practical to remove only the leaf litter that is easily scraped from the soil surface by hand, and to include the more heavily decayed material (humus, which is often more intimately associated with mineral soil) with the soil sample.

3. In the center of each litter quadrat, use another frame to cut the sides of a 30 cm × 30 cm block of mineral soil. Excavate the soil to a depth of 30 cm, and move this to another sheet of plastic for sorting in the field.
4. From both the litter and soil samples, pick out all invertebrates (arthropods and earthworms) by hand and immediately place them into 70%–95% ethanol.
5. Identify the specimens using keys mentioned for various groups in Chapter 4, Secondary Production: Activities of Heterotrophic Organisms—The Soil Fauna. For nonspecialists, it may be more practical to sort the specimens into morphospecies (but hopefully at least within the correct order for arthropods), and conduct community analyses on the resultant data. Examples of community measures that can be calculated include Shannon's diversity, evenness, percent similarity (as in, for example, Wenk et al., 2016), or more advanced measures such as indicator species analysis (Dufrene and Legendre, 1997).

References

Aanderud, Z.T., Jones, S.E., Fierer, N., Lennon, J.T., 2015. Resuscitation of the rare biosphere contributes to pulses of ecosystem activity. Front. Microbiol. 6, 24–36.

Abad, P., Gouzy, J., Aury, J.-M., Castagnone-Sereno, P., Danchin, E.G.J., Deleury, E., et al., 2008. Genome sequence of the metazoan plant-parasitic nematode Meloidogyne incognita. Nat. Biotechnol. 26, 909–915.

Abuzinadah, R.A., Read, D.J., 1989. The role of proteins in the nitrogen nutrition of ectomycorrhizal plants. V. Nitrogen transfer in birch (*Betula pendula*) grown in association with mycorrhizal and non-mycorrhizal fungi. New Phytol. 112 (1), 61–68.

Adejuyigbe, C.O., Tian, G., Adeoye, G.O., 1999. Soil microarthropod populations under natural and planted fallows in southwestern Nigeria. Agroforestry Syst. 47, 263–272.

Adl, S.M., 2003. The Ecology of Soil Decomposition. CABI Publishing, Wallingford.

Adl, S.M., Coleman, D.C., 2005. Dynamics of soil protozoa using a direct count method. Biol. Fertil. Soils 42, 168–171.

Adl, S.M., Acosta-Mercado, D., Lynn, D.H., 2008. Protozoa. In: Carter, M.R., Gregorich, E.G. (Eds.), Soil Sampling and Methods of Analysis, second ed. CRC Press, Boca Raton, FL, pp. 455–469.

Agerer, R., 2001. Exploration types of ectomycorrhizae: a proposal to classify ectomycorrhizal mycelial systems according to their patterns of differentiation and putative ecological importance. Mycorrhiza 11, 107–114.

Albers, B.P., Beese, F., Hartmann, A., 1995. Flow-microcalorimetry measurements of aerobic and anaerobic soil microbial activity. Biol. Fertil. Soils 19, 203–208.

Aldrete, A.N.G., 1990. Insecta: psocoptera. In: Dindal, D.L. (Ed.), Soil Biology Guide. Wiley, New York, pp. 1033–1052.

Alef, K., 1995. Heat output. In: Alef, K., Nannipieri, P. (Eds.), Methods in Applied Soil Microbiology and Biochemistry.. Academic Press, London, pp. 223–224.

Alef, K., Nannipieri, P., 1995a. Microbial biomass. In: Alef, K., Nannipieri, P. (Eds.), Methods in Applied Soil Microbiology and Biochemistry.. Academic Press, London, pp. 375–417.

Alef, K., Nannipieri, P., 1995b. Chapter 7, enzyme activities. In: Alef, K., Nannipieri, P. (Eds.), Methods in Applied Soil Microbiology and Biochemistry. Academic Press, London, pp. 311–373.

Allen, E.B., Allen, M.F., Helm, D.Y., Trappe, J.M., Molina, R., Rincon, E., 1995. Patterns and regulation of mycorrhizal plant and fungal diversity. Plant & Soil 70, 47–62.

Allen, M.F., 1991. The Ecology of Mycorrhizae. Cambridge University Press, Cambridge.

Allen, M.F. (Ed.), 1992. Mycorrhizal Functioning: An Integrative Plant-Fungal Process. Chapman and Hall, London.

Allen, R.T., 2003. Two new species of Epigean Litocampa (Insecta: Diplura: Campodeidae) from the Southeastern appalachians. Trans. Am. Entomol. Soc. 129, 549–559.

Allen-Morley, C.R., Coleman, D.C., 1989. Resilience of soil biota in various food webs to freezing perturbations. Ecology 70, 1127–1141.

Alphei, J., Bonkowski, M., Scheu, S., 1995. Application of the selective inhibition method to determine bacterial: fungal ratios in three beechwood soils rich in carbon-optimization of inhibitor concentrations. Biol. Fertil. Soils 19, 173–176.

Anderson, D.W., 1988. The effect of parent material and soil development on nutrient cycling in temperate ecosystems. Biogeochemistry 5, 71–97.

Anderson, D.W., Coleman, D.C., 1985. The dynamics of organic matter in grassland soils. J. Soil Water Conserv. 40, 211–216.

Anderson, J.M., 1973. Carbon dioxide evolution from two temperate, deciduous woodland soils. J. Appl. Ecol. 10, 361–378.

Anderson, J.M., 1975a. Succession, diversity and trophic relationships of some soil animals in decomposing leaf litter. J. Anim. Ecol. 44, 475–495.

Anderson, J.M., 1975b. The enigma of soil animal species diversity. In: Vanek, J. (Ed.), Progress in Soil Zoology. Academia, Prague, pp. 51–58.

Anderson, J.M., 1992. Responses of soils to climate change. Adv. Ecol. Res. 22, 163–210.

Anderson, J.M., 2000. Food web functioning and ecosystem processes: problems and perceptions of scaling. In: Coleman, D.C., Hendrix, P.F. (Eds.), Invertebrates as Webmasters in Ecosystems. CAB International, Wallingford, UK, pp. 3–24.

Anderson, J.M., Huish, S.A., Ineson, P., Leonard, M.A., Splatt, P.R., 1985. Interactions of invertebrates, micro-organisms, and tree roots in nitrogen and mineral element fluxes in deciduous woodland soils. In: Fitter, A.H., Atkinson, D., Read, D.J., Usher, M.B. (Eds.), Ecological Interactions in Soil: Plants, Microbes and Animals. Blackwell Scientific, Oxford, pp. 377–392.

Anderson, J.P.E., Domsch, K.H., 1978. A physiological method for the quantitative measurement of microbial biomass in soils. Soil Biol. Biochem. 10, 215–221.

Anderson, R.V., Coleman, D.C., Cole, C.V., 1981a. Effects of saprotrophic grazing on net mineralization. Terrestrial nitrogen cycles. Ecol. Bull. 33, 201–216.

Anderson, R.V., Coleman, D.C., Cole, C.V., Elliott, E.T., 1981b. Effect of the nematodes *Acrobeloides* sp. and *Mesodiplogaster lheritieri* on substrate utilization and nitrogen and phosphorus mineralization in soil. Ecology 62, 549–555.

Anderson, R.V., Gould, W.D., Woods, L.E., Cambardella, C., Ingham, R.E., Coleman, D.C., 1983. Organic and inorganic nitrogenous losses by microbivorous nematodes in soil. Oikos 40, 75–80.

Anderson, T.-H., Domsch, K.H., 1993. The metabolic quotient for CO_2 (qCO_2) as a specific activity parameter to assess the effects of environmental conditions, such as pH, on the microbial biomass of forest soils. Soil Biol. Biochem. 25, 393–395.

Anderson, T.-H., 1994. Physiological analysis of microbial communities in soil: applications and limitations. In: Ritz, K., Dighton, J., Giller, K.E. (Eds.), Beyond the Biomass. Wiley/Sayce, Chichester, pp. 67–76.

Andrade, G., Linderman, R.G., Bethlenfalvay, G.J., 1998. Bacterial associations with the mycorrhizosphere and hyphosphere of the arbuscular mycorrhizal fungus *Glomus mosseae*. Plant & Soil 202, 79–87.

André, H.M., Ducarme, X., Anderson, J., Crossley Jr., D.A., Koehler, H., Paoletti, M., et al., 2001. Skilled eyes are needed to go on studying the richness for the soil. Nature 409, 761.

André, H.M., Ducarme, X., Lebrun, P., 2002. Soil biodiversity: myth, reality or conning? Oikos 96, 3–24.

Andrén, O., Lindberg, T., Paustian, K., Rosswall, T. (Eds.), 1990. Ecology of Arable Land. Organisms, Carbon and Nitrogen Cycling. Munksgaard International, Copenhagen.

Andrén, O., Brussaard, L., Clarholm, M., 1999. Soil organism influence on ecosystem-level process - bypassing the ecological hierarchy. Appl. Soil Ecol. 11, 177–188.

Andrew, J.A., Harrison, K.G., Matamala, R., Schlesinger, W.H., 1999. Separation of root respiration from soil respiration using carbon-13 labeling during Free-Air Carbon Dioxide Enrichment (FACE). Soil Sci. Soc. Am. J. 63, 1429–1435.

Andrews, S.S., Carroll, C.R., 2001. Designing a soil quality assessment tool for sustainable agroecosystem management. Ecol. Appl. 11 (6), 1573–1585.

Anthony, K.W., Daanen, R., Anthony, P., von Deimling, T.S., Ping, C.-L., Chanton, J.P., et al., 2016. Methane emissions proportional to permafrost carbon thawed in Arctic lakes since the 1950s. Nat. Geosci. 9, 679–682.

Appelhoff, M., Fenton, M.F., Harris, B.L., 1993. Worms Eat Our Garbage: Classroom Activities for a Better Environment. Flowerfield Press, Kalamazoo, MI.

Arlian, L.G., Woolley, T.A., 1970. Observations on the biology of Liacarus cidarus (Acari: Cryptostigmata, Liacaridae). J. Kansas Entomol. Soc. 43, 297–301.

Arrhenius, S., 1896. On the influence of carbonic acid in the air upon the temperature of the ground. Philos. Mag. 41, 237–257.

Artursson, V., Jansson, J.K., 2003. Use of bromodeoxyuridine immunocapture to identify active bacteria associated with arbuscular mycorrhizal hyphae. Appl. Environ. Microbiol. 69, 6208–6215.

Auerbach, S.I., 1958. The soil ecosystem and radioactive waste disposal to the ground. Ecology 39, 522–529.

Ayres, E., Steltzer, H., Simmons, B.L., Simpson, R.T., Steinweg, J.M., Wallenstein, M.D., et al., 2009. Home-field advantage accelerates leaf litter decomposition in forests. Soil Biol. Biochem. 41, 606–610.

Azcón-Aguilar, C., Barea, J.M., 1995. Arbuscular mycorrhizas and biological control of soil-borne plant pathogens—an overview of the mechanisms involved. Mycorrhiza 6, 457–464.

Bååth, E., Lohm, U., Lundgren, B., Rosswall, T., Soderstrom, B., Sohlenius, B., 1981. Impact of microbial-feeding animals on total soil activity and nitrogen dynamics: a soil microcosm experiment. Oikos 37, 257–264.

Badri, D.V., Vivanco, J.M., 2009. Regulation and function of root exudates. Plant Cell Environ. 32, 666–681.

Baker, D.D., Schwintzer, C.R., 1990. Introduction. In: Schwintzer, C.R., Tjepkema, J.D. (Eds.), The Biology of *Frankia* and Actinorhizal Plants. Academic Press, San Diego, pp. 1–13.

Baker, E.W., Camin, J.H., Cunliffe, E., Woolley, T.A., Yunker, C.E., 1958. Guide to the Families of Mites. Institute of Acarology, University of Maryland, College Park, MD.

Baker, G.H., 1998. The ecology, management and benefits of earthworms in agricultural soils, with particular reference to Southern Australia. In: Edwards, C.A. (Ed.), Earthworm Ecology.. Lewis Publishers, Boca Raton, FL, pp. 229–258.

Bal, L., 1982. Zoological Ripening of Soils. Wageningen, Pudoc.

Baldock, J.A., 2002. Interactions of organic materials and microorganisms with minerals in the stabilization of soil structure. In: Huang, P.M., Bollag, J.-M., Senesi, N. (Eds.), Interactions Between Soil Particles and Microorganisms.. Wiley, Chichester, pp. 85–131.

Baldrian, P., Kolařík, M., Štursová, M., Kopecký, J., Valášková, V., Větrovský, T., et al., 2012. Active and total microbial communities in forest soil are largely different and highly stratified during decomposition. ISME J. 6, 248–258.

Balogh, J., Balogh, P. (Eds.), 1988. Soil Mites of the World: Oribatid Mites of the Neotropical Region, vol. 1. Elsevier Science Ltd., Amsterdam.

Balogh, J., Balogh, P. (Eds.), 1990. Soil Mites of the World: Oribatid Mites of the Neotropical Region, vol. 2. Kiadó Press, Budapest, Akadémiai.

Bamforth, S.S., 1980. Terrestrial protozoa. J. Protozool. 27, 33–36.

Bamforth, S.S., 1997. Protozoa: recyclers and indicators of agroecosystem quality. In: Benckiser, G. (Ed.), Fauna in Soil Ecosystems. Marcel Dekker, New York, pp. 63–84.

Banerjee, B., 1970. A mathematical model on sampling diplopods using pitfall traps. Oecologia 4, 102–105.

Barber, D.A., Martin, J.K., 1976. The release of organic substances by cereal roots into soil. New Phytol. 76, 69–80.

Bardgett, R.D., van der Putten, W., 2014. Belowground biodiversity and ecosystem functioning. Nature 515, 505–511.

Barois, I., 1999. Ecology of earthworm species with large environmental tolerance and/or extended distributions. In: Lavelle, P., Brussaard, L., Hendrix, P. (Eds.), Earthworm Management in Tropical Agroecosystems. CABI Publishing, New York, pp. 57–85.

Barra, J.A., Christiansen, K., 1975. Experimental study of aggregation during the development of *Pseudosinella impediens* (Collembola, Entomobryidae). Pedobiologia 15, 343–347.

Bater, J.E., 1996. Micro- and Macro-arthropods. In: Hall, G.S. (Ed.), Methods for the Examination of Organismal Diversity in Soils and Sediments. CAB International, New York, pp. 163–174.

Bates, S.T., Berg-Lyons, D., Caporaso, J.G., Walters, W.A., Knight, R., Fierer, N., 2011. Examining the global distribution of dominant archaeal populations in soil. ISME J. 5, 908–917.

Battley, E.H., 1987. Energetics of Microbial Growth. Wiley-Interscience, New York.

Beare, M.H., Neely, C.L., Coleman, D.C., Hargrove, W.L., 1990. A substrate-induced respiration (SIR) method for measurement of fungal and bacterial biomass on plant residues. Soil Biol. Biochem. 22, 585–594.

Beare, M.H., Neely, C.L., Coleman, D.C., Hargrove, W.L., 1991. Characterizations of a substrate-induced respiration method for measuring fungal, bacterial and total microbial biomass on plant residues. Agric., Ecosyst. Environ. 34, 65–73.

Beare, M.H., Parmelee, R.W., Hendrix, P.F., Cheng, W., Coleman, D.C., Crossley Jr., D.A., 1992. Microbial and faunal interactions and effects on litter nitrogen and decomposition in agroecosystems. Ecol. Monogr. 62, 569–591.

Beare, M.H., Hendrix, P.F., Coleman, D.C., 1994a. Water-stable aggregates and organic matter fractions in conventional and no-tillage soils. Soil Sci. Soc. Am. J. 58, 777–786.

Beare, M.H., Cabrera, M.L., Hendrix, P.F., Coleman, D.C., 1994b. Aggregate-protected and unprotected pools of organic matter in conventional and no-tillage Ultisols. Soil Sci. Soc. Am. J. 58, 787–795.

Beare, M.H., Coleman, D.C., Crossley Jr., D.A., Hendrix, P.F., Odum, E.P., 1995. A Hierarchical approach to evaluating the significance of soil biodiversity to biogeochemical cycling. Plant and Soil 170, 5–22.

Beare, M.H., Hu, S., Coleman, D.C., Hendrix, P.F., 1997. Influences of mycelial fungi on soil aggregation and organic matter storage in conventional and no-tillage soils. Appl. Soil Ecol. 5, 211–219.

Behan, V.M., Hill, S.B., 1978. Feeding habits and spore dispersal of oribatid mites in the North American arctic. Revue d'écologie et de biologie du sol 15, 497–516.

Behan-Pelletier, V., Newton, G., 1999. Linking soil biodiversity and ecosystem function—the taxonomic dilemma. BioScience 49, 149–153.

Behan-Pelletier, V.M., Bissett, B., 1993. Biodiversity of nearctic soil arthropods. Can. Biodiversity 2, 5–14.

Behan-Pelletier, V.M., Hill, S.B., 1983. Feeding habits of sixteen species of Oribatei (Acari) from an acid peat bog, Glenamoy, Ireland. Revue d'écologie et de biologie du sol 20, 221–267.

Behan-Pelletier, V.M., Norton, R.A., 1983. Epidamaeus (Acari: Damaeidae) of arctic western North America and extreme northeast U.S.S.R. Can. Entomol. 115, 1253–1289.

Behan-Pelletier, V.M., Walter, D.E., 2000. Biodiversity of oribatid mites (Acari: Oribatida) in tree canopies and litter. In: Coleman, D.C., Hendrix, P.F. (Eds.), Invertebrates as Webmasters in Ecosystems. CABI Publishing, Wallingford, pp. 187–202.

Behre, G.F., 1987. Die Sieb-Flotations-Methode. Bau und Erprobung eines Ökologischen Arbeitsgerätes zur mechanischen Auslese von Bodenarthropoden. Jahresberichte des Naturwissenschaftlichen Vereins in Wuppertal. 40, 52–55.

Bell, R.T., 1990. Insecta: coleoptera carabidae, adults and larvae. In: Dindal, D.L. (Ed.), Soil Biology Guide. John Wiley & Sons, New York, pp. 1053–1092.

Belnap, J., 2002. Nitrogen fixation in biological soil crusts from Southeastern Utah, USA. Biol. Fertil. Soils 35, 128–135.

Belnap, J., Phillips, S.L., 2001. Soil biota in an ungrazed grassland: response to annual grass (*Bromus tectorum*) invasion. Ecol. Appl. 11, 1261–1275.

Bending, G.D., Read, D.J., 1996. Nitrogen mobilization from protein-polyphenol complex by ericoid and ectomycorrhizal fungi. Soil Biol. Biochem. 28, 1603–1612.

Bengtsson, G., Hedlund, K., Rundgren, S., 1994. Food- and density- dependent dispersal: evidence from a soil Collembolan. J. Anim. Ecol. 63, 513–520.

Bengtsson, J., 1998. Which species? What kind of diversity? Which ecosystem function? Some problems in studies of relations between biodiversity and ecosystem function. Appl. Soil Ecol. 10, 191–199.

Bentham, H., Harris, J.A., Birch, P., Short, K.C., 1992. Habitat classification and soil restoration assessment using analysis of soil microbiological and physico-chemical characteristics. J. Appl. Ecol. 29, 711–718.

Berg, B., 1986. Nutrient release from litter and humus in coniferous forest soils—a mini review. Scand. J. For. Res. 1, 359–369.

Berg, B., 2014. Decomposition patterns for foliar litter—a theory for influencing factors. Soil Biol. Biochem. 78, 222–232.

Berg, B., Staaf, H., 1981. Leaching, accumulation and release of nitrogen in decomposing forest litter. In: Clark, F.E., Rosswall, T. (Eds.), Terrestrial Nitrogen Cycles. Processes, Ecosystem Strategies and Management Impacts. Ecological Bulletins, 33, 163–178.

Bernard, E.C., 1985. Two new species of *Protura* (Insecta) from North America. Proc. Biol. Soc. Wash. 98, 72–80.

Bever, J.D., 1999. Dynamics within mutualism and the maintenance of diversity: inference from a model of interguild frequency dependence. Ecol. Lett. 2, 52–61.

Bever, J.D., Schultz, P.A., Pringle, A., Morton, J.B., 2001. Arbuscular mycorrhizal fungi: more diverse than meets the eye, and the ecological tale of why. BioScience 51, 923–931.

Bignell, D.E., 1984. The arthropod gut as an environment for microorganisms. In: Anderson, J.M., Rayner, A.D.M., Walton, D.W.H. (Eds.), Invertebrate-Microbial Interactions. Cambridge University Press, Cambridge, pp. 205–227.

Bignell, D.E., 2000. Introduction to symbiosis. In: Abe, T., Bigell, D.E., Higashi, M. (Eds.), Termites: Evolution, Sociality, Symbioses, Ecology. Kluwer Academic, Dordrecht, pp. 189–208.

Bignell, D.E., Eggleton, P., 2000. Termites in ecosystems. In: Abe, T., Bignell, D.E., Higashi, M. (Eds.), Termites: Evolution, Sociality, Symbioses, Ecology. Kluwer Academic, Dordrecht, pp. 363–387.

Billings, S., Min, K., Ballantyne IV, F., Chen, Y., Sellers, M., 2016. Aging exo-enzymes can create temporally shifting, temperature-dependent resource landscapes for microbes. Biogeochemistry 131, 163–172.

Binkley, D., 2002. Ten-year decomposition in a loblolly pine forest. Can. J. For. Res. 32, 2231–2235.

Bintrim, S., Donohue, T., Handelsman, J., Roberts, G., Goodman, R., 1997. Molecular phylogeny of Archaea from soil. Proc. Natl. Acad. Sci. 94, 277–282.

Blackshaw, R.P., Hicks, H., 2013. Distribution of adult stages of soil insect pests across an agricultural landscape. J. Pest Sci. 86, 53–62.

Blagodatskaya, E., Kuzyakov, Y., 2013. Active microorganisms in soil: critical review of estimation criteria and approaches. Soil Biol. Biochem. 67, 192–211.

Blair, J.M., Crossley Jr., D.A., 1988. Litter decomposition, nitrogen dynamics and litter microarthropods in a southern Appalachian hardwood forest 8 years following clearcutting. J. Appl. Ecol. 25, 683–698.

Blair, J.M., 1988a. Nitrogen, sulfur and phosphorus dynamics in decomposing deciduous leaf litter in the southern appalachians. Soil Biol. Biochem. 20, 693–701.

Blair, J.M., 1988b. Nutrient release from decomposing foliar litter of three tree species with special reference to calcium, magnesium and potassium dynamics. Plant & Soil 110, 49–55.

Blair, J.M., Crossley Jr., D.A., Callaham, L.C., 1992. Effects of litter quality and microarthropods on N dynamics and retention of exogenous ^{15}N in decomposing litter. Biol. Fertil. Soils 12, 241–252.

Blair, W.F., 1977. Big Biology: The US/IBP. Dowden, Hutchinson, and Ross, Stroudsburg, PA.

Blakemore, R.J., 2002. Cosmopolitan earthworms: an eco-taxonomic guide to the peregrine species of the world. VermEcology, Kippax, Australia.

Bloem, J., de Ruiter, P., Bouwman, L., 1997. SoilFood webs and nutrient cycling in agroecosystems. In: Van Elsas, J.D., Trevors, J.T., Wellington, E.M.H. (Eds.), Modern Soil Microbiology. Marcel Dekker, New York, pp. 245–278.

Blum, M.S., Edgar, A.L., 1971. 4-Methyl-3-heptanone: identification and role in opilionid exocrine secretions. Insect Biochem. 1, 181–188.

Blumberg, A.Y., Crossley Jr., D.A., 1983. Comparison of soil surface arthropod populations in conventional tillage, no-tillage and old field systems. Agroecosystems 8, 247–253.

Boag, B., Yeates, G.W., 2001. The potential impact of the New Zealand flatworm, a predator of earthworms, in Western Europe. Ecol. Appl. 11, 1276–1286.

Bocock, K.L., Gilbert, O.J.W., 1957. The disappearance of leaf litter under different woodland conditions. Plant & Soil 9, 179–185.

Bohlen, P.J., Edwards, C.A., Zhang, Q., Parmelee, R.W., Allen, M., 2002. Indirect effects of earthworms on microbial assimilation of labile carbon. Appl. Soil Ecol. 20, 255–261.

Böhm, W., 1979. Methods of Studying Root Systems. Springer, Berlin.

Bolger, T.M., Heneghan, L.J., Neville, P., 2000. Invertebrates and nutrient cycling in coniferous forest ecosystems: spatial heterogeneity and conditionality. In: Coleman, D.C., Hendrix, P.F. (Eds.), Invertebrates as Webmasters in Ecosystems.. CABI Publishing, Wallingford, pp. 161–184.

Bolger, T., 2001. The functional value of species biodiversity—a review. Proc. R. Ir. Acad. 101B, 199–204.

Bolton, B., 1994. Identification Guide to the Ant Genera of the World. Harvard University Press, Cambridge.

Bomberg, M., Jurgens, G., Saana, A., Sen, R., Timonen, S., 2003. Nested PCR detection of Archaea in defined compartments of pine mycorrhizospheres developed in boreal forest humus microcosms. FEMS Microbiol. Ecol. 43, 163–171.

Bonfante, P., Genre, A., 2010. Mechanisms underlying beneficial plant fungus interactions in mycorrhizal symbiosis. Nat. Commun. 1, 48.

Bongers, T., 1990. The maturity index: and ecological measure of environmental disturbance based on nematode species composition. Oecologia 83, 14–19.

Bonkowski, M., Clarholm, M., 2012. Stimulation of plant growth through interactions of bacteria and protozoa: testing the auxiliary microbial loop hypothesis. Acta Protozool. 51, 237–247.

Bonkowski, M., Cheng, W., Griffiths, B.S., Alphei, J., Scheu, S., 2000. Microbial-faunal interactions in the rhizosphere and effects on plant growth. Eur. J. Soil Biol. 36, 135–147.

Bonkowski, M., Griffiths, B., Scrimgeour, C., 2000. Substrate heterogeneity and microfauna in soil organic "hot-spots" as determinants of nitrogen capture and growth of ryegrass. Appl. Soil Ecol. 14, 37–53.

Bornebusch, C.H., 1930. The Fauna of Forest Soil. Copenhagen.

Borneman, J., Triplett, E.W., 1997. Molecular microbial diversity in soils from Eastern Amazonia: evidence for unusual microorganisms and microbial population shifts associated with deforestation. Appl. Environ. Microbiol. 63, 2647–2653.

Borror, D.J., DeLong, D.M., Triplehorn, C.A., 1981. An Introduction to the Study of Insects (fifth edition). Saunders College Publishing, Philadelphia, PA.

Bouché, M.B., 1975. Action de la faune sur les etats de la matiére organique dans les ecosystemes. In: Kilbertus, G., Reisinger, O., Mourey, A., Cacela da Fonseca, J.A. (Eds.), Biodégradation et Humification. Pierron, Sarrugemines, pp. 157–168.

Bouché, M.B., 1977. Stratégies lombriciennes. In: Lohm, U., Persson, T. (Eds.), Soil organisms as components of ecosystems. Ecological Bulletins (Stockholm). 25, 122–133.

Bouché, M.B., 1983. The establishment of earthworm communities. In: Satchell, J.E. (Ed.), Earthworm Ecology from Darwin to Vermiculture.. Chapman and Hall, London, pp. 431–448.

Bouché, M.B., Beugnot, M., 1972. Contribution à l' approache éthodologique de l' étude des Biocenoses. II. L' extraction des macroéléments du sol par lavage-tamisage. Annales de Zoologie Ecologie Animale 4, 537–544.

Bouchon, D., Zimmer, M., Dittmer, J., 2016. The terrestrial isopod microbiome: an all-in-one toolbox for animal–microbe interactions of ecological relevance. Front. Microbiol. 7, 1472. Available from: https://doi.org/10.3389/fmicb.2016.01472.

Boutton, T.W., Yamasaki, S. (Eds.), 1996. Mass Spectrometry of Soils. Marcel Dekker, New York.

Bowen, H.J.M., 1979. Environmental Chemistry of the Elements. Academic Press, London, New York.

Box Jr., J.E., Johnson, J.W., 1987. Minirhizotron rooting comparisons of three wheat cultivars. In: Taylor, H.M. (Ed.), Minirhizotron Observation Tubes: Methods and Applications for Measuring Rhizosphere Dynamics. American Society of Agronomy, Special Publication No. 50, Madison, Wisconsin, pp. 123–130.

Box Jr., J.E., Hammond, L.C., 1990. Rhizosphere Dynamics. Westview Press, Boulder, CO.

Bradford, M.A., 2016. Editorial: re-visioning soil food webs. Soil Biol. Biochem. 102, 1–3.

Bradford, M.A., Jones, T.H., Bardgett, R.D., Black, H.I.J., Boag, B., Bonkowski, M., et al., 2002. Impacts of soil faunal community composition on model grassland ecosystems. Science 298, 615–618.

Bradford, M.A., Warren II, R.J., Baldrian, P., Crowther, T.W., Maynard, D.S., Oldfield, E.E., et al., 2014. Climate fails to predict wood decomposition at regional scales. Nat. Clim. Change 4, 625–630.

Bradford, M.A., Berg, B., Maynard, D.S., Wieder, W.R., Wood, S.A., 2016. Understanding the dominant controls on litter decomposition. J. Ecol. 104, 229–238.

Bradley, R.A., 2013. Common Spiders of North America. University of California Press, Berkeley, CA.

Brady, N.C., 1974. The Nature and Properties of Soils, eighth ed. MacMillan, New York.

Brady, N.C., Weil, R.R., 2000. Elements of the Nature and Properties of Soils.. Prentice Hall, Upper Saddle River, NJ.

Bragazza, L., Parisod, J., Buttler, A., Bardgett, R.D., 2013. Biogeochemical plant-soil microbe feed- back in response to climate warming in peatlands. Nat. Clim. Change 3, 273–277.

Brassard, B.W., Chen, H.Y.H., Bergeron, Y., Paré, D., 2011. Differences in fine root productivity between mixed- and single-species stands. Funct. Ecol. 25, 238–246.

Brauman, A., Bignell, D.E., Tayasu, I., 2000. Soil-feeding termites: biology, microbial associations and digestive mechanisms. In: Abe, T., Bignell, D.E., Higashi, M. (Eds.), Termites: Evolution, Sociality, Symbioses, Ecology. Kluwer Academic, Dordrecht, pp. 233–259.

Breznak, J.A., 1984. Biochemical aspects of symbiosis between termites and their intestinal microbiota. In: Anderson, J.M., Rayner, A.D.M., Walton, D.W.H. (Eds.), Invertebrate-Microbial Interactions.. Cambridge University Press, Cambridge, pp. 173–203.

Breznak, J., 2000. Ecology of prokaryotic microbes in the guts of wood- and litter-feeding termites. In: Abe, T., Bignell, D.E., Higashi, M. (Eds.), Termites: Evolution, Sociality, Symbioses, Ecology. Kluwer Academic, Dordrecht, pp. 209–231.

Briones, M.J.I., Ostle, N.J., Piearce, T.G., 2008. Stable isotopes reveal that the calciferous gland of earthworms is a CO_2-fixing organ. Soil Biol. Biochem. 40, 554–557.

Brockmeyer, V., Schmid, R., Westheide, W., 1990. Quantitative investigations of the food of two terrestrial enchytraeid species (Oligochaeta). Pedobiologia 34, 151–156.

Brown, G.G., 1995. How do earthworms affect microfloral and faunal community diversity? Plant & Soil 170, 247–269.

Brown, G.G., Callaham Jr., M.A., Niva, C.C., Feijoo, A., Sautter, K.D., James, S.W., et al., 2013. Terrestrial oligochaete research in Latin America: the importance of the Latin American meetings on oligochaete ecology and taxonomy. Appl. Soil Ecol. 69, 2–12.

Brown, S.P., Callaham Jr., M.A., Oliver, A.K., Jumpponen, A., 2013. Deep ion torrent sequencing identifies soil fungal community shifts after frequent prescribed fires in a southeastern forest ecosystem. FEMS Microbiol. Ecol. 86, 557–566.

Brundrett, M., Bougher, N., Dell, B., Grove, T., Malajczuk, N., 1996. Working with mycorrhizas in forestry and agriculture. ACIAR Monogr. 32, 374.

Bruno, J.F., Stachowicz, J.J., Bertness, M.D., 2003. Inclusion of facilitation into ecological theory. Trends Ecol. Evol. 18 (3), 119–125.

Brussaard, L., Bouwman, L.A., Geurs, M., Hassink, J., Zwart, K.B., 1990. Biomass, composition and temporal dynamics of soil organisms of a silt loam soil under conventional and integrated management. Neth. J. Agric. Sci. 38, 283–302.

Brussaard, L., Kooistra, M.J. (Eds.), 1993. Soil Structure/Soil Biota Interrelationships. Elsevier, Amsterdam.

Bryant, R.J., Woods, L.E., Coleman, D.C., Fairbanks, B.C., McClellan, J.F., Cole, C.V., 1982. Interactions of bacterial and amoebal populations in soil microcosms with fluctuating moisture content. Appl. Environ. Microbiol. 43, 7447–7752.

Brzostek, E.R., Dragoni, D., Brown, Z.A., Phillips, R.P., 2015. Mycorrhizal type determines magnitude and direction of root-induced changes in decomposition in a temperate forest. New Phytol. 206, 1274–1282.

Buckley, D., Graber, J., Schmidt, T., 1998. Phylogenetic analysis of nonthermophilic members of the kingdom Crenarchaeota and their diversity and abundance in soils. Appl. Environ. Microbiol. 64, 4333–4339.

Buckman, H.O., Brady, N.C., 1970. The Nature and Properties of Soil. Macmillan, New York.

Bue'e, M., Reich, M., Murat, C., Morin, E., Nilsson, R.H., Uroz, S., et al., 2009. 454 Pyrosequence analyses of forest soils reveal and unexpectedly high fungal diversity. New Phytol. 184, 449–456.

Bundt, M., Widmer, F., Pesaro, M., Zeyer, J., Blaser, P., 2001. Preferential flow paths: biological 'hot spots' in soils. Soil Biol. Biochem. 33, 729–738.

Burch, J.B., Pearce, T.A., 1990. Terrestrial Gastropoda. In: Dindal, D. (Ed.), Soil Biology Guide. Wiley, New York, pp. 201–309.

Burns, R.G., DeForest, J.L., Marxsen, J., Sinsabaugh, R.L., Stromberger, M.E., Wallenstein, M.D., et al., 2013. Soil enzymes in a changing environment: current knowledge and future directions. Soil Biol. Biochem. 58, 216–234.

Butcher, J.W., Snider, R., Snider, R.J., 1971. Bioecology of edaphic Collembola and Acarina. Annu. Rev. Entomol. 16, 249–288.

Butt, K.R., 2008. Earthworms in soil restoration: lessons learned from United Kingdom case studies of land reclamation. Restor. Ecol. 16, 637–641.

Butt, K.R., Callaham Jr., M.A., Loudermilk, E.L., Blaik, R., 2016. Action of earthworms on flint burial—a return to Darwin's estate. Appl. Soil Ecol. 104, 157–162.

Byers, J.E., Noonburg, E.G., 2003. Scale dependent effects of biotic resistance to biological invasion. Ecology 84, 1428–1433.

Byers, R.A., Barratt, B.I.P., Calvin, D., 1989. Comparison between defined-area traps and refuge traps for sampling slugs in conservation-tillage crop environments. In: Henderson, I. (Ed.), Slugs and Snails in World Agriculture. British Crop Protection Council Monograph No. 41, Thornton Heath, pp. 187–192.

Byrd, D.W.J., Barker, K.R., Ferris, H., Nusbaum, C.J., Griffin, W.E., Small, R.H., et al., 1976. Two semi-automatic elutriators for extracting nematodes and certain fungi from soil. J. Nematol. 8, 206–212.

Cabrera, M.L., Beare, M.H., 1993. Alkaline persulfate oxidation for determining total nitrogen in microbial biomass extracts. Soil Sci. Soc. Am. J 57, 1007–1012.

Caldwell, M.M., Camp, L.B., 1974. Belowground productivity of two cool desert communities. Oecologia 17, 123–130.

Callaham Jr., M.A., Hendrix, P.F., 1997. Relative abundance and seasonal activity of earthworms (Lumbricidae and Megascolecidae) as determined by hand-sorting and formalin extraction in forest soils on the southern Appalachian Piedmont. Soil Biol. Biochem. 29, 317–322.

Callaham Jr., M.A., Whiles, M.R., Meyer, K.C., Brock, B.L., Charlton, R.E., 2002. Feeding ecology and emergence production of annual cicadas (Homoptera, Cicadidae) in tallgrass prairie. Oecologia 123, 535–542.

Callaham Jr., M.A., Hendrix, P.F., Phillips, R.J., 2003. Occurrence of an exotic earthworm (*Amynthas agrestis*) in undisturbed soils of the southern Appalachian mountains, USA. Pedobiologia 47, 466–470.

Callaham Jr., M.A., González, G., Hale, C.M., Heneghan, L., Lachnicht, S.L., Zou, X., 2006. Policy and management responses to earthworm invasions in North America. Biol. Invasions 8, 1317–1329.

Callaham Jr., M.A., Richter, D.D., Coleman, D.C., Hofmockel, M., 2006. Long-term land use effects on soil invertebrate communities in Southern Piedmont soils. Eur. J. Soil Biol. 42, S150–S156.

Callaham, M.A., Snyder, B.A., James, S.W., Oberg, E.T., 2016. Evidence for ongoing introduction of non-native earthworms in the Washington, DC, metropolitan area. Biol. Invasions 18, 3133–3136.

Cameron, E.K., Bayne, E.M., 2009. Road age and its importance in earthworm invasion of northern boreal forests. J. Appl. Ecol. 46, 28–36.

Cameron, E.K., Bayne, E.M., Clapperton, M.J., 2007. Human-facilitated invasion of exotic earthworms into northern boreal forests. Ecoscience 14, 482–490.

Cameron, E.K., Shaw, C.H., Bayne, E.M., Kurz, W.A., Kull, S.J., 2015. Modelling interacting effects of invasive earthworms and wildfire on forest floor carbon storage in the boreal forest. Soil Biol. Biochem. 88, 189–196.

Campbell, E.E., Paustian, K., 2015. Current developments in soil organic matter modeling and the expansion of model applications: a review. Environ. Res. Lett. 10 (12), 1–36.

Campos-Herrera, R., El-Borai, F.E., Duncan, L.W., 2015. Modifying soil to enhance biological control of belowground dwelling insects in citrus groves under organic agriculture in Florida. Biocontrol 84, 53–63.

Catania, K.C., 2008. Worm grunting, fiddling, and charming—humans unknowingly mimic a predator to harvest bait. Plos One 3 (10), e3472.

Carapelli, A., Convey, P., Nardi, F., Frati, F., 2014. The mitochondrial genome of the antarctic springtail Folsomotoma octooculata (Hexapoda; Collembola), and an update on the phylogeny of collembolan lineages based on mitogenomic data. Entomologia 2, 190.

Carey, J.C., Tang, J., Templer, P.H., Kroeger, K.D., Crowther, T.W., Burton, A.J., et al., 2016. Temperature response of soil respiration largely unaltered with experimental warming. PNAS 113 (48), 13797–13802.

Carini, P., Marsden, P.J., Leff, J.W., Morgan, E.E., Strickland, M.S., Fierer, N., 2016. Relic DNA is abundant in soil and obscures estimates of soil microbial diversity. Nat. Microbiol.2, Article no. 16242.

Carrera-Martínez, R., Snyder, B.A., 2016. First report of *Amynthas carnosus* (Goto & Hatai 1899) (Oligochaeta: Megascolecidae) in the Western Hemisphere. Zootaxa 4111 (3), 297–300.

Catts, E.P., Haskell, N.H., 1990. Entomology and death: a procedural guide. Joyce's Print Shop, Clemson, South Carolina.

Cavigelli, M.A., Maul, J.E., Szlavecz, K., 2012. Managing soil biodiversity and ecosystem services. In: Wall, D.H., Bardgett, R.D., Behan-Pelletier, B., Herrick, J.E., Jones, H., Ritz, K., Six, J., Strong, D.R., van der Putten, W.H. (Eds.), Soil Ecology and Ecosystem Services. Oxford University Press, Oxford, pp. 337–356. Chapter 5.4.

Chakraborty, S., Warcup, J.H., 1983. Soil amoebae and saprophytic survival of *Gaeumannomyces graminis tritici* in a suppressive pasture soil. Soil Biol. Biochem. 15, 181–185.

Chakraborty, S., Old, K.M., Warcup, J.H., 1983. Amoebae from a take-all suppressive soil which feed on *Gaeumannomyces graminis tritici* and other soil fungi. Soil Biol. Biochem. 15, 17–24.

Chalot, M., Brun, A., 1998. Physiology of organic nitrogen acquisition by ectomycorrhizal fungi and ectomycorrhizas. FEMS Microbiol. Rev. 22, 21–44.

Chamberlain, P.M., Bull, I.D., Black, H.I.J., Ineson, P., Evershed, R.P., 2006. Collembolan trophic preferences determined using fatty acid distributions and compound-specific stable carbon isotope values. Soil Biol. Biochem. 38, 1275–1281.

Chapin III, F.S., 1980. The mineral nutrition of wild plants. Annu. Rev. Ecol. Evol. Syst. 11, 233–260.

Chapin III, F.S., Zavaleta, E.S., Eviner, V.T., Naylor, R.L., Vitousek, P.M., Reynolds, H.L., et al., 2000. Consequences of changing biodiversity. Nature 405, 234–242.

Cheeke, T.E., Rosenstiel, T.N., Cruzan, M.B., 2012. Evidence of reduced arbuscular mycorrhizal fungal colonization in multiple lines of Bt maize. Am. J. Bot. 99, 700–707.

Cheng, L., Booker, F.L., Tu, C., Burkey, K.O., Zhou, L., Shew, T.D., et al., 2012. Arbuscular mycorrhizal fungi increase organic carbon decomposition under elevated carbon dioxide. Science 337, 1084–1087.
Cheng, W., 1996. Measurement of rhizosphere respiration and organic matter decomposition using natural ^{13}C. Plant & Soil 183, 163–168.
Cheng, W., Coleman, D.C., 1990. Effect of living roots on soil organic matter decomposition. Soil Biol. Biochem. 22, 781–787.
Cheng, W., Coleman, D.C., Box Jr., J.E., 1990. Root dynamics, production and distribution in agroecosystems on the georgia piedmont using minirhizotrons. J. Appl. Ecol. 27, 592–604.
Cheng, W., Coleman, D.C., Carroll, C., Hoffman, C.A., 1993. In situ measurement of root respiration and soluble carbon concentrations in the rhizosphere. Soil Biol. Biochem. 25, 1189–1196.
Cheng, W., Parton, W.J., Gonzalez-Meler, M.A., Phillips, R., Asao, S., McNickle, G.G., et al., 2014. Synthesis and modeling perspectives of rhizosphere priming. New Phytol. 201, 31–44.
Cheshire, M., 1979. Soil Carbohydrates. Academic Press, London.
Cheshire, M.V., Sparling, G.P., Mundie, C.M., 1984. Influence of soil type, crop and air drying on residual carbohydrate content and aggregate stability after treatment with periodate and tetraborate. Plant Soil 76, 339–347.
Chiariello, N., Hickman, J.C., Mooney, H.A., 1982. Endomycorrhizal role for interspecific transfer of phosphorus in a community of annual plants. Science 217, 941–943.
Christiansen, K., 1970. Experimental studies on the aggregation and dispersion of Collembola. Pedobiologia 10, 180–198.
Christiansen, K.A., 1990. Insecta: collembola. In: Dindal, D.L. (Ed.), Soil Biology Guide.. Wiley, New York, pp. 965–995.
Christiansen, K.A., 1992. Springtails. The kansas school. Naturalist 39, 1–16.
Christiansen, K.A., Bellinger, P.F., 1998. The Collembola of North America North of the Rio Grande., second ed. Grinnell College, Grinnell, IA.
Christie, J.R., Perry, V.G., 1951. Removing nematodes from soil. Proc. Helminthol. Soc. Wash. 18, 106–108.
Clarholm, M., 1981. Protozoan grazing of bacteria in soil-impact and importance. Microb. Ecol. 7, 343–350.
Clarholm, M., 1985. Possible roles for roots, bacteria, protozoa and fungi in supplying nitrogen to plants. In: Fitter, A.H., Atkinson, D., Read, D.J., Usher, M.B. (Eds.), Ecological Interactions in Soil: Plants, Microbes and Animals.. Blackwell, Oxford, pp. 355–365.
Clarholm, M., 1994. The microbial loop in soil. In: Ritz, K., Dighton, J., Giller, K.E. (Eds.), Beyond the Biomass. Wiley/Sayce, Chichester, pp. 221–230.
Clemmensen, K.E., Finlay, R.D., Dahlberg, A., Stenlid, J., Wardle, D.A., Lindahl, B.D., 2015. Carbon sequestration related to mycorrhizal fungal community shifts during long-term succession in boreal forests. New Phytol. 205, 1525–1536.
Cock, M.J., Biesmeijer, J.C., Cannon, R.J., Gerard, P.J., Gillespie, D., Jiménez, J.J., et al., 2012. The positive contribution of invertebrates to sustainable agriculture and food security. CAB Reviews Perspectives in Agriculture Veterinary Science Nutrition and Natural Resources 7, 1–27.
Coe, K.K., Belnap, J., Sparks, J.P., 2012. Precipitation-driven carbon balance controls survivorship of desert biocrust mosses. Ecology 93, 1626–1636.
Coineau, Y., 1974. Introduction a l'Étude des microarthropodes du Sol et de ses Anexes. Documents pour l'Enseignement Practique de l'Écologie. Doin, Paris.
Čoja, T., Zehetner, K., Bruckner, A., Watzinger, A., Meyer, E., 2008. Efficacy and side effects of five sampling methods for soil earthworms (Annelida, Lumbricidae). Ecotoxicol. Environ. Saf. 71, 552–565.
Cole, L.C., 1946. The cryptozoa of an Illinois woodland. Ecol. Monogr. 16, 49–86.
Coleman, D.C., 1973. Soil carbon balance in a successional grassland. Oikos 24, 195–199.
Coleman, D.C., 1976. A review of root production processes and their influence on soil biota in terrestrial ecosystems. In: Anderson, J.M., Macfadyen, A. (Eds.), The Role of Terrestrial and Aquatic Organisms in Decomposition Processes. Blackwell, Oxford, pp. 417–434.
Coleman, D.C., 1985. Through a ped darkly: an ecological assessment of root-soil-microbial-faunal interactions. In: Fitter, A.H., Atkinson, D., Read, D.J., Usher, M.B. (Eds.), Ecological Interactions in Soil: Plants, Microbes and Animals. Blackwell, Oxford, pp. 1–21.
Coleman, D.C., 1994a. Compositional analysis of microbial communities. In: Ritz, K., Dighton, J., Giller, K. (Eds.), Beyond the Biomass. Wiley-Sayce, Chichester, pp. 201–220.

Coleman, D.C., 1994b. The microbial loop concept as used in terrestrial soil ecology studies. Microb. Ecol. 28, 245–250.

Coleman, D.C., 2001. SoilBiota, soil systems and processes. In: Levin, S. (Ed.), Encyclopedia of Biodiversity, Vol. 5. Academic Press, San Diego, CA, pp. 305–314.

Coleman, D.C., 2008. From Peds to Paradoxes: linkages between soil biota and their influences on ecological processes. Soil Biol. Biochem. 40, 271–289.

Coleman, D.C., 2010. BigEcology: the Emergence of Ecosystem Science. University of California Press, Berkeley, CA.

Coleman, D.C., Fry, B. (Eds.), 1991. Carbon Isotope Techniques in Plant, Soil, and Aquatic Biology. Academic Press, San Diego, CA.

Coleman, D.C., Sasson, A., 1980. Chapter 7: decomposers subsystem. In: Breymeyer, A., van Dyne, G. (Eds.), Grasslands, Systems Analysis, and Man. IBP Synthesis, vol. 19. Cambridge University Press, London, pp. 609–655.

Coleman, D.C., Schoute, J.F.T., 1993. Translation of soil features across levels of spatial resolution- Introduction to round table discussion. Geoderma 57, 171–181.

Coleman, D.C., Wall, D.H., 2015. Chapter 5: soil fauna: occurrence, biodiversity, and roles in ecosystem function. In: Paul, E.A. (Ed.), Soil Microbiology, Ecology and Biochemistry., fourth ed. Elsevier Academic Press, San Diego, CA, pp. 111–147.

Coleman, D.C., Andrews, R., Ellis, J.E., Singh, J.S., 1976. Energy flow and partitioning in selected man-managed and natural ecosystems. Agro Ecosyst. 3, 45–54.

Coleman, D.C., Cole, C.V., Anderson, R.V., Blaha, M., Campion, M.K., Clarholm, M., et al., 1977. Analysis of rhizosphere-saprophage interactions in terrestrial ecosystems. In: Lohm, U., Persson, T. (Eds.), Soil Organisms as Components of Ecosystems. Ecological Bulletins (Stockholm). 25, 299–309.

Coleman, D.C., Reid, C.P.P., Cole, C.V., 1983. Biological strategies of nutrient cycling in soil systems. Adv. Ecol. Res. 13, 1–55.

Coleman, D.C., Cole, C.V., Elliott, E.T., 1984. Decomposition, organic matter turnover and nutrient dynamics in agroecosystems. In: Lowrance, R., Stinner, B.R., House, G.J. (Eds.), Agricultural Ecosystems-Unifying Concepts. Wiley/lnterscience, New York, pp. 83–104.

Coleman, D.C., Ingham, E.R., Hunt, H.W., Elliott, E.T., Reid, C.P.P., Moore, J.C., 1990. Seasonal and faunal effects on decomposition in semiarid prairie, meadow and lodgepole pine forest. Pedobiologia 34, 207–219.

Coleman, D.C., Odum, E.P., Crossley Jr., D.A., 1992. Soil biology, soil ecology, and global change. Biol. Fertil. Soils 14, 104–111.

Coleman, D.C., Hendrix, P.F., Beare, M.H., Cheng, W., Crossley Jr., D.A., 1993. Microbial and faunal dynamics as they affect soil organic matter dynamics in subtropical Agroecosystems. In: Paoletti, M.G., Foissner, W., Coleman, D.C. (Eds.), Soil Biota and Nutrient Cycling Farming Systems. Lewis Publishing Company, Chelsea, MI, pp. 1–14.

Coleman, D.C., Hendrix, P.F., Beare, M.H., Crossley Jr., D.A., Hu, S., van Vliet, P.C.J., 1994a. The impacts of management and biota on nutrient dynamics and soil structure in sub-tropical agroecosystems: impacts on detritus food webs. In: Pankhurst, C.E., Doube, B.M., Gupta, V.V.S.R., Grace, P.R. (Eds.), Soil Biota Management in Sustainable Farming Systems. CSIRO, Melbourne, pp. 133–143.

Coleman, D.C., Dighton, J., Ritz, K., Giller, K.E., 1994b. Perspectives on the compositional and functional analysis of soil communities. In: Ritz, K., Dighton, J., Giller, K.E. (Eds.), Beyond the Biomass. Wiley-Sayce, Chichester, pp. 261–271.

Coleman, D.C., Hendrix, P.F., Odum, E.P., 1998. Ecosystem health: an overview. In: Huang, P.M. (Ed.), Soil Chemistry and Ecosystem Health. Soil Science Society of America Special Publication No. 52, Madison, WI, pp. 1–20.

Coleman, D.C., Blair, J.M., Elliott, E.T., Wall, D.H., 1999. Soil invertebrates. In: Robertson, G.P., Coleman, D.C., Bledsoe, C.S., Sollins, P. (Eds.), Standard Soil Methods for Long-Term Ecological Research. Oxford University Press, New York, pp. 349–377.

Coleman, D.C., Fu, S., Hendrix, P.F., Crossley Jr., D.A., 2002. Soil foodwebs in agroecosystems: impacts of herbivory and tillage management. Eur. J. Soil Biol. 38, 21–28.

Coleman, D.C., Hunter, M.D., Hendrix, P.F., Crossley Jr., D.A., Simmons, B., Wickings, K., 2006. Long-term consequences of biochemical and biogeochemical changes in the horseshoe bend agroecosystem, Athens, GA. Eur. J. Soil Biol. 42, S79–S84.

Coleman, D.C., Vadakattu, G., Moore, J.C., 2012. Soil ecology and agroecosystem studies: a dynamic world. In: Cheeke, T., Coleman, D.C., Wall, D.H. (Eds.), Microbial Ecology in Sustainable Agroecosystems. CRC Press, Boca Raton, FL, pp. 1–21.

Collins, C.G., Wright, S.J., Wurzburger, N., 2016. Root and leaf traits reflect distinct resource acquisition strategies in tropical lianas and trees. Oecologia 180 (4), 1037–1047.

Connell, J.H., Lowman, M.D., 1989. Low diversity tropical rain forests: some possible mechanisms for their existence. Am. Nat. 122, 661–696.

Conrad, R., 2009. The global methane cycle: recent advances in understanding the microbial processes involved. Environ. Microbiol. Rep. 1, 285–292.

Copeland, T.P., Imadaté, G., 1990. Insecta: protura. In: Dindal, D.L. (Ed.), Soil Biology Guide. Wiley, New York, pp. 911–933.

Cornelissen, J.H.C., Aerts, R., Cerabolini, B., Werger, M.J.A., van der Heijden, M.G.A., 2001. Carbon cycling traits of plant species are linked with mycorrhizal strategy. Oecologia 129, 611–619.

Costanza, R., d'Arge, R., de Groot, R., Farber, S., Grasso, M., Hannon, B., et al., 1997. The value of the world's ecosystem services and natural capital. Nature 387, 253–260.

Costello, D.M., Tiegs, S.D., Lamberti, G.A., 2011. Do non-native earthworms in Southeast Alaska use streams as invasional corridors in watersheds harvested for timber? Biol. Invasions 13, 177–187.

Cotrufo, M.F., Wallenstein, M.D., Boot, C.M., Denef, K., Paul, E., 2013. The Microbial Efficiency Matrix Stabilization (MEMS) Framework integrates plant litter decomposition with soil organic matter stabilization: do labile plant inputs form stable soil organic matter? Global Change Biol. 19, 988–995.

Coûteaux, M.-M., 1972. Distribution des thécamoebiens de la litiére et de l'humus de deux sols forestier d'humus brut. Pedobiologia 12, 237–243.

Coûteaux, M.-M., 1985. Relation entre la densité apparente d'un humus et l'aptitude a la croissance de ses ciliés. Pedobiologia 28, 289–303.

Coûteaux, M.-M., Mousseau, M., Celerier, M.L., Bottner, P., 1991. Increased atmospheric CO_2 and litter quality: decomposition of sweet chestnut leaf litter with animal food webs of different complexities. Oikos 61, 54–64.

Coyle, D.R., Duman, J.G., Raffa, K.F., 2011. Temporal and species variation in cold hardiness among invasive rhizophagous weevils (Coleoptera: Curculionidae) in a northern hardwood forest. Ann. Entomol. Soc. Am. 104, 59–67.

Crawford, C.S., 1981. Biology of Desert Invertebrates. Springer-Verlag, New York.

Crawford, C.S., 1990. Scorpiones, solfugae, and associated desert taxa. In: Dindal, D.L. (Ed.), Soil Biology Guide. Wiley, New York, pp. 421–475.

Crocker, R.L., 1952. Soil genesis and the pedogenic factors. Quart. Rev. Biol. 27, 139–168.

Crocker, T.L., Hendrick, R.L., Ruess, R.W., Pregitzer, K.S., Burton, A.J., Allen, M.F., et al., 2003. Substituting root numbers for length: improving the use of minirhizotrons to study fine root dynamics. Appl. Soil Ecol. 23, 127–135.

Cromack, K., 1973. Litter Production and Litter Decomposition in a Mixed Hard-Wood Watershed and in a White Pine Watershed at Coweeta Hydrologic Station, North Carolina. Ph.D. Dissertation. University of Georgia, Athens, Georgia, USA.

Cromack Jr., K., Fichter, B.L., Moldenke, A.M., Entry, J.A., Ingham, E.R., 1988. Interactions between soil animals and ectomycorrhizal fungal mats. Agric., Ecosyst. Environ. 24, 161–168.

Crossley Jr., D.A., 1960. Comparative external morphology and taxonomy of nymphs of the Trombiculidae (Acarina). Univ. Kans. Sci. Bull. 40, 135–321.

Crossley Jr, D.A., Blair, J.M., 1991. A high efficiency, 'low technology' Tullgren-type extractor for soil microarthropods. Agric., Ecosyst. Environ. 34, 182–187.

Crossley Jr., D.A., Hoglund, M.P., 1962. A litter-bag method for the study of microarthropods inhabiting leaf litter. Ecology 43, 571–573.

Crossley Jr., D.A., Mueller, B.R., Perdue, J.C., 1992. Biodiversity of microarthropods in agricultural soils: relations to functions. Agric., Ecosyst. Environ. 40, 37–46.

Crotty, F.V., Adl, S.M., Blackshaw, R.P., Murray, P.J., 2012. Using stable isotopes to differentiate trophic feeding channels within soil food webs. J. Eukaryotic Microbiol. 59, 520–526.

Crotty, F.V., Blackshaw, R.P., Adl, S.M., Inger, R., Murray, P.J., 2014. Divergence of feeding channels within the soil food web determined by ecosystem type. Ecol. Evol. 4, 1–13.

Crotty, F.V., Fychan, R., Benefer, C.M., Allen, D., Shaw, P., Marley, C.L., 2016. First documented pest outbreak of the herbivorous springtail *Sminthurus viridis* (Collembola) in Europe. Grass Forage Sci. 71, 699–704.

Crowder, D.W., Northfield, T.D., Strand, M.R., Snyder, W.E., 2010. Organic agriculture promotes evenness and natural pest control. Nature 466, 109–112.

Crowe, J.H., 1975. The physiology of cryptobiosis in tardigrades. Memoirs of the Institute of Italian Hydrobiology 32 (Suppl.), 37–59.

Crowe, J.H., Cooper Jr., A.F., 1971. Cryptobiosis, 225. Sci. Am.30–36.

Curl, E.A., Truelove, B., 1986. The Rhizosphere.. Springer-Verlag, Berlin, NY.

Currie, W.S., 2003. Relationships between carbon turnover and bioavailable energy fluxes in two temperate forest soils. Global Change Biol. 9, 919–929.

Curry, J.P., 1994. Grassland Invertebrates. Ecology, Influence on Soil Fertility and Effects on Plant Growth. Chapman & Hall, London.

Curry, J.P., Good, J.A., 1992. Soil fauna degradation and restoration. Adv. Soil Sci. 17, 171–215.

Curry, J.P., Byrne, D., Boyle, K.E., 1995. The earthworm population of a winter cereal field and its effects on soil and nitrogen turnover. Biol. Fertil. Soils 19, 166–172.

Cutler, D.W., Crump, L.M., Sandon, H., 1923. A quantitative investigation of the bacterial and protozoan population of the soil. Philos. Trans. R. Soc., B 211, 317–350.

Daily, G. (Ed.), 1997. Nature's Services. Societal Dependence on Natural Ecosystems.. Island Press, Washington, DC.

Daniel, O., Anderson, J.M., 1992. Microbial biomass and activity in contrasting soil materials after passage through the gut of the earthworm *Lumbricus rubellus* Hoffmeister. Soil Biol. Biochem. 24, 465–470.

Darbyshire, J.F., Greaves, M., 1967. Protozoa and bacteria in the rhizosphere of *Sinapis alba* (L), *Trifolium repens* (L.), and *Lolium perenne* (L). Can. J. Microbiol. 13, 1057–1068.

Darwin, C., 1837. On the formation of mould. Proc. Geol. Soc. 2, 574–576.

Darwin, C., 1881. The Formation of Vegetable Mould, Through the Action of Worms, With Observations on Their Habits.. John Murray, London.

David, J.-F., Gillon, D., 2002. Annual feeding rate of the millipede *Glomeris marginata* on holm oak (*Quercus ilex*) leaf litter under Mediterranean conditions. Pedobiologia 46, 42–52.

Davidson, D.A., Bruneau, P.M.C., Grieve, I.C., Young, I.M., 2002. Impacts of fauna on an upland grassland soil as determined by micromorphological analysis. Appl. Soil Ecol. 20, 133–143.

Davis, E.L., Hussey, R.S., Baum, T.J., Bakker, J., Schots, A., Rosso, M., et al., 2000. Nematode parasitism genes. Annu. Rev. Phytopathol. 38, 365–396.

Davis, J.P., Haines, B.L., Coleman, D.C., Hendrick, R.L., 2004. Fine root dynamics along an elevational gradient in the southern Appalachian Mountains, USA. For. Ecol. Manage. 187, 19–34.

Dayrat, B., Conrad, M., Balayan, S., White, T.R., Albrecht, C., Golding, R., et al., 2011. Phylogenetic relationships and evolution of pulmonate gastropods (Mollusca): new insights from increased taxon sampling. Mol. Phylogenet. Evol. 59, 425–437.

De Angelis, D.L., 1992. Dynamics of Nutrient Cycling and Food Webs. Chapman & Hall, London.

deDeyn, G.B., Cornelissen, J.H.C., Bardgett, R.D., 2008. Plant functional traits and soil carbon sequestration in contrasting biomes. Ecol. Lett. 11, 516–531.

de Deyn, G.B., Quirk, H., Yi, Z., Oakley, S., Ostle, N.J., Bardgett, R.D., 2009. Vegetation composition promotes carbon and nitrogen storage in model grassland communities of contrasting soil fertility. J. Ecol 97, 864–875.

Del Grosso, S.J., Ojima, D.S., Parton, W.J., Mosier, A.R., Peterson, G.A., Schimel, D.S., 2002. Simulated effects of dryland cropping intensification on soil organic matter and greenhouse gas exchanges using the DAYCENT ecosystem model. Environ. Pollut. 116, S75–S83.

Del Grosso, S.J., Mosier, A.R., Parton, W.J., Ojima, D.S., 2005. DAYCENT model analysis of past and contemporary soil N_2O and net greenhouse gas flux for major crops in the USA. Soil Tillage Res. 83, 9–24.

Delsuc, F., Phillips, M.J., Penny, D., 2003. Comment on hexapod origins: monophyletic or paraphyletic? Date accessed: May 3, 2017, Science, http://science.sciencemag.org/content/sci/301/5639/1482.4.full.pdf.

Denef, K., Six, J., Bossuyt, H., Frey, S.D., Elliott, E.T., Merckx, R., et al., 2001. Influence of dry-wet cycles on the interrelationship between aggregate, particulate organic matter, and microbial community dynamics. Soil Biol. Biochem. 33, 1599–1611.

Denis, R., 1949. Super-ordre des Ectotrophes. In: Grassé, P.P. (Ed.), Traité de Zoologie: Anatomie, Systématique, Biologie., vol. IX. Masson, Paris, pp. 209–275.

De Ruiter, P.C., Moore, J.C., Zwart, K.B., Bouwman, L.A., Hassink, J., Bloem, J., et al., 1993. Simulation of nitrogen mineralization in the below-ground food webs of two winter wheat fields. J. Appl. Ecol. 30, 95–106.

De Ruiter, P.C., Neutel, A.-M., Moore, J.C., 1998. Biodiversity in soil ecosystems: the role of energy flow and community stability. Appl. Soil Ecol. 10, 217–228.

de Vries, F.T., Liiri, M.E., Bjørnlund, L., Bowker, M.A., Christensen, S., Setälä, H.M., et al., 2012. Land use alters the resistance and resilience of soil food webs to drought. Nat. Clim. Change 2, 276–280.

De Wit, H.A., 1978. Soils and Grassland Types of the Serengeti Plain (Tanzania). Ph.D. Thesis. Agricultural University, Wageningen, The Netherlands.

Diaz, S., Cabido, M., 2001. Vive la différence: plant functional diversity matters to ecosystem processes. Trends Ecol. Evol. 16, 646–655.

Didden, W.A.M., 1990. Involvement of enchytraeidae (Oligochaeta) in soil structure evolution in agricultural fields. Biol. Fertil. Soils 9, 152–158.

Didden, W.A.M., 1993. Ecology of enchytraeidae. Pedobiologia 37, 2–29.

Didden, W.A.M., 1995. The effect of nitrogen deposition on enchytraeid-mediated decomposition and mobilization- a laboratory experiment. Acta Zoologica Fennica 196, 60–64.

Dighton, J., 2003. Fungi in Ecosystem Processes. Dekker, New York.

Dillon, E.S., Dillon, L.S., 1961. A Manual of Common Beetles of Eastern North America. Row, Peterson, and Co., Evanston, IL.

Dindal, D.L. (Ed.), 1990. Soil Biology Guide. John Wiley & Sons, New York.

Domene, X., Chelinho, S., Campana, P., Natal-da-Luz, T., Alcañiz, J.M., Andrés, P., et al., 2011. Influence of soil properties on the performance of *Folsomia candida*: implications for its use in soil ecotoxicology testing. Environ. Toxicol. Chem. 30, 1497–1505.

Dongying, W., Hugenholtz, P., Mavromatis, K., Pukall, R., Dalin, E., Ivanova, N.N., et al., 2009. A phylogeny driven genomic encyclopaedia of Bacteria and Archaea. Nature 462, 1056–1060.

Donner, J., 1966. Rotifers. Warne, London.

Doran, J.W., 1980a. Microbial changes associated with residue management with reduced tillage. Soil Sci. Soc. Am. J. 44, 518–524.

Doran, J.W., 1980b. Soil microbial and biochemical changes associated with reduced tillage. Soil Sci. Soc. Am. J. 44, 765–771.

Doran, J.W., 2002. Soil health and global sustainability: translating science into practice. Agric., Ecosyst. Environ. 88, 119–127.

Doran, J.W., Parkin, T.B., 1994. Defining and assessing soil quality. In: Doran, J.W., Coleman, D.C., Bezdicek, D.F., Stewart, B.A. (Eds.), Defining Soil Quality for a Sustainable Environment. SSSA Special Publication 35, ASA, Madison, WI, pp. 3–21.

Doran, J.W., Coleman, D.C., Bezdicek, D.F., Stewart, B.S. (Eds.), 1994. Defining Soil Quality for a Sustainable Environment. SSSA Special Publication, ASA, Madison, WI.

Dornbush, M.E., Isenhart, T.M., Raich, J.W., 2002. Quantifying fine-root decomposition: an alternative to buried litterbags. Ecology 83, 2985–2990.

Dósza-Farkas, K., 1996. Reproduction strategies in some enchytraeid species. In: Dósza-Farkas, K. (Ed.), Newsletter on Enchytraeidae No. 5. Eötvös Loránd University, Budapest, pp. 25–33.

Drake, H.L., Horn, M.A., 2007. As the worm turns: The earthworm gut as a transient habitat for soil microbial biomes. Annu. Rev. Microbiol. 61, 169–189.

Draney, M.S., 1997. Ground-layer spiders (Araneae) of a Georgia Piedmont floodplain agroecosystem: species list, phenology, and habitat selection. J. Arachology 25, 333–351.

Drigo, B., Pijl, A.S., Duyts, H., Kielak, A.M., Gamper, H.A., Houtekamer, M.J., et al., 2010. Shifting carbon flow from roots into associated microbial communities in response to elevated atmospheric CO_2. Proc. Natl. Acad. Sci. 107, 10938–10942.

Duarte, A.P., Melo, V.F., Brown, G.G., Pauletti, V., 2014. Earthworm (*Pontoscolex corethrurus*) survival and impacts on properties of soils from a lead mining site in Southern Brazil. Biol. Fertil. Soils 50, 851–860.

Dufrene, M., Legendre, P., 1997. Species assemblages and indicator species: the need for a flexible asymmetrical approach. Ecol. Monogr. 67 (3), 345–366.

Dungait, J.A., Hopkins, D.W., Gregory, A.S., Whitmore, A.P., 2012. Soil organic matter turnover is governed by accessibility not recalcitrance. Global Change Biol. 18, 1781–1796.

Dunger, W., 1983. Fauna in Soils. Ziemsen Verlag, Wittenberg Lutherstadt.

Dyer, M.I., Bokhari, U.G., 1976. Plant-animal interactions: studies of the effects of grasshopper grazing on blue grama grass. Ecology 57, 762–772.

Eash, N.S., Karlen, D.L., Parkin, T.B., 1994. Fungal contributions to soil aggregation and soil quality. In: Doran, J., Coleman, D.C., Bezdicek, D.F., Stewart, B.A. (Eds.), Defining Soil Quality for a Sustainable Environment. SSSA Special Publication No. 35, Madison, WI, pp. 221–228.

Edgar, A.L., 1990. Opiliones (Phalangida). In: Dindal, D.L. (Ed.), Soil Biology Guide. Wiley, New York, pp. 529–581.

Edgecombe, G.D., Giribet, G., 2007. Evolutionary biology of centipedes (Myriapoda: Chiolopoda). Annu. Rev. Entomol. 52, 151–170.

Edwards, C.A., 1959. The ecology of Symphyla. II. Seasonal Soil Migrations. Entomologia Experimentalis et Applicata 2, 257–267.

Edwards, C.A., 1990. Symphyla. In: Dindal, D.L. (Ed.), Soil Biology Guide. Wiley, New York, pp. 891–910.

Edwards, C.A., 1991. The assessment of populations of soil-inhabiting invertebrates. Agric., Ecosyst. Environ. 34, 145–176.

Edwards, C.A., 1998. Earthworm Ecology. St. Lucie Press, Boca Raton.

Edwards, C.A., 2000. Soil invertebrate controls and microbial interactions in nutrient and organic matter dynamics in natural and agroecosystems. In: Coleman, D.C., Hendrix, P.F. (Eds.), Arthropods as Webmasters in Ecosystems. CABI Publishing, Wallingford, pp. 141–159.

Edwards, C.A. (Ed.), 2004. EarthwormEcology. second ed. St. Lucie Press, Boca Raton.

Edwards, C.A., Bohlen, P.J., 1996. Earthworm Biology and Ecology, third ed. Chapman and Hall, London.

Edwards, C.A., Heath, G.W., 1963. The role of soil animals in breakdown of leaf material. In: Doeksen, J., van der Drift, J. (Eds.), Soil Organisms. North Holland Publishing Co., Amsterdam, pp. 76–80.

Edwards, C.A., Lofty, J.R., 1977. Biology of Earthworms., second ed. Chapman and Hall, London.

Edwards, W.M., Shipitalo, M.J., 1998. Consequences of earthworms in agricultural soils: aggregation and porosity. In: Edwards, C.A. (Ed.), Earthworm Ecology. Lewis Publishers, Boca Raton, FL, pp. 147–161.

Ehrenfeld, J.G., Kourtev, P., Huang, W., 2001. Changes in soil functions following invasions of understory plants in deciduous forests. Ecol. Appl. 11, 1287–1300.

Eigenberg, R.A., Doran, J.W., Nienaber, J.A., Ferguson, R.B., Woodbury, B.L., 2002. Electrical conductivity monitoring of soil condition and available N with animal manure and a cover crop. Agric., Ecosyst. Environ. 88, 183–193.

Eisenbeis, G., Wichard, W., 1987. Atlas on the Biology of Soil Arthropods. Springer-Verlag, Stuttgart.

Ekblad, A., Wallander, H., Godbold, D.L., Cruz, C., Johnson, D., Baldrian, P., et al., 2013. The production and turnover of extramatrical mycelium of ectomycorrhizal fungi in forest soils: role in carbon cycling. Plant Soil 366, 1–27.

Elliott, E.T., 1986. Hierarchic aggregate structure and organic C, N, and P in native and cultivated grassland soils. Soil Sci. Soc. Am. J. 50, 627–633.

Elliott, E.T., Coleman, D.C., 1977. Soil protozoan dynamics in a shortgrass prairie. Soil Biol. Biochem. 9, 113–118.

Elliott, E.T., Coleman, D.C., 1988. Let the soil work for Us. Ecol. Bull. 39, 23–32.

Elliott, E.T., Anderson, R.V., Coleman, D.C., Cole, C.V., 1980. Habitable pore space and microbial trophic interactions. Oikos 35, 327–335.

Elliott, E.T., Horton, K., Moore, J.C., Coleman, D.C., Cole, C.V., 1984. Mineralization dynamics in fallow dryland wheat plots, Colorado. Plant & Soil 76, 149–155.

Elliott, E.T., Janzen, H.H., Campbell, C.A., Cole, C.V., Myers, R.J.K., 1994. Principles of ecosystem analysis and their application to integrated nutrient management and assessment of sustainability, Proceedings of Sustainable Land Management for the 21st Century. Vol. 2: Plenary Papers. ISSS, Acapulco, Mexico.

Elton, C.S., 1958. The Ecology of Invasions by Animals and Plants. Methuen-Wiley, London, NY.

Emerson, A.E., 1956. Regenerative behavior and social homeostasis of termites. Ecology 27, 248–258.

Entry, J.A., Rose, C.L., Cromack, K., 1991. Litter decomposition and nutrient release in ectomycorrhizal mat soils of a Douglas fir ecosystem. Soil. Biol. Biochem. 23, 285–290.

Entry, J.A., Rose, C.L., Cromack, K., 1992. Microbial biomass and nutrient concentrations in hyphal mats of the ectomycorrhizal fungus *Hysterangium setchellii* in a coniferous forest soil. Soil Biol. Biochem. 24, 447–453.

Ettema, C.H., Bongers, T., 1993. Characterization of nematode colonization and succession in disturbed soil using the Maturity Index. Biol. Fertil. Soils 16, 79–85.

Ettema, C.H., Yeates, G.W., 2003. Nested spatial biodiversity patterns of nematode genera in a New Zealand forest and pasture soil. Soil Biol. Biochem. 35, 339–342.

Evans, R.D., Belnap, J., 1999. Long-term consequences of disturbance on nitrogen dynamics in an arid ecosystem. Ecology 80, 150–160.

Evans, R.D., Rimer, R., Sperry, L., Belnap, J., 2001. Exotic plant invasion alters nitrogen dynamics in an arid grassland. Ecol. Appl. 11, 1301–1310.

Evans, R.D., Belnap, J., Garcia-Pichel, F., Phillips, S.L., 2003. Global change and the future of biological soil crusts. In: Belnap, J., Lange, O.L. (Eds.), Biological Soil Crusts: Structure, Function, and Management, vol. 150. Springer-Verlag, Berlin, pp. 417–429.

Evans, A.W., 2014. Beetles of Eastern North America. Princeton University Press, Oxford.

FAO (United Nations Food and Agriculture Organization), 1990. Soilless Culture for Horticultural Crop Production. FAO, Rome.

Fahey, T.J., Bledsoe, C.S., Day, F.P., Ruess, R.W., Smucker, A.J.M., 1999. Fine root production and demography. In: Robertson, G.P., Coleman, D.C., Bledsoe, C., Sollins, P. (Eds.), Standard Soil Methods for Long-Term Ecological Research.. Oxford University Press, New York, pp. 437–455.

Falkowski, P.G., Fenchel, T., DeLong, E.F., 2008. The microbial engines that drive earth's biogeochemical cycles. Science 320, 1034–1039.

Farrah, S.R., Bitton, G., 1990. Viruses in the soil environment. In: Bollag, J.-M., Stotzky, G. (Eds.), Soil Biochemistry. Marcel Dekker, New York, pp. 529–556.

Farrar, J., Hawes, M., Jones, D., Lindow, S., 2003. How roots control the flux of carbon to the rhizosphere. Ecology 84, 827–837.

Feller, C., 1997. The concept of soil humus in the past three centuries. In: Yaalon, D.H., Berkowicz, S. (Eds.), History of Soil Science—International Perspectives. Catena Verlag, Reiskirchen, pp. 15–46.

Feller, C., Beare, M.H., 1997. Physical control of soil organic matter dynamics in the tropics. Geoderma 79, 69–116.

Fender, W.M., McKey-Fender, D., 1990. Oligochaeta: megascolecidae and other earthworms from western North America. In: Dindal, D. (Ed.), Soil Biology Guide. Wiley, New York, pp. 357–378.

Fenster, C.R., Peterson, G.A., 1979. Effects of No-Tillage Fallow as Compared to Conventional Tillage in a Wheat-Fallow System. Research Bulletin 289. Agricultural Experiment Station, University of Nebraska, Lincoln, Nebraska.

Ferguson, B.A., Dreisbach, T.A., Parks, C.G., Filip, G.M., Schmitt, C.L., 2003. Coarse-scale population structure of pathogenic Armillaria species in a mixed-conifer forest in the Blue Mountains of northeast Oregon. Can. J. Forest Res. 33, 612–623.

Ferguson, L.M., 1990a. Insecta: diplura. In: Dindal, D.L. (Ed.), Soil Biology Guide.. Wiley, New York, pp. 951–963.

Ferguson, L.M., 1990b. Insecta: microcoryphia and thysanura. In: Dindal, D.L. (Ed.), Soil Biology Guide. Wiley, New York, pp. 935–949.

Fernandez, C.W., Langley, J.A., Chapman, S., McCormack, M.L., Koide, R.T., 2016. The decomposition of ectomycorrhizal fungal necromass. Soil Biol. Biochem. 93, 38–49.

Ferris, H., Bongers, T., 2006. Nematode indicators of organic enrichment. J. Nematol. 38, 3–12.

Ferris, H., Bongers, T., de Goede, R.G.M., 2001. A framework for soil foodweb diagnostics: extension of the nematode faunal analysis concept. Appl. Soil 18, 13–29.

Fiera, C., 2014. Application of stable isotopes and lipid analysis to understand trophic interactions in springtails. North-Western J. Zool. 10, 227–235.

Fierer, N., Jackson, R., 2006. The diversity and biogeography of soil bacterial communities. Proc. Natl. Acad. Sci. U. S. A. 103, 626–631.

Fierer, N., Schimel, J.P., Holden, P.A., 2003. Variations in microbial community composition through two soil depth profiles. Soil Biol. Biochem. 35, 167–176.

Filser, J., 2002. The role of Collembola in carbon and nitrogen cycling in soil. Pedobiologia 46, 234–245.

Finlay, R., 2008. Ecological aspects of mycorrhizal symbiosis: with special emphasis on the functional diversity of interactions involving the extraradical mycelium. J. Exp. Bot. 59, 1115–1126.

Finlay, R.D., Frostegard, A., Sonnerfeldt, A.M., 1992. Utilization of organic and inorganic nitrogen sources by ectomycorrhizal fungi in pure culture and in symbiosis with *Pinus contorta* Dougl. Ex Loud. New Phytol. 120, 105–115.

Fissore, C., Jurgensen, M.F., Pickens, J., Miller, C., Page-Dumroese, D., Giardina, C.P., 2016. Role of soil texture, clay mineralogy, location, and temperature in coarse wood decomposition—a mesocosm experiment. Ecosphere 7 (11), 1–13.

Fitter, A.H., 1985. Functional significance of root morphology and root system architecture. In: Fitter, A.H., Atkinson, D., Read, D.J., Usher, M.B. (Eds.), Ecological Interactions in Soil; Plants, Microbes and Animals. Blackwell, Oxford, pp. 87–106.

Fitter, A.H., 1991. The ecological significance of root system architecture: an economic approach. In: Atkinson, D. (Ed.), Plant Root Growth: An Ecological Perspective. Blackwell, Oxford, pp. 229–243.

Fitter, A.H., 2005. Darkness visible, reflections on underground ecology. J. Ecol. 93, 231–243.

FitzPatrick, E.A., 1984. Micromorphology of Soils. Chapman and Hall, London.

Foelix, R.F., 1996. Biology of Spiders, second ed. Oxford University Press, Oxford.

Fogel, R., 1985. Roots as primary producers in below-ground ecosystems. In: Fitter, A.H., Atkinson, D., Read, D.J., Usher, M.B. (Eds.), Ecological Interactions in Soil: Plants, Microbes and Animals. Blackwell, Oxford, pp. 23–36.

Fogel, R., 1991. Root system demography and production in forest ecosystems. In: Atkinson, D. (Ed.), Plant Root Growth an Ecological Perspective. Blackwell, Oxford, pp. 89–101.

Fogel, R., Lussenhop, J., 1991. The University of Michigan soil biotron: a platform for soil biology research in a natural forest. In: Atkinson, D. (Ed.), Plant Root Growth an Ecological Perspective. Blackwell, Oxford, pp. 61–73.

Foissner, W., 1987a. Soil protozoa: fundamental problems, ecological significance, adaptations in ciliates and testaceans, bioindicators, and guide to the literature. Prog. Protistology 2, 69–212.

Foissner, W., 1987b. Global soil ciliate (Protozoa, Ciliophora) diversity: a probability based approach using large sample collections from Africa, Australia and Antarctica. Biodiversity Conserv. 6, 1627–1638.

Foissner, W., 1994. Soil protozoa as bioindicators in ecosystems under human influence. In: Darbyshire, J.F. (Ed.), Soil Protozoa. CABI Publishing, Wallingford, pp. 147–193.

Folke, C., Holling, C.S., Perrings, C., 1996. Biological diversity, ecosystems and the human scale. Ecol. Appl. 6, 1018–1024.

Fonte, S.J., Barrios, E., Six, J., 2010. Earthworms, soil fertility and aggregate-associated soil organic matter dynamics in the Quesungual agroforestry system. Geoderma 155, 320–328.

Forge, T.A., Kimpinski, J., 2008. Chapter 33: Nematodes. In: Carter, M.R., Gregorich, E.G. (Eds.), Soil Sampling and Methods of Analysis, second ed. CRC Press, Francis & Taylor, Boca Raton, FL, pp. 415–426.

Forterre, P., 2015. The universal tree of life: an update. Front. Microbiol. 6, 717.

Foster, R.C., 1985. In situ localization of organic matter in soils. Quaestiones Entomologicae 21, 609–633.

Foster, R.C., 1988. Microenvironments of soil microorganisms. Biol. Fertil. Soils 6, 189–203.

Foster, R.C., 1994. Microorganisms and soil aggregates. In: Pankhurst, C.E., Doube, B.M., Gupta, V.V.S.R., Grace, P.R. (Eds.), Soil Biota. CSIRO, Melbourne, pp. 144–155.

Foster, R.C., Dormaar, J.F., 1991. Bacteria-grazing amoebae in situ in the rhizosphere. Biol. Fertil. Soils 11, 83–87.

Foster, R.C., Rovira, A.D., Cock, T.W., 1983. Ultrastructure of the Root Soil Interface. America Phytopathology Society, St. Paul, MN.

Foth, H.D., 1990. Fundamentals of Soil Science. Wiley, New York.

Fouquet, D., Costa-Leonardo, A.M., Fournier, R., Blanco, S., Jost, C., 2014. Coordination of construction behavior in the termite *Procornitermes araujoi*: structure is a stronger stimulus than volatile marking. Insect. Soc. 61, 253–264.

Fragoso, C., Kanyonyo, J., Moreno, A., Senapati, B.K., Blanchart, E., Rodriguez, C., 1999. A survey of tropical earthworms: taxonomy, biogeography and environmental plasticity. In: Lavelle, P., Brussaard, L., Hendrix, P. (Eds.), Earthworm Management in Tropical Agroecosystems. CABI Publishing, Wallingford, pp. 1–26.

Francé, R.H., 1921. Das Edaphon. Arbeiten aus dem Biologischen Institut München, Nr. 2, Franckh'sche Verlashandlung, Stuttgart, Germany.

Frank, D.A., Kuns, M.M., Guido, D.R., 2002. Consumer control of grassland plant production. Ecology 83, 602–606.

Franklin, E., Santos, E.M.R., Albuquerque, M.I.C., 2007. Edaphic and arboricolous oribatid mites (Acari; Oribatida) in tropical environments: changes in the distribution of higher level taxonomic groups in the communities of species. Braz. J. Biol. 67 (3), 447–458.

Freckman, D.W., 1994. Life in the soil/soil biodiversity: its importance to ecosystem processes. Report on a workshop held at The Natural HistoryMuseum, London. NREL, Colorado State University, Ft. Collins, Colorado.

Freckman, D.W., Virginia, R.A., 1989. Plant-feeding nematodes in deep-rooting desert ecosystems. Ecology 70, 1665–1678.

Freckman, D.W., Kaplan, D.T., van Gundy, S.D., 1977. A comparison of techniques for extraction and study of anhydrobiotic nematodes from dry soils. J. Nematol. 9, 176–181.

Frelich, L.E., Hale, C.M., Scheu, S., Holdsworth, A.R., Heneghan, L., Bohlen, P.J., et al., 2006. Earthworm invasion into previously earthworm-free temperate and boreal forests. Biol. Invasions 8, 1235–1245.

Freschet, G.T., Aerts, R., Cornelissen, J.H., 2012. Multiple mechanisms for trait effects on litter decomposition: moving beyond home-field advantage with a new hypothesis. J. Ecol. 100 (3), 619–630.

Freschet, G.T., Cornwell, W.K., Wardle, D.A., Elumeeva, T.G., Liu, W., Jackson, B.G., et al., 2013. Linking litter decomposition of above- and below-ground organs to plant-soil feedbacks worldwide. J. Ecol. 101, 943–952.

Frey, S.D., Elliott, E.T., Paustian, K., Peterson, G.A., 2000. Fungal translocation as a mechanism for soil nitrogen inputs to surface residue decomposition in a no-tillage agroecosystem. Soil Biol. Biochem. 32 (5), 689–698.

Frost, S.W., 1942. General Entomology. McGraw-Hill, New York.

Frouz, J. (Ed.), 2015. Soil Biota and Ecosystem Development in Post Mining Sites. CRC Press, Taylor & Francis, Boca Raton, FL.

Fu, S., Kisselle, K.W., Coleman, D.C., Hendrix, P.F., Crossley Jr., D.A., 2001. Short-term impacts of aboveground herbivory (grasshopper) on the abundance and ^{14}C activity of soil nematodes in conventional tillage and no-till agroecosystems. Soil Biol. Biochem. 33, 1253–1258.

Furlong, M.A., Singleton, D.R., Coleman, D.C., Whitman, W.B., 2002. Molecular and culture-based analyses of prokaryotic communities from an agricultural soil and the burrows and casts of the earthworm Lumbricus rubellus. Appl. Environ. Microbiol. 68, 1265–1279.

Gadgil, R.L., Gadgil, P.D., 1975. Suppression of litter decomposition by mycorrhizal roots of *Pinus radiata*. N. Z. J. For. Sci. 5, 33–41.

Gange, A., 2000. Arbuscular mycorrhizal fungi, Collembola and plant growth. Trends Ecol. Evol. 15, 369–372.

Garbaye, J., 1991. Biological interactions in the mycorrhizosphere. Experientia 47, 370–375.

Garcia-Palacios, P., Maestre, F.T., Kattge, J., Wall, D.H., 2013. Climate and litter quality differently modulate the effects of soil fauna on litter decomposition across biomes. Ecol. Lett. 16, 1045–1053.

Garey, J.R., 2001. Ecdysozoa: the relationship between Cycloneuralia and Panarthropoda. Zool. Anz. 240, 321–330.

Garrett, C.J., Crossley Jr., D.A., Coleman, D.C., Hendrix, P.F., Kisselle, K.W., Potter, R.L., 2001. Impact of the rhizosphere on soil microarthropods in agroecosystems on the Georgia piedmont. Appl. Soil Ecol. 16, 141–148.

Garrison, N.L., Rodriguez, J., Agnarsson, I., Coddington, J.A., Griswold, C.E., Hamilton, C.A., et al., 2016. Spider phylogenomics: untangling the spider tree of life. Peer J. 4, e1719, https://doi.org/10.7717/peerj.1719.

Gates, G.E., 1967. On the earthworm fauna of the Great American Desert and adjacent areas. The Great Basin Naturalist 27, 142–1761.

Gaudinski, J.B., Trumbore, S.E., Davidson, E.A., Zheng, S., 2000. Soil carbon cycling in a temperate forest: radiocarbon-based estimates of residence times, sequestration rates and partitioning of fluxes. Biogeochemistry 51, 33–69.

Gaudinski, J.B., Trumbore, S.E., Davidson, E.A., Cook, A.C., Markewitz, D., Richter, D.D., 2001. The age of fine-root carbon in three forests of the eastern United States measured by radiocarbon. Oecologia 129, 420–429.

Geisen, S., 2016. The bacterial-fungal energy channel concept challenged by enormous functional versatility of soil protists. Soil Biol. Biochem. 102, 22–25.

Geisseler, D., Scow, K.M., 2014. Long-term effects of mineral fertilizers on soil microorganisms—a review. Soil Biol. Biochem. 75, 54–63.

Gerland, P., Raftery, A.E., Ševčíková, H., Li, N., Gu, D., Spoorenberg, T., et al., 2014. World population stabilization unlikely this century. Science 346 (6206), 234–237.

Gerson, U., Smiley, R.L., Ochoa, R., 2003. Mites (Acari) for Pest Control. Blackwell Science, Oxford.

Gilbert, G.S., 2002. Evolutionary ecology of plant diseases in natural ecosystems. Annu. Rev. Phytopathol. 40, 13–43.

Gill, H.S., Abrol, I.P., 1986. Salt affected soils and their amelioration through afforestation. In: Prinsley, R.T., Swift, M.J. (Eds.), Amelioration of Soil by Trees. Commonwealth Science Council, London, pp. 43–53.

Gillard, O., 1967. Coprophagous beetles in pasture ecosystems. J. Aust. Inst. Agric. Sci. 33, 30–34.

Giller, K.E., 2001. Nitrogen Fixation in Tropical Cropping Systems, second ed. CABI Publishing, Wallingford.

Gilmore, S.K., 1970. Collembola predation on nematodes. Search Agric. 1 (3), 1–12.

Gilmore, S.K., Potter, D.A., 1993. Potential role of Collembola as biotic mortality agents for entomopathogenic fungi nematodes. Pedobiologia 37, 30–38.

Gisin, H., 1962. Sur la fauna européen des Collemboles IV. Revue Suisse de zoologie 69, 1–23.

Gisin, H., 1963. Collemboles d'Europe V. Revue Suisse de zoologie 70, 77–101.

Gisin, H., 1964. Collemboles d'Europe VII. Revue Suisse de zoologie 71, 649–678.

Gist, C., Crossley Jr., D.A., Merchant, V.A., 1974. An analysis of life tables for *Sinella curviseta* (Collembola). Environ. Entomol. 3, 840–844.

Gist, C.S., Crossley Jr., D.A., 1973. A method for quantifying pitfall trapping. Environ. Entomol. 2, 951–952.

Gist, C.S., Crossley Jr., D.A., 1975. A model of mineral cycling for an arthropod food web in a southeastern hardwood forest litter community. In: Howell, F.G., Smith, M.H. (Eds.), Mineral Cycling in Southeastern Ecosystems. Energy Research and Development Administration, Washington, DC, pp. 84–106.

Gjelstrup, P., Petersen, P., 1987. Jordbundens mider og springhaler (Mites and springtails in the soil). Natur og Museum Århus, 26.41–32.

Glinka, K.D., 1927. Dokuchaiev's ideas in the development of pedology and cognate sciences. In: Academy of Sciences of the Union of the Soviet Socialist Republics. Russian pedological investigations, I. Leningrad, the Academy, Russia.

Golley, F.B., 1993. History of the Ecosystem Concept in Ecology: More Than the Sum of the Parts. Yale University Press, New Haven, CT.

González, G., Seastedt, T.R., 2001. Soil fauna and plant litter decomposition in tropical and subalpine forests. Ecology 82, 955–964.

González, G., Ley, R.L., Schmidt, S.K., Zou, X., Seastedt, T.R., 2001. Soil ecological interactions: comparisons between tropical and subalpine forests. Oecologia 128, 549–556.

Goodale, C.L., Apps, M.J., Birdsey, R.A., Field, C.B., Heath, L.S., Houghton, R.A., et al., 2002. Forest carbon sinks in the northern hemisphere. Ecol. Appl. 12 (3), 891–899.

Goodnight, C.J., Goodnight, M.L., 1960. Speciation among cave opilionids of the United States. Am. Midl. Nat. 64, 34–38.

Górny, M., Grüm, L., 1993. Methods in Soil Zoology. Polish Scientific Publishers, Warsaw.

Graefe, U., 1973. Systematische Untersuchungen and der Gattung *Achaeta* (Enchytraeidae, Oligochaeta). Diplomarbeit. Universität Hamburg, Hamburg.

Green, R.N., Trowbridge, R.L., Klinka, K., 1993. Towards a taxonomic classification of humus forms. For. Sci., Monogr. 29, 1–49.

Greenslade, P., 1964. Pitfall trapping as a method for studying populations of Carabidae (Coleoptera). J. Anim. Ecol. 33, 301–310.

Griffiths, B.S., 1994. In: Darbyshire, J.F. (Ed.), Soil Nutrient Flow. Soil Protozoa. CABI Publishing, Wallingford, pp. 65–91.

Griffiths, B.S., Caul, S., 1993. Migration of bacterial-feeding nematodes, but not protozoa, to decomposing grass residues. Biol. Fertil. Soils 15, 201–207.

Griffiths, B.S., Bonkowski, M., Dobson, G., Caul, S., 1999. Changes in soil microbial community structure in the presence of microbial-feeding nematodes and protozoa. Pedobiologia 43, 297–304.

Griffiths, B.S., Ritz, K., Bardgett, R.D., Cook, R., Christensen, S., Ekelund, F., et al., 2000. Ecosystem response of pasture soil communities to fumigation-induced microbial diversity reductions: an examination of the biodiversity-ecosystem function relationship. Oikos 90, 279–294.

Griffiths, B.S., Ritz, K., Wheatley, R., Kuan, H.L., Boag, B., Christensen, S., et al., 2001. An examination of the biodiversity-ecosystem function relationship in arable soil microbial communities. Soil Biol. Biochem. 33, 1713–1722.

Griffiths, B.S., Römbke, J., Schmelz, R.M., Scheffczyk, A., Faber, J.H., Bloem, J., et al., 2016. Selecting cost effective and policy-relevant biological indicators for European monitoring of soil biodiversity and ecosystem function. Ecol. Indic. 69, 213–223.

Griffiths, D.A., 1996. Mites. In: Hall, G.S. (Ed.), Methods for the Examination of Organismal Diversity in Soils and Sediments. CAB International, Wallingford, pp. 175–185.

Griffiths, E., 1965. Micro-organisms and soil structure. Biol. Rev. 40, 129–142.

Grodzinski, W., Yorks, T.P., 1981. Species and ecosystem-level bioindicators of airborne pollution: an analysis of two major studies. Water, Air, Soil Pollut. 16, 33–53.

Groffman, P.M., Bohlen, P.J., 1999. Soil and sediment biodiversity. BioScience 49, 139–148.

Guggenberger, G., Kaiser, K., 2003. Dissolved organic matter in soil: challenging the paradigm of sorptive preservation. Geoderma 113, 293–310.

Gundale, M.J., 2002. Influence of exotic earthworms on the soil organic horizon and the rare fern *Botrychium mormo*. Conserv. Biol. 16 (6), 1555–1561.

Gunn, A., Cherrett, J.M., 1993. The exploitation of food resources by soil meso- and macroinvertebrates. Pedobiologia 37, 303–320.

Gupta, V.V.S.R., Germida, J.J., 1988. Populations of predatory protozoa in field soils after 5 years of elemental S fertilizer application. Soil Biol. Biochem. 20, 787–791.

Gupta, V.V.S.R., Germida, J.J., 1989. Influence of bacterial-amoebal interactions on sulfur transformations in soil. Soil Biol. Biochem. 21, 921–930.

Gupta, V.V.S.R., Yeates, G.W., 1997. Soil microfauna as bioindicators of soil health. In: Pankhurst, C., Doube, B.M., Gupta, V.V.S.R. (Eds.), Biological Indicators of Soil Health. CABI Publishing, Wallingford, pp. 201–233.

Gyawaly, S., Koppenhofer, A.M., Wu, S., Kuhar, T.P., 2016. Biology, ecology, and management of masked chafer (Coleoptera: Scarabaeidae) grubs in turfgrass. J. Integr. Pest Manage. 7 (1), 1–11.

Haas, H.J., Evans, C.E., Miles, E.F., 1957. Nitrogen and carbon changes in Great Plains soils as influenced by cropping and soil treatments. Technical Bulletin No. 1164, USDA, Washington, DC.

Hadas, A., 1979. Heat capacity. In: Fairbridge, R.W., Finkl, C.W. (Eds.), The Encyclopedia of Soil Science, Part 1: Physics, Chemistry, Biology, Fertility, and Technology. Dowden, Hutchinson & Ross, Stroudsburg, PA, p. 189.

Hairston Jr., N.G., Hairston Sr., N.G., 1993. Cause-effect relationships in energy flow, trophic structure, and interspecific interactions. Am. Nat. 142, 379–411.

Hale, C.M., Frelich, L.E., Reich, P.B., Pastor, J., 2005. Effects of European earthworm invasion on soil characteristics in northern hardwood forests of Minnesota, USA. Ecosystems 8, 911–927.

Hall, G.S. (Ed.), 1996. Methods for the Examination of Organismal Diversity in Soils and Sediments. CAB International, New York.

Hall, S.J., Raffaelli, D.G., 1993. Food webs: theory and reality. Adv. Ecol. Res. 24, 187–239.

Hallsworth, E.G., Crawford, D.V., 1965. Experimental Pedology. Butterworths, London.

Handa, I.T., Aerts, R., Berendse, F., Berg, M.P., Bruder, A., Butenschoen, O., et al., 2014. Consequences of biodiversity loss for litter decomposition across biomes. Nature 509, 218–221.

Hansen, R.A., 1999. Red oak litter promotes a microarthropod functional group that accelerates its decomposition. Plant & Soil 209, 37–45.

Hansen, R.A., 2000. Diversity in the decomposing landscape. In: Coleman, D.C., Hendrix, P.F. (Eds.), Invertebrates as Webmasters in Ecosystems.. CABI Publishing, Wallingford, pp. 203–219.

Hanson, P.J., Edwards, N.T., Garten, C.T., Andrews, J.A., 2000. Separating root and soil microbial contributions to soil respiration. Biogeochemistry 48, 115–146.

Hansson, A., Andrén, O., Steen, E., 1991. Root production of four arable crops in Sweden and its effect on abundance of soil organisms. In: Atkinson, D. (Ed.), Plant Root Growth an Ecological Perspective. Blackwell, Oxford, pp. 247–266.

Hanula, J.L., 1995. Relationship of wood-feeding insects and coarse woody debris. In: McMinn, J.W., Crossley, D.A. (Eds.), Biodiversity and Coarse Woody Debris in Southern Forests. USDA Forest Service, Asheville, NC, pp. 55–81.

Hargrove, W.W., Crossley Jr., D.A., 1988. Video digitizer for the rapid measurement of leaf area lost to herbivorous insects. Ann. Entomol. Soc. Am. 81 (4), 593–598.

Harmon, M.D., Chen, H., 1991. Coarse woody debris dynamics in two old growth ecosystems. BioScience 41, 604–610.

Harris, R.F., Chesters, G., Allen, O.N., Attoe, O.J., 1964. Mechanisms involved in soil aggregate stabilization by fungi and bacteria. Soil Sci. Soc. Am. Proc. 28, 529–532.

Harris, W.F., Kinerson Jr., R.S., Edwards, N.T., 1977. Comparison of belowground biomass of natural deciduous forest and loblolly pine plantations. Pedobiologia 7, 369–381.

Harry, M., Jusseaume, N., Gambier, B., Garnier-Sillam, E., 2001. Use of RAPD markers for the study of microbial community similarity from termite mounds and tropical soils. Soil Biol. Biochem. 33, 417–427.

Hartnett, D.C., Wilson, G.W.T., 1999. Mycorrhizae influence plant community structure and diversity in tallgrass prairie. Ecology 80, 1187–1195.

Harvey, R.W., Kinner, N.E., Bunn, A., MacDonald, D., Metge, D., 1995. Transport behavior of groundwater protozoa and protozoan-sized micro-spheres in sandy aquifer sediments. Appl. Environ. Microbiol. 61, 209–271.

Hashimoto, T., Horikawa, D.D., Saito, Y., Kuwahara, H., Kozuka-Hata, H., Shin-I, T., et al., 2016. Extremotolerant tardigrade genome and improved radiotolerance of human cultured cells by tardigrade-unique protein. Nat. Commun. 7, 12808.

Hastings, J.M., Holliday, C.W., Long, A., Jones, K., Rodriguez, G., 2010. Size-Specific Provisioning by Cicada Killers, Sphecius speciosus, (Hymenoptera: Crabronidae) in North Florida. Fla. Entomol. 93, 412–421.

Hattori, T., 1994. Soil micro environment. In: Darbyshire, J.F. (Ed.), Soil Protozoa. CABI Publishing, Wallingford, pp. 43–64.

Hawksworth, D.L., 1991a. The Biodiversity of Microorganisms and Invertebrates. CABI Publishing, Wallingford.

Hawksworth, D.L., 1991b. The fungal dimension of biodiversity: magnitude, significance and conservation. Mycol. Res. 95, 641–655.

Hawksworth, D.L., 2001. The magnitude of fungal diversity: the 1.5 millions species estimate revisited. Mycol. Res. 105, 1422–1432.

He, Y., Trumbore, S.E., Torn, M.S., Harden, J.W., Vaughn, L.J.S., Allison, S.D., et al., 2016. Radiocarbon constraints imply reduced carbon uptake by soils during the 21st century. Science 353 (6306), 1419–1424.

Heal, O.W., Dighton, J., 1985. Resource quality and trophic structure in the soil system. In: Fitter, A.H., Atkinson, D., Read, D.J., Usher, M.B. (Eds.), Ecological Interactions in Soil: Plants, Microbes and Animals. Blackwell Scientific, Oxford, pp. 339–354.

Heal, O.W., Anderson, J.M., Swift, M.J., 1997. Plant litter quality and decomposition: an historical overview. In: Cadisch, G., Giller, K.E. (Eds.), Driven by Nature. Plant Litter Quality and Decomposition. CABI Publishing, Wallingford, pp. 3–30.

Hedley, M.J., Stewart, J.W.B., 1982. Method to measure microbial phosphate in soil. Soil Biol. Biochem. 14, 377–385.

Heidemann, K., Scheu, S., Ruess, L., Maraun, M., 2011. Molecular detection of nematode predation and scavenging in oribatid mites: laboratory and field experiments. Soil Biol. Biochem. 43, 2229–2236.

Heidemann, K., Hennies, A., Schakowske, J., Blumenberg, L., Ruess, L., Scheu, S., et al., 2014. Free-living nematodes as prey for higher trophic levels of forest soil food webs. Oikos 123, 1199–1211.

Helal, H.M., Sauerbeck, D., 1991. Short term determination of the actual respiration rate of intact plant roots. In: Michael, B.L., Perssons, H. (Eds.), Plant Roots and Their Environment. Elsevier Science Publishers, Amsterdam, pp. 88–92.

Henderson, L.J., 1913. The Fitness of the Environment. Beacon Hill Press, Boston, MA.

Henderson, L.S., 1952. Household insects. In: Stefferud, A. (Ed.), Insects, the Yearbook of Agriculture 1952. U.S. Government Printing Office, Washington, DC, pp. 469–475.

Hendrick, R.L., Pregitzer, K.S., 1992. The demography of fine roots in a Northern hardwood forest. Ecology 73, 1094–1104.

Hendrix, P.F. (Ed.), 1995. Earthworm Ecology and Biogeography in North America. Lewis Publishers, Boca Raton, FL.

Hendrix, P.F., 2000. Earthworms. In: Sumner, M.E. (Ed.), Handbook of Soil Science. CRC Press, Boca Raton, FL, pp. C-77–C-85.

Hendrix, P.F., Bohlen, P.J., 2002. Exotic earthworm invasions in North America: ecological and policy implications. BioScience 52, 801–811.

Hendrix, P.F., Parmelee, R.W., Crossley Jr., D.A., Coleman, D.C., Odum, E.P., Groffman, P., 1986. Detritus food webs in conventional and no-tillage agroecosystems. Bioscience 36, 374–380.

Hendrix, P.F., Crossley Jr., D.A., Coleman, D.C., Parmelee, R.W., Beare, M.H., 1987. Carbon dynamics in soil microbes and fauna in conventional and no-tillage agroecosystems. INTECOL Bull. 15, 59–63.

Hendrix, P.F., Crossley Jr., D.A., Blair, J.M., Coleman, D.C., 1990. Soil biota as components of sustainable agroecosystems. In: Edwards, C.A., Lal, R., Madden, P., Miller, R.H., House, G. (Eds.), Sustainable Agricultural Systems. Soil and Water Conservation Society, Ankeny, IA, pp. 637–654.

Hendrix, P.F., Coleman, D.C., Crossley Jr., D.A., 1992. Using knowledge of soil nutrient cycling processes to design sustainable agriculture. J. Sustainable Agric. 2, 63–82.

Hendrix, P.F., Callaham Jr., M.A., Kirn, L., 1994. Ecological studies of nearctic earthworms in the southern USA: II. Effects of bait harvesting on Diplocardia populations in Apalachicola National Forest in north Florida. Megadrilogica 5, 73–76.

Hendrix, P.F., Callaham Jr., M.A., Drake, J.M., Huang, C.-Y., James, S.W., Snyder, B.A., et al., 2008. Pandora's box contained bait: the global problem of introduced earthworms. Annu. Rev. Ecol. Evol. Syst. 39, 593–613.

Heneghan, L., Coleman, D.C., Zou, X., Crossley Jr., D.A., Haines, B.L., 1998. Soil microarthropod community structure and litter decomposition dynamics: a study of tropical and temperate sites. Appl. Soil Ecol. 9, 33–38.

Heneghan, L., Coleman, D.C., Zou, X., Crossley Jr., D.A., Haines, B.L., 1999. Soil microarthropod contributions to decomposition dynamics: tropical-temperate comparisons of a single substrate. Appl. Soil Ecol. 9, 33–38.

Hijii, N., 1987. Seasonal changes in abundance and spatial distribution of the soil arthropods in a Japanese cedar (*Cryptomeria japonica* D. Don) plantation, with special reference to Collembola and Acarina. Ecol. Res. 2, 159–173.

Hillel, D.J., 1991. Out of the Earth: Civilization and the Life of the Soil. Free Press, New York.

Hillel, D.J., 1998. Environmental Soil Physics. Academic Press, San Diego, CA.

Hiltner, L., 1904. Über neuere erfahrungen und probleme auf dem gebiet der bodenbakteriologie und unter besonderer berücksichtigung der gründüngung und brache. Arbeiten der Deutschen Landwirtschafts-Gesellschaft H. 98, 59–78.

Hobbie, E.A., Agerer, R., 2010. Nitrogen isotopes in ectomycorrhizal sporocarps correspond to belowground exploration types. Plant and Soil 327, 71–83.

Hobbie, E.A., Weber, N.S., Trappe, J.M., van Klinken, G.J., 2002. Using radiocarbon to determine the mycorrhizal status of fungi. New Phytol. 156, 129–136.

Hoff, C.C., 1949. The pseudoscorpions of Illinois. Ill. Nat. Hist. Surv. Bull. 24, 407–498.

Hoffman, R.L., 1990. Diplopoda. In: Dindal, D.L. (Ed.), Soil Biology Guide. Wiley, New York, pp. 835–860.

Hoffman, R.L., 1999. Checklist of the Millipedes of North and Middle America. Special Publication Number 8. Virginia Museum of Natural History, Martinsville, Virginia.

Högberg, M.N., Briones, M.J.I., Keel, S.G., Metcalfe, D.B., Campbell, C., Midwood, A.J., et al., 2010. Quantification of effects of season and nitrogen supply on tree below-ground transfers to ectomycorrhizal fungi and other soil organisms in a boreal pine forest. New Phytol. 187, 485–493.

Holland, E.A., Coleman, D.C., 1987. Litter placement effects on microbial and organic matter dynamics in an agroecosystem. Ecology 68, 425–433.

Holland, J.N., Cheng, W., Crossley Jr., D.A., 1996. Herbivore-induced changes in plant carbon allocation: assessment of below-ground C fluxes using carbon-14. Oecologia 107, 87–94.

Hölldobler, B., Wilson, E.O., 1990. The Ants. Belknap Press, Harvard University, Cambridge.

Holtkamp, R., van der Wal, A., Kardol, P., van der Putten, W., de Ruiter, P.C., Dekker, S.C., 2011. Modelling Carbon and Nitrogen mineralization in soil food webs during secondary succession on ex-arable land. Soil Biol. Biochem. 43, 251–260.

Hominick, W.M., 2002. Biogeography. In: Gaugler, R. (Ed.), Entomopathogenic Nematology. CABI Publishing, Wallingford, pp. 115–143.

Hongoh, Y., Ohkuma, M., Kudo, T., 2003. Molecular analysis of bacterial microbiota in the gut of the termite *Reticulitermes speratus* (Isoptera; Rhinotermitidae). FEMS Microbiol. Ecol. 44, 231–242.

Hooper, D.U., Bignell, D.E., Brown, V.K., Brussaard, L., Dangerfield, J.M., Wall, D.H., et al., 2000. Interactions between aboveground and belowground biodiversity in terrestrial ecosystems: patterns, Mechanisms, and Feedbacks. BioScience 50, 1049–1061.

Hoover, C., Crossley Jr., D.A., 1995. Leaf litter decomposition and microarthropod abundance along an altitudinal gradient. In: Collins, H.P., Robertson, G.P., Klug, M.J. (Eds.), The Significance and Regulation of Soil Biodiversity. Kluwer Academic, Dordrecht, pp. 287–292.

Hopkin, S.P., 1997. The Biology of Springtails (Insecta: Collembola). Oxford University Press, Oxford.
Hopkin, S.P., Read, H.J., 1992. The Biology of Millipedes. Oxford University Press, Oxford.
Horwath, W.R., Pregitzer, K.S., Paul, E.A., 1994. ^{14}C allocation in tree-soil systems. Tree Physilol. 14, 163–176.
Hotopp, K.P., 2002. Land snails and soil calcium in central appalachian mountain forest. Southeastern Naturalist 1, 27–44.
Houghton, R.A., 2003. Why are estimates of the terrestrial carbon balance so different? Global Change Biol. 9, 500–509.
Houghton, R.A., Hobbie, J.E., Melillo, J.M., Moore, B., Peterson, B.J., Shaver, G.R., et al., 1983. Changes in the carbon content of terrestrial biota and soils between 1860 and 1980: a net release of CO_2 to the atmosphere. Ecol. Monogr. 53, 235–262.
House, G.J., Stinner, B.R., Crossley Jr., D.A., 1984. Nitrogen cycling in conventional and no-tillage agroecosystems: analysis of pathways and processes. J. Appl. Ecol. 21, 991–1012.
House, G.J., Worsham, A.D., Sheets, T.J., Stinner, R.E., 1987. Herbicide effects on soil arthropod dynamics and wheat straw decomposition in a North Carolina no-tillage agroecosystem. Biol. Fertil. Soils 4, 109–114.
Hubricht, L., 1985. The distributions of the native land mollusks of the eastern United States. Fieldiana, Zoology, New Series 24. Field Museum of Natural History, Chicago, IL.
Hugenholtz, P., Goebel, B.M., Pace, N.R., 1998. Impact of culture-independent studies on the emerging phylogenetic view of bacterial diversity. J. Bacteriol. 180, 4765–4774.
Huhta, V., Wright, D.H., Coleman, D.C., 1989. Characteristics of defaunated soil. I. A comparison of three techniques applied to two different forest soils. Pedobiologia 33, 417–426.
Humphreys, W.F., 1979. Production and respiration in animal populations. J. Anim. Ecol. 48, 427–453.
Hunt, H.W., Wall, D.H., 2002. Modelling the effects of loss of soil biodiversity on ecosystem function. Global Change Biol. 8, 33–50.
Hunt, H.W., Coleman, D.C., Ingham, E.R., Ingham, R.E., Elliott, E.T., Moore, J.C., et al., 1987. The detrital food web in a shortgrass prairie. Biol. Fertil. Soils 3, 57–68.
Hunt, H.W., Ingham, E.R., Coleman, D.C., Elliott, E.T., Reid, C.P.P., 1988. Nitrogen limitation of production and decomposition in prairie, mountain meadow and pine forest. Ecology 69, 1009–1016.
Hunter, P.E., Rosario, R.M.T., 1988. Associations of Mesostigmata with other arthropods. Annu. Rev. Entomol. 33, 393–413.
Husband, R., Herre, E.A., Turner, S.L., Gallery, R., Young, J.P.W., 2002. Molecular diversity of arbuscular mycorrhizal fungi and patterns of host association over time and space in a tropical forest. Mol. Ecol. 11, 2669–2678.
Hutchinson, G.E., 1957. Concluding remarks. Cold Spring Harbor Symp. Quan. Biol. 22, 415–427.
Iannone III, B.V., Umek, L.G., Wise, D.H., Heneghan, L., 2012. A simple, safe, and effective sampling technique for investigating earthworm communities in woodland soils: implications for citizen science. Nat. Areas J. 32 (3), 283–292.
Ingham, E.R., Klein, D.A., 1982. Relationship between fluorescein diacetate-stained hyphae and oxygen utilization, glucose utilization, and biomass of submerged fungal batch cultures. Appl. Environ. Microbiol. 44, 363–370.
Ingham, E.R., Trofymow, J.A., Ames, R.N., Hunt, H.W., Morley, C.R., Moore, J.C., et al., 1986a. Trophic interactions and nitrogen cycling in a semi-arid grassland soil. I. Seasonal dynamics of the natural populations, their interactions and effects on nitrogen cycling. J. Appl. Ecol. 23, 597–614.
Ingham, E.R., Trofymow, J.A., Ames, R.N., Hunt, H.W., Morley, C.R., Moore, J.C., et al., 1986b. Trophic interactions and nitrogen cycling in a semiarid grassland soil. II. System responses to removal of different groups of soil microbes or fauna. J. Appl. Ecol. 23, 615–630.
Ingham, E.R., Horton, K.A., 1987. Bacterial, fungal and protozoan responses to chloroform fumigation in stored soil. Soil Biol. Biochem. 19, 545–550.
Ingham, E.R., Coleman, D.C., Moore, J.C., 1989. An analysis of food web structure and function in a shortgrass prairie, mountain meadow and lodgepole pine forest. Biol. Fertil. Soils 8, 29–37.
Ingham, R.E., Trofymow, J.A., Ingham, E.R., Coleman, D.C., 1985. Interactions of bacteria, fungi, and their nematode grazers: effects on nutrient cycling and plant growth. Ecol. Monogr. 55, 119–140.
Insam, H., 1990. Are the soil microbial biomass and basal respiration governed by the climatic regime? Soil Biol. Biochem. 22, 525–532.
Insam, H., Domsch, K.H., 1988. Relationship between soil organic carbon and microbial biomass on chronosequences of reclamation sites. Microb. Ecol. 15, 177–188.

Intergovernmental Panel on Climate Change, 2013. Climate Change 2013: The Physical Science Basis. Cambridge University Press, New York, NY.

Ito, M., Abe, W., 2001. Micro-distribution of soil inhabiting tardigrades (Tardigrada) in a sub-alpine coniferous forest of Japan. Zool. Anzeiger 240, 403–407.

Jackman, J.A., 1997. A Field Guide to Spiders and Scorpions of Texas. Gulf Publication, Houston, TX.

Jackson, R.R., Willey, M.B., 1994. The comparative study of the behavior of *Myrmarachne*, ant-like jumping spiders (Araneae: Salticidae). Zool. J. Linn. Soc. 110, 77–91.

Jacot, A.P., 1936. Soil structure and soil biology. Ecology 17, 359–379.

James, S.W., 1990. Oligochaeta: megascolecidae and other earthworms from Southern and Midwestern North America. In: Dindal, D.L. (Ed.), Soil Biology Guide. Wiley, New York, pp. 379–386.

James, S.W., 2004. Planetary processes and their interactions with earthworm distributions and ecology. In: Edwards, C.A. (Ed.), Earthworm Ecology, second ed. CRC Press, Boca Raton, FL, pp. 53–62.

James, S.W., Davidson, S.K., 2012. Molecular phylogeny of earthworms (Annelida: Crassiclitellata) based on 28S, 18S and 16S gene sequences. Invertebrate Systematics 26, 213–229.

Jamieson, B.G.M., 1988. On the phylogeny and higher classification of the Oligochaeta. Cladistics 4, 367–401.

Janssens, F., Christiansen, K.A., 2011. Class Collembola Lubbock, 1870. In: Zhang, Z.-Q. (Ed.) Animal biodiversity: an outline of higher-level classification and survey of taxonomic richness. Zootaxa, 3148, 192–193.

Jastrow, J.D., Miller, R.M., 1991. Methods for assessing the effects of biota on soil structure. Agric., Ecosyst. Environ. 34, 279–303.

Jastrow, J.D., Miller, R.M., Lussenhop, J., 1998. Contributions of interacting biological mechanisms to soil aggregate stabilization in restored prairie. Soil Biol. Biochem. 30, 905–916.

Jeffery, S., Gardi, C., Jones, A., Montanarella, L., Marmo, L., Miko, L., et al., 2010. European Atlas of Soil Biodiversity. European Commission, Publications Office of the European Union, Luxembourg.

Jenkinson, D.S., 1966. Studies on the decomposition of plant material in soil, II. Partial sterilization of soil and the soil biomass. J. Soil Sci. 17, 280–302.

Jenkinson, D.S., 1988. Determination of microbial biomass carbon and nitrogen in soil. In: Wilson, J.R. (Ed.), Advances in Nitrogen Cycling in Agricultural Ecosystems. CABI Publishing, Wallingford.

Jenkinson, D.S., Parry, L.C., 1989. The nitrogen cycle in the broadbalk wheat experiment: a model for the turnover of nitrogen through the soil microbial biomass. Soil Biol. Biochem. 21, 535–541.

Jenkinson, D.S., Powlson, D.S., 1976. The effects of biocidal treatments on metabolism in soil. V. A method for measuring soil biomass. Soil Biol. Biochem. 8, 209–213.

Jenkinson, D.S., Adams, D.E., Wild, A., 1991. Model estimates of CO_2 emissions from soil in response to global warming. Nature 351, 304–306.

Jenkinson, D.S., Brookes, P.C., Powlson, D.S., 2004. Measuring microbial biomass. Soil Biol. Biochem. 36, 5–7.

Jennings, T.J., Barkham, J.P., 1975. Food of slugs in mixed deciduous woodland. Oecologia 26, 211–221.

Jenny, H., 1941. Factors of Soil Formation. McGraw-Hill, New York.

Jenny, H., 1980. The soil resource: origin and behavior. Ecological Studies 37. Springer-Verlag, New York.

Jenny, H., Grossenbacher, K., 1963. Root-soil boundary zones as seen in the electron microscope. Soil Sci. Soc. Am. Proc. 27, 273–277.

Joergensen, R.G., Anderson, T.-H., Wolters, V., 1995. Carbon and nitrogen relationships in the microbial biomass of soils in beech (*Fagus sylvatica* L) forests. Biol. Fertil. Soils 19, 141–147.

Johnson, D., Leake, J.R., Ostle, N., Ineson, P., Read, D.J., 2002. In situ $^{13}CO_2$ pulse-labelling of upland grassland demonstrates a rapid pathway of carbon flux from arbuscular mycorrhizal mycelia to the soil. New Phytol. 153, 327–334.

Johnson, D., Krsek, M., Wellington, E.M.H., Stott, A.W., Cole, L., Bardgett, R.D., et al., 2005. Soil invertebrates disrupt carbon flow through fungal networks. Science 309 (5737), 1047.

Johnson, D.L., 1990. Biomantle evolution and the redistribution of earth materials and artifacts. Soil Sci. 149, 84–102.

Johnson, D.L., Ambros, S.H., Bassett, T.J., Bowen, M.L., Crummey, D.E., Isaacson, J.S., et al., 1997. Meaning of environmental terms. J. Environ. Qual. 26 (3), 581–589.

Johnson, D.L., Johnson, D.N., 2010. A holistic and universal view of soil. In: Proceedings of the 19th World Soil Congress, Brisbane, Queensland, Australia. pp. 1-4. ISBN: 978-0-646-53783-2.

Johnson, D.L., Schaetzl, R.J., 2015. Differing views of soil and pedogenesis by two masters: Darwin and Dokuchaev. Geoderma 237–238, 176–189.

Jones, C.G., Lawton, J.H., Shachak, M., 1994. Organisms as ecosystem engineers. Oikos 69, 373–386.

Jones, D.A., Nguyen, C., Finlay, R.D., 2009. Carbon flow in the rhizosphere: carbon trading at the soil-root interface. Plant Soil 321, 5–33.

Jones, D.L., Darrah, P.R., 1996. Re-sorption of organic compounds by roots of *Zea mays* L. and its consequences in the rhizosphere. Plant Soil 178 (1), 153–160.

Jones, D.L., Farrar, J., Giller, K.E., 2003. Associative nitrogen fixation and root exudation--what is theoretically possible in the rhizosphere? Symbiosis 35, 19–38.

Jones, D.L., Hodge, A., Kuzyakov, Y., 2004. Plant and mycorrhizal regulation of rhizodeposition. New Phytol. 163, 459–480.

Jones, P.C.T., Mollison, J.E., 1948. A technique for the quantitative estimation of soil microorganisms. J. Gen. Microbiol. 2, 54–69.

Jongerius, A., 1964. Soil Micromorphology. Elsevier, Amsterdam.

Jouquet, P., Traoré, S., Choosai, C., Hartmann, C., Bignell, D., 2011. Influence of termites on ecosystem functioning. Ecosystem services provided by termites. Eur. J. Soil Biol. 47, 215–222.

Jouquet, P., Bottinelli, N., Kerneis, G., Henry-des-Tureaux, T., Doan, T.T., Planchon, O., et al., 2013. Surface casting of the tropical *Metaphire posthuma* increases soil erosion and nitrate leaching in a laboratory experiment. Geoderma 204–205, 10–14.

Jurgens, G., Lindström, K., Saano, A., 1997. Novel group within the kingdom Crenarchaeota from boreal forest soil. Appl. Environ. Microbiol. 63, 803–805.

Kaczmarek, M., 1993. Apparatus and tools for the extraction of animals from the soil. In: Górny, M., Grüm, L. (Eds.), Methods in Soil Zoology. Polish Scientific Publishers, Warsaw, Poland, pp. 112–141.

Kaiser, C., Kilburn, M.R., Clode, P.L., Fuchslueger, L., Koranda, M., Cliff, J.B., et al., 2015. Exploring the transfer of recent plant photosynthates to soil microbes: mycorrhizal pathway vs. direct root exudation. New Phytol. 205, 1537–1551.

Kalisz, P.J., Dotson, D.B., 1989. Land-use history and the occurrence of exotic earthworms in the mountains of eastern Kentucky. Am. Midl. Nat. 122, 288–297.

Kalisz, P.J., Wood, H.B., 1995. Native and exotic earthworms in wildland ecosystems. In: Hendrix, P. (Ed.), Earthworm Ecology and Biogeography in North America. Lewis Publishers, Boca Raton, FL, pp. 117–126.

Kaneko, N., 1988. Feeding habits and cheliceral size of oribatid mites in cool temperate forest soils in Japan. Revue d'écologie et de biologie du sol 25 (3), 353–363.

Kanters, C., Anderson, I.C., Johnson, D., 2015. Chewing up the wood-wide web: selective grazing on ectomycorrhizal fungi by collembola. Forests 6, 2560–2570.

Karg, W., 1982. Investigations of habitat requirements, geographical distributions and origin of predatory mite genera in the cohort Gamasina for use as bioindicators. Pedobiologia 24, 241–247.

Karim, N., Jones, J.T., Okada, H., Kikuchi, T., 2009. Analysis of expressed sequence tags and identification of genes encoding cell-wall-degrading enzymes from the fungivorous nematode Aphelenchus avenae. BMC Genomics 10, 525.

Karlen, D.L., Andrews, S.S., Doran, J.W., 2001. Soil quality: current concepts and applications. Adv. Agron. 74, 1–40.

Kasprzak, K., 1982. Review of enchytraeid (Oligochaeta, Enchytraeidae) community structure and function in agricultural ecosystems. Pedobiologia 23, 217–232.

Kaston, B.J., 1978. How to Know the Spiders. McGraw Hill, Boston.

Keilin, D., 1959. The problem of anabiosis or latent life: history and current concept. Proc. R. Soc. B 150, 149–191.

Kemmpf, W.W., 1964. On the number of species of ants in the Neotropical region. Studia Entomologica 8, 161–200.

Kent, A.D., Triplett, E.W., 2002. Microbial communities and their interactions in soil and rhizosphere systems. Annu. Rev. Microbiol. 56, 211–236.

Kethley, J., 1990. Acarina: Prostigmata (Actenidida). In: Dindal, D.L. (Ed.), Soil Biology Guide.. Wiley, New York, pp. 667–756.

Khazanehdari, M., Mirshamsi, O., Aliabadian, M., 2016. Contribution to the solpugid (Arachnida: Solifugae) fauna of Iran. Turk. J. Zool. 40 (4), 608–614.

Kilbertus, G., 1980. Études des microhabitats contenus dans les agrégats du sol leur relation avec la biomasse bacterienne et la taille des procaryotes presents. Revue d'écologie et de biologie du sol 17, 543–557.

Kilbertus, G., Vannier, G., 1981. Relations microflore-microfaune dans la grotte de Sainte-Catherine (Pyrenées ariegeoises) II.—Le regime alimentaire de Tomocerus minor (Lubbock) et Tomocerus problematicus Cassagnau (Insectes Collemboles). Revue d'écologie et de biologie du sol 18 (3), 319–338.

Kisselle, K.W., Garrett, C.J., Fu, S., Hendrix, P.F., Crossley Jr., D.A., Coleman, D.C., et al., 2001. Budgets for root-derived C and Litter-derived C: comparison between conventional tillage and no tillage soils. Soil Biol. Biochem. 33, 1067–1075.

Kjøller, A., Struwe, S., 1994. Analysis of fungal communities on decomposing beech litter. In: Ritz, K., Dighton, J., Giller, K.E. (Eds.), Beyond the Biomass. Wiley-Sayce, Chichester, pp. 191–200.

Klarner, B., Maraun, M., Scheu, S., 2013. Trophic diversity and niche partitioning in a species rich predator guild: natural variations in stable isotope ratios ($^{13}C/^{12}C$, $^{15}N/^{14}N$) of mesostigmatid mites (Acari, Mesostigmata) from Central European beech forests. Soil Biol. Biochem. 57, 327–333.

Klironomos, J.N., Kendrick, W.B., 1995. Palatability of microfungi to soil arthropods in relation to the functioning of arbuscular mycorrhizal fungi. Biol. Fertil. Soils 21, 43–52.

Klironomos, J.N., Hart, M.M., 2001. Animal nitrogen swap for plant carbon. Nature 410, 651–652.

Knoepp, J.D., Coleman, D.C., Crossley Jr., D.A., Clark, J., 2000. Biological indices of soil quality: an ecosystem case study of their use. For. Ecol. Manage. 138, 357–368.

Koch, A., McBratney, A., Adams, M., Field, D., Hill, R., Crawford, J., et al., 2013. Soil security: solving the global soil crisis. Glob. Policy 4, 434–441.

Koller, R., Rodriguez, A., Robin, C., Scheu, S., Bonkowski, M., 2013. Protozoa enhance foraging efficiency of arbuscular mycorrhizal fungi for mineral nitrogen from organic matter in soil to the benefit of host plants. New Phytol. 199, 203–211.

Kourtev, P.S., Ehrenfeld, J.G., Häggblom, M., 2002. Exotic plant species alter the microbial community structure and function in the soil. Ecology 83, 3152–3166.

Krantz, G.W., 1978. A Manual of Acarology, second ed. Oregon State University Book Stores, Corvallis, OR.

Krantz, G.W., Ainscough, B.D., 1990. Acarina: Mesostigmata (Gamasida). In: Dindal, D.L. (Ed.), Soil Biology Guide.. Wiley, New York, pp. 583–665.

Krantz, G.W., Walter, D.E. (Eds.), 2009. A Manual of Acarology.. third ed. University of Texas Press, Lubbock, TX.

Kristufek, V., Ravasz, K., Pizl, V., 1992. Changes in densities of bacteria and microfungi during gut transit in Lumbricus rubellus and Aporrectodea caliginosa (Oligochaeta: Lumbricidae). Soil Biol. Biochem. 24, 1499–1500.

Krool, S., Bauer, T., 1987. Reproduction, development and pheromone secretion in *Heteromurus nitidus* Templeton, 1835 (Collembola, Entomobryidae). Revue d'écologie et de biologie du sol 24, 187–195.

Kubiëna, W.L., 1938. Micropedology.. Collegiate Press, Ames, IA.

Kühnelt, W., 1958. Zoogenic crumb-formation in undisturbed soils (in German). Sonderdruck aus Tagungsberichte 13, 193–199.

Kühnelt, W., 1976. Soil Biology.. Michigan State University, East Lansing, MI.

Kuikman, P., van Veen, J.A., 1989. The impact of protozoa on the availability of bacterial nitrogen to plants. Biol. Fertil. Soils 8, 13–18.

Kuikman, P.J., Jansen, A.G., van Veen, J.A., Zehnder, A.J.B., 1990. Protozoan predation and the turnover of soil organic carbon and nitrogen in the presence of plants. Biol. Fertil. Soils 10, 22–28.

Kulman, H.M., 1974. Comparative ecology of North American carabidae with special reference to biological control. Entomophaga Memoirs 7, 61–70.

Kurcheva, G.F., 1960. Role of soil organisms in the breakdown of oak litter. Pochvovedeniye 4, 16–23.

Kurcheva, G.F., 1964. Wirbellose Tiere als Faktor der Zersetzung von Waldstreu. Pedobiologia 4, 7–30.

Kuske, C.R., Barns, S.M., Busch, J.D., 1997. Diverse uncultivated bacterial groups from soils of the arid Southwestern United States that are present in many geographic regions. Appl. Environ. Microbiol. 63, 3858–3865.

Kuzyakov, Y., 2002. Separating microbial respiration of exudates from root respiration in non-sterile soils: a comparison of four methods. Soil Biol. Biochem. 34, 1621–1631.

Kuzyakov, Y., Blagodatskaya, E., 2015. Microbial hotspots and hot moments in soil: concept & review. Soil Biol. Biochem. 83, 184–199.

Kuzyakov, Y., Domanski, G., 2000. Carbon input by plants into the soil. Review. J. Plant Nutr. Soil Sci. 163, 421–431.

Kuzyakov, Y., Xu, X., 2013. Competition between roots and microorganisms for nitrogen: mechanisms and ecological relevance. New Phytol. 198, 656–669.

Labandeira, C.C., Phillips, L.T., Norton, R.A., 1997. Oribatid mites and the decomposition of plant tissues in Palaeozoic coal-swamp forests. Palaios 12, 319–353.

Ladygina, N., Henry, F., Kant, M.R., Koller, R., Reidinger, S., Rodriguez, A., et al., 2010. Additive and interactive effects of functionally dissimilar soil organisms on a grassland plant community. Soil Biol. Biochem. 42 (12), 2266–2275.

Lal, R., 2002. The potential of soils of the tropics to sequester carbon and mitigate the green house effect. Adv. Agron. 76, 1–30.

Lambers, H., Raven, J.A., Shaver, G.R., Smith, S.E., 2008. Plant nutrient-acquisition strategies change with soil age. Trends Res. Ecol. Evol. 25, 95–103.

Lambers, H., Brundrett, M.C., Raven, J.A., Hopper, S.D., 2010. Plant mineral nutrition in ancient landscapes. Plant Soil 334, 11–31.

Lamoncha, K.L., Crossley Jr., D.A., 1998. Oribatid mite diversity along an elevation gradient in a southeastern appalachian forest. Pedobiologia 42, 43–55.

Landeweert, R., Hoffland, E., Finlay, R.D., Kuyper, T.W., van Breemen, N., 2001. Linking plants to rocks: ectomycorrhizal fungi mobilize nutrients from minerals. Trends Ecol. Evol. 16, 248–254.

Landeweert, R., Leeflang, P., Kuyper, T.W., Hoffland, E., Rosling, A., Wernars, K., et al., 2003. Molecular identification of ectomycorrhizal mycelium in soil horizons. Appl. Environ. Microbiol. 69, 327–333.

Lane, D.J., 1991. 16S/23S rRNA sequencing. In: Stackebrandt, T., Goodfellow, M. (Eds.), Nucleic Acid Techniques in Bacterial Systematics. Wiley, New York, pp. 115–175.

Larink, O., 1997. Springtails and Mites. Important Knots in the Food Web of Soils. In: Benckiser, G. (Ed.), Fauna in Soil Ecosystems. Recycling Processes, Nutrient Fluxes, and Agricultural Production. Marcel Dekker, New York, pp. 225–264.

Larsen, K.J., Work, T.T., Purrington, F.F., 2003. Habitat use patterns by ground beetles (Coleoptera: Carabidae) of northeastern Iowa. Pedobiologia 47 (3), 288–299.

Lartey, R.T., Curl, E.A., Peterson, C.M., 1994. Interactions of mycophagous collembola and biological control fungi in the suppression of *Rhizoctonia solani*. Soil Biol. Biochem. 26, 81–88.

Lauber, C.L., Hamady, M., Knight, R., Fierer, N., 2009. Pyrosequencing-based assessment of soil pH as a predictor of soil bacterial community structure at the continental scale. Appl. Environ. Microbiol. 75, 5111–5120.

Lavelle, P., 1978. Les vers de terre de la savane de Lamto (Ivory Coast): peuplements, populations et fonctions dans l'écosystème. Thèse de Doctorat. Université Paris VI.

Lavelle, P., 1983. The structure of earthworm communities. In: Satchell, J.E. (Ed.), Earthworm Ecology from Darwin to Vermiculture. Chapman & Hall, London, pp. 449–466.

Lavelle, P., 2000. Ecological challenges for soil science. Soil Sci. 165, 73–86.

Lavelle, P., Lapied, E., 2003. Endangered earthworms of Amazonia: an homage to Gilberto Righi. Pedobiologia 47, 419–427.

Lavelle, P., Martin, A., 1992. Small-scale and large-scale effects of endogeic earthworms on soil organic matter dynamics in soils of the humid tropics. Soil Biol. Biochem. 24, 1491–1498.

Lavelle, P., Spain, A.V., 2001. Soil Ecology. Kluwer, Dordrecht.

Lavelle, P., Blanchart, E., Martin, A., 1992. Impact of soil fauna on the properties of soils in the humid tropics. In: Lal, R., Sanchez, P. (Eds.), Myths and Science of Soils of the Tropics.. Soil Science Society of America, Madison, WI, pp. 157–185.

Lavelle, P., Lattaud, D.T., Barois, I., 1995. Mutualism and biodiversity in soils. Plant Soil 170, 23–33.

Lavelle, P., Pashanasi, B., Charpentier, F., Gilot, C., Rossi, J., Derouard, L., et al., 1998. Influence of earthworms on soil organic matter dynamics, nutrient dynamics and microbiological ecology. In: Edwards, C.A. (Ed.), Earthworm Ecology. Lewis Publisher, Boca Raton, FL, pp. 103–122.

Lavelle, P., Brussaard, L., Hendrix, P. (Eds.), 1999. Earthworm Management in Tropical Agroecosystems. CABI Publishing, Wallingford.

Lavelle, P., Spain, A., Blouin, M., Brown, G., Decaëns, T., Grimaldi, M., et al., 2016. Ecosystem engineers in a self-organized soil: a review of concepts and future research questions. Soil Sci. 181, 91–109.

Lawrence, K.L., Wise, D.H., 2000. Spider predation on forest-floor Collembola and evidence for indirect effects on decomposition. Pedobiologia 44, 33–39.

Leake, J.R., 2004. Myco-heterotroph/epiparasitic plant interactions with ectomycorrhizal and arbuscular mycorrhizal fungi. Curr. Opin. Cell Biol. 7, 422–428.

Leake, J.R., Read, D.J., 1989. The biology of mycorrhiza in the Ericaceae. 13. Some characteristics of the extracellular proteinase activity of the ericoid endophyte *Hymenoscyphus ericae*. New Phytol. 112, 69–76.

Lee, K.E., 1959. The Earthworm fauna of New Zealand. New Zealand, Department of Scientific and Industrial Research, Bulletin 130. Government Printer, Wellington, New Zealand.

Lee, K.E., 1985. Earthworms: Their Ecology and Relationships with Soils and Land Use. Academic Press, Sydney.

Lee, K.E., 1995. Earthworms and sustainable land use. In: Hendrix, P.F. (Ed.), Earthworm Ecology and Biogeography. Lewis Publishers, Boca Raton, FL, pp. 215–234.

Lee, K.E., Wood, T.G., 1971. Termites and Soils. Academic Press, London and New York.

Lee, K.E., Foster, R.C., 1991. Soil fauna and soil structure. Aust. J. Soil Res. 29, 745–775.

Leetham, J.W., McNary, T.J., Dodd, J.L., Lauenroth, W.K., 1982. Response of soil nematodes, rotifers, and tardigrades to three levels of season-long sulphur dioxide exposure. Water, Air, Soil Pollut. 18, 343–356.

Lehmann, J., Kleber, M., 2015. The contentious nature of soil organic matter. Nature 528, 60–68.

Leigh, J., Hodge, A., Fitter, A.H., 2009. Arbuscular mycorrhizal fungi can transfer substantial amounts of nitrogen to their host plant from organic material. New Phytol. 181, 199–207.

Liiri, M., Setälä, H., Haimi, J., Pennanen, T., Fritze, H., 2002. Soil processes are not influenced by the functional complexity of soil decomposer food webs under disturbance. Soil Biol. Biochem. 34, 1009–1020.

Lindeman, R.L., 1942. The trophic-dynamic aspect of ecology. Ecology 23, 399–418.

Linden, D.R., Hendrix, P.F., Coleman, D.C., van Vliet, P.C.J., 1994. Faunal indicators of soil quality. In: Doran, J.W., Coleman, D.C., Bezdicek, D.F., Stewart, B.A. (Eds.), Defining Soil Quality for a Sustainable Environment. SSSA Special Publication No. 35.. American Society of Agronomy, Madison, WI, pp. 91–106.

Lindo, Z., Winchester, N.N., 2006. A comparison of microarthropod assemblages with emphasis on oribatid mites in canopy suspended soils and forest floors associated with ancient western red cedar trees. Pedobiologia 50, 31–41.

Lloyd, M., Dybas, H.S., 1966. The periodical cicada problem. I. Population ecology. Evolution 20, 133–149.

Lobe, J.W., Callaham Jr., M.A., Hendrix, P.F., Hanula, J.L., 2014. Removal of an invasive shrub (Chinese privet: Ligustrum sinense Lour) reduces exotic earthworm abundance and promotes recovery of native North American earthworms. Appl. Soil Ecol. 83, 133–139.

Lock, K., Dekoninck, W., 2001. Centipede communities on the inland dunes of eastern Flanders (Belgium). Eur. J. Soil Biol. 37, 113–116.

Lotka, A.J., 1925. Elements of Physical Biology. Williams & Wilkins, Baltimore, MD.

Lousier, J.D., Bamforth, S.S., 1990. Soil protozoa. In: Dindal, D.L. (Ed.), Soil Biology Guide. Wiley, New York, pp. 97–136.

Lousier, J.D., Parkinson, D., 1981. Evaluation of a membrane filter technique to count soil and litter Testacea. Soil Biol. Biochem. 13, 209–213.

Lousier, J.D., Parkinson, J., 1984. Annual population dynamics and production ecology of Testacea (Protozoa, Rhizopoda) in an aspen woodland soil. Soil Biol. Biochem. 16, 103–114.

Lovelock, J.E., 1979. Gaia: A New Look at Life on Earth. Oxford University Press, Oxford and London.

Lovelock, J.E., 1988. The Ages of Gaia. W.W. Norton Co, New York.

Lubbers, I.M., van Groenigen, J., Fonte, S.J., Six, J., Brussaard, L., van Groenigen, J.W., 2013. Greenhouse gas emissions from soils increased by earthworms. Nat. Clim. Change 3, 187–194.

Luo, Y., Zhou, X., 2006. Soil Respiration and the Environment. Academic Press, Waltham, MA.

Lussenhop, J., 1976. Soil arthropod response to prairie burning. Ecology 57, 88–98.

Lussenhop, J., 1992. Mechanisms of microarthropod-microbial interactions in soil. Adv. Ecol. Res. 23, 1–33.

Luxton, M., 1967. The ecology of saltmarsh Acarina. J. Anim. Ecol. 36, 257–277.

Luxton, M., 1972. Studies on the oribatid mites of a Danish beech forest. I. Nutritional biology. Pedobiologia 12, 434–463.

Luxton, M., 1975. Studies on the oribatid mites of a Danish beech forest. II. Biomass, calorimetry and respirometry. Pedobiologia 15, 161–200.

Luxton, M., 1979. Food and energy processing by oribatid mites. Revue d'écologie et de biologie du sol 16, 103–111.

Luxton, M., 1981a. Studies on the astigmatic mites of a Danish beech wood soil. Pedobiologia 22, 29–38.

Luxton, M., 1981b. Studies on the prostigmatic mites of a Danish beech wood soil. Pedobiologia 22, 277–303.

Lynch, J.M. (Ed.), 1990. The Rhizosphere. Wiley-Interscience, Chichester.

MacArthur, R.H., 1972. Geographical Ecology: Patterns in the Distribution of Species. Harper & Row, New York.

Macfadyen, A., 1969. The systematic study of soil ecosystems. In: Sheals, J.G. (Ed.), The Soil Ecosystem. The Systematics Association, London, pp. 191–197.

Machmuller, M.B., Mohan, J.E., Minucci, J.M., Phillips, C.A., Wurzburger, N., 2016. Season, but not experimental warming, affects the activity and temperature sensitivity of extracellular enzymes. Biogeochemistry 131, 255–265.

Mackie-Dawson, L.A., Atkinson, D., 1991. Methodology for the study of roots in field experiments and the interpretation of results. In: Atkinson, D. (Ed.), Plant Root Growth: An Ecological Perspective. Blackwell, London, pp. 25–47.

Makkonen, M., Berg, M.P., Handa, I.T., Haettenschweiler, S., van Ruijven, J., van Bodegom, P.M., et al., 2012. Highly consistent effects of plant litter identity and functional traits on decomposition across a latitudinal gradient. Ecol. Lett. 15, 1033–1041.

Malloch, D.W., Pirozynski, K.A., Raven, P.A., 1980. Ecological and evolutionary significance of mycorrhizal symbioses in vascular plants (a review). Proc. Natl. Acad. Sci. 77, 2113–2118.

Mann, L.K., 1986. Changes in soil organic carbon storage after cultivation. Soil Sci. 142, 279–288.

Maraun, M., Erdmann, G., Fischer, B.M., Pollierer, M.M., Norton, R.A., Schneider, K., et al., 2011. Stable isotopes revisited: their use and limits for oribatid mite trophic ecology. Soil Biol. Biochem. 43, 877–882.

Maraun, M., Schatz, H., Scheu, S., 2007. Awesome or ordinary? Global diversity patterns of oribatid mites. Ecography 30, 209–216.

Marinissen, J.C.Y., Dexter, A.R., 1990. Mechanisms of stabilization of earthworm casts and artificial casts. Biol. Fertil. Soils 9, 163–167.

Marshall, V.G., 1977. Effects of Manures and Fertilizers on Soil Fauna: A Review. Special Publication 3. Commonwealth Bureau of Soils, Slough.

Marshall, V.G., Reeves, R.M., Norton, R.A., 1987. Catalogue of the Oribatida (Acari) of the continental United States and Canada. Mem. Entomol. Soc. Can. 139, 418.

Martin, M.M., 1984. The role of ingested enzymes in the digestive processes of insects. In: Anderson, J.M., Rayner, A.D.M., Walton, D.W.H. (Eds.), Invertebrate-Microbial Interactions. Cambridge University Press, Cambridge, pp. 155–172.

Martin, J.K., Kemp, J.R., 1986. The measurement of carbon transfers within the rhizosphere of wheat grown in field plots. Soil Biol. Biochem. 18, 103–107.

Mason, C.F., 1974. Mollusca. In: Dickinson, C.H., Pugh, G.J.F. (Eds.), Biology of Plant Litter Decomposition., vol. 2. Academic Press, London, pp. 555–591.

Matamala, R., Gonzàlez-Meler, M.A., Jastrow, J.D., Norby, R.J., Schlesinger, W.H., 2003. Impacts of fine root turnover on Forest NPP and Soil C sequestration potential. Science 302, 1385–1387.

Matsuko, K., 1994. Specialized predation on oribatid mites by two species of the ant genus *Myrmecina* (Hymenoptera, Formicidae). Psyche 101, 159–173.

McAlpine, J.F., 1990. Insecta: diptera adults. In: Dindal, D.L. (Ed.), Soil Biology Guide. Wiley, New York, pp. 1211–1252.

McBrayer, J.F., 1973. Exploitation of deciduous leaf litter by *Apheloria montana*. Pedobiologia 13, 90–98.

McCormack, M.L., Dickie, I.A., Eissenstat, D.M., Fahey, T.J., Fernandez, C.W., Guo, D., et al., 2015. Redefining fine roots improves understanding of belowground contributions to terrestrial biosphere processes. New Phytol. 207 (3), 505–518.

McDowell, W.H., 2003. Dissolved organic matter in soils—future directions and unanswered questions. Geoderma 113, 179–186.

McFarlane, K.J., Tornwilliams, M.S., Hanson, P.J., Porras, R.C., Swanston, C.W., Callaham Jr., M.A., et al., 2013. Comparison of soil organic matter dynamics at five temperate deciduous forests with physical fractionation and radiocarbon measurements. Biogeochemistry 112, 457–476.

McGill, W.B., Cole, C.V., 1981. Comparative aspects of cycling of organic C, N, S, and P through soil organic matter. Geoderma 26, 267–286.

McGonigle, T.P., Miller, M.H., Evans, D.G., Fairchild, G.L., Swan, J.A., 1990. A new method which gives an objective measure of colonization of roots by vesicular—arbuscular mycorrhizal fungi. New Phytol. 115 (3), 495–501.

McNaughton, S.J., 1976. Serengeti migratory wildebeest: facilitation of energy flow by grazing. Science 191, 92–94.

McNaughton, S.J., Banyikawa, F.F., McNaughton, M.M., 1998. Root biomass and productivity in a grazing ecosystem: the Serengeti. Ecology 79, 587–592.

McSorley, R., Frederick, J.J., 2004. Effects of extraction method on perceived composition of the soil nematode community. Appl. Soil Ecol. 27, 55–63.

Meentemeyer, V., 1978. Macroclimate and lignin control of decomposition. Ecology 59, 465–472.

Melillo, J.M., Aber, J.D., Muratore, J.F., 1982. Nitrogen and lignin control of hardwood leaf litter decomposition dynamics. Ecology 63, 621–626.

Meyer, E., 1996. Mesofauna. In: Schinner, F., Öhlinger, R., Kandeler, E., Margesin, R. (Eds.), Methods in Soil Biology. Springer-Verlag, Berlin, pp. 338–345.

Michener, C.D., Michener, M.H., 1951. American Social Insects. Van Nostrand, New York.

Mikhail, W.Z.A., 1993. Effect of soil structure on soil fauna in a desert wadi in southern Egypt. J. Arid Environ. 24, 321–331.

Milchunas, D.G., Lauenroth, W.K., Singh, J.S., Cole, C.V., Hunt, H.W., 1985. Root turnover and production by ^{14}C dilution: implications of carbon partitioning in plants. Plant & Soil 88, 353–365.

Mills, A.L., 2003. Keeping in touch: microbial life on soil particle surfaces. Adv. Agron. 78, 1–43.

Mitchell, M.J., Parkinson, D., 1976. Fungal feeding of oribatid mites (Acari: Cryptostigmata) in an aspen woodland soil. Ecology 57, 302–312.

Mitra, O., Callaham Jr., M.A., Smith, M.L., Yack, J.E., 2009. Grunting for worms: Seismic vibrations cause *Diplocardia* earthworms to emerge from the soil. Biol. Lett. 5, 16–19.

Møbjerg, N., Halberg, K.A., Jørgensen, A., Persson, D., Bjørn, M., Ramløv, H., et al., 2011. Survival in extreme environments—on the current knowledge of adaptations in tardigrades. Acta Physiol. 202, 409–420.

Molleman, F., Walter, D.E., 2001. Niche separation and can-openers: scydmaenid beetles as predators of armoured mites in Australia. In: Halliday, R.B., Walter, D.E., Proctor, H.C., Norton, R.A., Collof, M.J. (Eds.), Acarology: Proceedings of the 10th International Congress. CSIRO Publications, Melbourne, Australia, pp. 283–288.

Monz, C.A., Reuss, D.E., Elliott, E.T., 1991. Soil microbial biomass carbon and nitrogen estimates using 2450 MHz microwave irradiation or chloroform fumigation followed by direct extraction. Agric., Ecosyst. Environ. 34, 55–63.

Moore, J.C., de Ruiter, P.C., 1991. Temporal and spatial heterogeneity of trophic interactions within below-ground food webs. Agric., Ecosyst. Environ. 34, 371–397.

Moore, J.C., de Ruiter, P.C., 2000. Invertebrates in detrital food webs along gradients of productivity. In: Coleman, D.C., Hendrix, P.F. (Eds.), Invertebrates as Webmasters in Ecosystems. CABI Publishing, Wallingford, pp. 161–184.

Moore, J.C., de Ruiter, P.C., 2012. Energetic Food Webs: An analysis of real and model ecosystems. Oxford University Press, Oxford.

Moore, J.C., St. John, T.V., Coleman, D.C., 1985. Ingestion of vesicular arbuscular mycorrhizal hyphae and spores by soil microarthropods. Ecology 66, 1979–1981.

Moore, J.C., Ingham, E.R., Coleman, D.C., 1987. Inter- and intraspecific feeding selectivity of *Folsomia candida* (Willem) (Collembola, Isotomidae) on fungi. Biol. Fertil. Soils 5, 6–12.

Moore, J.C., Walter, D.E., Hunt, H.W., 1988. Arthropod regulation of micro- and mesobiota in belowground food webs. Annu. Rev. Entomol. 33, 419–439.

Moore, J.C., de Ruiter, P.C., Hunt, H.W., 1993. Influence of productivity on the stability of real and model ecosystems. Science 261, 906–908.

Moore, J.C., de Ruiter, P.C., Hunt, H.W., Coleman, D.C., Freckman, D.W., 1996. Microcosms and soil ecology: critical linkages between field studies and modeling food webs. Ecology 77, 694–705.

Moore, J.C., McCann, K., Setälä, H., de Ruiter, P.C., 2003. Top-down is bottom-up: does predation in the rhizosphere regulate aboveground dynamics? Ecology 84, 846–857.

Moorhead, D.L., Lashermes, G., Sinsabaugh, R.L., 2012. A theoretical model of C-and N-acquiring exoenzyme activities, which balances microbial demands during decomposition. Soil Biol. Biochem. 53, 133–141.

Morita, R.Y., 1997. Bacteria in Oligotrophic Environments. Chapman & Hall, New York.

Morriën, E., 2016. Understanding soil food web dynamics, how close do we get? Soil Biol. Biochem. 102, 10–13.

Mosier, A., Schimel, D., Valentine, D., Bronson, K., Parton, W., 1991. Methane and nitrous oxide fluxes in native, fertilized and cultivated grasslands. Nature 350, 330–332.

Muchmore, W.B., 1990. Pseudoscorpionida. In: Dindal, D.L. (Ed.), Soil Biology Guide. Wiley, New York, pp. 503–527.

Mueller, B.R., Beare, M.H., Crossley Jr., D.A., 1990. Soil mites in detrital food webs of conventional and no-tillage agroecosystems. Pedobiologia 34, 389–401.

Muma, M.H., 1966. Feeding behavior of North American Solpugida (Arachnida). Fla. Entomol. 49, 199–216.

Mundel, P., 1990. Chilopoda. In: Dindal, D.L. (Ed.), Soil Biology Guide. Wiley, New York, pp. 819–833.

Murphy, K.L., Klopatek, J.M., Klopatek, C.C., 1998. The effects of litter quality and climate on decomposition along an elevation gradient. Ecol. Appl. 8, 1061–1071.

Nabholz, J.V., Reynolds, L.J., Crossley Jr., D.A., 1977. Range extension of *Pardosa lapidicina* Emerton (Araneida: Lycosidae) to Georgia. J. Ga. Entomol. Soc. 12, 241–243.

Nadelhoffer, K.J., Raich, J.W., 1992. Fine root production estimates and belowground carbon allocation in forest ecosystems. Ecology 73, 1139–1147.

Nadelhoffer, K.J., Aber, J.D., Melillo, J.M., 1985. Fine roots, net primary production, and soil nitrogen availability: a new hypothesis. Ecology 66, 1377–1390.

Nadkarni, N., Schaefer, D., Matelson, T.J., Solano, R., 2002. Comparison of arboreal and terrestrial soil characteristics in a lower montane forest, Monteverde, Costa Rica. Pedobiologia 46, 24–33.

Nannipieri, P., 1994. The potential use of soil enzymes as indicators of productivity, sustainability and pollution. In: Pankhurst, C.E., Doube, B.M., Gupta, V.V.S.R., Grace, P.R. (Eds.), Soil Biota: Management in sustainable farming. CSIRO, Melbourne, pp. 238–244.

Nannipieri, P., Grego, S., Ceccanti, B., 1990. Ecological significance of the biological activity in soil. In: Bollag, J.-M., Stotzky, G. (Eds.), Soil Biochemistry, vol. 6. Marcel Dekker, New York, pp. 293–356.

Nannipieri, P., Kandeler, E., Ruggiero, P., 2002. Enzyme activities and microbiological and biochemical processes in soil. In: Burns, R.G., Dick, R.P. (Eds.), Enzymes in the Environment. Marcel Dekker, New York, pp. 1–33.

Nekola, J., 2003. Large-scale terrestrial gastropod community composition patterns in the great lakes region of North America. Diversity Distrib. 9, 55–71.

Nekola, J.C., 2014. Overview of the North American terrestrial gastropod fauna. Am. Malacological Bull. 32 (2), 225–235.

Nelson, D.R., Adkins, R.G., 2001. Distribution of tardigrades within a moss cushion: do tardigrades migrate in response to changing moisture conditions? Zool. Anz. 240, 493–500.

Nelson, D.R., Higgins, R.P., 1990. Tardigrada. In: Dindal, D.L. (Ed.), Soil Biology Guide.. Wiley, New York, pp. 393–419.

Neuhauser, E.F., Hartenstein, P., 1978. Phenolic content and palatability of leaves and wood to soil isopods and diplopods. Pedobiologia 18, 99–109.

Neutel, A.-M., Heesterbeek, J.A.P., de Ruiter, P.C., 2002. Stability in real food webs: weak links in long loops. Science 296, 1120–1123.

Newell, K., 1984a. Interaction between two decomposer basidiomycetes and a collembolan under Sitka spruce: distribution, abundance and selective grazing. Soil Biol. Biochem. 16, 227–233.

Newell, K., 1984b. Interaction between two basidiomycetes and Collembola under Sitka spruce: grazing and its potential effects on fungal distribution and litter decomposition. Soil Biology and Biochem. 16, 235–240.

Newell, S.Y., Fallon, R.D., 1991. Toward a method for measuring instantaneous fungal growth rates in field samples. Ecology 72, 1547–1559.

Newman, A.S., Norman, A.G., 1943. An examination of thermal methods for following microbiological activity in soil. Soil Sci. Soc. Am. Proc. 8, 250–253.

Newman, R.H., Tate, K.R., 1980. Soil phosphorus characterization by 31P nuclear magnetic resonance. Commun. Soil Sci. Plant Anal. 11, 835–842.

Newton Jr., A.F., 1990. Insects: Coleoptera Staphylinidae adults and larvae. In: Dindal, D.L. (Ed.), Soil Biology Guide. Wiley, New York, pp. 1137–1174.

Nielsen, C.O., Christensen, B., 1959a. The Enchytraeidae: critical revision and taxonomy of European species. Natura Jutlandica 10 (Suppl. 1), 1–23, 1961.

Nielsen, C.O., Christensen, B., 1959b. The Enchytraeidae: critical revision and taxonomy of European species. Natura Jutlandica 8–9, 1–160.

Nielsen, C.O., Christensen, B., 1961. The enchytraeidae: critical revision and taxonomy of European species. Natura Jutlandica 10 (Suppl. 2), 1–23.

Nielsen, C.O., Christensen, B., 1963. The enchytraeidae: critical revision and taxonomy of European species. Nat. Jutlandica 10 (Suppl. 2), 1–19.

Nielson, G.A., Hole, F.D., 1964. Earthworms and the development of coprogenous A1-horizons in forest soils of Wisconsin. Soil Sci. Soc. Am. Proc. 28, 426–430.

Northup, R.R., Yu, Z., Dahlgren, R.A., Vogt, K.A., 1995. Polyphenol control of nitrogen release from pine litter. Nature 377, 227–229.

Norton, R.A., 1984. Monphyletic groups in the Enarthronota (Sarcoptiformes. In: Griffiths, D.A., Bowman, C.E. (Eds.), Acarology VI, vol. 1. Ellis Harwood, Chichester, pp. 233–240.

Norton, R.A., Behan-Pelletier, V.M., 1991. Calcium carbonate and calcium oxalate as cuticular hardening agents in oribatid mites (Acari: Oribatida). Can. J. Zool. 69, 1504–1511.

Norton, R.A., Bonamo, P.M., Grierson, J.D., Shear, W.A., 1987. Fossil mites from the Devonian of New York state. In: Channabasavanna, G.P., Viraktamath, C.A. (Eds.), Prog. Acarol., vol. 1. Oxford & IBM Publishing Co., New Delhi, pp. 271–277.

Norton, R.A., Alberti, G., Weigmann, G., St. Woas, S., 1997. Porose integumental organs of oribatid mites (Acari, Oribatida). 1. Overview of types and distribution. Zoologica (Stuttgart) 146, 1–31.

Nosek, J., 1973. The European Protura. Their taxonomy, ecology and distribution with keys for determination. Muséum d'Histoire Naturelle, Genéve.

Noti, M.-I., André, H.M., Ducarme, X., Lebrun, P., 2003. Diversity of soil oribatid mites (Acari: Oribatida) from High Katanga (Democratic Republic of Congo): a multiscale and multifactor approach. Biodiversity Conserv. 12, 767–785.

Nunan, N., Wu, K., Young, I.M., Crawford, J.W., Ritz, K., 2002. In situ spatial patterns of soil bacterial populations, mapped at multiple scales, in an arable soil. Microb. Ecol. 44, 296–305.

Oades, J.M., 1984. Soil organic matter and structural stability: mechanisms and implications for management. Plant & Soil 76, 319–337.

Oades, J.M., Gillman, G.P., Uehara, G., Hue, N.V., van Noordwijk, M., Robertson, G.P., et al., 1989. Interactions of soil organic matter and variable-charge clays. In: Coleman, D.C., Oades, J.M., Uehara, G. (Eds.), Dynamics of Soil Organic Matter in Tropical Ecosystems. University of Hawaii Press, Honolulu, Hawaii, pp. 69–95.

Oades, J.M., Waters, A.G., 1991. Aggregate hierarchy in soils. Aust. J. Soil Res. 29, 815–828.

O'Connor, F.B., 1955. Extraction of enchytraeid worms from a coniferous forest soil. Nature 175, 815–816.

Odum, E.P., 1971. Fundamentals of Ecology, third ed. W.B. Saunders Company, Philadelphia, PA.

Odum, E.P., Biever, L.J., 1984. Resource quality, mutualism and energy partitioning in food chains. Am. Nat. 124, 360–376.

Oepik, M., Vanatoa, A., Vanatoa, E., Moora, M., Davison, J., Kalwij, J.M., et al., 2010. The online database MaarjAM reveals global and ecosystemic distribution patterns in Arbuscular Mycorrhizal Fungi (Glomeromycota). New Phytol. 188, 223–241.

Ogram, A., Sharma, K., 2002. Methods of soil microbial community analysis. In: Hurst, C.J., Crawford, R.L., Knudsen, G.R., McInerney, M.J., Stetzenbach, L.D. (Eds.), Manual of Environmental Microbiology, second ed. American Society forMicrobiology, Washington, DC, pp. 554–563.

Oldroyd, G.E.D., 2013. Speak, friend, and enter: signalling systems that promote beneficial symbiotic associations in plants. Nat. Rev./Microbiol. 11, 252–263.

O'Lear, H.A., Blair, J.M., 1999. Responses of soil microarthropods to changes in soil water availability in tallgrass prairie. Biol. Fertil. Soils 29, 207–217.

Olson, J.S., 1963. Energy storage and the balance of producers and decomposers in ecological systems. Ecology 44, 322–331.

Orgiazzi, A., Bardgett, R.D., Barrios, E., Behan-Pelletier, V., Briones, M.J.I., Chotte, J.-L., et al., 2016. Global soil biodiversity atlas. Luxembourg, Publications Office of the European Union.

Pace, N.R., 1997. A molecular view of microbial diversity and the biosphere. Science 276, 735–740.

Pace, N.R., 1999. Microbial ecology and diversity. ASM News 65, 328–333.

Padmanabhan, P., Padmanabhan, S., DeRito, C., Gray, A., Gannon, D., Snape, J.R., et al., 2003. Respiration of ^{13}C-labeled substrates added to soil in the field and subsequent 16S rRNA gene analysis of ^{13}C-labeled soil DNA. Appl. Environ. Microbiol. 69, 1614–1622.

Pallant, D., 1974. Assimilation of the grey field slug (*Agriolimax reticulatus* (Mueller)). Proc. Malacological Soc. London 41, 99.

Palm, C.A., Gachengo, C.N., Delve, R.J., Cadisch, G., Giller, K.E., 2001. Organic inputs for soil fertility management in tropical agroecosystems: application of an organic resource database. Agric., Ecosyst. Environ. 83, 27–42.

Pantastico-Caldas, M., Duncan, K.E., Istock, C.A., Bell, J.A., 1992. Population dynamics of bacteriophage and *Bacillus subtilis* in soil. Ecology 73, 1888–1902.

Park, O., 1947. Observations on *Batrisodes* (Coleoptera: Pselaphidae), with particular reference to the American species east of the Rocky Mountains. Bull. Chicago Acad. Sci. 8, 45–132.

Parker, L.W., Santos, P.F., Phillips, J., Whitford, W.G., 1984. Carbon and nitrogen dynamics during the decomposition of litter and roots of a Chihuahuan desert annual *Lepidium lasiocarpum*. Ecol. Monogr. 54, 339–360.

Parkinson, D., Coleman, D.C., 1991. Microbial populations, activity and biomass. Agric., Ecosyst. Environ. 34, 3–33.

Parkinson, D., Gray, T.R.G., Williams, S.T., 1971. Methods for Studying the Ecology of Soil Microorganisms. IBP Handbook No. 19. Blackwell, Oxford.

Parmelee, R.W., Beare, M.H., Cheng, W., Hendrix, P.F., Rider, S.J., Crossley Jr., D.A., et al., 1990. Earthworms and enchytraeids in conventional and no-tillage agroecosystems: a biocide approach to assess their role in organic matter breakdown. Biol. Fertil. Soils 10, 1–10.

Parmelee, R.W., Bohlen, P.J., Blair, J.M., 1998. Earthworms and nutrient cycling processes: integrating across the ecological hierarchy. In: Edwards, C. (Ed.), Earthworm Ecology. St. Lucie Press, Boca Raton, FL, pp. 123–143.

Parton, W.J., Schimel, D.A., Cole, C.V., Ojima, D.S., 1987. Analysis of factors controlling soil organic matter levels in Great Plains grasslands. Soil Sci. Soc. Am. J 51, 1173–1179.

Parton, W.J., Cole, C.V., Stewart, J.W.B., Schimel, D.S., Ojima, D., 1989a. Simulating the long-term dynamics of C, N and P in soils. In: Clarholm, M., Bergström, L. (Eds.), Arable Land—Perspectives and Challenges. Martinus Nijhoff, Dordrecht.

Parton, W.J., Sanford, R.L., Sanchez, P.A., Stewart, J.W.B., 1989b. Modeling soil organic matter dynamics in tropical soils. In: Coleman, D.C., Oades, J.M., Uehara, G. (Eds.), Dynamics of Soil Organic Matter in Tropical Ecosystems. University of Hawaii, Honolulu, Hawaii, pp. 153–171.

Parton, W.J., Ojima, D.S., Cole, C.V., Schimel, D.S., 1994. A general model for soil organic matter dynamics: sensitivity to litter chemistry, texture and management. In: Bryant, R.B., Arnold, R.W. (Eds.), Quantitative Modeling of Soil Forming Processes. Soil Science Society of America, Madison, WI, pp. 147–167.

Parton, W.J., Del Grosso, S.J., Plante, A.F., Adair, E.C., Lutz, S.M., 2016. Modeling the dynamics of soil organic matter and nutrient cycling. In: Paul, E.A. (Ed.), Soil Microbiology, Ecology, and Biochemistry, fourth ed. Elsevier Academic Press, Amsterdam, pp. 505–537.

Pastor, J., Post, W.M., 1988. Responses of northern forests to CO_2- induced climate change. Nature 334, 55–58.

Paul, E.A. (Ed.), 2015. Soil Microbiology, Ecology, and Biochemistry. fourth ed. Elsevier/Academic Press, Amsterdam and London.

Paul, E.A., 2016. The nature and dynamics of soil organic matter: plant inputs, microbial transformations, and organic matter stabilization. Soil Biol. Biochem. 98, 109–126.

Paul, E.A., Clark, F.E., 1996. Soil Microbiology and Biochemistry, second ed. Academic Press, San Diego.

Paustian, K., Andrén, O., Clarholm, M., Hansson, A.-C., Johansson, G., Lagerlöf, J., et al., 1990. Carbon and nitrogen budgets of four agro-ecosystems with annual and perennial crops, with and without N fertilization. J. Appl. Ecol. 27, 60–84.

Paustian, K., Ågren, G.I., Bosatta, E., 1997. Modeling litter quality effects on decomposition and soil organic matter dynamics. In: Cadisch, G., Giller, K.E. (Eds.), Driven by Nature. Plant Litter Quality and Decomposition. CABI Publishing, Wallingford, pp. 313–335.

Pawlowski, J., Audic, S., Adl, S., Bass, D., Belbahri, L., Berney, C., et al., 2012. CBOL Protist Working Group: barcoding eukaryotic richness beyond the animal, plant and fungal kingdoms. PLOS Biol. 10 (11), 1–5.

Pawluk, S., 1987. Faunal micromorphological features in moder humus of some western Canadian soils. Geoderma 40, 3–16.

Payne, J.A., 1965. A summer carrion study of the baby pig, *Sus scrofa* L. Ecology 46, 592–602.

Payne, W.J., 1970. Energy yields and growth of heterotrophs. Annu. Rev. Microbiol. 24, 17–52.

Pearce, M.J., 1997. Termites: Biology and Pest Management. CABI Publishing, Wallingford.

Perdomo, G., Evans, A., Maraun, M., Sunnucks, P., Thompson, R., 2012. Mouthpart morphology and trophic position of microarthropods from soils and mosses are strongly correlated. Soil Biol. Biochem. 53, 56–63.

Perdue, J.C., 1987. Population Dynamics of Mites (Acari) in Conventional and Conservation Tillage Agroecosystems. PhD Dissertation. University of Georgia, Athens, Georgia.

Perdue, J.C., Crossley Jr., D.A., 1989. Seasonal abundance of soil mites (Acari) in experimental agroecosystems: effects of drought in no-tillage and conventional tillage. Soil Tillage Res. 15, 117–124.

Perez-Moreno, J., Read, D.J., 2001. Nutrient transfer from soil nematodes to plants: a direct pathway provided by the mycorrhizal mycelial network. Plant Cell Environ. 24, 1219–1226.

Persson, T. (Ed.), 1980. Structure and function of northern coniferous forests- an ecosystem study. Ecol. Bull. 32.

Petersen, H., 2002. Collembolan ecology at the turn of the millennium. Pedobiologia 46, 246–260.

Petersen, H.A., Luxton, M., 1982. A comparative analysis of soil fauna populations and their role in decomposition processes. Oikos 39, 287–388.

Philips, J.R., 1990. Acarina: Astigmata (Acaridida). In: Dindal, D.L. (Ed.), Soil Biology Guide. Wiley, New York, pp. 757–778.

Phillips, R.E., Phillips, S.H., 1984. No-tillage Agriculture: Principles and Practices. Van Nostrand Reinhold, New York.

Piearce, T.G., Phillips, M.J., 1980. The fate of ciliates in the earthworm gut: an in vitro study. Microb. Ecol. 5, 313–320.

Pimm, S.L., 1982. Food Webs. Chapman and Hall, London, UK.

Pimm, S.L., Lawton, J.H., 1980. Are food webs divided into compartments? J. Anim. Ecol. 49, 879–898.

Pirozynski, K.A., Hawksworth, D.L., 1988. Coevolution of Fungi with Plants and Animals. Academic Press, London.

Pirozynski, K.A., Malloch, D.W., 1975. The origin of land plants: a matter of mycotrophism. BioSystems 6, 153–164.

Plass, G.N., 1956. Carbon dioxide and the climate. Am. Sci. 44, 302–319.

Poinar Jr., G.O., 1983. The natural history of nematodes. Prentice Hall, Englewood Cliffs, NJ.

Pokarzhevskii, A.D., van Straalen, N.M., Zaboev, D.P., Zaitsev, A.S., 2003. Microbial links and element flows in nested detrital food-webs. Pedobiologia 47, 213–224.

Polis, G.A., 1991. Complex trophic interactions in deserts: an empirical critique of food-web theory. Am. Nat. 138, 123–155.

Pollierer, M.M., Langel, R., Scheu, S., Maraun, M., 2009. Compartmentalization of the soil animal food web as indicated by dual analysis of stable isotope ratios (N-15/N-14 and C-13/C-12). Soil Biol. Biochem. 41, 1221–1226.

Pomeroy, L.R., 1974. The ocean's food web, a changing paradigm. Bioscience 24, 499–504.

Ponge, J.-F., 1991. Food resources and diets of soil animals in a small area of Scots pine litter. Geoderma 49, 33–62.

Ponge, J.-F., 2003. Humus forms in terrestrial ecosystems: a framework to diversity. Soil Biol. Biochem. 35, 935–945.

Ponge, J.-F., 2013. Plant-soil feedbacks mediated by humus—a review. Soil Biol. Biochem. 57, 1048–1060.

Ponge, J.-F., Chevalier, R., Loussot, P., 2002. Humus index: an integrated tool for the assessment of forest floor and topsoil properties. Soil Sci. Soc. Am. J. 66, 1996–2001.

Porazinska, D.L., Bardgett, R.D., Blaauw, M.B., Hunt, H.W., Parsons, A.N., Seastedt, T.R., et al., 2003. Relationships at the aboveground-belowground interface: plants, soil biota, and soil processes. Ecol. Monogr. 73, 377–395.

Porazinska, D.L., Creer, S., Caporaso, J.G., Knight, R., Thomas, W.K., 2012. Sequencing our way towards understanding global eukaryotic biodiversity. Trends Ecol. Evol. 27, 234–244.

Porter, K.G., 1975. Enhancement of algal growth and productivity by grazing zooplankton. Science 192, 1332–1334.

Post, W.H., Peng, T.H., Emanuel, W.R., King, A.W., Dale, V.H., De Angelis, D.L., 1990. The global carbon cycle. Am. Sci. 78, 310–326.

Post, W.H., King, A.W., Wullschleger, S.D., 1996. Soil organic matter models and global estimates of soil organic carbon. In: Powlson, D.S., Smith, P., Smith, J.U. (Eds.), Evaluation of Soil Organic Matter Models. NATO ASI Series, 38, 201–222.

Postgate, J.R., 1987. Nitrogen fixation, Studies in Biology, No. 92, second ed. Arnold, London.

Postma, J., Altemüller, H.-J., 1990. Bacteria in thin soil sections stained with the fluorescence brightener Calcofluor M2R. Soil Biol. Biochem. 22, 89–96.

Powers, J.S., Montgomery, R.A., Adair, E.C., Brearley, F.Q., DeWalt, S.J., Castanho, C.T., et al., 2009. Decomposition in tropical forests: a pan-tropical study of the effects of litter type, litter placement and mesofaunal exclusion across a precipitation gradient. J. Ecol. 97, 801–811.

Powlson, D.S., 1975. Effects of biocidal treatments on soil organisms. In: Walker, N. (Ed.), Soil Microbiology. Butterworths, London, pp. 193–224.

Powlson, D.S., 1994. The soil microbial biomass: before, beyond and back. In: Ritz, K., Dighton, J., Giller, K.E. (Eds.), Beyond the Biomass. Wiley/Sayce, Chichester, pp. 3–20.

Pregitzer, K.S., DeForest, J.L., Burton, A.J., Allen, M.F., Ruess, R.W., Hendrick, R.L., 2002. Fine root architecture of nine North American trees. Ecol. Monogr. 72, 293–309.

Price, D.W., 1975. Vertical distribution of small arthropods in a California pine forest soil. Ann. Entomol. Soc. Am. 68, 174–180.

Pries, C.E.H., Castanha, C., Porras, R.C., Torn, M.S., 2017. The whole-soil carbon flux in response to warming. Science 355, 1420–1423.

Prosser, J.I., 2002. Molecular and functional diversity in soil micro-organisms. Plant & Soil 244, 9–17.

Publicover, D.A., Vogt, K.A., 1993. A comparison of methods for estimating forest fine root production with respect to sources of error. Can. J. For. Res. 23, 1179–1186.

Punzo, F., 1998. The Biology of Camel Spiders. Kluwer Academic Publication, Boston, MA.

Purvis, G., Fadel, A., 2002. The influence of cropping rotations and soil cultivation practice on the population ecology of carabids (Coleoptera: Carabidae) in arable land. Pedobiologia 46, 452–474.

Radajewski, S., Ineson, P., Parekh, N.R., Murrell, J.C., 2000. Stable-isotope probing as a tool in Microbial Ecology. Nature 403, 646–649.

Rao, V.R., Ramakrishnan, B., Adhya, T.K., Kanungo, P.K., Nayak, D.N., 1998. Current status and future prospects of associative nitrogen fixation in rice. World J. Microbiol. Biotechnol. 14, 621–633.

Raupach, M.J., Hannig, K., Moriniere, J., Hendrich, L., 2016. A DNA barcode library for ground beetles (Insecta, Coleoptera, Carabidae) of Germany: the genus Bembidion Latreille, 1802 and allied taxa. ZooKeys 592, 121.

Read, D.J., 1991. Mycorrhizas in ecosystems. Experientia 47, 376–3911.

Read, D.J., Francis, R., Finlay, R.D., 1985. Mycorrhizal mycelia and nutrient cycling in plant communities. In: Fitter, A.H., Atkinson, D., Read, D.J., Usher, M.B. (Eds.), Ecological Interactions in Soil: Plants, Microbes and Animals. Blackwell, Oxford, pp. 193–217.

Read, D.J., Duckett, J.G., Francis, R., Ligrone, R., Russell, A., 2000. Symbiotic fungal associations in "lower" land plants. Philos. Trans. R. Soc., B 355, 815–830.

Reavy, B., Swanson, M.M., Taliansky, M., 2014. Viruses in soil. In: Dighton, J., Krumins, J.A. (Eds.), Interactions in Soil: Promoting Plant Growth. Springer, pp. 163–180.

Redeker, D., 2002. Molecular identification and phylogeny of arbuscular mycorrhizal fungi. Plant & Soil 244, 67–73.

Redeker, D., Morton, J.B., Bruns, T.D., 2000. Ancestral lineages of arbuscular mycorrhizal fungi (Glomales). Mol. Phylog. Evol. 14, 276–284.

Reeves, R.M., 1967. Seasonal distribution of some forest soil Oribatei. In: Evans, G.O. (Ed.), Proceedings 2nd International Congress of Acarology. Akademiai Kiado, Budapest, pp. 23–30.

Reichle, D.E., Ausmus, B.S., McBrayer, J.F., 1975. Ecological energetics of decomposer invertebrates in a deciduous forest and total respiration budget. In: Vanek, J. (Ed.), Progress in Soil Zoology. Proceedings of the 5th International Colloquium on Soil Zoology, Prague, Czech Republic, pp. 283–292.

Reid, J.B., Goss, M.J., 1982. Suppression of decomposition of ^{14}C- labelled plant roots in the presence of living roots of maize and perennial ryegrass. Eur. J. Soil Sci. 33 (3), 387–395.

Rengasamy, P., 2006. World salinization with emphasis on Australia. J. Exp. Bot. 57 (5), 1017–1023.

Reynolds, J.W., 1977. The earthworms (Lumbricidae and Sparganophilidae) of Ontario. Life Sciences Miscellaneous Publications, Royal Ontario Museum, Toronto.

Reynolds, J.W., 1995. The status of exotic earthworm systematics and biogeography in North America. In: Hendrix, P.F. (Ed.), Ecology and Biogeography of Earthworms in North America. Lewis Publications, Boca Raton, FL, pp. 1–27.

Richter Jr., D.D., Markewitz, D., 2001. Understanding Soil Change: Soil Sustainability over Millennia, Centuries, and Decades. Cambridge University Press, Cambridge, MA.

Richter Jr., D.D., Hofmockel, M., Callaham Jr., M.A., Powlson, D.S., Smith, P., 2007. Long-term soil experiments: keys to managing earth's rapidly changing ecosystems. Soc. Sci. Soc. Am. J. 71, 266–279.

Rillig, M.C., Wright, S.F., Eviner, V.T., 2002. The role of arbuscular mycorrhizal fungi and glomalin in soil aggregation: comparing effects of five plant species. Plant Soil 238, 325–333.

Rizzo, A.M., Altiero, T., Corsetto, P.A., Montorfano, G., Guidetti, R., Rebecchi, L., 2015. Space flight effects on antioxidant molecules in dry tardigrades: the TARDIKISS experiment. BioMed Research International, vol. 2015, Article ID 167642, 7 pages.

Robertson, G.P., Gross, K.L., 1994. Assessing the heterogeneity of belowground resources: quantifying pattern and scale. In: Caldwell, M.M., Pearcy, R.W. (Eds.), Exploitation of Environmental Heterogeneity by Plants. Academic Press, San Diego, CA, pp. 237–253.

Robinson, D., Scrimgeour, C.M., 1995. The contribution of plant C to soil CO_2 measured using ^{13}C. Soil Biol. Biochem. 27, 1653–1656.

Rogers, J.E., Whitman, W.B., 1991. Microbial Production and Consumption of Greenhouse Gasses: Methane, Nitrogen Oxides, and Halomethanes. American Society for Microbiology, Washington, DC.

Rosomer, W.S., Stoffolano Jr., J.G., 1994. The Science of Entomology, third ed. William C. Brown Publishers, Dubuque. IA.

Rothwell, F.M., 1984. Aggregation of surface mine soil by interaction between VAM fungi and lignin degradation products of lespedeza. Plant & Soil 80, 99–104.

Rovira, A.D., Foster, R.C., Martin, J.K., 1979. Note on terminology: origin, nature and nomenclature of the organic materials in the rhizosphere. In: Harley, J.L., Russell, R.S. (Eds.), The Soil-Root Interface. Academic Press, London, pp. 1–4.

Ruess, L., Häggblom, M.M., Zapata, E.J.G., Dighton, J., 2002. Fatty acids of fungi and nematodes—possible biomarkers in the soil food chain? Soil Biol. Biochem. 34, 745–756.

Ruess, L., Häggblom, M.M., Reinhard, L., Scheu, S., 2004. Nitrogen isotope ratios and fatty acid composition as indicators of animal diets in belowground systems. Oecologia 139 (3), 336–346.

Rumpel, C., Koegel-Knabner, I., 2011. Deep soil organic matter—a key but poorly understood component of the terrestrial Carbon cycle. Plant Soil 338, 143–158.

Rusek, J., 1975. Die bodenbildende Funktion von Collembolen und Acarina. Pedobiologia 15, 299–308.

Rusek, J., 1985. Soil microstructures—contributions on specific soil organisms. Quaestiones Entomologicae 21, 497–514.

Russo, R.O., 2005. Chapter 8: actinorhiza in forestry and agroforestry. In: Werner, D., Newton, W.E. (Eds.), Nitrogen Fixation in Agriculture, Forestry, Ecology, and the Environment. Springer, Amsterdam, pp. 143–171.

Russell, E.J., 1973. Soil Conditions and Plant Growth. Longmans, London.

Russell, E.J., Hutchinson, H.B., 1909. The effect of partial sterilization of soil on the production of plant food. J. Agric. Sci. 3, 111–144.

Saano, A., Lindström, K., 1995. Isolation and identification of DNA from soil. In: Alef, K., Nannipieri, P. (Eds.), Methods in Applied Soil Microbiology and Biochemistry. Academic Press, London, pp. 440–451.

Sanchez, P.A., Palm, C.A., Buol, S.W., 2003. Fertility capability soil classification: a tool to help assess soil quality in the tropics. Geoderma 114, 157–185.

Sanchez de Leon, Y., Johnson-Maynard, J., 2009. Dominance of an invasive earthworm in native and non-native grassland ecosystems. Biol. Invasions 11, 1393–1401.

Santos, F., Torn, M.S., Bird, J.A., 2012. Biological degradation of pyrogenic organic matter in temperate forest soils. Soil Biol. Biochem. 51, 115–124.

Santos, P.F., Whitford, W.G., 1981. The effects of microarthropods on litter decomposition in a Chihuahuan Desert ecosystem. Ecology 62, 654–663.

Santos-González, J.C., Finlay, R.D., Tehler, A., 2007. Seasonal dynamics of arbuscular mycorrhizal root colonization in a semi- natural grassland. Appl. Environ. Microbiol. 73, 5613–5623.

Satchell, J.E., 1969. Methods of sampling earthworm populations. Pedobiologia 9, 20–25.

Satchell, J.E. (Ed.), 1983. Earthworm Ecology: From Darwin to Vermiculture. Chapman and Hall, London.

Scharlemann, J.P.W., Tanner, E.V.J., Hiederer, R., Kapos, V., 2014. Global soil carbon: understanding and managing the largest terrestrial carbon pool. Carbon Manage. 5 (1), 81–91.

Scharpenseel, H.W., Schomaker, M., Ayoub, A., 1990. Soils on a Warmer Earth. Elsevier, Amsterdam.

Schauermann, J., 1983. Eine Verbesserung der Extraktionsmethode für terrestrische Enchytraeiden. In: Lebrun, P., André, H.M., De Medts, A., Grégoire-Wibo, C., Wauthy, G. (Eds.), New Trends in Soil Biology. Dieu-Brichart, Louvain-la-Neuve, Belgium, pp. 669–670.

Scheller, U., 1988. The Pauropoda of the Savanna River Plant, Aiken, South Carolina. SRO-NERP, 17, 1–99.

Scheller, U., 1990. Pauropoda. In: Dindal, D.L. (Ed.), Soil Biology Guide. Wiley, New York, pp. 861–890.

Scheller, U., 2002. Pauropods—the little ones among the Myriapods. All Taxa Biodiversity Inventory 3, 3.

Schenker, R., 1984. Spatial and seasonal distribution patterns of oribatid mites (Acari: Oribatei) in a forest soil ecosystem. Pedobiologia 27, 133–149.

Scheu, S., 2002. The soil food web: structure and perspectives. Eur. J. Soil Biol. 38, 11–20.

Scheu, S., Setälä, H., 2002. Multitrophic interactions in decomposer food-webs. In: Tscharntke, B., Hawkins, B.A. (Eds.), Multitrophic Level Interactions. Cambridge University Press, Cambridge, MA, pp. 223–264.

Schimel, D.S., 1993. Population and community processes in the response of terrestrial ecosystems to global change. In: Kareiva, P.M., Kingsolver, J.G., Huey, R.B. (Eds.), Biotic Interactions and Global Change. Sinauer Associates, Sunderland, MA, pp. 45–54.

Schimel, D.S., Braswell, B.H., Holland, E.A., McKeown, R., Ojima, D.S., Painter, T.H., et al., 1994. Climatic, edaphic, and biotic controls over storage and turnover of carbon in soils. Global Biogeochem. Cycles 8, 279–293.

Schimel, J.P., Gulledge, J.M., 1998. Microbial community structure and global trace gases. Global Change Biol. 4, 745–758.

Schimel, J.P., Weintraub, M.N., 2003. The implications of exoenzyme activity on microbial carbon and nitrogen limitation in soil: a theoretical model. Soil Biol. Biochem. 35, 549–563.

Schinner, F., Öhlinger, R., Kandeler, E., Margesin, R. (Eds.), 1996. Methods in Soil Biology.. Springer-Verlag, Berlin.

Schlesinger, W.H., 1996. Biogeochemistry: An Analysis of Global Change, second ed. Academic Press, San Diego, CA.

Schmelz, R.M., Collado, R., 2012. An updated checklist of currently accepted species of Enchytraeidae (Oligochaeta, Annelida). Landbauforschung–vTI Agriculture and Forestry Research, Special Issue, 357, 67–87.

Schmelz, R.M., Collado, R., 2015. Checklist of taxa of Enchytraeidae (Oligochaeta): an update. Soil Org. 87, 149–152.

Schmidt, O., 2001. Appraisal of the electrical octet method for estimating earthworm populations in arable land. Ann. Appl. Biol. 138, 231–241.

Schmidt, M.W.I., Torn, M.S., Abiven, S., Dittmar, T., Guggenberger, G., Janssens, I.A., et al., 2011. Persistence of soil organic matter as an ecosystem property. Nature 478 (7367), 49–56.

Schneider, K., Migge, S., Norton, R.A., Scheu, S., Langel, R., Reineking, A., et al., 2004. Trophic niche differentiation in soil microarthropods (Oribatida, Acari): evidence from stable isotope ratios ($^{15}N/^{14}N$). Soil Biol. Biochem. 36, 1769–1774.

Schoenholzer, F., Hahn, D., Zeyer, J., 1999. Origins and fate of fungi and bacteria in the gut of *Lumbricus terrestris* L. studied by image analysis. FEMS Microbiol. Ecol. 28, 235–248.

Schouten, A.J., Arp, K.K.M., 1991. A comparative study on the efficiency of extraction methods for nematodes from different forest litters. Pedobiologia 35, 393–400.

Schrader, H.S., Schrader, J.O., Walker, J.J., Bruggeman, N.B., Vanderloop, J.M., Shaffer, J.M., et al., 1997. Effects of host starvation on bacteriophage dynamics. In: Morita, R.Y. (Ed.), Bacteria in Oligotrophic Environments. Chapman & Hall, New York, pp. 368–385.

Schultz, P.A., 1991. Grazing preferences of two collembolan species, *Folsomia candida* and *Proisotoma minuta*, for ectomycorrhizal fungi. Pedobiologia 35, 313–325.

Schuster, R., 1956. Der anteil der Oribatiden and den Zersetzungsvorgängen im Böden. Zeitschrift für Morphologie und Ökologie der Tiere 45, 1–33.

Schwert, D.P., 1990. Oligochaeta: lumbricidae. In: Dindal, D.L. (Ed.), Soil Biology Guide. Wiley, New York, pp. 341–356.

Scow, K.M., 1997. Soil microbial communities and carbon flow in agroecosystems. In: Jackson, L.E. (Ed.), Ecology in Agriculture. Academic Press, San Diego, CA, pp. 367–413.

Seastedt, T.R., 1984a. Microarthropods of burned and unburned tallgrass prairie. J. Kansas Entomol. Soc. 57, 468–476.

Seastedt, T.R., 1984b. The role of microarthropods in decomposition and mineralization processes. Annu. Rev. Entomol. 29, 25–46.

Setälä, H., Rissanen, J., Markkola, M., 1997. Conditional outcomes in the relationship between pine and ectomycorrhizal fungi in relation to biotic and abiotic environment. Oikos 80, 112–122.

Sgardelis, S., Stamou, G., Margaris, N.S., 1981. Structure and spatial distribution of soil arthropods in a Phryganic (East Mediterranean) ecosystem. Revue d'écologie et de biologie du sol 18, 221–230.

Shain, D.H., Carter, M.R., Murray, K.P., Maleski, K.A., Smith, N.R., McBride, T.R., Michalewicz, L.A., Saidel, W. M., 2000. Morphologic characterization of the ice worm *Mesenchytraeus solifugus*. J. Morphol. 246, 192–197.

Shamoot, S., McDonald, L., Bartholomew, W.V., 1968. Rhizo-deposition of organic debris in soil. Soil Sci. Soc. Am. Proc. 32, 817–820.

Shanks, R.E., Olson, J.S., 1961. First-year breakdown of leaf litter in southern Appalachian forests. Science 134, 194–195.

Shaw, M.R., Zavaleta, E.S., Chiariello, N.R., Cleland, E.E., Mooney, H.A., Field, C.B., 2002. Grassland responses to global environmental changes suppressed by elevated CO_2. Science 298, 1987–1990.

Shelley, R.M., 2002. A Synopsis of the North American Centipedes of the Order Scolopendromorpha (Chilopoda). Memoir 5, Virginia Museum of Natural History, Martinsville, Virginia.

Shelley, R.M., Sisson, W.D., 1995. Distributions of the scorpions *Centruroides vittatus* (Say) and *Centruroides hentzi* (Banks) in the United States and Mexico (Scorpiones, Buthidae). J. Arachnology 23, 100–110.

Shimmel, S.M., Darley, W.M., 1985. Productivity and density of soil algae in an agricultural system. Ecology 66, 1439–1447.

Shipton, P.J., 1986. Infection by foot and root rot pathogens and subsequent damage. In: Wood, R.K.S., Jellis, G.K. (Eds.), Plant Diseases, Infection, Damage and Loss. Blackwell Scientific, Oxford, pp. 139–155.

Siepel, H., de Ruiter-Dijkman, E.M., 1993. Feeding guilds of oribatid mites based on their carbohydrase activities. Soil Biol. Biochem. 25, 1491–1497.

Sileshi, G.W., Arshad, M.A., Konaté, S., Nkunika, P.O.Y., 2010. Termite-induced heterogeneity in African savanna vegetation: mechanisms and patterns. J. Veg. Sci. 21, 923–937.

Sims, R.W., Gerard, B.M., 1985. Earthworms: Keys and Notes for the Identification and Study of the Species. The Linnaean Society of London, London.

Sinclair, J.L., Ghiorse, W.C., 1989. Distribution of aerobic bacteria, protozoa, algae, and fungi in deep subsurface sediments. Geomicrobiol. J. 7, 15–31.

Singh, B.P., Cowie, A.L., Smernik, R.J., 2012. Biochar carbon stability in a clayey soil as a function of feedstock and pyrolysis temperature. Environ. Sci. Technol. 46, 11770–11778.

Singh, J., Singh, S., Vig, A.P., 2016. Extraction of earthworm from soil by different sampling methods: a review. Environ. Develop. Sustainability 18, 1521–1539.

Singh, J.S., Lauenroth, W.K., Hunt, H.W., Swift, D.M., 1984. Bias and random errors in estimators of net root production: a simulation approach. Ecology 65, 1760–1764.

Singleton, D.R., Furlong, M.A., Peacock, A.D., White, D.C., Coleman, D.C., Whitman, W.B., 2003. *Solirubrobacter pauli* gen. nov., sp. nov., a mesophilic bacterium within the Rubrobacteridae related to common soil clones. Int. J. Syst. Evol. Microbiol. 53, 485–490.

Sinsabaugh, R.L., 2010. Phenol oxidase, peroxidase and organic matter dynamics of soil. Soil Biol. Biochem. 42, 391–404.

Sinsabaugh, R.L., Moorhead, D.L., Linkins, A.E., 1994. The enzymic basis of plant litter decomposition: emergence of an ecological process. Appl. Soil Ecol. 1, 97–111.

Six, J., Elliott, E.T., Paustian, K., Doran, J.W., 1998. Aggregation and soil organic matter accumulation in cultivated and native grassland soils. Soil Sci. Soc. Am. J. 62, 1367–1377.

Six, J., Elliott, E.T., Paustian, K., 1999. Aggregate and soil organic matter dynamics under conventional and no-tillage systems. Soil Sci. Soc. Am. J. 63, 1350–1358.

Six, J., Conant, R.T., Paul, E.A., Paustian, K., 2002a. Stabilization mechanisms of soil organic matter: implications for C-saturation of soils. Plant Soil 241, 155–176.

Six, J., Feller, C., Denef, K., Ogle, S.M., de Moraes Sa, J.C., Albrecht, A., 2002b. Soil organic matter, biota and aggregation in temperate and tropical soils—effects of no-tillage. Agronomie 22 (7–8), 755–775.

Six, J., Bossuyt, H., Degryze, S., Denef, K., 2004. A history of the research on the link between (micro) aggregates, soil biota, and soil organic matter dynamics. Soil Tillage Res. 79, 7–31.

Smith, M.D., Hartnett, D.C., Wilson, G.W.T., 1999. Interacting influence of mycorrhizal symbiosis and competition on plant diversity in tallgrass prairie. Oecologia 121, 574–582.

Smith, M.L., Bruhn, J.N., Anderson, J.B., 1992. The root-infecting fungus *Armillaria bulbosa* may be among the largest and oldest living organisms. Nature 356, 428–443.

Smith, P., Andrén, O., Brussaard, L., Dangerfield, M., Ekschmitt, K., Lavelle, P., et al., 1998. Soil biota and global change at the ecosystem level: describing soil biota in mathematical models. Global Change Biol. 4, 773–784.

Smith, S.J., Read, D.J., 1997. Mycorrhizal Symbiosis, second ed. Academic Press, Cambridge.

Smucker, A.J.M., Ferguson, J.C., De Bruyn, W.P., Belford, R.K., Ritchie, J.T., 1987. Image analysis of video-recorded plant root systems. In: Taylor, H.M. (Ed.), Minirhizotron Observation Tubes: Methods and Applications for Measuring Rhizosphere Dynamics. American Society of Agronomy Special Publication No. 50, Madison, WI, pp. 67–80.

Snider, R.J., 1967. An annotated list of the Collembola (Springtails) of Michigan. Michigan Entomologist 1, 178–234.

Snider, R.J., Snider, R.M., Smucker, A.J.M., 1990. Collembolan populations and root dynamics in Michigan agroecosystems. In: Box Jr., J.E., Hammonds, L.C. (Eds.), Rhizosphere Dynamics. Westview Press, Boulder, CO, pp. 168–191.

Snider, R.M., 1973. Laboratory observations on the biology of *Folsomia candida* (Willem) Collembola: Isotomidae. Revue d'écologie et de biologie du sol 10, 103–124.

Snider, R.M., Snider, R.J., 1997. Efficiency of arthropod extraction from soil cores. Entomol. News 108 (3), 203–208.

Söderström, B.E., 1977. Vital staining of fungi in pure cultures and in soil with fluorescein diacetate. Soil Biol. Biochem. 9, 59–63.

Sojka, R.E., Upchurch, D.R., 1999. Reservations regarding the soil quality concept. Soil Sci. Soc. Am. J. 63, 1039–1054.

Sommer, R.J., Streit, A., 2011. Comparative genetics and genomics of nematodes: genome structure, development, and lifestyle. Annu. Rev. Genet. 45, 1–20.

South, A., 1992. Terrestrial Slugs: Biology. Ecology, and Control. Chapman and Hall, London, UK.

Southwood, T.R.E., 1978. Ecological Methods With Particular Reference to the Study of Insect Populations., second ed. Chapman and Hall, London.

Southey, J., 1986. Laboratory Methods for Work With Plant and Soil Nematodes. Ministry of Agriculture, Fisheries & Food, Ref. Bk. 402. H.M.S.O. Books Norwich, Norfolk, U.K.

Sparling, G.P., 1981. Microcalorimetry and other methods to assess biomass and activity in soil. Soil Biol. Biochem. 13, 93–98.

Sparling, G.P., Ross, D.J., 1993. Biochemical methods to estimate soil microbial biomass: current developments and applications. In: Mulongoy, K., Merckx, R. (Eds.), Soil Organic Matter Dynamics and Sustainability of Tropical Agriculture. John Wiley, New York.

Speiser, B., 2001. Food and feeding behaviour. In: Barker, G.M. (Ed.), The Biology of Terrestrial Molluscs. CAB International, Wallingford, pp. 259–288.

Staaf, H., Berg, B., 1982. Accumulation and release of plant nutrients in decomposing Scots pine needle litter. Long-term decomposition in a Scots pine forest II. Can. J. Bot. 60, 1561–1568.

Staddon, P.L., Ramsey, C.B., Ostle, N., Ineson, P., Fitter, A.H., 2003. Rapid turnover of hyphae of mycorrhizal fungi determined by AMS microanalysis of C14. Science 300, 1138–1140.

Stark, J.M., 1994. Causes of soil nutrient heterogeneity at different scales. In: Caldwell, M.M., Pearcy, R.W. (Eds.), Exploitation of Environmental Heterogeneity by Plants.. Academic Press, San Diego, CA, pp. 255–284.
Stary, J., Block, W., 1998. Distribution and biogeography of oribatid mites (Acari: Oribatida) in Antarctica, the sub-Antarctic islands and nearby land areas. J. Nat. Hist. 32, 861–894.
Steen, E., 1984. Variation of root growth in a grass ley studied with a mesh bag technique. Swed. J. Agric. Res. 14, 93–97.
Steen, E., 1991. Usefulness of the mesh bag method in quantitative root studies. In: Atkinson, D. (Ed.), Plant Root Growth: An ecological Perspective. Blackwell, Oxford, pp. 75–86.
Steinaker, D.F., Wilson, S.D., 2008. Scale and density dependent relationships among roots, mycorrhizal fungi and collembola in grassland and forest. Oikos 117, 703–710.
Steinberger, Y., Wallwork, J.A., 1985. Composition and vertical distribution patterns of the microarthropod fauna in a Negev desert soil. J. Zool. 206 (3), 329–339.
Sterner, R.W., Elser, J.J., 2002. Ecological Stoichiometry. Princeton University Press, Princeton and Oxford.
Stewart, J.W.B., McKercher, R.B., 1982. Phosphorus cycle. In: Burns, R.G., Slater, J.H. (Eds.), Experimental Microbial Ecology. Blackwell, Oxford, pp. 221–238.
Stewart, J.W.B., Anderson, D.W., Elliott, E.T., Cole, C.V., 1990. The use of models of soil pedogenic processes in understanding changing land use and climatic change. In: Scharpenseel, H.W., Schomaker, M., Ayoub, A. (Eds.), Soils on a Warmer Earth.. Elsevier, Amsterdam, pp. 121–131.
Stinner, B.R., Crossley Jr., D.A., 1980. Comparison of mineral elements cycling under till and no-till practices: an experimental approach to agroecosystems analysis. In: Dindal, D. (Ed.), Soil Biology as Related to Land Use Practices. U.S. Environmental Protection Agency, Washington, DC, pp. 180–288.
St. John, T.V., Coleman, D.C., 1983. The role of mycorrhizae in plant ecology. Can. J. Bot. 61, 1005–1014.
Stockdill, S.M.J., 1966. The effect of earthworms on pastures. Proc. N. Z. Ecol. Soc. 13, 68–74.
Stocker, B.D., Prentice, I.C., Cornell, S.E., Davies-Barnard, T., Finzi, A.C., Franklin, O., et al., 2016. Terrestrial nitrogen cycling in Earth system models revisited. New Phytol. 210, 1165–1168.
Störmer, K., 1908. Ueber die Wirkung des Schwefelkohlenstoffs und aehnlicher Stoffe auf den Boden. Zentralblatt für Bakteriologie 20, 282–286.
Stout, J.D., 1963. The terrestrial plankton. Tuatara 11, 57–65.
Stout, J.D., Goh, K.M., Rafter, T.A., 1981. Chemistry and turnover of naturally occurring resistant organic compounds in soil. In: Paul, E.A., Ladd, J.N. (Eds.), Soil Biochemistry, vol. 5. Dekker, New York, pp. 1–73.
Strandtmann, R.W., 1967. Terrestrial Prostigmata (Trombidiform mites). In: Gressitt, J.L. (Ed.), Entomology of Antarctica. Antarctic Research Series, vol. 10. American Geophysical Union, Washington, DC, pp. 51–80.
Strandtmann, R.W., Crossley Jr., D.A., 1962. A new species of soil-inhabiting mite, *Hypoaspis marksi* (Acarina, Laelaptidae). J. Kansas Entomol. Soc. 35, 180–185.
Strong, D.R., 2002. Populations of entomopathogenic nematodes in foodwebs. In: Gaugler, R. (Ed.), Entomopathogenic Nematology. CABI Publishing, Wallingford, pp. 225–240.
Strong, D.R., Kaya, H.K., Whipple, A.V., Child, A.L., Kraig, S., Bondonno, M., et al., 1996. Entomopathogenic nematodes: natural enemies of root-feeding caterpillars on bush lupine. Oecologia 108, 167–173.
Strong, D.R., Whipple, A.V., Child, A.L., Dennis, B., 1999. Model selection for a subterranean trophic cascade: root-feeding caterpillars and entomopathogenic nematodes. Ecology 80, 2750–2761.
Sturm, H., 1959. Die Nährung der Proturen. Beobachtungen an Acerentomon doderoiSilv. und Eosentomon transitoriumBERL. Naturwissenschaften 46 (2), 90–91.
Summerhayes, V.S., Elton, C.S., 1923. Contributions to the ecology of Spitsbergen and Bear Island. J. Ecol. 11, 214–286.
Sutter, P., 2015. Let Us Now Praise Famous Gullies: Providence Canyon and the Soils of the South. University of Georgia Press, Athens, Georgia, USA, p. 288.
Swank, W.T., Crossley Jr., D.A., 1988. Forest Hydrology and Ecology at Coweeta. Springer-Verlag, New York.
Swift, M.J., 1999. Integrating soils, systems and society. Nat. Resou. 35, 12–20.
Swift, M.J., Heal, O.W., Anderson, J.M., 1979. Decomposition in Terrestrial Ecosystems. University of California Press, Berkeley, CA.
Syers, J.K., Sharpley, A.N., Keeney, D.R., 1979a. Cycling of nitrogen by surface-casting earthworms in a pasture ecosystem. Soil Biol. Biochem. 11, 181–185.

Tandarich, J.P., Darmody, R.G., Follmer, L.R., Johnson, D.L., 2002. Historical development of soil and weathering profile concepts from Europe to the United States of America. Soil Sci. Soc. Am. J. 66, 335–346.
Tansley, A.G., 1935. The use and abuse of vegetational concepts and terms. Ecology 16, 284–307.
Targulian, V.O., Krasilnikov, P.V., 2007. Soil System and Pedogenic processes: self-organization, time scales, and environmental significance. Catena 71, 373–381.
Tashiro, H., 1990. Insecta: coleoptera scarabaeidae larvae. In: Dindal, D.L. (Ed.), Soil Biology Guide.. Wiley, New York, pp. 1191–1209.
Tate, K.R., Newman, R.H., 1982. Phosphorus fractions of a climosequence of soils in New Zealand tussock grassland. Soil Biol. Biochem. 14, 191–196.
Taylor, A.F.S., Alexander, I., 2005. The ectomycorrhizal symbiosis: life in the real world. Mycologist 19, 102–112.
Taylor, D.L., Hollingsworth, T.N., McFarland, J.W., Lennon, N.J., Nusbaum, C., Ruess, R.W., 2014. A first comprehensive census of fungi in soil reveals both hyperdiversity and fine-scale niche partitioning. Ecol. Monogr. 84, 3–20.
Taylor, H.M., 1987. Minirhizotron Observation Tubes: Methods and applications for Measuring Rhizosphere Dynamics. American Society of Agronomy Special Publication No. 50, Madison, WI.
Tedersoo, L., Bahram, M., Põlme, S., Kõljalg, U., Yorou, N.S., Wijesundera, R., et al., 2014. Global diversity and geography of soil fungi. Science 346 (6213), 1256688–1-1256688-10.
Teskey, H.J., 1990. Insecta: diptera larvae. In: Dindal, D.L. (Ed.), Soil Biology Guide. Wiley, New York, pp. 1253–1276.
Tevis Jr., L., Newell, I.M., 1962. Studies on the biology and seasonal cycle of the giant red velvet mite, *Dinothrombium pandorae* (Acarina: Trombidiidae). Ecology 43, 797-505.
Theng, B.K.G., 1979. Formation and Properties of Clay-Polymer Complexes. Elsevier, Amsterdam.
Theng, B.K.G., Tate, K.R., Sollins, P., Moris, N., Nadkarni, N., Tate III, R.L., 1989. Constituents of organic matter in temperate and tropical soils. In: Coleman, D.C., Oades, J.M., Uehara, G. (Eds.), Dynamics of Soil Organic Matter in Tropical Ecosystems. University of Hawaii Press, Honolulu, Hawaii, pp. 5–32.
Thielemann, U., 1986. Elektrischer Regenwurmfang mit der Oktett-Metode. Pedobiologia 29, 296–302.
Thompson, R.M., Brose, U., Dunne, J.A., Hall Jr., R.O., Hladyz, S., Kitching, R.L., et al., 2012. Food webs: reconciling the structure and function of biodiversity. Trends Ecol. Evol. 27, 689–697.
Tian, G., Brussaard, L., Kang, B.T., 1995. Breakdown of plant residues with chemically contrasting compositions: effect of earthworms and millipedes. Soil Biol. Biochem. 27, 277–280.
Thimm, T., Hoffmann, A., Borkott, H., Munch, J.C., Tebbe, C.C., 1998. The gut of the soil microarthropod *Folsomia candida* (Collembola) is a frequently changeable but selective habitat and a vector for microorganisms. Appl. Environ. Microbiol. 64, 2660–2669.
Tiedje, J.M., Cho, J.C., Murray, A., Treves, D., Xia, B., Zhou, J., 2001. Soil teeming with life: new frontiers for soil science. In: Rees, R.M., Ball, B.C., Campbell, C.D., Watson, C.A. (Eds.), Sustainable Management of Soil Organic Matter. CABI Publishing, Wallingford, pp. 393–412.
Tinker, B., Goudriaan, J., Teng, P., Swift, M., Linder, S., Ingram, J., et al., 1996. Global change impacts on agriculture, forestry and soils: the programme of the global change and terrestrial ecosystems core project of IGBP. In: Bazzaz, F., Sombroek, W. (Eds.), Global Climate Change and Agricultural Production. FAO and Wiley & Sons, Chichester, pp. 295–318.
Tippkötter, R., Ritz, K., Darbyshire, J.F., 1986. The preparation of soil thin sections for biological studies. Eur. J. Soil Sci. 37 (4), 681–690.
Tisdall, J.M., 1991. Fungal hyphae and structural stability of soil. Aust. J. Soil Res. 29, 729–743.
Tisdall, J.M., Oades, J.M., 1979. Stabilization of soil aggregates by the root systems of ryegrass. Aust. J. Soil Res. 17, 429–441.
Tisdall, J.M., Oades, J.M., 1982. Organic matter and waterstable aggregates in soils. Eur. J. Soil Sci. 33, 141–163.
Titlyanova, A.A., 1987. Ecosystem succession and biological turnover. Vegetatio 50, 43–51.
Tiunov, A.V., Scheu, S., 2005. Facilitative interactions rather than resource partitioning drive diversity-functioning relationships in laboratory fungal communities. Ecol. Lett. 8, 618–625.
Todd, R.L., Crossley Jr., D.A., Stormer Jr., J.A., 1974. Chemical composition of microarthropods by electron microprobe analysis: A preliminary report. Proceedings of 32nd Annual Proceedings of the Electron Microscopy Society of America, St. Louis, Missouri.

Tomlin, A.D., 1977. Pipeline construction—impact on soil micro- and mesofauna (Arthropoda and Annelida) in Ontario. Proc. Entomol. Soc. Ont. 108, 13–17.

Tomlin, A.D., Shipitalo, M.J., Edwards, W.M., Protz, R., 1995. Earthworms and their influence on soil structure and infiltration. In: Hendrix, P.F. (Ed.), Earthworm Ecology and Biogeography in North America. Lewis Publisher, Boca Raton, FL, pp. 159–183.

Torsvik, V., Øvreås, L., 2002. Microbial diversity and function in soil: from genes to ecosystems. Curr. Opin. Microbiol. 5, 240–245.

Torsvik, V., Salte, K., Sørheim, R., Goksøyr, J., 1990a. Comparison of phenotypic diversity and DNA heterogeneity in a population of soil bacteria. Appl. Environ. Microbiol. 56, 776–781.

Torsvik, V., Goksøyr, J., Daae, F.L., 1990b. High diversity in DNA of soil bacteria. Appl. Environ. Microbiol. 56, 782–787.

Torsvik, V., Goksøyr, J., Daae, R.L., Sørheim, R., Michalsen, J., Salte, K., 1994. Use of DNA analysis to determine the diversity of microbial communities. In: Ritz, K., Dighton, J., Giller, K.E. (Eds.), Beyond the Biomass. Wiley-Sayce, Chichester, pp. 39–49.

Touchot, F., Kilbertus, G., Vannier, G., 1983. Role d'un collembole (*Folsomia candida*) au cours de la degradation des litiéres de charme et de chêne, en presence au en absence d'argile. In: Lebrun, P., André, H.M., Demedts, A., Grégoire-Wibo, C., Wauthy, G. (Eds.), New Trends in Soil Biology. Dieu-Brichart, Ottignies-Louvain-la-Neuve, Belgium, pp. 269–280.

Travé, J., André, H.M., Taberly, G., Bernini, F., 1996. Les Acariens Oribates. Editions AGAR, Wavre, Belgium.

Treonis, A.M., Wall, D.H., Virginia, R.A., 1999. Invertebrate biodiversity in Antarctic Dry Valley soils and sediments. Ecosystems 2, 483–492.

Treseder, K.K., 2008. Nitrogen additions and microbial biomass: a meta-analysis of ecosystem studies. Ecol. Lett. 11, 1111–1120.

Tripplehorn, N.F., Johnson, C.A., 2004. Borror and DeLong's Introduction to the Study of Insects, seventh ed. Brooks Cole Publishing Company, Boston, MA.

Trofymow, J.A., Coleman, D.C., 1982. The role of bacterivorous and fungivorous nematodes in cellulose and chitin decomposition in the context of a root/rhizosphere/ soil conceptual model). In: Freckman, D.W. (Ed.), Nematodes in Soil Systems. University of Texas Press, Austin, TX, pp. 117–137.

Trumbore, S., 2006. Carbon respired by terrestrial ecosystems. Recent Progress and Challenges. Global Change Biol. 12, 141–153.

Trumbore, S., 2009. Radiocarbon and soil carbon dynamics. Annu. Rev. Earth Planet. Sci. 37, 47–66.

Trumbore, S.E., Davidson, E.A., Barbosa de Carnago, P., Nepstad, D.C., Martinelli, L.A., 1995. Belowground cycling of carbon in forests and pastures of Eastern Amazonia. Global Biogeochem. Cycles 9, 515–528.

Tunlid, A., White, D.C., 1992. Biochemical analysis of biomass, community structure, nutritional status, and metabolic activity of microbial communities in soil. In: Stotzky, G., Bollag, J.-M. (Eds.), Soil Biochemistry, vol. 7. Marcel Dekker, New York, pp. 229–262.

Turner, B.L., Mahieu, N., Condron, L.M., 2003. The phosphorus composition of temperate pasture soils determined by NaOH—EDTA extraction and solution 31P NMR spectroscopy. Org. Geochem. 34, 1199–1210.

Tynen, M.J., 1972. Ice-worms (Oligochaeta: Enchytraeidae) from Western British Columbia. National Museum of Natural Sciences, Ottawa.

Ubick, D., Paquin, P., Cushing, C.E., Roth, V. (Eds.), 2005. Spiders of North America: An Identification Manual. American Arachnological Society, Columbia, Missouri.

Uksa, M., Schloter, M., Endesfelder, D., Kublik, S., Engel, M., Kautz, T., et al., 2015. Prokaryotes in subsoil—evidence for a strong spatial separation of different phyla by analysing co-occurrence networks. Front. Microbiol. 6 (1269), 1–13.

Upchurch, D.R., Taylor, H.M., 1990. Tools for studying rhizosphere dynamics. In: Box Jr., J.E., Hammond, L.C. (Eds.), Rhizosphere Dynamics. Westview Press, Boulder, Colorado, pp. 83–115.

Usher, M.B., Balogun, R.A., 1966. A defense mechanism in *Onychiurus* (Collembola, Onychiuridae). Entomologist's Monthly Mag. 102, 237–238.

Van Bree, G.W., Ross, C.W., Shepherd, T.G., 1999. Populations of terrestrial planarians affected by crop management: implications for long-term management. Pedobiologia 42, 360–363.

van Breemen, N., 1993. Soils as biotic constructs favouring net primary productivity. Geoderma 57, 183–211.

van der Heijden, M.G.A., Klironomos, J.N., Ursic, M., Moutoglis, P., Streitwolf-Engel, R., Boller, T., et al., 1998. Mycorrhizal fungal diversity determines plant biodiversity, ecosystem variability and productivity. Nature 396, 69–72.

van der Heijden, M.G.A., Horton, T.R., 2009. Socialism in Soil? The importance of mycorrhizal fungal networks for facilitation in natural ecosystems. J. Ecol. 97, 1139–1150.

van der Heijden, M.G., Bardgett, R.D., van Straalen, N.M., 2008. The unseen majority: soil microbes and drivers of plant diversity and production in terrestrial ecosystems. Ecol. Lett. 11, 296–310.

van der Putten, W.H., Bardgett, R.D., Bever, J.D., Bezemer, T.M., Casper, B.B., Fukami, T., et al., 2013. Plant–soil feedbacks: the past, the present and future challenges. J. Ecol. 101, 265–276.

Van der Wal, A., Geydan, T.D., Kuyper, T.W., De Boer, W., 2013. A thready affair: linking fungal diversity and community dynamics to terrestrial decomposition processes. FEMS Microbiol. Rev. 37, 477–494.

Van Groenigen, J.W., Stein, A., 1998. Constrained optimisation of spatial sampling using continuous simulated annealing. J. Environ. Qual. 27, 1078–1086.

van Noordwijk, M., de Ruiter, P.C., Zwart, K.B., Bloem, J., Moore, J.C., van Faassen, H.G., et al., 1993. Synlocation of biological activity, roots, cracks and recent organic inputs in a sugar beet field. Geoderma 56, 265–276.

van Noordwijk, M., Hairiah, K., Woomer, P.L., Murdiyarso, D., 1998. Criteria and indicators of forest soils used for slash-and-burn agriculture and alternative land uses in Indonesia. In: Bigham, J.M., Kral, D.M., Viney, M.K., Adams, M.B., Ramakrishna, K., Davidson, E.A. (Eds.), The Contributions of Soil Science to the Development and Implementation of Criteria and Indicators of Sustainable Forest Management. SSSA Special Publication. Soil Science Society of America, Madison, WI, pp. 137–153.

van Vliet, P.C.J., 2000. Enchytraeids. In: Sumner, M. (Ed.), Handbook of Soil Science.. CRC Press, Boca Raton, FL, pp. C70–C77.

van Vliet, P.C.J., Hendrix, P.F., 2007. Role of fauna in soil physical processes. In: Abbott, L.K., Murphy, D.V. (Eds.), Soil Biological Fertility—A Key to Sustainable Land Use in Agriculture. Springer, New York, pp. 61–80.

van Vliet, P.C.J., West, L.T., Hendrix, P.F., Coleman, D.C., 1993. The influence of Enchytraeidae (Oligochaeta) on the soil porosity of small microcosms. Geoderma 56, 287–299.

van Vliet, P.C.J., Hendrix, P.F., Callaham Jr., M.A., 2012. Soil fauna: earthworms. In: Huang, P.M., Li, Y., Sumner, M.P. (Eds.), Handbook of Soil Sciences: Properties and Processes, second ed. CRC Press, Boca Raton, FL, pp. 25.35–25.44.

Vance, E.D., Brookes, P.C., Jenkinson, D.S., 1987. An extraction method for measuring microbial biomass C. Soil Biol. Biochem. 19, 703–707.

Vanlauwe, B., Diels, J., Sanginga, N., Merckx, R. (Eds.), 2002. Integrated Plant Nutrient Management in Sub-SaharanAfrica: From Concept to Practice. CABI Publishing, Wallingford.

Vannier, G., 1973. Originalité des conditions de vie dans le sol due a la presence de I'eau: importance thermodynamique et biologique de la porosphere. Annales de la Société royale zoologique de Belgique 103, 157–167.

Vannier, G., 1987. The porosphere as an ecological medium emphasized in Professor Ghilarov's work on soil animal adaptations. Biol. Fertil. Soils 3, 39–44.

Vargas, R., Hattori, T., 1986. Protozoan predation of bacterial cells in soil aggregates. FEMS Microbiol. Ecol. 38, 233–242.

Veloz, S.D., Williams, J.W., Blois, J.L., He, F., Otto-Bliesner, B., Liu, Z., 2012. No-analog climates and shifting realized niches during the late quaternary: implications for 21st-century predictions by species distribution models. Global Change Biol. 18, 1698–1713.

Venette, R.C., Ferris, H., 1998. Influence of bacterial type and density on population growth of bacterial-feeding nematodes. Soil Biol. Biochem. 30, 949–960.

Verbruggen, E., Pena, R., Fernandez, C.W., Soong, J.L., 2017. Mycorrhizal interactions with saprotrophs and impact on soil carbon storage. In: Johnson, N.C., Gehring, C., Jansa, J. (Eds.), Mycorrhizal Mediation of Soil—Fertility, Structure and Carbon Storage.. Elsevier, Amsterdam, pp. 441–460.

Veresoglou, S.D., Chen, B., Rillig, M., 2012. Arbuscular mycorrhiza and soil nitrogen cycling. Soil Biol. Biochem. 46, 53–62.

Verhoef, H.A., DeGoede, R.G.M., 1985. Effects of collembolan grazing on nitrogen dynamics in a coniferous forest. In: Fitter, A.H., Atkinson, D., Read, D.J., Usher, M.B. (Eds.), Ecological Interactions in Soil: Plants, Microbes and Animals.. Blackwell, Oxford, pp. 367–376.

Vernadsky, V.I., 1944. Problems of biogeochemistry. II. The fundamental matter-energy difference between the living and the inert natural bodies of the biosphere. Trans. Conn. Acad. Arts Sci. 35, 483–512.

Vernadsky, V.I., 1998. The Biosphere. Copernicus. Springer-Verlag, New York.

Versteegh, E.A.A., Black, S., Hodson, M.E., 2014. Environmental controls on the production of calcium carbonate by earthworms. Soil Biol. Biochem. 70, 159–161.

Vervoort, M.T.W., Vonk, J.A., Mooijman, P.J.W., Van den Elsen, S.J.J., Van Megen, H.H.B., Veenhuizen, P., et al., 2012. SSU ribosomal DNA-based monitoring of nematode assemblages reveals distinct seasonal fluctuations within evolutionary heterogeneous feeding guilds. PLOS ONE 7 (10), e47555.

Vincent, C., Rowland, D., Na, C., Schaffer, B., 2017. A high-throughput method to quantify root hair area in digital images taken *in situ*. Plant and Soil 412, 61–80.

Vitousek, P.M., 1994. Beyond global warming: ecology and global change. Ecology 75, 1861–1876.

Vitousek, P.M., Hattenschwiler, S., Olander, L., Allison, S., 2002. Nitrogen and nature. Ambio 31, 97–101.

Vitousek, P.M., Menge, D.N.L., Reed, S.C., Cleveland, C.C., 2013. Research article: biological nitrogen fixation: rates, patterns and ecological controls in terrestrial ecosystems. Philos. Trans. R. Soc. B 368 (20130119), 1–9.

Vogt, K.A., Grier, C.C., Meier, C.E., Edmonds, R.L., 1982. Mycorrhizal role in net primary production and nutrient cycling in *Abies amabilis* ecosystems in western Washington. Ecology 63, 370–380.

Vogt, K.A., Grier, C.C., Gower, S.T., Sprugel, D.G., Vogt, D.J., 1986. Overestimation of net root production: a real or imaginary problem? Ecology 67, 577–579.

Volobuev, V.R., 1964. Ecology of Soils. Israel Program for Science Translations. Davey & Co., New York.

Voroney, R.P., Paul, E.A., 1984. Determination of Kc and Kn in situ for calibration of the chloroform fumigation-incubation method. Soil Biol. Biochem. 16, 9–14.

Voroney, R.P., Brookes, P.C., Beyaert, R.P., 2007. Soil microbial biomass C N P and S. In: Carter, M.R., Gregorich, E.G. (Eds.), Soil Sampling and Methods of Analysis, second ed. CRC Press, Boca Raton, FL, pp. 637–651.

Voroney, R.P., Winter, J.P., Gregorich, E.G., 1991. Microbe, plant, soil interactions. In: Coleman, D.C., Fry, B. (Eds.), Carbon Isotope Techniques. Academic Press, Inc, San Diego, CA, pp. 77–99.

Vos, W., Stortelder, A.H.F., 1988. Vanishing Tuscan Landscapes. Landscape ecology of a Submediterranean -Montane area (Solano Basi, Tuscany, Italy). Ph. D. Thesis. University of Amsterdam, the Netherlands.

Vossbrinck, C.R., Coleman, D.C., Woolley, T.A., 1979. Abiotic and biotic factors in litter decomposition in a semi-arid grassland. Ecology 60, 265–271.

Waldorf, E.S., 1974. Sex pheromone in the springtail, *Sinella curviseta*. Environ. Entomol. 3, 916–918.

Walker, T.W., 1965. The significance of phosphorus in pedogenesis. In: Hallsworth, E.G., Crawford, D.V. (Eds.), Experimental Pedology. Butterworths, London, pp. 295–315.

Wall, D.H., Moore, J.C., 1999. Interactions underground: soil biodiversity, mutualism, and ecosystem processes. Bioscience 49, 109–117.

Wall, D.H., Virginia, R.A., 1999. Controls on soil biodiversity: insights from extreme environments. Appl. Soil Ecol. 13, 137–150.

Wall, D.H., Bradford, M.A., St. John, M.G., Trofymow, J.A., Behan-Pelletier, V., Bignell, D.E., et al., 2008. Global decomposition experiment shows soil animal impacts on decomposition are climate-dependent. Global Change Biol. 14, 2661–2677.

Wall, D.H., Bardgett, R.D., Behan-Pelletier, V., Herrick, J.E., Jones, H., Ritz, K., et al., 2012. Soil Ecology and Ecosystem Services. Oxford University Press, Oxford.

Wall, D.H., Nielsen, U.H., Six, J., 2015. Soil biodiversity and human health. Nature 528, 69–76.

Wallwork, J.A., 1970. Ecology of Soil Animals. McGraw-Hill, London.

Wallwork, J.A., 1976. The Distribution and Diversity of Soil Fauna. Academic Press, London.

Wallwork, J.A., 1982. Desert Soil Fauna. Praeger Scientific, New York.

Walsh, M.I., Bolger, T., 1990. Effects of diet on the growth and reproduction of some Collembola in laboratory cultures. Pedobiologia 34, 161–171.

Walter, D.E., 1988. Predation and mycophagy by endeostigmatid mites (Acariformes: Prostigmata). Exp. Appl. Acarology 4, 159–166.

Walter, D.E., Ikonen, E.K., 1989. Species, guilds and functional groups: taxonomy and behavior in nematophagous arthropods. J. Nematol. 21, 315–327.

Walter, D.E., Proctor, H.C., 1999. Mites. Ecology, Evolution and Behavior. CABI Publishing, Wallingford.

Walter, D.E., Proctor, H.C., 2000. Life at the Microscale: Mites and the Study of Ecology, Evolution and Behaviour. New South Wales Press, Sydney, Australia.

Walter, D.E., Hunt, H.W., Elliott, E.T., 1987. The influence of prey type on the development and reproduction of some predatory soil mites. Pedobiologia 30, 419–424.

Walter, D.E., Kaplan, D.T., Permar, T.A., 1991. Missing links: a review of methods used to estimate trophic links in soil food webs. Agric., Ecosyst. Environ. 34, 399–405.

Wang, G.M., Coleman, D.C., Freckman, D.W., Dyer, M.I., McNaughton, S.J., Acra, M.A., et al., 1989. Carbon partitioning patterns of mycorrhizal versus non-mycorrhizal plants. Real-time dynamic measurements using $^{11}CO_2$. New Phytol. 112, 489–493.

Wardle, D.A., 1998. Controls of temporal variability of the soil microbial biomass: a global-scale synthesis. Soil Biol. Biochem. 30, 1627–1637.

Wardle, D.A., 2002. Communities and Ecosystems: Linking the Aboveground and Belowground Components. Princeton University Press, Princeton.

Wardle, D.A., Yeates, G.W., 1993. The dual importance of competition and predation as regulatory forces in terrestrial ecosystems: evidence from decomposer food-webs. Oecologia 93, 303–306.

Wardle, D.A., Bonner, K.I., Barker, G.M., Yeates, G.W., Nicholson, K.S., Bardgett, R.D., et al., 1999. Plant removals in perennial grassland: vegetation dynamics, decomposers, soil biodiversity, and ecosystem properties. Ecol. Monogr. 69, 535–568.

Warnock, A.J., Fitter, A.H., Usher, M.B., 1982. The influence of a springtail *Folsomia candida* on the mycorrhizal association of leek *Allium porrum* and arbuscular mycorrhizal endophyte *Glomus fasciculatus*. New Phytol. 90, 285–292.

Wasylik, A., 1995. Indicatory groups of Acarina in the processes of degradative succession. In: Boczek, J., Ignatowicz, S. (Eds.), Advances of Acarology in Poland. Polish Academy of Science, Siedlec, pp. 90–93.

Waters, A.G., Oades, J.M., 1991. Organic matter in water stable aggregates. In: Wilson, W.S. (Ed.), Advances in Soil Organic Matter Research. The Impact on Agriculture and the Environment. Woodhead Publishing Limited, Cambridge, pp. 163–174.

Webb, D.P., 1977. Regulation of deciduous forest litter decomposition by soil arthropod feces. In: Mattson, W.J. (Ed.), The Role of Arthropods in Forest Ecosystems. Springer, New York, pp. 57–69.

Wenk, E.S., Callaham Jr., M.A., Hanson, P.J., 2016. Soil macroinvertebrate communities across a productivity gradient in deciduous forests of eastern North America. Northeastern Naturalist 23, 25–44.

Wheeler, G.C., Wheeler, J., 1990. Insecta: hymenoptera formicidae. In: Dindal, D.L. (Ed.), Soil Biology Guide.. Wiley, New York, pp. 1277–1294.

Whiles, M.R., Callaham Jr., M.A., Meyer, C.K., Brock, B.L., Charlton, R.E., 2001. Emergence of periodical cicadas from a Kansas riparian forest: densities, biomass and nitrogen flux. Am. Midl. Nat. 145, 176–187.

White, G., 1789. The Natural History of Selborne. Benjamin, White, London.

White, R.E., 1983. A Field Guide To the Beetles. Houghton Mifflin, Boston.

Whitehead, A.G., Hemming, J.R., 1965. A comparison of some quantitative methods of extracting small vermiform nematodes from soil. Ann. Appl. Biol. 55, 25–38.

Whitford, W.G., 2000. Keystone arthropods as webmasters in desert ecosystems. In: Coleman, D.C., Hendrix, P.F. (Eds.), Invertebrates as Webmasters in Ecosystems. CAB International, Wallingford, pp. 25–41.

Whitford, W.G., Santos, P.F., 1980. Arthropods and detritus decomposition in desert ecosystems. In: Dindal, D.L. (Ed.), Soil Biology as Related to Land Use Practices. U.S. Environmental Protection Agency, Washington, DC, pp. 770–778.

Whitford, W.G., Freckman, D.W., Parker, L.W., Schaefer, D., Santos, P.F., Steinberger, Y., 1983. The contributions of soil fauna to nutrient cycles in desert systems. In: Lebrun, P., André, H.M., de Medts, A., Grégoire-Wibo, C., Wauthy, G. (Eds.), New Trends in Soil Biology. Dieu-Brichart, Ottignies-Louvain-la-Neuve, Belgium, pp. 49–59.

Whitman, W.B., Coleman, D.C., Wiebe, W.J., 1998. Perspective: prokaryotes: the unseen majority. Proc. Natl. Acad. Sci. 95, 6578–6583.

Whitney, M., 1925. Soil and Civilization. Van Nostrand, New York.

Wickings, K., Grandy, A.S., 2011. The oribatid mite *Scheloribates moestus* (Acari: Oribatida) alters litter chemistry and nutrient cycling during decomposition, Soil Biol. Biochem., 43. pp. 351–358.

Wieder, R.K., Lang, G.E., 1982. A critique of the analytical methods used in examining decomposition data obtained from litterbags. Ecology 63, 1636–1642.

Wilkinson, M.T., Richards, P.J., Humphreys, G.S., 2009. Breaking ground: pedological, geological, and ecological implications of soil bioturbation. Earth Sci. Rev. 97, 257–272.

Wilkinson, D.M., Koumoutsaris, S., Mitchell, E.A.D., Bey, I., 2012. Modelling the effect of size on the aerial dispersal of microorganisms. J. Biogeogr. 39 (1), 89–97.

Willerslev, E., Hansen, A.J., Binladen, J., Brand, T.B., Gilbert, M.T.P., Shapiro, B., et al., 2003. Diverse plant and animal genetic records from Holocene and Pleistocene sediments. Science 300, 791–795.

Williams, J.W., Jackson, S.T., 2007. Novel climates, no-analog communities, and ecological surprises. Front. Ecol. Environ. 5 (9), 475–482.

Williams, S.C., 1987. Scorpion bionomics. Annu. Rev. Entomol. 32, 275–295.

Williamson, M., 1996. Biological invasions. Chapman & Hall, London.

Wilson, A.T., 1978. Pioneer agriculture explosion and CO_2 levels in the atmosphere. Nature 273, 40–41.

Wilson, D.S., 1980. The Natural Selection of Populations and Communities. Benjamin/Cummings, Menlo Park, CA.

Wilson, E.O., 1987. Causes of ecological success: the case of the ants. J. Anim. Ecol. 56, 1–9.

Wilson, E.O., 1992. The Diversity of Life. Norton, New York.

Wilson, K.J., Sessitsch, A., Akkermans, A., 1994. Molecular markers as tools to study the ecology of microorganisms. In: Ritz, K., Dighton, J., Giller, K.E. (Eds.), Beyond the Biomass. Wiley-Sayce, Chichester, pp. 149–156.

Wilson, M.J., Kakouli-Duarte, T. (Eds.), 2009. Nematodes as Environmental Indicators. CABI International, Cambridge, MA.

Winchester, N.N., 1997. Canopy arthropods of coastal Sitka spruce on Vancouver Island, British Columbia, Canada. In: Stork, N.E., Adis, J., Didham, R.K. (Eds.), Canopy Arthropods. Chapman & Hall, London, pp. 151–168.

Winsome, T., Epstein, L., Hendrix, P.F., Horwath, W.R., 2006. Competitive interactions between native and exotic earthworm species as influenced by habitat quality in a California grassland. Appl. Soil Ecol. 32, 38–53.

Winter, J.P., Voroney, R.P., Ainsworth, D.A., 1990. Soil microarthropods in long-term no-tillage and conventional corn production. Can. J. Soil Sci. 70, 641–653.

Wise, D.H., 1993. Spiders in Ecological Webs. Cambridge University Press, Cambridge.

Withington, J.M., Elkin, A.D., Bulaj, B., Olesinski, J., Tracy, K.N., Bouma, T.J., et al., 2003. The impact of material used for minirhizotron tubes for root research. New Phytol. 160, 533–544.

Witkamp, M., van der Drift, J., 1961. Breakdown of forest litter in relation to environmental factors. Plant & Soil 15, 295–311.

Woese, C.R., Kandler, O., Wheelis, M.L., 1990. Towards a natural system of organisms: proposal for the domains Archaea, Bacteria, and Eucarya. Proc. Natl. Acad. Sci. U. S. A. 87, 4576–4579.

Wolters, V., 1988. Effects of *Mesenchytraeus glandulosus* (Oligochaeta, Enchytraeidae) on decomposition processes. Pedobiologia 32, 387–398.

Wolters, V., 1991. Soil invertebrates—effects on nutrient turnover and soil structure—a review. Zeitschrift für Pflanzenernährung und Bodenkunde 154, 389–402.

Wolters, V., Ekschmitt, K., 1997. Gastropods, isopods, diplopods and chilopods: neglected groups of the decomposer food web. In: Benckiser, G. (Ed.), Fauna in Soil Ecosystems.. Dekker, New York, pp. 265–306.

Wolters, V., Silver, W.H., Bignell, D.E., Coleman, D.C., Lavelle, P., van der Putten, W.H., et al., 2000. Effects of global changes on above- and below-ground biodiversity in terrestrial ecosystems: implications for ecosystem functioning. BioScience 50, 1089–1098.

Wood, T.G., Johnson, R.A., Anderson, J.M., 1983. Modification of soils in Nigerian Savanna by soil-feeding *Cubitermes* (Isoptera, Termitidae). Soil Biol. Biochem. 15, 575–579.

Woomer, P.L., Swift, M.J., 1994. Report of the Tropical Soil Biology and Fertility Programme. Tropical Soil Biology and Fertility Programme, Nairobi, Kenya.

Wright, D.H., 1988. Inverted microscope methods for counting soil mesofauna. Pedobiologia 31, 409–411.

Wright, D.H., Huhta, V., Coleman, D.C., 1989. Characteristics of defaunated soil. II. Effects of reinoculation and the role of the mineral soil. Pedobiologia 33, 427–435.

Wright, J.C., 2001. Cryptobiosis 300 years on from van Leeuwenhoek: what have we learned about tardigrades? Zool. Anz. 240, 563–582.

Wu, P., Liu, X., Liu, S., Wang, J., Wang, Y., 2014. Composition and spatio-temporal variation of soil microarthropods in the biodiversity hotspot of northern Hengduan Mountains, China. Eur. J. Soil Biol. 62, 30–38.

Wu, T., Ayres, E., Li, G., Bardgett, R.D., Wall, D.H., Garey, J.R., 2009. Molecular profiling of soil animal diversity in natural ecosystems: incongruence of molecular and morphological results. Soil Biol. Biochem. 41, 849–857.

Wu, T., Ayres, E., Bardgett, R.D., Wall, D.H., Gareya, J.R., 2011. Molecular study of worldwide distribution and diversity of soil animals. Proc. Natl. Acad. Sci. 108, 7720–17725.

Xiong, Y., Gao, Y., Yin, W.-Y., Luan, Y.-X., 2008. Molecular phylogeny of Collebmola inferred from ribosomal RNA genes. Mol. Phylogenet. Evol. 49, 728–735.

Yang, L., 2004. Periodical cicadas as resource pulses in North American forests. Science 306, 1565–1567.

Yang, N., Schuetzenmeister, K., Grubert, D., Jungkunst, H.F., Gansert, D., Scheu, S., et al., 2015. Impacts of earthworms on nitrogen acquisition from leaf litter by arbuscular mycorrhizal ash and ectomycorrhizal beech trees. Environ. Exp. Bot. 120, 1–7.

Yeates, G.W., 1981. Soil nematode populations depressed in the presence of earthworms. Pedobiologia 22, 191–195.

Yeates, G.W., 1998. Feeding in free-living soil nematodes: a functional approach. In: Perry, R.N., Wright, D.J. (Eds.), The Physiology and Biochemistry of Free-living and Plant-parasitic Nematodes.. CABI Publishing, Wallingford, pp. 245–269.

Yeates, G.W., 1999. Effects of plants on nematode community structure. Annu. Rev. Phytopathol. 37, 127–149.

Yeates, G.W., Coleman, D.C., 1982. Role of nematodes in decomposition. In: Freckman, D.W. (Ed.), Nematodes in soil ecosystems.. University Texas Press, Austin, TX, pp. 55–80.

Yeates, G.W., Bongers, T., de Goede, R.G.M., Freckman, D.W., Georgieva, S.S., 1993. Feeding habits in soil nematode families and genera-An outline for soil ecologists. J. Nematol. 25, 101–313.

Yeates, G.W., Ross, C.W., Shepherd, T.G., 1999. Populations of terrestrial planarians affected by crop management: implications for long-term management. Pedobiologia 42, 360–363.

Yeates, G.W., Dando, J.L., Shepherd, T.G., 2002. Pressure plate studies to determine how moisture affects access of bacterial-feeding nematodes to food in soil. Eur. J. Soil Sci. 53, 355–365.

Youssef, N.H., Couger, M.B., McCully, A.L., Criado, A.E.G., Elshahed, M.S., 2015. Assessing the global phylum level diversity within the bacterial domain: a review. J. Adv. Res. 6, 269–282.

Zaborski, E.R., 2003. Allyl isothiocyanate: an alternative chemical expellant for sampling earthworms. Appl. Soil Ecol. 22, 87–95.

Zachariae, G., 1963. Was leisten Collembolen für den Waldhumus? In: Van der Drift, J., Doeksen, J. (Eds.), Soil Organisms. North Holland, Amsterdam, pp. 109–114.

Zachariae, G., 1964. Welche Bedeutung haben Enchyträus in Waldboden? In: Jongerius, A. (Ed.), Soil Micromorphology.. Elsevier, Amsterdam, pp. 57–68.

Zachariae, G., 1965. Spuren tierischer Tätigkeit im Boden des Buchenwaldes. Forstwissenschaftliche Forschungen 20, 1–68.

Zangerlé, A., Renard, D., Iriarte, J., Suarez Jimenez, L.E., Adame Montoya, K.L., Juilleret, J., et al., 2016. The surales, self-organized Earth-mound landscapes made by Earthworms in a seasonal tropical Wetland. Plos One 11 (5), e0154269.

Zechmeister-Boltenstern, S., Keiblinger, K.M., Mooshammer, M., Peñuelas, J., Richter, A., Sardans, J., et al., 2015. The application of ecological stoichiometry to plant–microbial–soil organic matter transformations. Ecol. Monogr. 85 (2), 133–155.

Zelles, L., Alef, K., 1995. Biomarkers. In: Alef, K., Nannipieri, P. (Eds.), Methods in Applied Soil Microbiology and Biochemistry. Academic Press, London, pp. 422–439.

Zhou, J., Bruns, M.A., Tiedje, J.M., 1996. DNA recovery from soils of diverse composition. Appl. Environ. Microbiol. 62, 316–322.

Zhou, L., Dai, L.M., Gu, H.Y., Zhong, L., 2007. Review on the decomposition and influence factors of coarse woody debris in forest ecosystem. J. For. Res. 18 (1), 48–54.

Zimmer, M., 2002. Nutrition in terrestrial isopods (Isopoda: Oniscidea): an evolutionary-ecological approach. Biol. Rev. 77, 455–493.

Zimmer, M., Kautz, G., Topp, W., 1996. Olfaction in terrestrial isopods (Isopoda: Oniscidea): responses of *Porcellio scaber* to the odour of litter. Eur. J. Soil Biol. 32, 141–147.

Zonn, S.V., Eroshkina, A.N., 1996. V.V. Dokuchaev, his disciples and followers. Eurasian Soil Sci. 29, 111–120.

Zou, K.J., Thebault, E., Lacroix, G., Barot, S., 2016. Interactions between the green and brown food web determine ecosystem functioning. Funct. Ecol. 30, 1454–1465.

Zwart, K.B., Kuikman, P.J., van Veen, J.A., 1994. Rhizosphere protozoa: their significance in nutrient dynamics. In: Darbyshire, J.F. (Ed.), Soil Protozoa. CABI Publishing, Wallingford, pp. 93–121.

Zwart, K.B., Darbyshire, J.F., 1992. Growth and nitrogenous excretion of a common soil flagellate, Spumella sp.—a laboratory experiment. Eur. J. Soil Sci. 43 (1), 145–157.

Index

Note: Page numbers followed by "*f*" and "*t*" refer to figures and tables, respectively.

N

Q

R

S

T

Printed in the United States
By Bookmasters